SPRINGER SERIES IN PHOTONICS 10

Springer
Berlin
Heidelberg
New York
Hong Kong
London
Milan
Paris
Tokyo

SPRINGER SERIES IN PHOTONICS

Series Editors: T. Kamiya B. Monemar H. Venghaus Y. Yamamoto

The Springer Series in Photonics covers the entire field of photonics, including theory, experiment, and the technology of photonic devices. The books published in this series give a careful survey of the state-of-the-art in photonic science and technology for all the relevant classes of active and passive photonic components and materials. This series will appeal to researchers, engineers, and advanced students.

1 **Advanced Optoelectronic Devices**
By D. Dragoman and M. Dragoman

2 **Femtosecond Technology**
Editors: T. Kamiya, F. Saito, O. Wada, H. Yajima

3 **Integrated Silicon Optoelectronics**
By H. Zimmermann

4 **Fibre Optic Communication Devices**
Editors: N. Grote and H. Venghaus

5 **Nonclassical Light from Semiconductor Lasers and LEDs**
By J. Kim, S. Somani, and Y. Yamamoto

6 **Vertical-Cavity Surface-Emitting Laser Devices**
By H. Li and K. Iga

7 **Active Glass for Photonic Devices**
Photoinduced Structures and Their Application
Editors: K. Hirao, T. Mitsuyu, J. Si, and J. Qiu

8 **Nonlinear Photonics**
Nonlinearities in Optics, Optoelectronics and Fiber Communications
By Y. Guo, C.K. Kao, E.H. Li, and K.S. Chiang

9 **Optical Solitons in Fibers**
Third Edition.
By A. Hasegawa and M. Matsumoto

10 **Nonlinear Photonic Crystals**
Editors: R.E. Slusher and B.J. Eggleton

Series homepage – http://www.springer.de/phys/books/ssp/

R.E. Slusher B.J. Eggleton (Eds.)

Nonlinear Photonic Crystals

With 174 Figures

Springer

Richard E. Slusher
Lucent Technologies
Optical Physics Research Department
600 Mountain Avenue
Murray Hill, NJ 07974-0000, USA
E-mail: res@lucent.com

Benjamin J. Eggleton
OFS, Laboratories
& Specialty Photonics Division
19 Schoolhouse Rd.
Somerset, NJ 08873-0000, USA
egg@ofsoptics.com

Series Editors:

ISSN 1437-0379
ISBN 3-540-43900-5 Springer-Verlag Berlin Heidelberg New York

Library of Congress Cataloging-in-Publication Data
Nonlinear photonic crystals / R.E. Slusher, B.J. Eggleton (eds.).
p. cm. - - (Springer series in photonics, ISSN 1437-0379 ; 10)
Includes bibliographical references and index.
ISBN 3-540-43900-5 (acid-free paper)
1. Photons. 2. Crystal optics. 3. Nonlinear optics. I. Slusher, R. . (Richart E.), 1938- II. Eggleton, B. J. (Benjamin J.), 1970- III. Springer series in photonics ; v. 10.
QC703.5.P427 N66 2003
621.36'94- -dc21 2002030637

Springer-Verlag Berlin Heidelberg New York
a member of BertelsmannSpringer Science+Business Media GmbH

http://www.springer.de

Typesetting: Camera ready by the Authors using a Springer LaTeX macro package
Cover concept: eStudio Calamar Steinen
Cover production: *design & production* GmbH, Heidelberg

Printed on acid-free paper SPIN: 10885614 57/3141/yu 5 4 3 2 1 0

Preface

Photonic crystals, periodic dielectric arrays in one, two or three dimensions that selectively transmit or reflect light at various wavelengths, are the physical basis of beautiful natural phenomena. For example, colors in butterfly wings result from the selective reflection of the spectral components of sunlight from periodic dielectric structures that form on the surface of the butterfly wing. In the laboratory periodic optical arrays in the form of diffraction gratings have been used for over a hundred years to separate the various colors in a light beam. In this book we will concentrate on nonlinear phenomena in photonic crystals. The light wavelengths of interest are in the visible and infrared region of the spectrum. The dielectric periodicity scale size in photonic crystals is a fraction of the light wavelength, periods of a few tenths of a micron. A range of wavelengths where a photonic crystal exhibits strong reflection is called a photonic bandgap.

The name "photonic crystal" and the exciting growth of photonic crystal research began with the work of Eli Yablonovitch and his group in the late 1980s. They began with basic concepts and experiments in the microwave regime where the 3D structures are easily fabricated. At present fabrication techniques with spatial resolutions in the sub-micron regime are flourishing, resulting in an explosion of new photonic bandgap structures in the visible and infrared.

This book describes the initial research in nonlinear photonic crystals. This new research direction results from the growth, intersection and overlap of research in the fields of nonlinear optics and photonic crystals. Intense laser sources and confinement of light to small spatial regions in photonic crystals allow us to generate optical fields that make significant nonlinear changes in the dielectric constants in photonic crystals. This means that we can wavelength tune the photonic crystal reflection and transmission bands by simply varying the intensity of the incident light.

Light with wavelengths near the period of a photonic crystal slows down since it is reflected back and forth between the periodic dielectric layers instead of being transmitted directly through the material. Since this slowing down depends strongly on the light wavelength, different colors propagate with much different velocities, i.e. there is strong dispersion near a photonic bandgap. The combination of this linear dispersion and nonlinear change in dielectric constant with the light intensity combine to form a new class of soliton phenomena. Soliton light pulses retain their shape as they propagate in spite of the strong dispersion that quickly broadens the pulse in time at low intensities. In fact the new solitons, called "gap solitons" or "Bragg solitons" have many unique properties. For example, gap solitons can propagate at any velocity, even zero velocity. A related nonlinear spatial soliton can form in periodic waveguide arrays where the nonlinear change in the dielectric constant at high light intensities can compensate for the diffractive spreading of the light beam.

Introducing dispersion at different wavelengths in a photonic crystal can be used to control and enhance phase mating of a number of nonlinear processes. For example, second harmonic generation can be enhanced by forming photonic bandgaps near either the fundamental or second harmonic wavelengths in order to achieve phase matching of the fundamental and second harmonic waves over much longer distance than is possible in the bare material.

The new physical phenomena being discovered in nonlinear photonic crystals almost certainly will lead to advances in optical devices and applications in optical systems. This will require advances in fabrication of the photonic crystals and improvements in the nonlinear response of the composite materials. Examples of optical device function include pulse shaping, pulse compression, and pulse regeneration. Optical buffers that store light for extended periods and allow random acccess to the stored data are important optical devices needed for many applications. Optical buffers are far from competitive with electronic memory devices at present. Nonlinear photonic crystals may allow for significant advances in optical buffering devices by using resonantly stored light in dielectric array defects. Switching of optical pulses and optical data streams is most often accomplished today by converting the optical pulses to electrical pulses, processing the electrical pulse trains with electronic circuitry and then converting the processed electrical pulses back to light. All-optical switches based on nonlinear photonic crystal are now being explored for low power, low cost alternatives to the optical-electronic-optical techniques. Optical parametric amplification enhanced by dielectric arrays is another important possible application.

Semiconductor optical amplifiers (SOAs) are one of the leading materials for nonlinear photonic crystal applications. SOAs have large nonlinearities for wavelengths near the semiconductor electronc bandgap and they have gain to compensate for losses in the optical circuits. As described in Chap. 13 a number of very important applications are being developed based on SOAs with distributed feedback periodic dielectric arrays including optical switches, optical regeneration, wavelength conversion and optical memories. Alternate nonlinear materials and new glass microstructed fibers described in this book offer new solutions the problem of achieving low power all-optical devices with a broad range of applications.

The book is divided into four parts that deal with different aspects of nonlinear photonic crystals. After the introduction, Part 1 (*Nonlinear Photonic Crystal Theory*) presents the theoretical framework for describing nonlinear optics in Bragg gratings and photonic crystal structures. Chapter 2 presents the general theoretical formulation for nonlinear pulse propagation in periodic media, focusing on third-order nonlinearity in photonic crystals with weak index modulations. In this weak modulation limit the well known nonlinear coupled mode equations are derived. Chapter 3 explores polarization effects in nonlinear periodic structures. Intriguing nonlinear phenomena that rely on the interplay of the underlying periodicity with the nonlinearity and intrinsic birefringence of the material are described. Raman gap solitons are theoretically investigated in Chap. 4. Raman gap solitons are stable, long-lived quasistationary excitation that can exist within the grating even after the pump pulse has passed. By solving the modified nonlinear coupled mode equations numerically it is shown that slow Raman gap solitons can propagate with velocities as low as 1 percent of the speed of light. In Chap. 5 self-transparency and localization in gratings with quadratic nonlinearity are described. In Chap. 6 photonic band edge effects in finite grating structures are described and applications involving $\chi^{(2)}$ interactions are suggested. Chapter 7 introduces the theory of parametric photonic crystals along with the concept of the simultons, i.e. simultaneous solitary wave solutions.

Part 2 (*Nonlinear Fiber Grating Experiments*) describes experiments on nonlinear pulse propagation effects in one-dimensional photonic crystals formed as Bragg gratings in optical fibers. In Chap. 8 some of the initial experiments are described that demonstrate optical pulse compression and soliton propagation in nonlinear Bragg gratings. Chapter 9 describes experiments demonstrating gap soliton propagation in optical fiber Bragg grating. In Chap. 10 theoretical and experimental aspects of cross-phase modulation effects in nonlinear Bragg grating structures are reviewed. This includes experimental demonstrations of the optical "push-broom" effect where a light pulse can be compressed by using cross-phase modulation from a second pulse.

Part 3 (*Novel Nonlinear Periodic Systems*) deals with novel nonlinear periodic structures and materials. Chapter 11 reviews recent progress in developing new material systems for nonlinear periodic devices. In particular this chapter deals with chalcogenide glasses that exhibit a third-order optical nonlinearity between 100 and 1000 times larger than that of silica. Chapter 12 reviews experimental and theoretical aspects of nonlinear pulse propagation in air-silica microstructured optical fibers. Although these microstructured fibers are not necessarily periodic and may not exhibit a photonic bandgap, they represent an exciting new class of nonlinear material that is opening many new research directions. Chapter 13 reviews experimental studies of nonlinear effects in periodic structures integrated within semiconductor optical amplifiers. Chapter 14 looks to the future possibility of atomic solitons in optical lattices where the role of light and material are reversed. The basic formalism leading to the formulation of nonlinear atom optics is described. Dark and bright soliton solutions are discussed, the chapter then concludes with a theoretical description of gap solitons (atomic wavefunctions) in optical lattices.

Part 4 (*Spatial Solitons in Photonic Crystals*) describes spatial soliton effects in waveguide arrays and nonlinear photonic crystals. Experiments demostrating spatial solitons in nonlinear array waveguides are described in Chap. 15. Finally, in Chapter 16 nonlinear propagation effects in 2D photonic crystal structures are studied theoretically including spatially localized light states.

We are delighted to have been able to participate in the early experiments in this field and edit this book. This book has contributions from many of the world leaders in this field. We appreciate their time and resources that went into this book. There are many others that contributed ideas to this book and helped us to understand the concepts in this field. In particular we would like to thank Demetrios Christoduolides and John Joannopolis for their inspiration and leadership in this field. Preparation of the manuscript for this book was greatly facilitated by Adelheid Duhm at Springer. We would also like to thank Shirley Slusher and Susan Eggleton for their loving support and encouragement.

Murray Hill, New Jersey, *Richart Slusher*
July 2002 *Benjamin Eggleton*

Contents

List of Contributors

Alejandro Aceves
Univ. of New Mexico
Mathematics and Statistics Dept.
Albuquerque, NM 87131-0000, USA
aceves@math.unm.edu

Govind P. Agrawal
Univ. of Rochester
Inst. of Optics
Wilmot 206
Rochester, NY 14627-0186, USA
gpa@optics.rochester.edu

Mario Bertolotti
Istituto Nazionale di Fisica
della Materia
at Dipartimento di Energetica
Univertita' di Roma, "La Sapienza"
Via A. Scarpa, 16
00161 Roma, Italy
mario.bertolotti@uniroma1.it

Mark J. Bloemer
U.S. Army Aviation and
Missile Command
AMSAM-RD-WS-CM
Huntsville, AL 35898-0000, USA
mark.bloemer
@ws.redstone.army.mil

Charles M. Bowden
US Army Aviation and
Missile Command
US Army RD+E Center, Bldg. 7804
AMSAM-RD-WS-ST
Huntsville, AL 35898-5000, USA
cmbowden@ws.redstone.army.mil

Neil G.R. Broderick
Univ. of Southampton
Optoelectonics Research Ctr.
University Rd.
Southampton, SO171BJ, UK
ngb@orc.soton.ac.uk

M. Centini
Istituto Nazionale di Fisica
della Materia
at Dipartimento di Energetica
Universita' di Roma, "La Sapienza"
Via Scarpa 16
00161 Rome, Italy
and
Weapons Sciences Directorate
Research Development and
Engineering Center
U.S. Army Aviation and
Missile Command
Building 7804, Redstone Arsenal
AL 35898-5000, USA
marco.centini@uniroma1.it

C. Conti
National Institute
for the Physics of Matter
INFM-RM3
Terza University of Rome
84 Via della Vasca Navale
00146 Roma, Italy
claudio_conti71@hotmail.com

G. D'Aguanno
Istituto Nazionale di Fisica
della Materia
at Dipartimento di Energetica
Universta' di Roma "La Sapienza"
Via Scarpa 16
00161 Rome, Italy
and
Weapons Sciences Directorate
Research Development and
Engineering Center
U.S. Army Aviation and
Missile Command
Building 7804, Redstone Arsenal
AL 35898-5000, USA
giuseppe.daguanno@uniroma1.it

Peter D. Drummond
Department of Physics
(Rm. 2.14)
Centre for Laser Science
The University of Queensland
Brisbane, Queensland 4072
Australia
drummond@physics.uq.edu.au

Benjamin J. Eggleton
OFS
Labs. and Specialty Phot. Div.
19 Schoolhouse Rd.
Somerset, NJ 08873-0000, USA
egg@ofsoptics.com

Hagai Eisenberg
Weizmann Inst. of Science
Dept. of Complex Systems
Rehovot, 76100 Israel
feisen@wisemail.weizmann.ac.il

Alexander L. Gaeta
Cornell University
School of Applied and Engineering
Physics
224 Clark Hall
Ithaca, NY 14853-2501, USA
a.gaeta@cornell.edu

Joseph W. Haus
Univ of Dayton
Kettering Labs KL441
Electro-Optics Program
Dayton, OH 45469-0245, USA
jwhaus@udayton.edu

H. He
42 Epping Av.
Eastwood, NSW 2122, Australia
h.he@physics.usyd.edu.au

Yuri S. Kivshar
Australian National University
Nonlinear Physics Group
Research School of Physical Science
and Engineering
Canberra, ACT, 0200, Australia
yuri@cyberone.com.au

Gadi Lenz
Kodeos Communications
111 Corporate Blvd.
South Plainfield, NJ 07080-0000,
USA
g.lenz@kodeos.com

Drew N. Maywar
Lucent Technologies
1908 Knollwood Drive
Middletown, NJ 07748, USA
maywar@lucent.com

Pierre Meystre
Univ. of Arizona
Optical Sciences Ctr.
Tucson, AZ 85721-0001, USA
pierre.meystre
@optics.arizona.edu

S.F. Mingaleev
Australian National University
Nonlinear Physics Group
Research School of Physical Science
and Engineering
Canberra, ACT, 0200, Australia
sfm124@rsphysse.anu.edu.au

Suresh Pereira
Univ. of Toronto
Physics Dept.
60 St. George St.
Toronto, ON M5S-1A7, Canada
pereira@physics.utoronto.ca

Victor E. Perlin
Univ. of Michigan
EECS Bldg.
1301 Beal Ave.
Ann Arbor, MI 48109-0000, USA
vperlin@umich.edu

Sierk Pötting
Optical Sciences Center
University of Arizona
Tucson, AZ 85721, USA
and
Max-Planck Institute for Quantum Optics
D-85748 Garching, Germany
sierk.poetting
@optics.arizona.edu

Jinendra K. Ranka
Sycamore Networks
150 Apollo Dr.
Chelmsford, MA 01824-0000, USA
jinendra.ranka@sycamorenet.com

Michael Scalora
US Army Aviation and
Missile Command
Bldg. 7804
AMSAM-RD-WS-ST
Huntsville, AL 35898-0000, USA
mscalora@ws.redstone.army.mil

Concita Sibilia
Istituto Nazionale di Fisica
della Materia
at Dipartimento di Energetica
Universta' di Roma "La Sapienza"
Via Scarpa 16
00161 Rome, Italy
concita.sibilia@uniroma1.it

Yaron Silberberg
Weizmann Inst. of Science
Phys. of Complex Systems Dept.
Rehovot, 76100 Israel
yaron.silberberg@
weizmann.ac.il

John E. Sipe
Univ. of Toronto
Physics Dept.
60 St. George St.
Toronto, ON M5S-1A7, Canada
sipe@physics.utoronto.ca

Richart E. Slusher
Lucent Technologies
Optical Physics Res. Dept.
600 Mountain Ave.
Murray Hill, NJ 07974-0000, USA
res@lucent.com

S. Spalter
Siemens AG
ICN ON PNE AT 3
Hofmannstr. 51
81359 Munich, Germany
stefan.spaelter@icn.siemens.de

Michael J. Steel
RSoft, Inc.
Ste. 255
830 Hillview Ct.
Milpitas, CA 95035-0000, USA
mike@rsoftinc.com

C. Martijn de Sterke
Univ. of Sydney
School of Physics A28
Sydney, NSW 2006, Australia
desterke@physics.usyd.edu.au

Stefano Trillo
Univ. of Ferrara
Via Saragat 1
44100 Ferrara, Italy
strillo@ing.unife.it

Michael I. Weinstein
Lucent Technologies
600 Mountain Ave.
Murray Hill, NJ 07974-0000, USA
miw@research.bell-labs.com

Herbert G. Winful
Univ. of Michigan
EECS Department
1301 Beal Ave.
Ann Arbor, MI 48109-2122, USA
winful@eecs.umich.edu

Ewan M. Wright
Univ. of Arizona
Optical Science Ctr.
Tucson, AZ 85721-0001, USA
ewan.wright@optics.arizona.edu

1 Introduction

R.E. Slusher and B.J. Eggleton

Nonlinear photonic crystals, periodic arrays of dielectric materials along with light beams strong enough to change the dielectric constants, may bring to reality the vision of light controlling light in microscale photonic circuits, the analog of present day electronic integrated circuits where electrons control electrons. These integrated all-optical microphotonic circuits could perform a multitude of functions in optical communications systems, optical interconnections and optical processing.

Nonlinear photonic crystals are also being used to explore new physical phenomena based on their ability to control and enhance light fields in unique new geometries. For example, nonlinear light pulses propagating in photonic crystals can form soliton-like [1–3] pulses that propagate without change in the pulse shape at velocities that can vary between zero and the speed of light in the material. Another example of a new phenomenon is found in coupled waveguide arrays that support "discrete" solitons [4] . In the linear intensity regime light will rapidly spread throughout the waveguide array, whereas in the high intensity nonlinear regime light remains concentrated in a discrete soliton localized in a narrow range of waveguides or a single waveguide.

This book is an introduction to the vision of light controlling light in periodic photonic structures. The advent of the nonlinear photonic crystal research field has progressed rapidly during the past two decades. Microscale photonic circuits are now being formed in many materials using high refractive index contrast photonic crystals. Features in these microphotonic circuits must be controlled to an accuracy of a small fraction of the optical wavelength in the material. These sub-micron dimensions are now routinely achieved using phtolithography and electron beam lithography. This is a new field with many developments in progress and a continuing potential for surprising new results. We hope to capture some of the basic concepts in this book and also describe some of the exciting new directions for progress.

A photonic crystal is a periodic array of dielectric materials where the period is chosen so that light is strongly reflected, refracted and confined. The period of the array is often chosen so that light is controlled through the Bragg reflection effects. Bragg reflection requires a wavelength of incident light

$$\lambda_{\mathrm{B}} = 2 n_0 \Lambda \,, \qquad (1.1)$$

where λ_B is the Bragg wavelength, n_0 is the average linear index of refraction in the material and Λ is the period of the index grating in the photonic

Fig. 1.1. A photograph of a cross-section of a glass fiber that has been fabricated with a photonic crystal structure

crystal along one of its axes. The photonic crystal can be three, two or one-dimensional. The amplitude of the index contrast determines the width of the wavelength region in which light is excluded, e.g. reflected, from the photonics crystal. This region is called the photonic bandgap width and, for the one-dimensional case, is given by

$$\Delta\lambda = \lambda_{\mathrm{B}}\Delta n/n_0 \,, \tag{1.2}$$

where Δn is the amplitude of the index modulation in the grating. These relatively simple concepts have evolved into beautifully complex photonic crystals that include resonator structures, defects, prisms and waveguides. As a result light can be guided through highly functional circuits that include optical filters, efficient couplers, and dispersion elements. The basic linear properties of photonic crystals are described in a recent book [5]. A picture of a photonic crystal formed with air holes in a glass fiber is shown in Fig. 1.1.

In this book the photonic crystal research directions are expanded to include nonlinear optical interactions in the materials that form the crystals. Highly concentrated light fields within the very small spatial features in photonic crystal structures allow us to optimize nonlinear interactions and approach the limit where one can control the pathway, pulse shape, spectrum and phase of one light beam with another light beam. The basic optical element size in photonic crystals is of the order of the optical wavelength in the material, e.g. a few hundred nanometers for optical wavelengths near a micron and refractive indices near two.

Optical nonlinearities described in this volume include both second-order and third-order nonlinear susceptibilities. The nonlinear polarization [6,7] in a material can be written as

$$P_{\mathrm{NL}} = 2dE^2 + 4\chi^{(3)}E^3 \,, \tag{1.3}$$

where d is the second order nonlinear coefficient and $\chi^{(3)}$ is the third order nonlinear coefficient. The third-order nonlinearities, called Kerr nonlinearities, are often described by a nonlinear refractive index, n_2. The index in a nonlinear one-dimensional photonic crystal can then be expressed as

$$n(z) = n_0 + n_2 I + \Delta n(z) \,, \tag{1.4}$$

where n_0 is the linear refractive, n_2 is the nonlinear Kerr refractive index, I is the light intensity and $\Delta n(z)$ is the linear periodic refractive index modulated along the z axis.

Second-order nonlinearities typically are much larger than the third-order nonlinearities. A typical second-order nonlinear material is lithium niobate where the second-order susceptibility for a particular crystalline axis is $d_{13} = 4.3 \times 10^{-23}$ MKS units. Optical nonlinear processes in $\chi^{(2)}$ materials often require phase matching of light at widely different wavelengths. For example, second harmonic generation efficiency is often limited by the different propagation velocities of the fundamental and second harmonic waves that results from normal material dispersion. Photonic crystal bandgaps can be introduced at the fundamental or second harmonic wavelengths to compensate for normal materials dispersion and strongly enhance the second harmonic generation efficiency. Another periodic dielectric technique, called quasi-phase matching, introduces a periodic reversal of the axis of the $\chi^{(2)}$ material in order to achieve phase matching. Quasi-phase matching in materials like $LiNbO_3$ are being used to cascade the $\chi^{(2)}$ process in order to achieve large effective third order nonlinearities.

Silica glass is the prototypical third-order nonlinear material, although its susceptibility is one of the smallest at a value of $n_2 = 2.6 \times 10^{-16} cm^2/W$ at wavelengths near $1.5\mu m$. Other third-order material susceptibilities are often given relative to the silica value. Semiconductors like AlGaAs and glasses like the chalcogenides (e.g. As_2Se_3) can be optimized so that the third order response is nearly a thousand times that of glass. New composite materials are being designed to enhance the internal light fields in order to achieve even higher effective nonlinear susceptibilites.

Semiconductors with electronic bandgaps near the operating light wavelength have even larger suceptibilites nearly 10^5 times that of silica, however, free carriers are often formed at the light intensities required for nonlinear effects. The excited carrier lifetimes limit the response time to the range from picoseconds to nanoseconds. Semiconductors can also be electrically excited to provide optical gain that can compensate for material losses and control the nonlinear interactions.

Many of the interesting nonlinear phenomena, for example, formation of soliton pulses [1–3] that propagate without distortion in the presence of strong dispersion, require nonlinear phase shifts in the structure near π, i.e.

$$\Delta\phi_{\mathrm{NL}} = 2\pi L n_2 I/\lambda = \pi\,, \tag{1.5}$$

where L is the length of the photonic crystal interaction region. For example, in a one-dimensional photonic crystal comprised of a Bragg grating in a silica fiber with a core diameter near ten microns and a grating length of a few centimeters, the required intensities for a π phase shift are near $5\,\mathrm{GW/cm^2}$ and the peak powers are near $1\,\mathrm{kW}$. These very high intensities are near the damage thresholds for many materials. The ideal third-order nonlinear material should have nonlinear susceptibilities a million or even billion times that of silica. These extremely high susceptibilities would allow achieving a π phase shift with peak powers as low as a milliwatt over propagation lengths of only millimeters. At present these extremely nonlinear materials are still a vision, but they are not necessarily out of reach with future material engineering advances.

Nonlinear material response times are important for characterizing optical pulse propagation. For example, sub-picosecond response times are required for soliton pulse propagation where the pulse widths are in the picosecond regime. Nonresonant nonlinearities are usually required for these ultra-fast response times. Silica glass is an example of a materials with ultra-fast nonlinear response times. However, the nonlinear index for silica is very small since the operating wavelengths are typically nearly an order of magnitude longer than those resonant with the silica electronic bandgap. This trade-off between nonlinear strength and response time is described in more detail in Chap. 11.

Nonlinear photonic crystals exhibit a wide range of interesting, often unique phenomena. For example, the gap soliton is an optical pulse that propagates at wavelengths within the photonic bandgap for long distances without distortion at velocities that can vary between zero and the speed of light in the material. Gap solitons have recently been reviewed by de Sterke and Sipe [40]. A more general type of soliton, the Bragg soliton, forms when the pulse wavelengths are either inside, partially inside and partially outside the gap or even for wavelengths near but primarily outside the gap. Bragg solitons[22] can travel at velocities well below the speed of light in the medium even when their wavelengths are primarily outside the gap. Optical dispersion for wavelengths near the edges of the photonic bandgap is typically nearly six orders of magnitude larger than for propagation in the uniform material. This large dispersion along with nonlinear changes in the refractive index caused by intense optical pulses results in Bragg soliton formation lengths that are only centimeters, compared with lengths over a hundred meters required for soliton formation in bare fiber cores. An example of a Bragg soliton forming in only a few centimeters is shown in Fig. 1.2.

At present these interesting solitons are just beginning to be explored experimentally as described in Chaps. 8 and 9. Other interesting phenom-

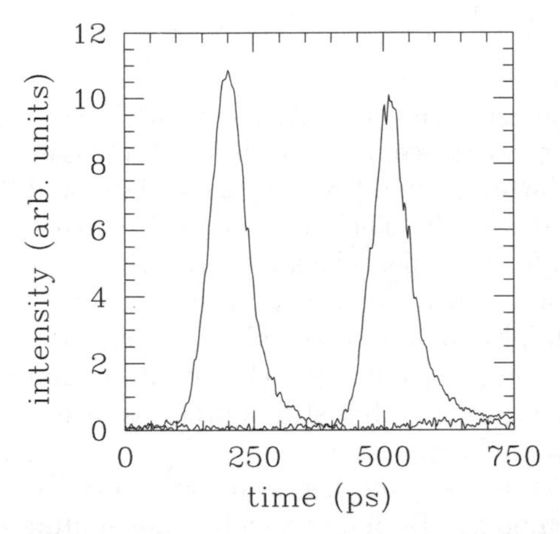

Fig. 1.2. A Bragg soliton has formed at the right from the inident pulse at the left. The soliton pulse wavelength is tuned near the short wavelength edge of a 6 cm long fiber grating. If the nonlinearities had not compensated for disperive pulse broadening, the strong dispersion of the fiber grating would have broadened the pulse by over a factor of two. The propagation velocity in this experiment is nearly half of the normal propagation speed in the glass fiber core. The peak pulse intensity is approximately 2 GW/cm^2

ena include the modulational instability [23], enhanced polarization instabilities [24], and pulse compression [25,26].

Quantum phenomena can play an important role in photonic crystals. The crystal structure can be fabricated so that emitters or aborbers localized in the structure experience enhanced or suppressed vacuum field fuctuations due to peaks or nulls in the electric field distribution in the unit cell of the photonic crystal. This allows control of the fundamental coupling between the optically active material and the light field. Although this is an important topic, it is not covered in this book due to space limitations.

Birefringence associated with photonic crystals can be much larger than that found in the uniform materials. This large birefringence leads to a rich class of nonlinear polarization pulse propagation phenomena [27].

The ability to tailor the amplitude of the periodic index profile, the phase of the photonic crystal period and the crystal birefringent properties allow us to engineer photonic crytals that control both linear and nonlinear propagation phenomena in great detail. For example, by slowly tapering the grating index amplitude at the beginning and end of the photonic crystal, a process called apodization [31–33], we can control the reflection of light at wavelengths near the edges of the photonic bangap. Other examples include phase or amplitude defects in photonic crystals that act as light traps and resonators capable of enhancing the nonlinear interaction by increasing the light intensity near the defect [51]. Photonic crystal defects can also be used to slow or even stop light pulses propagating in the crystal [54].

1.1 History

Nonlinear photonic crystal research began in the late 1970s and early 1980s. The first papers to describe nonlinear properties of periodic dielectric crystals included those authored by Winful, Marburger and Garmire [28], Winful and Cooperman [29] and Winful and Stegeman [30]. These authors realized that a rich new set of phenomena could be found in periodic structures with Kerr nonlinearities including optical bistability, switching and optical limiting.

In the late 1980s Eli Yablonovitch [8] popularized the photonic bandgap concept. This concept led to a spectacular growth in research and fabrication of photonic crystal materials. This explosive growth is still in progress. It has produced a wealth of new science and applications.

Nonlinear effects in photonic crystals took another leap forward when Chen and Mills [9] and Mills and Trullinger [10] found steady state soliton solutions for periodic dielectric structures. This was followed by propagating gap soliton theories and analytical solutions with work by Sipe and Winful [11], de Sterke and Sipe [14], Christodoulides and Joseph [12] and Aceves and Wabnitz [13]. These solitons can be further catagorized by their wavelength relative to the photonic bandgap. Solitons that have frequencies within the linear gap are called gap solitons and have the facinating property that they can in principle travel at any velocity. Physically the nonlinear index shift induced by the pulse light field shifts the photonic bandgap spectrum sufficiently that the pulse can propagate at a wavelength where it would only be an evanescent field in the linear regime. The Bragg soliton [2] regime includes those cases where the incident pulse wavelengths are near the bandgap, but predominantly outside the bandgap.

Numerical modeling has played a major role in this research field. For example, de Sterke and Sipe [15–18] often used numerical modeling in their studies of pulse propagation behavior as a function of pulse energy and detuning of the central pulse frequency from the edge of the photonic bandgap. Broderick, de Serke and Jackson [19] used numerical modeling to better simulate experimental conditions.

Much of the initial work in this field is focused on specific one-dimensional cases, often relating to fiber Bragg gratings. The analysis of the more general class of nonlinear photonic crystals began with a paper by Scalora, et al. [46] in 1994. The number of papers using the more general concept has grown rapidly and many of these ideas are presented in this book. A good example, is a Bloch wave approach for understanding propagation and stability of oblique waves in photonic crystals by Russell and Archambault [34]. It is also important to remember that much of this work is related to previous work in related fields, for example, a paper by Radic, et. al. [20] shows the close analogy between nolinear photonic crystals and distributed feedback lasers.

Discrete solitons are another important class of nonlinear phenomena in periodic dielectric structures. For optical sytems, discrete solitons were first studied theoretically in nonlinear waveguide arrays by Christodoulides and

Joseph [4] in 1988 and later by Aceves, et al. [58]. The periodic structures used to study discrete solitons are waveguide arrays where the waveguide width is typically a few microns and the waveguide separation is slightly larger than the waveguide width. This periodic arrangement allows single mode propagation in each waveguide with a small evanescent coupling between the waveguides. In the linear case, light launched into one of the waveguides in the array of many waveguides evanescently couples between the guides and spreads quickly throughout the array. In the nonlinear case where there is a π nonlinear phase shift in a distance of the order of the evanescent waveguide coupling length, the nonlinear phase shift acts to prevent evanescent spreading and the light remains concentrated in a narrow range of waveguides or a single waveguide forming a "discrete" spatial soliton. The description of this system involves a discrete set of coupled nonlinear propagation equations, each equation in the set corresponding to one of the waveguides. The period of the waveguide array is chosen so that the evanescent coupling between waveguides is small enough to allow a large nonlinear interaction length, i.e. nonlinear phase shifts near π over the distance required to couple light from one guide to another. This coupling distance is typically of the order of millimeters or centimeters, much larger than the waveguide width.

Experimental work in nonlinear photonic crystal began in the early 1990s. The first experimental paper by Larochelle, Hinino, Mizrahi and Stegeman [47] demonstrated all-optical switching in a fiber grating using cross-phase modulation. Brown's group [48,49] made a detailed study of optical switching using a corrugated silicon-on-insulator optical waveguide. Kerr nonlinearities in the silicon waveguide result when carriers are generated by radiation with wavelengths near the silicon electronic bandedge. They observed switching, hysteresis, and instabilities. The nonlinear response times in this system are comparable to the optical pulse widths, making propagation phenomena difficult to interpret.

Two groups pioneered the experimental studies of nonlinear pulse propagation in photonic crystals. The Bragg soliton regime was observed for wavelengths near the photonic bandgap edge in fiber Bragg gratings by Eggleton, et. al. [2,22,36]. Gap soliton formation, where a dominant portion of the pulse energy is within the linear photonic bandgap was first opbserved by the Southhampton group led by N. Broderick [3,38] using fiber Bragg gratings with relatively narrow bandgaps and long optical pulse in the nanosecond range. The long pulse lengths and associated narrow optical spectrum allow lower powers for initiating the soliton formation. Again this is understood from the simple physical model, i.e. the nonlinear index shift must shift the bandgap by at least the width of the incident pulse spectrum, thus longer duration pulses with smaller spectral width require less index shift and smaller incident light intensities. Pulse compression and switching were also observed by the Southhampton group in the important 1.5 μm communications wavelength band.

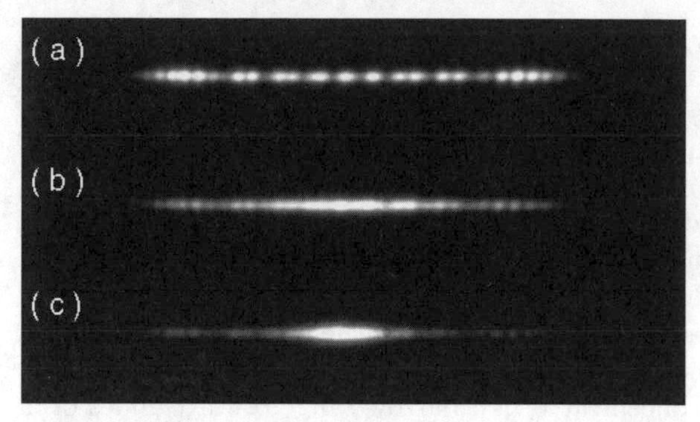

Fig. 1.3. Discrete soliton forming in a waveguide array. An array of parallel waveguides fabricated on a planar guiding structure is viewed at the edge of the plane where the waveguides terminate. In (**a**) an excitation of the guides in the linear intensity regime results in a spreading of the light to many waveguides as seen by the many discrete points of light with a variety of intensities. As the exciting light intensity is increased in (**b**) and (**c**) a soliton forms that results in a concentration of the light in a small number of waveguides

Discrete solitons have been observed experimentally [55] using AlGaAs planar waveguide arrays. Strong confinement in a narrow range of waveguides is observed in the nonlinear regime as shown in Fig. 1.3.

Discrete solitons in either one or two-dimensional waveguide arrays can move with high mobilities throughout the arrays by applying phase "chirps" to optical field across the waveguide mode. These mobile localized light solitons can be used for optical processing. Christodoulides and his group have recently described a variety of techniques to manipulate discrete solitons [56,57]. For example, a soliton moving through the waveguide array can be "blocked" by a second control soliton located in a single waveguide and can then be redirected toward a new location in the array. These innovative, optically controlled spatial switching techniques could find applications in future photonic circuits.

1.2 Applications

Device dimensions of the order of an optical wavelength are easily achieved using photonic crystals. For example, a laser resonator has been demonstrated [50] using a single defect in a planar photonic crystal. The dimension of the defect is only a few tenth of a micron on a side. Highly compact multifunctional photonic circuits may well be in the future for nonlinear photonic crystals.

For applications in optical communications, photonic crystals can form a narrow band filter for selecting a particular wavelength from the hundreds of different wavelength channels that comprise a wavelength division multiplexed (WDM) system. A nonlinear shifting of the central filter frequency could then allow dynamic control of WDM channel routing. Pulse shaping and compression using Bragg soliton effects [44] are additional examples of possible applications. Pulse compression can be achieved by tapering the amplitude of the photonic crystal to force the soliton pulse width to nonlinearly narrow as it propagates. This pulse compression could be used to generate ultrashort optical pulses for a variety of applications. These two examples illustrate only a few of the many possible future applications.

Ultimately one can conceive of a multi-functional photonic network that could rival the functionality of electronic ICs. However, the photonic IC will probably address different functions than those commonly used in digital computation. One important device that has not been achieved with the required range of parameters is an optical buffer that can delay optical pulses by times up to the microsecond range without significant pulse distortion. Another unsolved problem is efficient coupling from optical fiber modes (typically near 10 microns in diameter) or free space optical modes into the submicron sized waveguides and resonators in photonic crystals.

1.3 Future Directions and Challenges

A very strong optical nonlinearity is required to achieve all-optical control of light at very small dimensions. For example, in order to achieve a π phase shift in a 10 micron long arm of a Mach-Zehnder interferometer that could route light from one of its output ports to the other requires a third-order nonlinear response nearly ten orders of magnitude larger than that of glass for control light powers near 10 mW. These super-nonlinear materials have not been developed at present. These extreme nonlinearities can in principle be obtained, possibly with resonant enhancement over narrow bandwidths. Both classical resonator [51] and electromagnetically induced transparency [52] schemes have been proposed. Collective effects are another interesting possibility for enhanced nonlinearities. Light fields in nonlinear materials have also been enhanced by small particle plasmon resonances [53].

Another important consideration for future materials development is the ratio of nonlinear phase shift to nonlinear absorption. For third-order nonlinearities, multiphoton loss and the nonlinear coefficient are fundamentally related so that nonlinear losses are unavoidable. These issues are discussed in more detail in Chap. 11. Linear losses due to material absorption or scattering in the phtonic crystal are also important. They can easily limit processes including slow light propagation and soliton formation.

Photonic crystal fabrication is progressing rapidly into two and three dimensions. The two dimensional nonlinear photonic crystal case is discussed in Chap. 16. In both two and three dimension spatial and temporal propagation properties are very intersting. Light bullets, as spatio-temporal solitons are sometimes called, are being explored in this regime. Many exciting results are expected in this area.

Defects in photonic crystals act as resonant structures that can enhance the nonlinear effects and lead to interesting new phenomena. Nonlinear trapping of light at defects has recently been analyzed theoretically [54]. Experimental work in this area is just beginning. Photonic crystals with gain media is still another area. Gain in semiconductor optical amplifiers with periodic structures, an important topic for applications, is considered in Chap. 13. Second-order nonlinear effects are very interesting and are described theoretically in Chaps. 5, 6 qnd 7. Again experimental work is just beginning.

1.4 Book Outline

This book is organized in five parts. After the introduction, Part 2 (Nonlinear Photonic Crystal Theory) describes the theory of nonlinear photonic crystals including polarization effects, second-order nonlinearities and new work on Raman gap solitons and simultons. Part 3 (Nonlinear Fiber Grating Experiments) describes experiments in fiber sytems where the photonic crystal is one dimensional. These experiments where the pioneering propagation experiments, primarily because of the simplicity of the photonic bandgap and the availbility of good quality fiber gratings. Part 4 (Novel Nonlinear Periodic Systems) explores new materials and structures, the chalcogenide glasses, microstructred fibers, and semiconductor optical amplifiers. There is an exciting chapter on atomic solitons in optical lattices. One finds that similar nonlinear photonic crystal phenomena are found when atomic waves replace light waves and light lattices replace phtonoic crystals. Part 5 (Spatial Solitons in Photonic Crystals) describes higher dimension structures and spatial "discrete" solitons in periodic dielectric structures.

References

1. L.F. Mollenauer, R.H. Stolen and J.P. Gordon: Phys. Rev. Lett. **45**, 1095 (1980)
2. B.J. Eggleton, R.E. Slusher, C.M. de Sterke, P.A. Krug and J.E. Sipe: Phys. Rev. Lett. **76**, 1627 (1996)
3. D. Taverner N.G.R. Broderick, D.J. Richardson, R.I. Laming, and M. Ibsen: Opt. Lett. **23**, 328 (1998); N.G.R. Broderick, D.J. Richardson, and M. Ibsen: Opt. Lett. 25, 536 (2000)
4. D.N. Christodoulides and R.I. Joseph: Opt. Lett. **13**, 794 (1988)
5. J.D. Joannopoulos, R.D. Meade, J.N. Winn: *Photonic Crystals*, (Princeton University Press, 1995)
6. G.P. Agrawal: *Nonlinear fiber optics*, 2nd Edn. (Academic Press, San Diego, 1995)
7. B.E.A. Saleh and M.C. Teich: *Fundamentals of Photonics*, (Wiley & Sons, Ney York Chichester 1991)
8. E. Yablonovitch, T.J. Gmitter: Phys. Rev. Lett. **63**, 160 (1989)

9. W. Chen and D.L. Mills: Phys. Rev. Lett. **58**, 160 (1987)
10. D.L. Mills and S.E. Trullinger: Phys. Rev. B **36**, 947 (1987)
11. J.E. Sipe and H.G. Winful: Opt. Lett. **13**, 132 (1988)
12. D.N. Christodoulides and R.I. Joseph: Phys. Rev. Lett. **62**, 1746 (1989)
13. A.B. Aceves and S. Wabnitz: Phys. Lett. A **141**, 37 (1989)
14. C.M. de Sterke and J.E. Sipe: Phys. Rev. A **39**, 5163 (1989)
15. C.M. de Sterke and J.E. Sipe: Phys. Rev. A **39**, 5163 (1989)
16. C.M. de Sterke and J.E. Sipe: Phys. Rev. A **42**, 550 (1990)
17. C.M. de Sterke and J.E. Sipe: Phys. Rev. A **42**, 2858 (1990)
18. C.M. de Sterke and J.E. Sipe: Phys. Rev. A **42**, 2858 (1990)
19. N.G.R. Broderick, C.M. de Sterke and K.R. Jackson: Opt. Quantum Elec. **26**, S219 (1994)
20. S. Radic, N. George and G.P. Agrawal: Opt. Lett. **19**, 1789 (1994)
21. H.G. Winful: Opt. Lett. **11**, 33 (1986)
22. B.J. Eggleton, C.M. de Sterke, and R.E. Slusher: J. Opt. Soc. Am. B **16**, 587 (1999)
23. B.J. Eggleton, C.M. de Sterke, A.B. Aceves, J.E. Sipe, T.A. Strasser, and R.E. Slusher: Opt. Comm. **149**, 267 (1998)
24. R.E. Slusher, S. Spalter, B.J. Eggleton, S. Pereira and J.E. Sipe: Opt. Lett. **25**, 749 (2000)
25. G. Lenz, B.J. Eggleton and N. Litchinitser: J. Opt. Soc. Am. B **15**, 715 (1997)
26. G. Lenz and B.J. Eggleton: J. Opt. Soc. Am. B **15**, 2979 (1998)
27. S. Pereira, J.E Sipe, and R.E. Slusher: submitted to JOSA B.
28. H.G. Winful et. al.: Appl. Phys. Lett. **35**, 379 (1979)
29. H.G. Winful and C.D. Cooperman: Appl. Phys. Lett.**40**, 298 (1982)
30. H.G. Winful and G.I. Stegeman: Proc. SPIE **517**, 214 (1984)
31. T.A. Strasser, P.J. Chandonnet, J. DeMarko, C.E. Soccolich, J.R. Pedrazzani, D.J. DiGiovanni, M.J. Andrejco, and D.S. Shenk, in *Optical Fiber Communication Conference.* Vol. 2 of 1996 OSA Technical Digest Series (Optical Society of America, Washington, D.C., 1996): postdeadline paper PD8-1
32. P.S. Cross and H. Kogelnik: Opt. Lett. **1**, 43 (1977)
33. B. Malo, D.C. Johnson, F. Bilodeau, J. Albert, and K.O. Hill: Electronic Letters **31**, 223 (1995)
34. P.S.J. Russell: J. Mod. Opt. **38**, 1599 (1991)
35. B.J. Eggleton, C.M. de Sterke, and R.E. Slusher: Opt. Lett. **21**, 1223 (1996)
36. B.J. Eggleton, C.M. de Sterke, and R.E. Slusher: J. Opt. Soc. Am. B **14**, 2980 (1997)
37. C.M. de Sterke, B.J. Eggleton, and P.A. Krug: J. Lightwave Technology **15**, 2908 (1997)
38. D. Taverner, N.G.R. Broderick, D.J. Richardson, M. Ibsen, and R.I. Laming: Opt. Lett. **23**, 259 (1998)
39. H.G. Winful: Appl. Phys. Lett. **46**, 527 (1985)
40. C.M. de Sterke and J.E. Sipe: in *Progress in Optics XXXIII*, E. Wolf, ed., (Elsevier, Amsterdam 1994), Chap. III-Gap Solitons
41. C.M. de Sterke, K.R. Jackson, and B.D. Robert: J. Opt. Soc. Am B **8**, 403 (1991)
42. M.J. Steel and C.M. de Sterke: Phys. Rev. A **49**, 5048 (1994)
43. C.M. de Sterke: Opt. Express **3**, 405 (1998)
44. B.J. Eggleton, G. Lenz, R.E. Slusher, and N.M. Litchinitser: Appl. Opt. **37**, 7055 (1998)

45. B.J. Eggleton, C.M. de Sterke, A.B. Aceves, J.E. Sipe, T.A. Strasser, and R.E. Slusher: Optics Comm. **149**, 267 (1998)
46. M. Scalora, R.J. Flynn, S.B. Reinhardt, R.L. Fork, M.J. Bloemer, M.D. Tocci, C.M. Bowden, H.S. Ledbetter, J.M. Bendickson, J.P. Dowling, and R.P. Leavitt: Phys. Rev. E. **54**, 1078 (1996)
47. S. Larochelle, V. Mizrahi, and G. Stegeman: Electron. Lett. **26**, 1459 (1990)
48. N.D. Sankey, D.F. Prelewitz, and T.G. Brown: Appl. Phys. Lett. **60**, 1427 (1992)
49. N.D. Sankey, D.F. Prelewitz, and T.G. Brown: J. App. Phys. **73**, 1 (1993)
50. O. Painter, R.K. Lee, A. Scherer, A. Yariv, J.D. O'Brien, P.D. Dapkus, I. Kim: Science **284**, 1819 (1999)
51. J. Heebner and R.W. Boyd: Optics Lett. **24**, 847 (1999)
52. S.E. Harris and L.V. Hau: Phys. Rev. Lett. **82**, 4611 (1999)
53. Y. Hamanaka, A. Nakamura, S. Omi, N. Del Fatti, F. Vallee, and C. Flytzanis: Appl. Phys. Lett. **75**, 1712 (1999)
54. R.H. Goodman, R.E. Slusher, and M.I. Weinstein: to be published in Jour. Opt. Soc. Am. B, (2002)
55. H.S. Eisenberg, Y. Silberberg, R. Morandotti, A.R. Boyd and J.S. Aitchson: Phys. Rev. Lett. **81**, 3383 (1998); R. Morandotti, U. Peschel, J.S. Aitchson, H.S. Eisenberg, and Y. Silberberg: Phys. Rev. Lett. **83**, 2726 (1999)
56. D.N. Christodoulides and E.D. Eugenieva: Phys. Rev. Lett. **87**, (2001).
57. E.D. Eugenieva, N.K. Efremidis, and D.N. Christodoulides: Optics Lett. **26**, (2001)
58. A.B. Aceves, C. Deangelis, S. Trillo, and S. Wabnitz: Optics Lett. **19**, 332 1994

Part I

Nonlinear Photonic Crystal Theory

2 Theory of Nonlinear Pulse Propagation in Periodic Structures

A. Aceves, C.M de Sterke, and M.I. Weinstein

2.1 Introduction

We consider here the propagation of intense light in a periodic structure. The models that we discuss are one-dimensional; this means that only a single spatial direction, that in which the light propagates, is relevant, and that the refractive index only depends on this spatial coordinate. Of course such a geometry is highly idealized, and, strictly speaking, only applies to a plane wave propagating through a periodically stratified medium, in a direction orthogonal to the layers. However, it is a very good approximation for light propagating through an optical fiber with a grating written in the core, or for a ridge waveguide with a periodic variation in its thickness. The reason for this is that the light is in a mode of the guided-wave structure and that the perturbation is so weak that coupling into other modes, including radiation modes, can often be neglected. A one-dimensional treatment is thus an excellent approximation, provided that the relevant mode is sufficiently far from cut-off. In practice the geometries are designed such that this is always true.

Assuming then that we can use a one-dimensional model, we need to solve the wave equation for the electric field $\boldsymbol{E}(z,t)$

$$\frac{\partial^2 \boldsymbol{E}}{\partial z^2} - \frac{\varepsilon(z)}{c^2}\frac{\partial^2 \boldsymbol{E}}{\partial t^2} = \mu_0 \frac{\partial^2}{\partial t^2}\left(\boldsymbol{P}_{\mathrm{NL}}(\boldsymbol{E})\right), \tag{2.1}$$

where ε is the dielectric function, c is the speed of light in vacuum, μ_0 is the permeability of vacuum, t is time, and z is the propagation direction; the dielectric function depends periodically on z, and may also depend on frequency, though we do not consider this case here. We thus take the dielectric function to satisfy

$$\varepsilon(z) = \varepsilon(z+d), \tag{2.2}$$

where d is the period. Finally, the term $\boldsymbol{P}_{\mathrm{NL}}(\boldsymbol{E})$ represents the medium's nonlinear polarization, which may be very complicated. Here, however, we consider a simple Kerr nonlinearity, according to which

$$\boldsymbol{P}_{\mathrm{NL}}(\boldsymbol{E}) = \chi^{(3)}\boldsymbol{E}\boldsymbol{E}\boldsymbol{E}, \tag{2.3}$$

where $\chi^{(3)}$ is the third order susceptibility. In practice $\chi^{(3)}$ may be a periodic function of z as well, but we neglect this here, though it can easily be included. The form of the nonlinearity in (2.3) assumes an instantaneous response of the medium to the field. In the physical contexts which interest us here (e.g. gratings that are written in the core of an optical fiber [2] and corrugated semiconductor channel waveguides [3]) nonlinearity is associated with the bound electrons of the medium. As a consequence, the nonlinearity is fairly weak and has a response time of the order of femtoseconds. The model (2.1)–(2.3) incorporates photonic band dispersion and nonlinearity. The envelope of slowly varying fields of appropriate carrier frequency and amplitude (neglecting coupling to higher harmonics) are governed by nonlinear coupled mode equations (2.28) (see Sect. 2.3), which give good agreement with certain experiments. However, as shown by Goodman et al. [1], temporal dispersion due to finite response time can play an important role. Due to the generation of *resonant* higher harmonics by the instantaneous nonlinearity, the nonlinear wave equation (2.1)–(2.3) is expected to exhibit shock formation, an unbounded growth of field gradients within a finite time, on the time (length) scale where dispersion and nonlinearity balance. The inclusion of temporal dispersion, where (2.3) is replaced by a nonlocal in time relation, modeling the delayed response of medium to field, is shown to be a stabilizing mechanism, precluding shock formation and ensuring the validity of the nonlinear coupled mode equations.

2.2 Gratings at Low Intensities

Before tackling the nonlinear problem, let us first consider the linear problem in which $\boldsymbol{P}_{\mathrm{NL}}(\boldsymbol{E}) = 0$. The wave equation (2.1) then reduces to the one dimensional linear wave equation with spatially periodic coefficients. This can be solved in principle analytically using the spectral theory of linear operators with periodic coefficients (Floquet-Bloch theory) or numerically. However, a more efficient approximate analytical approach is possible in the regimes which interest us. This is the regime where the electric field is nearly monochromatic and may be viewed as a slowly varying envelope of a highly oscillatory carrier wave. In the cases of interest to us, the wavelength of light and the medium periodicity are of the same order of magnitude and the envelope of the electric field varies slowly on this scale. Similarly, in the relevant structures [3], the electric field varies sinusoidally in space, with envelope variations in the amplitude and phase that vary slowly on the scale of a spatial period d. These slow modulations in space and time of the underlying plane waves of the uniform medium are induced by the weak periodic variations in the refractive index. Solution of the wave equation for the full field requires the resolution of temporal and spatial scales ranging over many orders of magnitude. Goodman et al. [1] present a rigorous and quantitative comparison between the full electric field predicted by a Maxwell model (the anharmonic Maxwell Lorentz equations) and the electric field envelope, governed by the nonlinear coupled mode equations. Test numerical simulations comparing full field to envelope computations were done by Goodman et al. [1]. Simulations of the full field took on the order of hours where for the envelope equations, simulations took on the order of minutes. These simulations were for wave fields with approximately 60 carrier oscillations under the dominant part of the envelope. Physical fields arising in experiments (e.g. wavelength $\lambda = 1.5\,\mu\text{m}$ with pulse width of about 30 picoseconds FWHM) have on the order 10,000 carrier oscillations. While the computation time for the envelope is unchanged, a calculation of the full field requires a huge computational effort. Due to these issues of efficiency as well as the fact that it is the envelope that contains the essential information on the location and dynamics of the field energy, it is natural to focus, as we do below, on a description in terms of the envelope. We should mention that gratings have been analyzed without using an envelope function approach. This includes particularly the pioneering work of Chen and Mills [4], who introduced the term *gap solitons* to describe self-localized pulses in nonlinear periodic media.

Since the dielectric function in the structures that we are considering is periodic, it may be written in a Fourier series as

$$\varepsilon(z) = \sum_l \varepsilon_l e^{2ilkz}, \tag{2.4}$$

where $k = k_B \equiv \pi/d$. Here, we shall consider lossless media. In this case the dielectric function is real-valued, so that $\varepsilon^\star_{-l} = \varepsilon_l$, and ε_0 is real. In this section we also only consider the experimentally relevant situation in which the grating is not too *deep*. This means that for all $l \neq 0$, we have $\varepsilon_l \ll \varepsilon_0$. In both the fiber and semiconductor geometries [3,5–7], for example, $|\varepsilon_{\pm 1}| < 0.001$, whereas the higher order coefficients are even smaller. We finally note that the phases of the various Fourier coefficients (with the exception of ε_0) depend our choice of the position of the origin. Using this freedom we now take $\varepsilon_{\pm 1}$ to be real and positive.

We now embark on an approximate solution to (2.1) in the linear case. First, note that in the case of a uniform medium ($\varepsilon(z) \equiv \varepsilon_0$), $E(z,t)$ satisfies the linear homogenous wave equation with plane wave solutions:

$$E(z,t) = \mathcal{E} e^{i(kz-\omega t)}, \quad \text{where} \quad \omega = \pm\sqrt{\varepsilon_0}k/c\,. \tag{2.5}$$

Here, $\mathcal{E}$ is an arbitrary constant. All solutions of the spatially homogeneous equation can be obtain by Fourier superposition.

For a weak grating (ε_l small), it is natural to seek solutions which are a superposition of such plane waves, with *slowly varying amplitude functions*. Thus we seek:

$$\boldsymbol{E}(z,t) = \sum_m \mathcal{E}_{2m+1}(z,t) e^{i[(2m+1)kz-\omega t]} + c.c.\,, \tag{2.6}$$

where *c.c.* denotes the complex conjugate, the $\mathcal{E}_{2m+1}$ are amplitudes and the frequency ω is as yet unspecified.

Here, however, we take a different approach [8]. We first note that the regime where backward and forward waves interact occurs when the frequency of the field is close to one of the Bragg resonances of the periodic medium. In fact, for the structures that are of relevance here, the field is tuned close to the lowest Bragg resonance, for which

$$\lambda = \lambda_{\mathrm{B}} \equiv 2\sqrt{\epsilon_0}d = \frac{2\pi\sqrt{\epsilon_0}}{k_{\mathrm{B}}} . \tag{2.7}$$

Considering here such a situation, we take

$$\omega_B = \sqrt{\varepsilon_0} k_B / c ,$$

which corresponds to the centre of the resonance; any small frequency differences can be included by the explicit time dependence of the envelopes $\mathcal{E}_m$. Under this condition, the envelopes are obviously slowly varying in time. However, for shallow grating they are also slowly varying in space and we therefore refer to them as *envelope functions.*

We are now also in a position to determine the dominant terms in (2.6). In the absence of the grating, the only term is that with $m = 0$ and so we expect this term to be large. Since the grating is in fact expected to reflect the incoming light by Bragg reflection, the other dominant term is that with the same wave number, but is backward propagating, i.e. that with $m = -1$. All other terms can be assumed to be smaller, as we demonstrate below.

With the above remarks in mind, we now substitute expressions (2.6) and (2.4) into the wave equation (2.1) to find (setting $k = k_B$ and $\omega = \omega_B$)

$$\begin{aligned} c^2 \sum_m & \left[-(2m+1)^2 k^2 \mathcal{E}_{2m+1} + 2i(2m+1)k \frac{\partial \mathcal{E}_{2m+1}}{\partial z} + \frac{\partial^2 \mathcal{E}_{2m+1}}{\partial z^2} \right] \\ & \times e^{i(2m+1)kz} + \sum_l \sum_m \varepsilon_l \left[\omega^2 \mathcal{E}_{2m+1} + 2i\omega \frac{\partial \mathcal{E}_{2m+1}}{\partial t} - \frac{\partial^2 \mathcal{E}_{2m+1}}{\partial t^2} \right] \\ & \times e^{i(2l+2m+1)kz} = 0 . \end{aligned} \tag{2.8}$$

We first consider the small envelopes $\mathcal{E}_m$ with $n \neq 0, -1$. Since, as verified later, the $\mathcal{E}_m$ are slowly varying, we ignore any of their derivatives. This then leads to

$$-c^2 k^2 (2n+1)^2 \mathcal{E}_{2n+1} + \varepsilon_0 \omega^2 \mathcal{E}_{2n+1} + \omega^2 \sum_{m \neq n} \varepsilon_{n-m} \mathcal{E}_{2m+1} = 0 . \tag{2.9}$$

Now in the summation the dominant terms are again those with $m = -1, 0$, and we thus find that

$$\mathcal{E}_{2n+1} = \frac{(\varepsilon_n/\varepsilon_0)\,\mathcal{E}_1 + (\varepsilon_{n+1}/\varepsilon_0)\,\mathcal{E}_{-1}}{(2n+1)^2 - 1} , \tag{2.10}$$

expressing the weak envelopes in terms of the two dominant ones.

We now return to general result (2.8) and consider the dominant amplitudes $\mathcal{E}_{0,-1}$. Since these terms are large, we cannot neglect their spatial derivatives. In fact we keep first derivatives in space and time associated with the dominant amplitudes. This leads to

$$\begin{aligned} i\left(+\frac{\partial \mathcal{E}_1}{\partial z}+\frac{1}{V}\frac{\partial \mathcal{E}_1}{\partial t}\right)+\kappa\mathcal{E}_{-1}+\frac{k}{V}\frac{\varepsilon_1}{\varepsilon_0}\frac{\partial \mathcal{E}_{-1}}{\partial t} \\ +\beta\mathcal{E}_1+\frac{1}{2k}\left(\frac{\partial^2 \mathcal{E}_1}{\partial z^2}-\frac{1}{V^2}\frac{\partial^2 \mathcal{E}_1}{\partial t^2}\right)=0\,, \\ i\left(-\frac{\partial \mathcal{E}_{-1}}{\partial z}+\frac{1}{V}\frac{\partial \mathcal{E}_{-1}}{\partial t}\right)+\kappa^\star\mathcal{E}_1+\frac{k}{V}\frac{\varepsilon_{-1}}{\epsilon_0}\frac{\partial E_1}{\partial t} \\ +\beta\mathcal{E}_{-1}+\frac{1}{2k}\left(\frac{\partial^2 \mathcal{E}_{-1}}{\partial z^2}-\frac{1}{V^2}\frac{\partial^2 \mathcal{E}_{-1}}{\partial t^2}\right)=0\,, \end{aligned} \tag{2.11}$$

where result (2.10) was used and

$$\begin{aligned} V &= \omega/k = \sqrt{\varepsilon_0}/c\,, \quad (\omega=\omega_B,\, k=k_B) \\ \kappa &= \frac{k\varepsilon_1}{2\varepsilon_0}+\frac{k}{2}\sum_{m\neq 0,-1}\frac{(\varepsilon_{-m}\varepsilon_{m+1}/\varepsilon_0^2)}{[(2m+1)^2-1]}\,, \\ \beta &= \frac{k}{2}\sum_{m\neq 0,-1}\frac{|\varepsilon_m|^2/\varepsilon_0^2}{[(2m+1)^2-1]}\,. \end{aligned} \tag{2.12}$$

Equations (2.11) are somewhat unwieldy, but they can be simplified considerably. First note that to leading order

$$\begin{aligned} \frac{\partial^2 \mathcal{E}_1}{\partial z^2}-\frac{1}{V^2}\frac{\partial^2 \mathcal{E}_1}{\partial t^2} &= \left(\frac{\partial}{\partial z}-\frac{1}{V}\frac{\partial}{\partial t}\right)\left(\frac{\partial}{\partial z}+\frac{1}{V}\frac{\partial}{\partial t}\right)\mathcal{E}_1 \\ &= i\kappa\left(\frac{\partial}{\partial z}-\frac{1}{V}\frac{\partial}{\partial t}\right)\mathcal{E}_{-1} = |\kappa|^2\mathcal{E}_1\,, \end{aligned} \tag{2.13}$$

which, since this expression enters (2.11) at the highest order, is sufficient. Using this and the equivalent expression for $\mathcal{E}_{-1}$, we thus find that

$$\begin{aligned} i\left(+\frac{\partial \mathcal{E}_1}{\partial z}+\frac{1}{V}\frac{\partial \mathcal{E}_1}{\partial t}\right)+\kappa\mathcal{E}_{-1}+\frac{k}{V}\frac{\varepsilon_1}{\varepsilon_0}\frac{\partial \mathcal{E}_{-1}}{\partial t}+\beta'\mathcal{E}_1=0\,, \\ i\left(-\frac{\partial \mathcal{E}_{-1}}{\partial z}+\frac{1}{V}\frac{\partial \mathcal{E}_{-1}}{\partial t}\right)+\kappa^\star\mathcal{E}_1+\frac{k}{V}\frac{\varepsilon_{-1}}{\epsilon_0}\frac{\partial \mathcal{E}_1}{\partial t}+\beta'\mathcal{E}_{-1}=0\,. \end{aligned} \tag{2.14}$$

where $\beta'=\beta+k|\kappa|^2/2$.

Note first of all that we have achieved our aim, since we have, in effect, replaced the wave equation by evolution equations for the envelopes. However, these are not the equations that are most commonly used. In fact, in (2.14) the terms with β', $\varepsilon_{\mp 1}$, and the second term in the definition of κ are smaller than the other terms if the grating is weak. These are therefore usually dropped leading to the well known *coupled mode equations*

$$
\begin{aligned}
&i\left(+\frac{\partial \mathcal{E}_+}{\partial z}+\frac{1}{V}\frac{\partial \mathcal{E}_+}{\partial t}\right)+\kappa\mathcal{E}_-=0\,,\\
&i\left(-\frac{\partial \mathcal{E}_-}{\partial z}+\frac{1}{V}\frac{\partial \mathcal{E}_-}{\partial t}\right)+\kappa\mathcal{E}_+=0\,,
\end{aligned}
\tag{2.15}
$$

where κ is now real and we have made the replacements $\mathcal{E}_{\pm 1} \to \mathcal{E}_{\pm}$. The parameter κ, which is proportional to $k\varepsilon_1/\varepsilon_0$, is small compared to k (as mentioned, by a factor of at least 10^{-3}) and therefore equations (2.15), which govern (for $V > 0$) respectively the right- and left- going directional derivatives of $\mathcal{E}_+$ and $\mathcal{E}_-$, assert that the envelope functions $\mathcal{E}_\pm$ are slowly varying and in fact are essentially functions of the variables: $Z = k\varepsilon_1/\varepsilon_0 z$ and $T = k\varepsilon_1/\varepsilon_0 t$. Note that in general, if the unperturbed dynamics governing propagation in a spatially homogeneous medium is dispersive, for example due to material or temporal dispersion, the factor $V = c/\sqrt{\epsilon_0}$ in (2.11) and (2.15) is replaced by the *group velocity* at the Bragg resonance.

We now discuss the plane waves and dispersion relation associated with (2.15). If we drop the terms in (2.15) that are proportional to κ we find that the solutions are non-interacting forward and backward propagating waves. For $\kappa \neq 0$, the forward and backward propagating waves are coupled. We find the dispersion relation of (2.15) by seeking plane wave solutions, i.e.

$$
\mathcal{E}_\pm = A_\pm e^{i(V\Delta t - Qz)}\,, \tag{2.16}
$$

where the $A_\pm$ are constant amplitudes and Q is the wave number of the envelope and $V\Delta$ is the frequency of the envelope. This corresponds to an electric field wave number and electric field frequency:

$$
k = \frac{\pi}{d} + Q\,, \quad \omega = V\left(\frac{\pi}{d} + \Delta\right). \tag{2.17}
$$

Thus, since $\omega_B = V\pi/d$ is the frequency at the Bragg resonance, the *detuning* Δ is, apart from a factor V, the difference of the frequency of the electric field and the Bragg frequency of the grating.

Substituting (2.16) into (2.15) we find the algebraic eigenvalue equations

$$
\begin{aligned}
-(\Delta - Q)A_+ + \kappa A_- &= 0\,,\\
\kappa A_+ - (\Delta + Q)A_- &= 0\,.
\end{aligned}
\tag{2.18}
$$

Eliminating the amplitudes $A_\pm$ we find the *dispersion relation*

$$
\Delta = \pm\sqrt{Q^2 + \kappa^2}\,, \tag{2.19}
$$

which is illustrated in Fig. 2.1. The dispersion clearly exhibits a frequency region in which plane wave solutions of the form (2.16) do not exist. Rather,

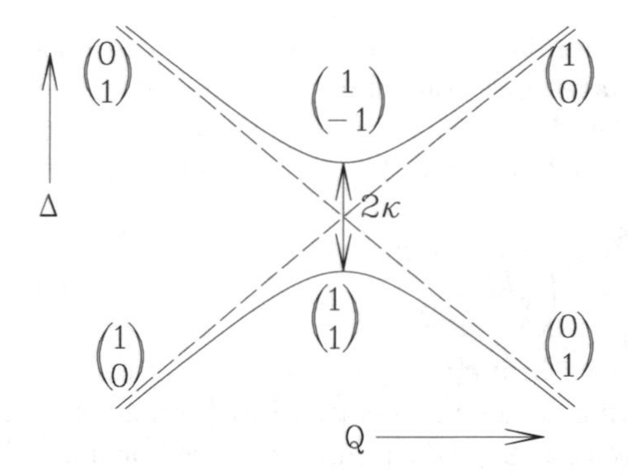

Fig. 2.1. Dispersion relation associated with coupled mode equations (2.15), showing the detuning Δ versus the local wave number Q. Also indicated are the (unnormalized) eigenstates (see text). The dashed lines indicate the dispersion relation in the limit $\kappa \to 0$

the solutions in this *one-dimensional band gap*[1] are evanescent, a behavior that is associated with Bragg reflection by the grating. Note that the width of the gap is $2V\kappa$, and is by (2.12) thus proportional to the lowest DC Fourier component of the grating. The dashed lines in Fig. 2.1 show the dispersion relation when $\kappa \to 0$, i.e. in the limit in which the grating disappears. In this case the solutions are simply forward and backward propagating plane waves, traveling at the group velocity of the medium.

From dispersion relation (2.19) we can easily find the group velocity in the presence of the grating. Defining v to be the group velocity in units of V we find

$$v_{\pm} = \frac{1}{V}\frac{\partial \omega}{\partial} = \frac{\partial \Delta}{\partial Q} = v = \pm\sqrt{1 - \kappa^2/\Delta^2}\,. \tag{2.20}$$

Thus, consistent with Fig. 2.1, $v = 0$ at the edges of the photonic band gap ($|\Delta| \to \kappa$), whereas far from the band gap ($|\Delta| \to \infty$), the group velocity is essentially that of the underlying uniform medium. For intermediate values of the detuning, the group velocity smoothly varies between these extreme values. Expression (2.20) allows us to rewrite dispersion relation (2.19) in a manner convenient for the discussion of Bloch functions below:

$$Q = \kappa\gamma|v|\,, \quad \Delta_{\pm} = \pm\kappa\gamma\,, \quad \gamma = 1 - v^{2-\frac{1}{2}}\,. \tag{2.21}$$

With each eigenvalue in (2.19) we can associate an eigenvector indicating the mixing of the forward and backward propagating mode by the grating[2];

[1] Though the band gap is understood to be one-dimensional, for convenience we henceforth simply refer to it simply as the band gap.

[2] The eigenvectors indicate the eigenstates of the field inside the grating. Since the grating is periodic, the eigenstates are Bloch functions [9]. For this reason we refer to the eigenvectors as Bloch functions as well.

these are indicated in Fig. 2.1 for a number key positions on the dispersion relation. In general, the eigenvectors are most conveniently expressed in terms of the group velocity v. To do this, we define each state on the upper branch of Fig. 2.1 to be labeled by its associated velocity. The Bloch function is then given by $\boldsymbol{v}_+$, where

$$\boldsymbol{v}_+ = \left(\sqrt{1+\frac{v}{2}}, \sqrt{1-\frac{v}{2}}\right) \quad \boldsymbol{v}_- = \left(\sqrt{1-\frac{v}{2}}, -\sqrt{1+\frac{v}{2}}\right) . \tag{2.22}$$

Similarly, we label a state on the *lower* branch by the velocity v of the state on the *upper* branch that has the same Q. The associated Bloch function can then be written as $\boldsymbol{v}_-$ in (2.22). Note that the Bloch functions are normalized and that Bloch functions on the two different branches and at the same Q are orthogonal. Far from the Bragg resonance, where $|v| \to 1$, the Bloch functions approach $(1,0)$ and $(0,1)$ and at the band edge, where $v = 0$, the Bloch functions are $(1/\sqrt{2}, \pm 1/\sqrt{2})$, all consistent with Fig. 2.1. These results can be understood as follows. Far from the Bragg resonance the effect of the gratings is small, and the eigenstates of the system should then essentially be (uncoupled) forward and backward propagating waves. At the band edges, where $v = 0$, the Bloch functions are standing waves; at the bottom edge, the nodes are in the low-index medium since this energetically favourable [10], whereas at the top edge the nodes are in the high-index medium, consistent with (2.22). We return to the Bloch functions when we discuss the solutions to the nonlinear coupled mode equations.

Now that we have explored some of the properties of infinite gratings, let us turn to grating of finite length. The reflection properties of gratings are well known [11] and are briefly reviewed here. To do this, consider a grating of finite length L, as shown in Fig. 2.2a. Then set

$$\mathcal{E}_\pm(z,t) = E_\pm(z,t)\, e^{-iV\Delta t} , \tag{2.23}$$

corrsponding to CW fields, and apply the boundary conditions

$$E_+(-L,t) = 1 , \quad E_-(0,t) = 0 , \tag{2.24}$$

corresponding to radiation incident from $-\infty$. Under these conditions the solutions to the coupled mode equations are found to be

$$\begin{aligned} E_+(z) &= \frac{i\alpha \cosh \alpha z - \Delta \sinh \alpha z}{i\alpha \cosh \alpha L + \Delta \sinh \alpha L} , \\ E_-(z) &= \frac{\kappa \sinh \alpha z}{i\alpha \cosh \alpha L + \Delta \sinh \alpha L} , \end{aligned} \tag{2.25}$$

where $\alpha = \sqrt{\kappa^2 - \Delta^2}$, and $|\Delta| \leq \kappa$. The corresponding result for frequencies outside the photonic bandgap (i.e. $|\Delta| > \kappa$), can be found by analytic continuation, and involves trigonometric functions. From (2.25) we can straightforwardly find the grating's reflectivity $R \equiv |E_+(-L)/E_-(-L)|^2$,

$$R = \frac{\kappa^2 \sinh^2 \alpha L}{\kappa^2 \cosh^2 \alpha L - \Delta^2} , \tag{2.26}$$

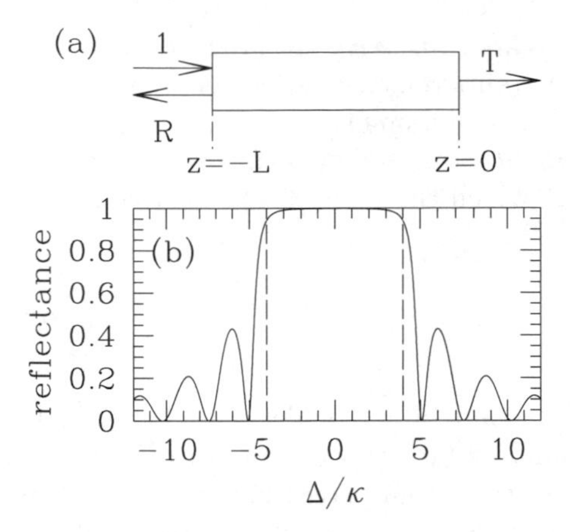

Fig. 2.2. a Geometry used for grating cw calculations. **b** Reflectivity versus δ/κ for a uniform grating with $\kappa L = 4$. The vertical dashed lines indicate the edges of the photonic band gap

and the transmissivity $T = 1 - R$, since the structure is taken to be lossless. This results is illustrated in Fig. 2.2b, showing R versus Δ/κ for a grating with $\kappa L = 4$. The vertical dashed lines indicate the edges of the photonic band gap. The reflectivity is clearly highest inside the band gap, reaching its maximum values $\tanh^2 \kappa L$ at the Bragg resonance, where $\Delta = 0$. Outside the bandgap the reflectivity exhibits prominent sidelobes, the envelope of which decreases as Δ^{-2} [12]. Thus, for frequencies very far from the Bragg resonance the reflectivity is negligible as expected. However, the reflectivity for frequencies outside the photonic band gap, but close to it, is highly undesirable for the type of work we are interested in, and in fact for almost all telecommunications applications of fibre gratings. One way to think about this out-of-band reflectivity is that it is due to a mismatch of the *effective impedance* [13,12] at the edges of the grating.[3] Now it is well known that such reflections may be reduced by smoothing out the abrupt effective impedance jump. Indeed, this is the main idea behind the use of *apodized* gratings, in which the grating strength κ smoothly disappears at the edges of the grating [14,2]. In this way the strong out of band reflection is significantly reduced and can essentially be neglected. Henceforth we assume that all gratings are of this type.

Being justified to drop the grating reflectivity for all detunings $|\Delta| > \kappa$ we now turn to grating dispersion, the key grating property for the work discussed here. We concluded earlier that in a grating the group velocity

[3] Note that the mismatch of the actual impedance is small because of the small refractive index variations involved. The effective impedance can be considered to be an impedance for the field envelopes, the discontiuity of which may be much larger.

depends on frequency, and that gratings are thus *dispersive.* In fact, for typical gratings that are of interest here, the group velocity varies between 0 and v within a wavelength range of less than a nanometer; for wavelengths just outside the photonic band gap, the grating dispersion is thus very large and dwarfs that of the underlying medium, which therefore can be neglected. The quadratic dispersion is given by:

$$\omega_2 \equiv \frac{\partial^2 \omega}{\partial k^2} = V \frac{\partial^2 \Delta}{\partial Q^2} = \pm V \frac{\kappa^2}{\Delta^3} , \tag{2.27}$$

where $\Delta = \Delta(Q) = \sqrt{\kappa^2 + Q^2}$. Thus, again in agreement with Fig. 2.1, the magnitude of the group velocity attains its largest value at the band edges, and decreases smoothly when tuning away from them. For a fiber grating with $\kappa = 10\,\mathrm{cm}^{-1}$, we thus find that $\omega_2 \approx 2 \times 10^5\,\mathrm{m}^2\,\mathrm{s}^{-1}$. This should be compared with the dispersion of standard optical fiber, which at $\lambda = 1.55\,\mu\mathrm{m}$ has a dispersion of roughly $\omega_2 \approx 2 \times 10^{-1}\,\mathrm{m}^2\,\mathrm{s}^{-1}$, six orders of magnitude smaller than the dispersion of a typical fiber grating. Similar conclusions can be drawn for higher order dispersion [15], but we do not discuss this here further. Finally, note that (2.27) implies that the quadratic dispersion becomes arbitrarily small far away from the Bragg resonance. Of course this is not quite true since at such frequencies the dispersion of untreated fiber dominates.

As mentioned, the coupled mode equations (2.15) are valid only when the grating is shallow. Equations (2.11) show that for deep gratings one obtains extra terms in the coupled mode equations (for example the β terms in (2.11)) and the size of existing terms is modified (for example the cross terms in κ in (2.15)). This is consistent with other deep grating treatments that have been reported. These include a more thorough version of the treatment given here [8], a treatment based on the Bloch function of the grating [16] and one that is somewhat similar to that given here, but which applies to nonuniform grating, albeit without nonlinear effects [13]. In the linear regime there are two types of deep grating effects. First, the width of the photonic band gap cannot increase indefinitely and cannot be proportional to the Fourier component $|\varepsilon_1|$ as in the first term in (2.12). Secondly, the Bragg resonance may shift away from the frequency $\omega = \pi c/\sqrt{\varepsilon_0} d$.

2.3 Gratings at High Intensities

Here we generalize the treatment in Sect. 2.2 to high intensities, where the nonlinearity of the medium is important. As mentioned, we consider only the Kerr effect, in which the nonlinear polarization is given by 2.3, which is now included in the wave equation (2.1). In the discussion below we will also neglect all effects due to deep gratings and we thus search for generalizations of (2.15). Under these simplified conditions, we formally obtain the following nonlinear coupled mode equations for the high intensity regime:

$$i\left(+\frac{\partial \mathcal{E}_+}{\partial z}+\frac{1}{V}\frac{\partial \mathcal{E}_+}{\partial t}\right)+\kappa\mathcal{E}_- + \Gamma(|\mathcal{E}_+|^2+2|\mathcal{E}_-|^2)\,\mathcal{E}_+ = 0\,,$$
$$i\left(-\frac{\partial \mathcal{E}_-}{\partial z}+\frac{1}{V}\frac{\partial \mathcal{E}_-}{\partial t}\right)+\kappa\mathcal{E}_+ + \Gamma(|\mathcal{E}_-|^2+2|\mathcal{E}_+|^2)\,\mathcal{E}_- = 0\,. \tag{2.28}$$

Here, the nonlinear coefficient Γ is given by

$$\Gamma = \frac{3\omega^2\chi^{(3)}}{2kc^2}\,. \tag{2.29}$$

A rigorous derivation of nonlinear coupled mode equations with analytic estimates on the deviation of this approximation from a nonlinear Maxwell model is presented in [1]. The new terms in (2.28) describe how the wave propagation is affected by nonlinear changes in the grating's refractive index. In this way, waves affect their own propagation (*self-phase modulation*), and the that of the wave propagating in the opposite direction (*cross-phase modulation*). Winful et al. were the first to derive these equations and to find CW and numerical solutions [17,18]. Pulse-like solutions were later reported by Christodoulides and Joseph [19], and by Aceves and Wabnitz [20]. Yet more general solutions, including dark solutions, were found by Feng and Kneubühl [21]. We now briefly discuss the key aspects of these solutions.

Winful et al. [17] were the first to consider the properties of nonlinear periodic media, initially in the CW regime. Their calculations, the initial steps of which, follow the linear regime treatment in Sect. 2.2, show that the field inside the grating can be written in terms in Jacobi elliptic functions [17]. We do not write the general expression here, but only give the result for the center of the band gap, where $\Delta = 0$, for which one can derive that

$$I/S = \mathrm{nd}\left[2\cosh(\chi)\kappa z|1/\cosh^2\chi\right]\,, \tag{2.30}$$

where I, the intensity and S, the energy flow through the grating, are given by

$$I \;=\; |E_+|^2+|E_-|^2,\quad S=|E_+|^2-|E_-|^2\,. \tag{2.31}$$

The function nd is one of the Jacobi elliptic functions [22], and χ follows from $\sinh\chi = 3\Gamma S/(4\kappa)$. Note that in the linear limit when $S\to 0$, the parameter $m = 1/\cosh^2\chi \to 1$, and in this limit the nd function reduces to a hyperbolic cosine, consistent with result (2.25). In general, however, for real argument nd is a real, positive periodic function that varies between $\mathrm{nd}=1$ and $\mathrm{nd}=1/\tanh\chi$ and has a period $2K$, where K is the complete elliptic integral of the first kind. Now we saw earlier that at the back of the grating, $E_-(0)=0$, and so $\mathrm{nd}(0)=1$. But if the field's envelope function is periodic, then so must be the electric field.[4] If, therefore, the period of the

[4] The periodicity of the envelopes cannot be gleaned from (2.30) alone, since it only demonstrates the periodicity of the envelope's magnitude.

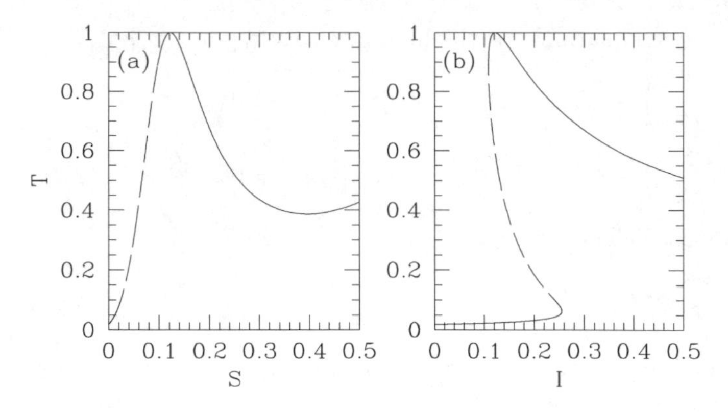

Fig. 2.3. **a** Transmissivity T versus energy flow S through a uniform grating with $\kappa = 5$, $L = 1$, $\Delta = 4.75$, $\Gamma = 1$. **b** Same as **a**, but as a function of the incident intensity. Dashed lines correspond to solutions that are guaranteed to be unstable; solid lines may indicate stable or unstable solutions

field envelope equals the length of the grating, then the field at the front and at the back of the grating are identical, and the grating is then thus perfectly transmitting, even though the light is tuned to the grating's Bragg resonance. The conclusion that the reflectivity vanishes at certain powers is general and is true for any frequency inside the band gap.

The behavior described above is illustrated in Fig. 2.3, which shows the nonlinear transmission through a uniform grating. Figure 2.3a shows the transmission T versus the energy flow S through the grating. When $S \approx 0.12$, the structure is perfectly transmitting, and, because of this, the energy flow through the structure equals the incident energy I, since $S = TI$. For other values of the power, $T < 1$, and thus $I > S$. This is consistent with Fig. 2.3b, which shows T versus I. This figure shows a low-transmission branch corresponding to the linear regime. More generally, the system is expected to demonstrate bistability, indicating that the state of the system is not uniquely determined by the input. The parts of the curve where $\partial T/\partial I < 0$ are guaranteed to be unstable, as indicated by the dashed lines [23]. The solid lines indicate solutions for which stability must be established using the full time-dependent equations. This shows that the lower branch is always stable, but that the upper branch is often unstable against the formation of sidebands which manifests itself as a time-varying output, which may be periodic or chaotic. [18,24,25]. For cases in which the upper branch is stable, the system exhibits bistable switching. Though this has not been demonstrated in this geometry due to high switching powers, switching has been observed in related geometries [26,27].

Another class of solutions to coupled mode equations (2.28) are pulse-like solutions, first reported by Christodoulides and Joseph [19], and, in more a more general context, by Aceves and Wabnitz [20]. The solutions, to which we will refer as *Bragg grating solitons*, have two free parameters: a velocity ν ($0 \leq \nu < 1$) and an amplitude δ ($0 < \delta < \pi$). Using the notation of Aceves and Wabnitz, these solutions can be written as

$$\mathcal{E}_+(z,t) = \sqrt{\frac{\kappa/\Gamma}{2+\frac{1+\nu^2}{1-\nu^2}}}\left(\frac{1+\nu}{1-\nu}\right)^{1/4} \sin\delta\,\mathrm{sech}(\vartheta - i\delta/2)\,e^{i\sigma}\,e^{i\eta}$$
$$\mathcal{E}_-(z,t) = -\sqrt{\frac{\kappa/\Gamma}{2+\frac{1+\nu^2}{1-\nu^2}}}\left(\frac{1-\nu}{1+\nu}\right)^{1/4} \sin\delta\,\mathrm{sech}(\vartheta + i\delta/2)\,e^{i\sigma}\,e^{i\eta}\,, \quad (2.32)$$

where the nonlinearity was assumed to be positive, and where

$$\vartheta = \gamma\kappa\sin\delta\,(z - \nu V t)\,, \quad \sigma = \gamma\kappa\cos\delta\,(\nu z - V t)\,,$$
$$\gamma = \frac{1}{\sqrt{1-\nu^2}}\,, \quad e^{i\eta} = \left(\frac{e^{2\vartheta} + e^{-i\delta}}{e^{2\vartheta} + e^{+i\delta}}\right)^{2\nu/(3-\nu^2)}.$$

These represent single-peaked solutions that can travel at any speed, νV between 0 and the speed of light in the medium V. At $\nu = 0$, this result is easy to interpret. In this case $|\mathcal{E}_+| = |\mathcal{E}_-|$ and the detuning from the center of the photonic band gap is $\Delta = \kappa\cos\delta$, thus as $\delta \to 0$, the field is tuned close to the upper edge of the gap, whereas when $\delta \to \pi$, it is tuned close to its lower edge. By the $\sin\delta$ term, the former limit is that of low intensities, whereas the latter corresponds to high intensities. When $\nu \neq 0$, then the solutions attains a nontrivial phase due to the presence of the $\exp(i\eta)$ term, and $|\mathcal{E}_+| \neq |\mathcal{E}_-|$. The velocity dependence of the latter can be understood by a comparison with the Bloch function, discussed in Sect. 2.2. Note that $|E_+/E_-|^2 = (1+\nu)/(1-\nu)$. This is precisely the ratio of the Bloch functions of the upper branch in (2.22) if we identify the soliton velocity ν with the group velocity v in a grating (see (2.20)). The stability of Bragg grating solitons has been studied by a number of authors [28,29]. They find that Bragg grating solitons exhibit oscillation, that cause them to be unstable at high intensities.

2.4 Nonlinear Schrödinger Limit

The nonlinear Schrödinger equation (NLSE) is well-known to arise as a universal envelope equation governing systems characterized by weak nonlinearity and strong dispersion [30]. In this section we demonstrate how the NLSE emerges in an appropriate limit from the nonlinear coupled mode equations. We first encode the regime of weak nonlinearity by introducing a small dimensionless parameter μ which measures the amplitude of the electric field: $\mathcal{E}_\pm = \mathcal{O}(\mu)$. With this scaling of the electric field, the nonlinear coupled mode equations (2.28) are seen formally to be an $\mathcal{O}\mu^2$ perturbation of the linear coupled equations (2.15). The latter have plane wave solutions of the form (see Sect. 2.2):

$$\mathcal{E} = e^{i(Qz - V\Delta_{\pm}t)}\boldsymbol{v}_{\pm}\,, \tag{2.33}$$

where $\boldsymbol{v}_{\pm}$ given by (2.22).

In analogy with the multiple scales procedure alluded to in obtaining the evolution of the slowly varying envelope functions $\mathcal{E}_{\pm}$, we construct solutions which are slow modulations of these *Bloch waves* on appropriately long spatial and temporal scales. The derivation of the NLSE can have two different starting points: one can either start from the Maxwell equations [16,31], or from the coupled mode equations [32]. In the former approach, one works with the exact generalized Bloch eigenfunctions of the linear wave equation, and therefore with an expansion which is valid for the case of deep gratings. In the latter approach, the expansion is in terms of plane wave approximations to the Bloch eigenfunctions, valid only for shallow gratings. Here, we adopt the latter approach and start from (2.28).

Introduce the slow time and spatial scales:

$$t_j = \mu^j t\,, \quad z_j = \mu^j z\,, \quad j = 0, 1, 2, ... \tag{2.34}$$

We seek solutions of nonlinear coupled mode equations in the form:

$$\begin{aligned}\bar{\mathcal{E}} = &\left(\mu a(z_1, z_2; \ldots, t_1, t_2, \ldots)\boldsymbol{v}_+ + \mu^2 b(z_1, z_2, \ldots; t_1, t_2, \ldots)\boldsymbol{v}_-\right)\\ &\times e^{i(V\Delta_+ t - Qz)}\,.\end{aligned} \tag{2.35}$$

where $\bar{\mathcal{E}}$ is the vector with components $\mathcal{E}_{\pm}$, a and b are envelope functions that modulate the Bloch function $\boldsymbol{v}_{\pm}$ defined in (2.22).

In our Ansatz, we have imposed that the $\boldsymbol{v}_+$ mode dominate the $\boldsymbol{v}_-$ mode. Note that the time dependence is that of the dominant Bloch function. The carrier (plane) wave varies rapidly on the spatial and temporal scales $z_0 = z \sim Q^{-1}$ and $t_0 = t =\sim (V\Delta_+)^{-1}$, while the envelope modulation functions a and b are taken to be independent of z and t. The assumption that they are slowly varying is incorporated by taking them to be functions of z_1, t_1, z_2, t_2, ... which vary on increasingly longer length and time scales. In the calculation below we treat the z_i and t_i as independent variables and we write the nonlinear coupled mode equations with the replacements $\partial/\partial z \to \partial/\partial z_0 + \mu\partial/\partial z_1 + \ldots$, and $\partial/\partial t \to \partial/\partial t_0 + \mu\partial/\partial t_1 + \ldots$.

Use of the *ansatz* (2.35) is justified under two conditions. The first of these is that the pulse spectrum does not extend deep into the gap, since otherwise the effect of Bloch functions associated with the upper and lower branches of the dispersion relation would be comparable. Since we have chosen $\boldsymbol{v}_+$ to dominate, the pulse spectrum must in fact have a positive detuning. Second, the spectrum should not be too wide, since otherwise the description by the single dominant Bloch function is again not valid.

To order μ^1, where a is a constant and b does not enter, we find that (2.35) satisfies (2.28) via relations (2.21) as required. To level μ^2 we find

$$\begin{aligned} \frac{i}{V}\frac{\partial a}{\partial t_1} + \frac{\partial a}{\partial z_1} + \frac{2\kappa}{1+v} b = 0 \\ \frac{i}{V}\frac{\partial a}{\partial t_1} - \frac{\partial a}{\partial z_1} - \frac{2\kappa}{1-v} b = 0\,, \end{aligned} \tag{2.36}$$

where (2.22) was used. By eliminating b we find that

$$\frac{\partial a}{\partial t_1} + vV\frac{\partial a}{\partial z_1} = 0\,, \tag{2.37}$$

and so $a = a(z - vVt)$. To this order, therefore, the a travels at the group velocity associated with the dominant Bloch function. From (2.36) we also find that

$$b = -\frac{i}{2\kappa\gamma^2}\frac{\partial a}{\partial z_1}\,, \tag{2.38}$$

expressing b in terms of a.

To order μ^3 we find expressions that are more similar to (2.36) but more complicated, and we therefore do not give them here. However, from these expression it is easy to show that

$$\frac{i}{V}\frac{\partial a}{\partial \tau_2} + \frac{i}{\gamma}\frac{\partial b}{\partial \zeta_1} + \frac{\Gamma}{2}(3 - v^2)|a|^2 a = 0\,. \tag{2.39}$$

where $\zeta_i = z_i - vVt_i$ and $\tau_i = t_i$. Using (2.38) this then leads to the NLSE

$$\frac{i}{V}\frac{\partial a}{\partial \tau_2} + \frac{\omega_2}{2}\frac{\partial^2 a}{\partial \zeta_1^2} + \frac{\Gamma}{2}(3 - v^2)|a|^2 a = 0\,, \tag{2.40}$$

which is the result that we were after. The dispersion ω_2 is given by (2.27) and is due to the presence of the grating. The nonlinear coefficient is affected by the grating through its dependence on v. The $(3 - v^2)/2$ term is unity far from the Bragg resonance and is 3/2 at the edge of the band gap. This can be understood as an enhancement due to the standing wave nature of the Bloch function.

The properties and solutions of the NLSE are well known, and include integrability and solitons, respectively. It can be shown that the fundamental soliton solution to (2.40) corresponds to the Aceves and Wabnitz solution discussed earlier, with agreement to second order in δ [32]. More generally, (2.40) is a good approximation to the full coupled mode equations (2.28) in the low-intensity limit.

In fact, most of the experimental results that have reported to date can be understood within the context of the NLSE, particularly the the generation and propagation of fundamental Bragg grating solitons [6]. However, at the highest incident powers, where higher order solitons are generated and modulational instability occurs, can deviations be observed [6,33].

Acknowledgement. This work was partly supported by the Australian Research Council.

References

1. R.H. Goodman, M.I. Weinstein. P.J. Holmes: J. Nonlinear Science **11**, 123 (2001)
2. R. Kashyap, *Fiber Bragg gratings*, 1st Edn. (Academic Press, San Diego 1999)
3. P. Millar, R.M. De La Rue, T.F. Krauss, J.S. Aitchison: Opt. Lett. **24**, 685 (1999)
4. W. Chen, and D.L. Mills: Phys. Rev. Lett. **58**, 160 (1987)
5. B.J. Eggleton, R.E. Slusher, C.M. de Sterke, P.A. Krug, and J.E. Sipe: Phys. Rev. Lett. **76**, 1627 (1996)
6. B.J. Eggleton, C.M. de Sterke, and R.E. Slusher: J. Opt. Soc. Am. B **16**, 587 (1999)
7. D. Taverner, N.G.R. Broderick, D.J. Richardson, R.I. Laming, and M. Ibsen, Opt. Lett. **15**, 328 (1998)
8. T. Iizuka and C.M. de Sterke: Phys. Rev. E **61**, 4491 (2000)
9. C. Kittel, *Introduction to solid state physics* 5th Edn. (Wiley & Sons, New York Chichester 1976), Ch. 7
10. J.D. Joannopoulos, R.D. Meade, J.N. Winn, *Photonic crystals* (Princeton University Press, Princeton 1995)
11. H. Kogelnik and C.V. Shank: Appl. Phys. Lett. **18**, 152 (1971)
12. A. Arraf, C.M. de Sterke, L. Poladian, and T.G. Brown: J. Opt. Soc. Am. A **14**, 1137 (1997)
13. J.E. Sipe, L. Poladian, and C.M. de Sterke: J. Opt. Soc. Am. A **11**, 1307 (1994)
14. P.S. Cross and H. Kogelnik: Opt. Lett. **1**, 43 (1977)
15. P.S.J. Russell: J. Mod. Opt. **38**, 1599 (1991)
16. C.M. de Sterke, D.G. Salinas, and J.E. Sipe: Phys. Rev. E **54**, 1969 (1996)
17. H.G. Winful, J.H. Marburger and E. Garmire: Appl. Phys. Lett. **35**, 379 (1979)
18. H.G. Winful and G.D. Cooperman: Appl. Phys. Lett. **40**, 298 (1982)
19. D.N. Christodoulides and R.I. Joseph: Phys. Rev. Lett. **62**, 1746 (1989)
20. A.B. Aceves and S. Wabnitz: Phys. Lett. A **141**, 37 (1989)
21. J. Feng and F.K. Kneubühl, J. Quantum Electron. **29**, 590 (1993)
22. L.M. Milne-Thomson, "Jacobi Elliptic functions and theta functions," in *Handbook of Mathematical functions*, edited by M. Abramowitz and I.A. Stegun (Dover, New York 1965)
23. C.M. de Sterke and J.E. Sipe, "Gap solitons," in *Progress in Optics XXXIII*, E. Wolf, ed., (Elsevier, Amsterdam 1994), Chap. III-Gap Solitons
24. C.M. de Sterke and J.E. Sipe: Phys. Rev. A **42**, 2858 (1990)
25. H.G. Winful, R. Zamir, and S. Feldman: Appl. Phys. Lett. **58**, 1001 (1991)
26. D. Taverner, N.G.R. Broderick, D.J. Richardson, M. Ibsen, and R.I. Laming: Opt. Lett. **23**, 259 (1998)
27. S. Lee and S.-T. Ho: Opt. Lett. **18**, 962 (1993)
28. I.V. Barashenkov, D.E. Pelinovsky, and E.V. Zemlyanaya: Phys. Rev. Lett. **80**, 5117 (1998)
29. A.D. Rossi, C. Conti, and S. Trillo: Phys. Rev. Lett. **81**, 85 (1998)

30. R.K. Dodd, J.C. Eilbeck, J.D. Gibbon, H.C. Morris, *Solitons and nonlinear wave equations* (Academic Press, London 1982), Ch. 8
31. C.M. de Sterke and J.E. Sipe: Phys. Rev. A **38**, 5149 (1988)
32. C.M. de Sterke and B.J. Eggleton: Phys. Rev. E **59**, 1267 (1999)
33. B.J. Eggleton, C.M. de Sterke, A.B. Aceves, J.E. Sipe, T.A. Strasser, and R.E. Slusher: Opt. Comm. **149**, 267 (1998)

3 Polarization Effects in Birefringent, Nonlinear, Periodic Media

S. Pereira and J.E. Sipe

3.1 Introduction

In recent years there have been many studies of one-dimensional photonic band-gap materials in the presence of a Kerr nonlinearity [1–5]. A great deal of the experimental work in this field has concentrated on fiber Bragg gratings, which typically have refractive index variations on the order of 10^{-4} [6,7]. With such small index variations, it is reasonable to apply the heuristic nonlinear coupled mode equations, or the appropriate nonlinear Schrödinger equation, to analyze experimental results. However, it is well known that the process of UV-writing a grating in a nominally isotropic optical fiber introduces a birefringence of the order of 10^{-6} [8]. For grating lengths of around 10 cm, this has observable consequences.

In this chapter we extend the isotropic nonlinear coupled mode equations, and the associated nonlinear Schrödinger equation, to account for birefringence in a grating system [9,10]. Our equations describe a system with one effective spatial dimension, in which both the birefringence and the grating index contrast are small relative to the background index. Although there do exist equations that describe stronger gratings [10], derived by considering the underlying Bloch functions of the periodic medium, many qualitative insights into birefringent, nonlinear gratings can be discussed using the weak-grating equations. Furthermore, the experimental results have concentrated on optical fibers where the weak grating equations are adequate.

Birefringence has the effect of separating the photonic band gaps of the two orthogonal polarizations so that, in certain frequency ranges, light of one polarization can propagate freely, while the other is blocked. This has immediate deleterious consequences for proposed devices based on circularly polarized light, where the linearly polarized signals are mixed; further, the robustness of nonlinear effects, such as soliton formation and propagation through grating structures, will be affected by birefringence. The analysis of nonlinear effects in a birefringent grating can be guided by the large literature on birefringent effects in bare optical fibers [11–14], which uses a set of coupled nonlinear Schrödinger equations similar to the ones we present below. However, even when the coupled nonlinear Schrödinger equations provide a good description of light propagating in the grating system, the dynamics are more complicated than in a bare fiber. In a bare fiber the group velocity and group velocity dispersion can be assumed equal for the two orthogonal polarizations, while this approximation often fails in gratings. Nevertheless, since the pulse propagation parameters in a grating are strong functions of frequency, one can always find a frequency where the group velocity and group velocity dispersion are similar for the two polarizations, so that the results of the bare optical fiber literature will be directly applicable.

For light with frequency content outside the photonic band gap, for which the coupled nonlinear Schrödinger equations are considered applicable, the grating system gives us access to three distinct regimes of nonlinear pulse propagation. We understand these regimes in terms of two intensities: I_{sol}, the intensity required to form a soliton along one of the principal axes of the system [15]; and I_{PI}, the intensity required to phase-match the nonlinear energy exchange between polarizations [15]. Far from the photonic band gap, we find $I_{\mathrm{sol}} < I_{\mathrm{PI}}$, so that an isotropic soliton can form without the complication of energy exchange. Very close to the gap we find $I_{\mathrm{PI}} < I_{\mathrm{sol}}$, and instability occurs at intensities far below those required to form a soliton. In between these two extremes we have a regime where $I_{\mathrm{PI}} \simeq I_{\mathrm{sol}}$. In this last regime the nonlinear dynamics are complicated, and it is as yet unclear what novel effects might emerge.

Light with its frequency components inside the photonic band gap is reflected. However, it is well known that if the light has sufficient intensity, then the change in the effective background index due to nonlinear self phase modulation can allow the light to be transmitted [5]. In a birefringent system, cross phase modulation and energy exchange between the polarizations arise in addition to the familiar self phase modulation. This makes the dynamics of the situation far more complicated than switching in the isotropic system. It also suggests an AND gate geometry, observed experimentally in 1998 [7], and simulated below.

This chapter is divided into four sections. In Sect. 3.2 we derive the coupled mode equations for a linear system; we then use those equations to determine the dispersion relation of the system, characterizing the effective birefringence. Next we include a Kerr nonlinearity, and derive a set of nonlinear coupled mode equations. Finally, we show how the nonlinear coupled mode equations (CME) are related to a set of coupled nonlinear Schrödinger equation (CNLSE), and discuss their regions of validity. In Sect. 3.3 we present numerical simulations and experimental results. We begin by modifying the nonlinear coupled mode equations to account for the finite length and apodization profile of the grating. We then give the three regimes of nonlinear pulse propagation outside the photonic band gap, and present some experimental and numerical results in these regimes. Finally, we turn to propagation inside the photonic bandgap, and simulate the AND gate discussed above. In Sect. 3.4 of the chapter we conclude.

3.2 Derivation of the Equations

In this section we derive two sets of equations to describe light propagating in the system shown in Fig. 3.1. There we have a dielectric medium, weakly periodic in the z direction, and extending to infinity in the transverse direction. We assume a birefringence both in the sense that the background index is different for the x and y polarizations, and in the sense that the index contrast is different for the two polarizations. Although our model is strictly one dimensional, it can be used to describe experiments in optical fibers, and can be used as a guide to experiments in dielectric waveguide systems with etched gratings. In the absence of both nonlinearity and a grating, the light will have four components – two polarizations, each of which can travel in either the forward or backward direction – and we can write down four uncoupled equations to describe the propagation. In the presence of a grating, the forward and backward travelling components of each polarization are coupled via Bragg scattering. Thus our four uncoupled equations become two sets of two coupled equations. The two polarizations, though, can still be considered independently.

With the addition of a Kerr nonlinear response in the presence of a grating, the nonlinearity couples the two polarizations, and the grating couples the two directions, so that our four propagation equations are now entirely coupled together. The effect of the grating is still the scattering of a forward travelling component of a given polarization into a backward travelling component of the same polarization (and vice versa). The effects of the nonlinearity include phase modulation, where the light sees a higher index of refraction due to its own intensity and the intensity of the other polarization; and phase conjugation, whereby, under certain conditions, energy can shift from one polarization to the other [10,15].

Returning to the linear situation in the presence of a grating, it is well known that the coupling between forward and backward travelling waves leads to the opening of photonic band gaps in the dispersion relation. The lowest frequency photonic band gap – typically the largest, and the one in which we are interested – is centered at the Bragg wave number of the system, $k_0 = \pi/d$, where d is the period of the system. This Bragg wave number corresponds to Bragg frequencies $\omega_{0\mathrm{x}} = ck_0/\overline{n}_\mathrm{x}$, $\omega_{0\mathrm{y}} = ck_0/\overline{n}_\mathrm{y}$ for the two

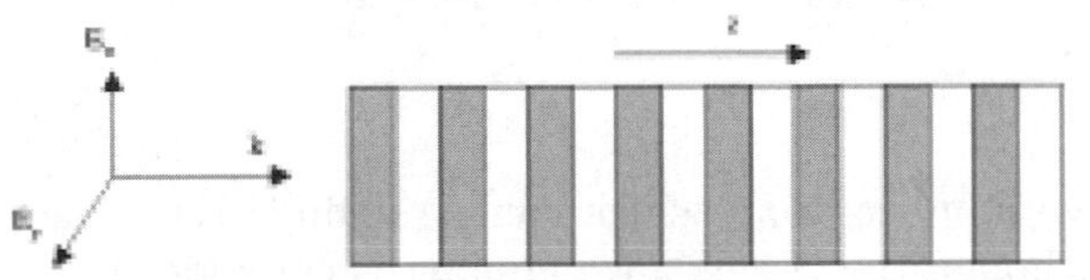

Fig. 3.1. Schematic of the one dimensional system being used in this paper. The periodic medium is birefringent, so we must explicitly consider both the x- and y-polarized light. We assume that the two polarizations see both a different background index of refraction, and a different index contrast

polarizations, where $\overline{n}_{\mathrm{x}}, \overline{n}_{\mathrm{y}}$ are the background indices of refraction seen by the two polarizations. Light at a Bragg frequency cannot propagate in the system. Futhermore, the stronger the index contrast of the grating, the larger the width of the photonic bandgap – that is, the larger the range of frequencies for which light cannot propagate. At frequencies outside but near the photonic bandgap light can propagate, but the dispersion relation is strongly modified by the grating, and a pulse will experience a greatly reduced group velocity, a greatly enhanced group velocity dispersion, and higher order dispersion [5,10].

If we choose to describe pulses whose frequency content is sufficiently far from the photonic band gap, then there will be negligible coupling to the backward going wave from light initially travelling in the forward direction. Thus, instead of using four equations, we can use only two. However, although the grating is no longer explicitly coupling the two directions of propagation, its effect is still felt in the modified dispersion relation. The inclusion of nonlinearity at this point will, of course, couple the two polarizations. If we further restrict ourselves to pulses with sufficiently narrow frequency content, then we can ignore third and higher order dispersion. In such a situation we find that the light is well described by a set of two coupled nonlinear Schrödinger equations.

3.2.1 The Linear Coupled Mode Equations

We begin with the linear Maxwell equations in the presence of a dielectric tensor that is a function of only one Cartesian component, $\boldsymbol{\varepsilon} = \boldsymbol{\varepsilon}(z)$. We assume that the (x, y) coordinates can be chosen such that for all z the tensor is diagonal, $\varepsilon = diag\left(\varepsilon_{\mathrm{xx}}(z), \varepsilon_{\mathrm{yy}}(z)\right)$. Neglecting magnetic effects by setting the permeability, μ, equal to that of free space, $\mu = \mu_0$, we can then define indices of refraction associated with the x and y axes, $n_{\mathrm{i}}^2(z) = \varepsilon_{\mathrm{ii}}(z)/\varepsilon_0$, where ε_0 is the permittivity of free space and where, for the remainder of the text, the index i runs over the polarizations x and y. We seek fields $\mathbf{E}(\mathbf{r}, t), \mathbf{H}(\mathbf{r}, t)$ that depend only on the coordinate z. In such a case, the Maxwell equations can be expressed as two uncoupled wave equations

$$\frac{\partial^2 E_{\mathrm{i}}}{\partial z^2} - \frac{n_{\mathrm{i}}^2(z)}{c^2}\frac{\partial^2 E}{\partial t^2} = 0\,. \tag{3.1}$$

Now suppose that the index of refraction varics according to

$$n_{\mathrm{i}}(z) = \overline{n}_{\mathrm{i}} + \delta n_{\mathrm{i}} \cos\left(2k_0 z\right)\,, \tag{3.2}$$

where k_0 is the Bragg wave number, from which can be defined the Bragg frequency $\omega_{0\mathrm{i}} = ck_0/\overline{n}_{\mathrm{i}}$. Although the Bragg wave number is common to the two polarizations, the Bragg frequencies are, in general, unequal. We assume that the grating index contrast, δn_i, is small relative to the background index $\overline{n}_{\mathrm{i}}$: $\delta n_{\mathrm{i}} \ll \overline{n}_{\mathrm{i}}$. As well, both the linear birefringence and grating contrast mismatch,

$$\begin{aligned} n_{\mathrm{b}} &\equiv \overline{n}_{\mathrm{y}} - \overline{n}_{\mathrm{x}}\,, \\ \delta n_{\mathrm{b}} &\equiv \delta n_{\mathrm{y}} - \delta n_{\mathrm{x}}\,, \end{aligned} \tag{3.3}$$

are assumed small relative to $\overline{n}_{\mathrm{i}}$. To first order,

$$n_{\mathrm{i}}^2(z) = \overline{n}_{\mathrm{i}}^2 \left[1 + 2\frac{\delta n_{\mathrm{i}}}{\overline{n}_{\mathrm{i}}} \cos(2k_0 z) + O\left(\frac{\delta n_{\mathrm{i}}^2}{\overline{n}_{\mathrm{i}}^2}\right)\right]. \tag{3.4}$$

If δn_{i} were zero, then a solution to the wave equation for the ith polarization would be $E_{\mathrm{i}}(z,t) = E_{\mathrm{i}+} e^{-i(\omega_{0\mathrm{i}} t - k_0 z)} + E_{\mathrm{i}-} e^{-i(\omega_{0\mathrm{i}} t + k_0 z)} + c.c.$, where the *c.c.* denotes complex conjugation, and the constants $E_{\mathrm{i}+}$ and $E_{\mathrm{i}-}$ label the electric field amplitudes associated with propagation in the forward and backward z direction respectively. The basis of coupled mode analysis is that, if the δn_{i} are small, we can use this same form for the solution to the perturbed wave equations; but, because the perturbation can scatter a forward going wave into a backward going wave, we must allow $E_{\mathrm{i}+}$ and $E_{\mathrm{i}-}$ to vary weakly with z and t. We thus seek solutions to (3.1) of the form

$$E(z,t) = \widetilde{E}_{\mathrm{i}+}(z,t)\, e^{-i(\omega_{0\mathrm{i}} t - k_0 z)} + \widetilde{E}_{\mathrm{i}-}(z,t)\, e^{-i(\omega_{0\mathrm{i}} t + k_0 z)} + c.c.\,, \tag{3.5}$$

where $\widetilde{E}_{\mathrm{i}\pm}(z,t)$ are slowly varying envelope functions, carried by the Bragg frequency associated with the given polarization. We now use equations (3.2) and (3.5) in (3.1). As is common in the heuristic derivation of coupled mode equations [5], we ignore 2nd derivatives of the slowly varying $\widetilde{E}_{\mathrm{i}\pm}$; a more rigorous approach relies on the method of multiple scales [4,10]. Collecting terms of matching phase, and keeping terms only to 1st order in our perturbations, we find the coupled mode equations in the absence of nonlinearity [5]. However, instead of presenting the equations in terms of the electric field, we choose to re-write our fields as

$$\widetilde{A}_{\mathrm{i}\pm} = \sqrt{\frac{2\overline{n}_{\mathrm{i}}}{Z_0}}\,\widetilde{E}_{\mathrm{i}\pm}\,, \tag{3.6}$$

where $Z_0 = \sqrt{\mu_0/\varepsilon_0}$ is the impedance of free space, and where $\left|\widetilde{A}_{\mathrm{i}\pm}\right|^2$ represents the intensity in the field. This gives a set of coupled mode equations

$$0 = \pm i\frac{\partial \widetilde{A}_{\mathrm{i}\pm}}{\partial z} + i\frac{\overline{n}_{\mathrm{i}}}{c}\frac{\partial \widetilde{A}_{\mathrm{i}\pm}}{\partial z} + \kappa_{\mathrm{i}} \widetilde{A}_{\mp}\,, \tag{3.7}$$

with

$$\kappa_{\mathrm{i}} = \frac{1}{2}\frac{\delta n_{\mathrm{i}}}{\overline{n}_{\mathrm{i}}} k_0\,. \tag{3.8}$$

3.2.2 Dispersion Relation for the Linear CME

By labelling the index of refraction and the electric fields according to polarization we have formally accounted for two types of birefringence in (3.7): First, the normal linear birefringence, n_{b}, leads to a mismatch in the group and phase velocity of the ith polarization in the absence of a grating ($\kappa_{\mathrm{i}} = 0$); second, according to (3.8) the grating strength, κ_{i}, will in general be unequal for the two polarizations. Thus, although the equations for the two polarizations are of the same form, their dispersion relations will be different. Because we expect the coupled mode equations to give a good description of light near the Bragg wave number, we can use them to approximate the dispersion relation in the region close to k_0. In fact, although we have stipulated that $\delta n_{\mathrm{i}} \ll \overline{n}_{\mathrm{i}}$, we could relax that restriction and the resulting coupled mode equations, based on the underlying Bloch functions of the periodic medium, would still provide a good approximation for the dispersion relation of the system near the Bragg wave number [16].

We re-write the linear coupled mode equations for a given polarization as a matrix equation

$$\left[i \begin{pmatrix} 1 & 0 \\ 0 & -1 \end{pmatrix} \frac{\partial}{\partial z} + i \frac{\overline{n}_{\mathrm{i}}}{c} \begin{pmatrix} 1 & 0 \\ 0 & 1 \end{pmatrix} \frac{\partial}{\partial t} + \begin{pmatrix} 0 & 1 \\ 1 & 0 \end{pmatrix} \kappa_{\mathrm{i}} \right] \begin{bmatrix} A_{\mathrm{i}+} \\ A_{\mathrm{i}-} \end{bmatrix} = 0 \,. \tag{3.9}$$

We seek solutions to (3.9) of the form

$$\begin{bmatrix} A_{\mathrm{i}+} \\ A_{\mathrm{i}-} \end{bmatrix} = \begin{bmatrix} r_{\mathrm{i}+} \\ r_{\mathrm{i}-} \end{bmatrix} e^{-i(\Omega_{\mathrm{i}\pm} t - Q_{\mathrm{i}} z)} \,, \tag{3.10}$$

where the wavevector detuning is

$$Q_{\mathrm{i}} = k_{\mathrm{i}} - k_0 \,. \tag{3.11}$$

If the full frequency $\omega_{\mathrm{i}} > \omega_{0\mathrm{i}}$, then we call the frequency detuning parameter $\Omega_{\mathrm{i}+}$ and otherwise we call it $\Omega_{\mathrm{i}-}$, with

$$\Omega_{\mathrm{i}\pm} = \omega_{\mathrm{i}} - \omega_{0\mathrm{i}} \,. \tag{3.12}$$

Substitution of the trial solution into (3.9) leads to the matrix equation

$$\begin{pmatrix} \frac{\overline{n}_{\mathrm{i}}}{c} \Omega_{\mathrm{i}\pm} - Q_{\mathrm{i}} & \kappa_{\mathrm{i}} \\ \kappa_{\mathrm{i}} & \frac{\overline{n}_{\mathrm{i}}}{c} \Omega_{\mathrm{i}\pm} + Q_{\mathrm{i}} \end{pmatrix} \begin{bmatrix} r_{\mathrm{i}+} \\ r_{\mathrm{i}-} \end{bmatrix} = 0 \,, \tag{3.13}$$

which has non-trivial solutions only when the determinant of the matrix is identically zero. This gives the dispersion relation,

$$\frac{\overline{n}_{\mathrm{i}}}{c} \Omega_{\mathrm{i}\pm}(Q) = \pm \sqrt{\kappa_{\mathrm{i}}^2 + Q^2} \,. \tag{3.14}$$

Clearly the frequency detuning is restricted so that $|\Omega_{\mathrm{i}\pm}| \geq c\kappa_{\mathrm{i}}/\overline{n}_{\mathrm{i}}$. Frequencies for which this is not the case are said to lie in the photonic band gap,

Table 3.1. Material parameters used in the simulations

Index of Refraction ($\overline{n}_x$)	1.46
Birefringence (n_b)	3.6×10^{-6}
Index Modulation (δn_x)	1.8×10^{-4}
Nonlinear Index (n_2; W/cm^2)	2.8×10^{-16}
grating length	7.7 cm
λ_{vac}	1052 nm
M	0.6
ν	0.02

where no propagating solutions exist. For frequencies not in the photonic band gap, we can use (3.14) to determine the associated group velocity and group velocity dispersion:

$$\Omega'_{i\pm}(Q) \equiv \frac{d\Omega_{i\pm}}{dQ} = \left(\frac{c}{\overline{n}_i}\right)^2 \frac{Q}{\Omega_{i\pm}},$$
$$\Omega''_{i\pm}(Q) \equiv \frac{d^2\Omega_{i\pm}}{dQ^2} = \left(\frac{c}{\overline{n}_i}\right)^2 \frac{1}{\Omega_{i\pm}} \left[1 - \rho_i^2(Q)\right], \tag{3.15}$$

where we have defined

$$\rho_i(Q) = \overline{n}_i \Omega'_{i+}(Q)/c, \tag{3.16}$$

the ratio of the group velocity at a given wave number for a point *above* the Bragg frequency, relative to the group velocity in the absence of the grating. From (3.15) it is clear that both the group velocity and dispersion are strong functions of frequency. In fact, at $Q = 0$, the group velocity vanishes, while $\Omega''_{i\pm}(0) = c/(\overline{n}_i \kappa_i)$. Notice that $\Omega''_{i\pm}(0)$ is inversely proportional to grating strength, so that a weaker grating leads to a higher group velocity dispersion. Using the parameters in Table 3.1, for the fiber grating used in experiments carried out by our colleagues [17,18], we find that $\left|\Omega''_{i\pm}(0)\right| = 0.38 \times 10^6\,\mathrm{m}^2/\mathrm{s}$, as compared to about $0.48\,\mathrm{m}^2/\mathrm{s}$ in a bare fiber – larger by six orders of magnitude! This immense difference between grating dispersion and material dispersion justifies our neglect of the latter in our theory.

In Fig. 3.2 we plot the dispersion relations (3.14) with $\kappa_x = \kappa_y \equiv \kappa$, but $\overline{n}_y > \overline{n}_x$. There we see that although the relations are similar for the two polarizations, they are offset because of the unequal Bragg frequencies. On the high frequency side, light can be inside the photonic band gap of the x polarization, while being outside the gap of the y polarization. The opposite holds true on the low frequency side. Furthermore, the horizontal distance between the curves at a given frequency is seen to be a strong function of frequency. This horizontal distance is equal to $(k_y - k_x)$, which gives the mismatch in phase accumulation for a monochromatic beam. Because this

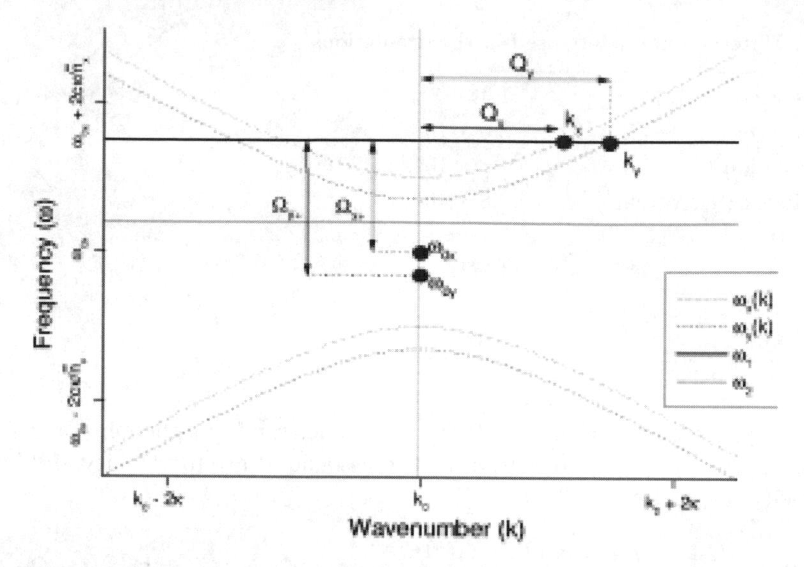

Fig. 3.2. Dispersion relations in the vicinity of the Bragg wave-vector. There are two situations of interest: **a** A carrier frequency ω_1, that lies on the dispersion relation of both polarizations. The frequency gives a different wave number for each polarization, which accounts for the effective birefringence; **b** A carrier frequency ω_2 that lies within the photonic bad gap of one or both of the polarizations. The quantities Q_x,Q_y and $\Omega_{\mathrm{x}+}$, $\Omega_{\mathrm{y}+}$ are detuning parameters defined in the text. The quantity κ is the grating strength parameter, defined in the text

mismatch is frequency dependent, we are led to the concept of an effective birefringence of the system [19].

3.2.3 Effective Birefringence

We assume that $n_\mathrm{b} > 0$, which in no way restricts the generality of what follows. We use the x polarization to define dimensionless detuning parameters

$$f_\mathrm{i} = \frac{\overline{n}_\mathrm{x}}{c\kappa_\mathrm{x}} (\omega_\mathrm{i} - \omega_{0\mathrm{x}}) , \tag{3.17}$$

$$\delta_\mathrm{i} = \frac{k_\mathrm{i} - k_0}{\kappa_\mathrm{x}} , \tag{3.18}$$

where f_i is the frequency detuning from the x Bragg frequency in units of the half-width of the x polarization photonic bandgap, and δ_i is the wavenumber detuning for the ith polarization in units of the x polarization grating strength; and where ω_i and k_i are the frequency and wavenumber of the ith polarization. In terms of these quantities the expressions for the dispersion relation of the x polarization are very simple:

$$f_\mathrm{x} = \pm\sqrt{\delta_\mathrm{x}^2 + 1}\,, \quad \rho_\mathrm{x} = \frac{\sqrt{f_\mathrm{x}^2 - 1}}{|f_\mathrm{x}|}\,, \quad \Omega''_{\mathrm{x}\pm} = \frac{c}{\overline{n}_\mathrm{x}\kappa_\mathrm{x}} \frac{1}{f_\mathrm{x}^3}\,, \tag{3.19}$$

where the $\pm$ refer to frequency detunings above and below $\omega_{0\mathrm{x}}$.

To find the normalized y dispersion relation we use the ratios

$$\nu = 2\frac{n_{\mathrm{b}}}{\delta n_{\mathrm{x}}}, \quad M = \frac{1}{2}\frac{\delta n_{\mathrm{b}}}{n_{\mathrm{b}}}, \tag{3.20}$$

in terms of which

$$\begin{aligned} f_{\mathrm{y}} &= \frac{\overline{n}_{\mathrm{x}}}{\overline{n}_{\mathrm{y}}}\left\{-\nu \pm \sqrt{\delta_{\mathrm{y}}^2 + \left(\frac{\overline{n}_{\mathrm{x}}}{\overline{n}_{\mathrm{y}}}(1+M\nu)\right)^2}\right\} \\ &\simeq \left\{-\nu \pm \sqrt{\delta_{\mathrm{y}}^2 + (1+M\nu)^2}\right\}, \end{aligned} \tag{3.21}$$

where, since n_{b} is a small number, we have put $n_{\mathrm{x}}/n_{\mathrm{y}} \simeq 1$. We have shown elsewhere that the value of M in an optical fiber grating can be related to the well known induced birefringence associated with the growth of a grating by UV exposure [19]. We assume that the grating strengths, $\delta n_{\mathrm{x}}, \delta n_{\mathrm{y}}$ are sufficiently strong that the bandgaps for the two polarizations overlap for some values of $f = f_{\mathrm{x}} = f_{\mathrm{y}}$. This is not necessary, but makes the analysis more straightforward. The Bragg frequency is the average of the frequencies at the extreme ends of the photonic bandgap, so for x-polarized light it occurs at $f_{\mathrm{x}} = 0$, while for y-polarized light it occurs at $f_{\mathrm{y}} = -\nu$.

The dispersion relations in Fig. 3.2 used $M = 0$. In Fig. 3.3 we plot the x and y dispersion relations with $M = 0.6, 2.0$ and $\nu = 0.02$. The fact that $\nu \ll 1$ means that the width of either photonic bandgap is much larger than the difference in the Bragg frequencies, which ensures that the bandgaps overlap. Notice the large asymmetry between the upper band ($f > 0$) and the lower band ($f < 0$) for $M \neq 0$. Increasing M while fixing ν has the effect of increasing the grating strength for y-polarized light while keeping all other parameters constant. This, in turn, increases the bandgap width for y-polarized light. When $M = 0$ the grating strengths for x- and y-polarized light are the same, so the y band is just the x band shifted down by a factor ν (Fig. 3.2). For $M > 0$, the larger y band gap means that the upper y band approaches the upper x band; for $M > 1$ it actually crosses it. The situation for the lower bands is simpler: increasing M monotonically increases the distance between the bands.

The effective birefringence of the system is related to the difference in wave number at a given frequency $f = f_{\mathrm{x}} = f_{\mathrm{y}}$, which is shown by the arrows in Fig. 3.3. We define a birefringence enhancement factor,

$$\begin{aligned} F_{\mathrm{b}}(f) &\equiv \frac{\delta_{\mathrm{y}}(f) - \delta_{\mathrm{x}}(f)}{\nu} \\ &= \pm\frac{1}{\nu}\left[\sqrt{(f+\nu)^2 - (1+M\nu)^2} - \sqrt{f^2 - 1}\right], \end{aligned} \tag{3.22}$$

since ν is the appropriate dimensionless wave number mismatch far from the Bragg wave number, and where the plus (minus) sign is used above (below)

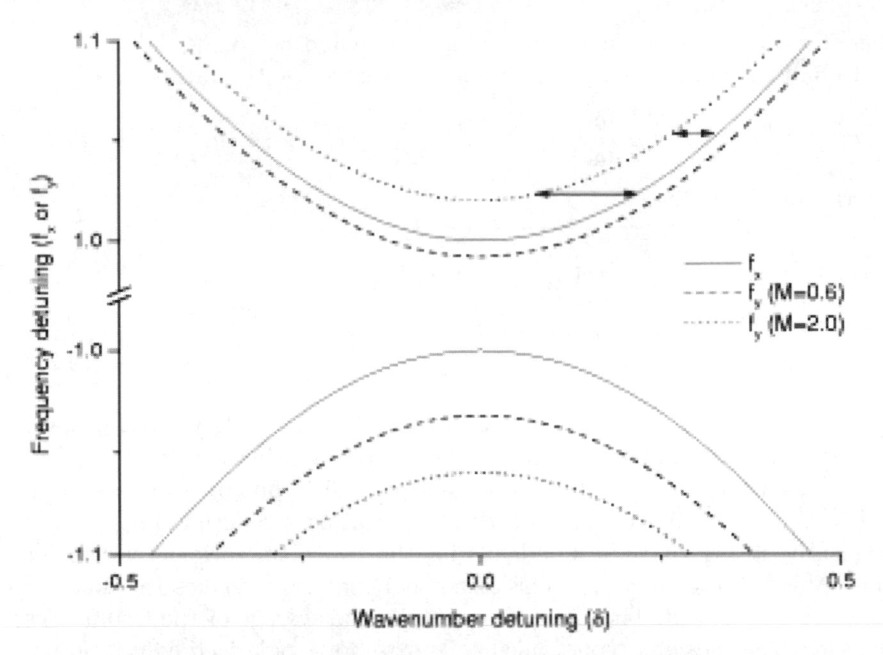

Fig. 3.3. Normalized dispersion relations with $\nu = 0.02$ and $M = 0.6, 2.0$; f_{x} refers to the solution (3.19) and f_{y} to the solution (3.21). The horizontal arrows give the wavenumber mismatch (or effective birefringence) between the two polarizations for $M = 2.0$. Closer to photonic band gap the effective birefringence is larger. Furthermore, the vertical distance between the x and y bands is different for positive and negative detuning. This asymmetry is not seen in Fig. 3.2, where $M = 0$

the Bragg frequencies. We also define an effective birefringence

$$n_{\mathrm{b}}^{\mathrm{eff}}(f) = F_{\mathrm{b}}(f)\, n_{\mathrm{b}} \,. \tag{3.23}$$

In Fig. 3.4 we plot $F_{\mathrm{b}}(f)$ with $\nu = 0.02$ and $M = 0, 1.5$ for positive and negative detunings. The $M = 0$ curve is symmetric for positive and negative detunings, while the $M = 1.5$ curve is strongly asymmetric. For $|M| \geq 1.0$ there are points where the wavenumber mismatch vanishes, which is seen in Fig. 3.4 for the $M = 1.5$ curve. This can be understood as follows. Referring back to Fig. 3.3, the frequencies on the upper bands associated with $\delta_{\mathrm{x}} = \delta_{\mathrm{y}} = 0$, which we denote F_{ui}, are $F_{\mathrm{ux}} = 1$, $F_{\mathrm{uy}} = (1 + (M-1)\,\nu)$. If $M \geq 1.0$, then $F_{\mathrm{uy}} > F_{\mathrm{ux}}$. However, we know that far from the grating the dispersion relation will approach its asymptotic value which, for a given value of k, has the x band frequency greater than the y band frequency. Since the upper bands of the dispersion relations are both monotonically increasing, they must cross, at which point the given frequency has the same wave number for both x and y polarizations. We set $F_{\mathrm{b}}(f) = 0$ in (3.22) to find

$$f_{\mathrm{no-bi}} = \frac{\nu}{2}\left(M^2 - 1\right) + M \,. \tag{3.24}$$

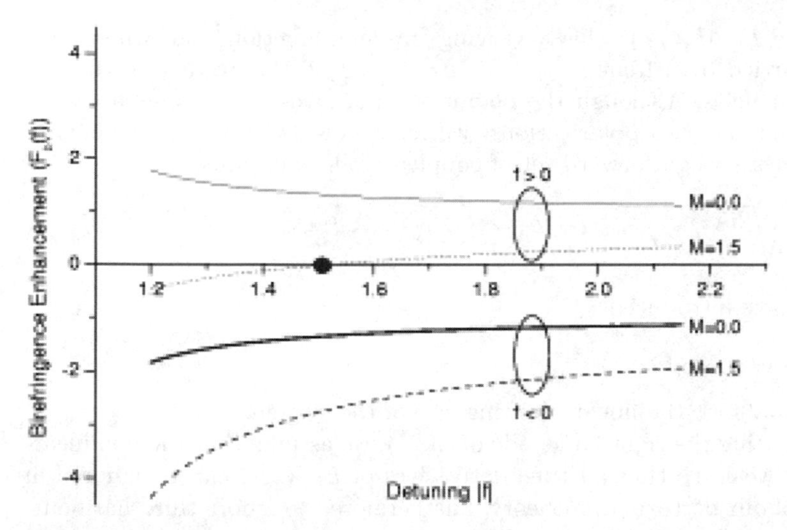

Fig. 3.4. Birefringence enhancement factor, $F_b(f)$, for positive and negative values of detuning, f, and for $M = 0, 1.5$. The $M = 0$ values are symmetric about the horizontal axis, while the $M = 1.5$ values are not. For $M = 1.5$ the birefringence vanishes for a particular a positive detuning. This is to be expected whenever $|M| > 1.0$

3.2.4 Nonlinearity

We now return to the wave equations (3.1) and include a nonlinear polarization $P_{\rm NL}$, which leads to the new wave equations

$$\frac{\partial^2 E_{\rm i}}{\partial z^2} - \frac{n_{\rm i}^2(z)}{c^2}\frac{\partial^2 E}{\partial t^2} = \mu_0 \frac{\partial^2 P_{\rm i}^{NL}}{\partial t^2} . \tag{3.25}$$

For simplicity, we treat the nonlinear properties of the medium as uniform, although this is not necessary. We adopt a nondispersive Kerr model for the nonlinearity

$$P_{\rm i}^{NL}(r,t) = \varepsilon_0 \chi_{\rm ijkl}^{(3)} E_{\rm j}(r,t) E_k(r,t) E_{\rm l}(r,t) , \tag{3.26}$$

with $i, j, k, l = x, y$. Clearly this form of the nonlinearity will couple the x and y polarizations of the electric field. We choose as a model for $\chi^{(3)}$ that of an isotroptic medium, so that only $\chi_{\rm xxxx}, \chi_{\rm yyyy}, \{\chi_{\rm xxyy}\}$, are non-zero, where $\{\chi\text{xxyy}\}$ is a set that includes all permutations of the subscripts, and $\chi_{\rm xxxx} = \chi_{\rm yyyy} = 3\{\chi_{\rm xxyy}\}$ [20]. In principle any model could be chosen, but because the birefringence of the system is considered small, the deviations in $\chi^{(3)}$ due to lack of isotropy will typically be of the next lowest order in the perturbation theory, and can thus be ignored. We now use an ansatz similar to (3.5) in the new wave equation (3.25). We assume

$$E(z,t) = E_{\rm i+}(z,t)\, e^{-i(\overline{\omega}t - k_0 z)} + E_{\rm i-}(z,t)\, e^{-i(\overline{\omega}t + k_0 z)} + c.c. , \tag{3.27}$$

where again $E_{\mathrm{i}\pm}(z,t)$ are slowly varying envelope functions, but where they are now carried by a frequency $\overline{\omega} = (\omega_{0\mathrm{x}} + \omega_{0\mathrm{y}})/2$, the average of the two Bragg frequencies. Although the definition of $\overline{\omega}$ gives us a common carrier frequency for our two polarizations, we must now redefine our A fields in order to get a straightforward set of coupled mode equations:

$$A_{\mathrm{x}\pm} = \sqrt{\frac{2\overline{n}_{\mathrm{x}}}{Z_0}} E_{\mathrm{x}\pm} e^{+i\delta t/4}, \quad A_{\mathrm{y}\pm} = \sqrt{\frac{2\overline{n}_{\mathrm{y}}}{Z_0}} E_{\mathrm{y}\pm} e^{-i\delta t/4}, \tag{3.28}$$

where we have introduced

$$\delta = 2(\omega_{0\mathrm{x}} - \omega_{0\mathrm{y}}),$$

which accounts for the linear birefringence of the system.

In evaluating the right hand side of (3.25), we assume that the nonlinearity is itself weak, so that all time derivatives of $E_{\mathrm{i}\pm}(z,t)$ can be ignored in the spirit of our perturbation theory. Furthermore, we ignore third harmonic generation on physical grounds: We have assumed that the underlying material is nondispersive, and while this may be valid for frequencies near $\overline{\omega}$, it will likely not be valid for frequency ranges extending to $\omega \simeq 3\overline{\omega}$; also, the assumption of no absorption at $\omega \simeq 3\overline{\omega}$ will likely be in error. We expect that in most cases the actual material dispersion and absorption will make any significant buildup of the third harmonic unlikely.

All of this leads to a set of four nonlinear coupled mode equations, which we show here explicitly for the x polarization [10],

$$\begin{aligned} 0 = {} & i\frac{\overline{n}_{\mathrm{i}}}{c}\frac{\partial A_{\mathrm{x}\pm}}{\partial t} \pm i\frac{\partial A_{\mathrm{x}\pm}}{\partial z} + \kappa_{\mathrm{x}} A_{\mathrm{x}\mp} + \alpha\left\{|A_{\mathrm{x}\pm}|^2 + 2|A_{\mathrm{x}\mp}|^2\right\} A_{\mathrm{x}\pm} \\ & + \beta\left\{|A_{\mathrm{y}\pm}|^2 + |A_{\mathrm{y}\mp}|^2\right\} A_{x\pm} + \beta A_{\mathrm{x}\mp} A^*_{\mathrm{y}\mp} A_{\mathrm{y}\pm} \\ & + \gamma\left\{A^*_{\mathrm{x}\pm} A^2_{\mathrm{y}\pm} + 2A^*_{\mathrm{x}\mp} A_{\mathrm{y}\pm} A_{\mathrm{y}\mp}\right\} e^{i\delta t}, \end{aligned} \tag{3.29}$$

where, introducing the familiar nonlinear index of refraction,

$$n_2 \equiv 3\chi^{(3)}_{\mathrm{xxxx}} Z_0 / 4\overline{n}^2_{\mathrm{x}}, \tag{3.30}$$

we find

$$\alpha = n_2 \frac{\overline{\omega}}{c}, \quad \beta = \frac{2}{3} n_2 \frac{\overline{\omega}}{c}, \quad \gamma = \frac{1}{3} n_2 \frac{\overline{\omega}}{c}. \tag{3.31}$$

The appropriate equations for the y polarization can be determined from (3.29) by switching $x \leftrightarrow y$ and $\delta \to -\delta$.

A careful derivation [10] shows that the nonlinear coefficients appearing the the equations for the x and y polarized light are actually different from each other. However, the difference in their values is entirely due to the linear birefringence, which we have assumed is small. To include such small variations in the nonlinear terms, which we have already assumed small, is unnecessary. Nevertheless, despite the simple ratio $\{\alpha : \beta : \gamma\} = \{3 : 2 : 1\}$, we persist in distinctly labelling the three coefficients because they have distinct interpretations. The coefficient α describes the phase modulation incurred by a given polarization component, travelling in a given direction, due to its intensity, and the intensity of the component with the same polarization but travelling in the opposite direction. The term in brackets involving β describes the phase modulation incurred by a given polarization component due to the intensity of the other polarization component. The second term involving β is a very weak energy exchange term. The terms involving γ would be strong energy exchange terms, except that because of the phase factors $e^{\pm i\delta t}$ they are not phase matched. In a later section we will discuss how phase matching can be achieved.

3.2.5 The Coupled Nonlinear Schrödinger Equations

If the frequency content of a pulse is sufficiently narrow, and lies outside the photonic band gaps of both polarizations, then the propagation of the pulse is well described by a set of coupled nonlinear Schrödinger equations [10]. This has been shown in two ways: First, from a direct consideration of the nonlinear wave equation (3.25); and second, as a consequence of the coupled mode equations (3.29). Both approaches require the use of the Bloch functions of the periodic medium. We here quote the results, and refer the reader to the appropriate literature for the details of the derivation [10].

We denote our nonlinear Schrödinger equation fields a_{i}, with $|a_{\mathrm{i}}|^2$ representing the intensity in the ith polarization. The fields are slowly varying envelope functions, with a common carrier frequency $\widetilde{\omega}$. The polarizations will have distinct wavenumbers, $\widetilde{k}_{\mathrm{i}}$, where the relationship between $\widetilde{k}_{\mathrm{i}}$ and $\widetilde{\omega}$ is given by (3.14), (3.11) and (3.12). As a shorthand we denote $\widetilde{\omega}_{\mathrm{i}}' \equiv (\partial\omega_{\mathrm{i}}/\partial k)\rfloor_{\widetilde{\omega}}$ and $\widetilde{\omega}_{\mathrm{i}}'' = (\partial^2\omega_{\mathrm{i}}/\partial k^2)\rfloor_{\widetilde{\omega}}$, and we note that although $\widetilde{\omega}$ is common to both polarizations, the derivatives will be different. Our nonlinear Schrödinger equations are

$$\begin{aligned} 0 = &\left(i\frac{\partial}{\partial t} + i\widetilde{\omega}_{\mathrm{x}}'\frac{\partial}{\partial z} + \frac{1}{2}\widetilde{\omega}_{\mathrm{x}}''\frac{\partial^2}{\partial z^2}\right)a_{\mathrm{x}} \\ &+ \left(\alpha_{spm}^{x}|a_{\mathrm{x}}|^2 + \alpha_{\mathrm{cpm}}^{x}|a_{\mathrm{y}}|^2\right)a_{\mathrm{x}} + \alpha_{\mathrm{pc}}^{x}a_{\mathrm{y}}^2 a_{\mathrm{x}}^* e^{i\Delta z}\,, \end{aligned} \quad (3.32)$$

$$\begin{aligned} 0 = &\left(i\frac{\partial}{\partial t} + i\widetilde{\omega}_{\mathrm{y}}'\frac{\partial}{\partial z} + \frac{1}{2}\widetilde{\omega}_{\mathrm{y}}''\frac{\partial^2}{\partial z^2}\right)a_{\mathrm{y}} \\ &+ \left(\alpha_{\mathrm{spm}}^{y}|a_{\mathrm{y}}|^2 + \alpha_{\mathrm{cpm}}^{y}|a_{\mathrm{x}}|^2\right)a_{\mathrm{y}} + \alpha_{\mathrm{pc}}^{y}a_{\mathrm{x}}^2 a_{\mathrm{y}}^* e^{-i\Delta z}\,, \end{aligned} \quad (3.33)$$

where

$$\Delta = 2\left(\widetilde{k}_{\mathrm{y}} - \widetilde{k}_{\mathrm{x}}\right) = 2n_{\mathrm{b}}^{\mathrm{eff}}\widetilde{\omega}/c\,, \quad (3.34)$$

and where the nonlinear coefficients are defined in terms of $\rho_{\mathrm{x/y}}$ (3.16) and the α coefficient for the CME (3.31),

$$\begin{aligned} \alpha^x_{\mathrm{spm}} &= \frac{c}{\overline{n}_{\mathrm{x}}}\alpha\left\{\frac{3-\rho_{\mathrm{x}}^2}{2}\right\}\frac{1}{\rho_{\mathrm{x}}}\,, \\ \alpha^x_{\mathrm{cpm}} &= \frac{2}{3}\frac{c}{\overline{n}_{\mathrm{x}}}\alpha\left\{\frac{2+\sqrt{(1-\rho_{\mathrm{x}}^2)(1-\rho_{\mathrm{y}}^2)}}{2}\right\}\frac{1}{\rho_{\mathrm{y}}}\,, \\ \alpha^x_{\mathrm{pc}} &= \frac{1}{3}\frac{c}{\overline{n}_{\mathrm{x}}}\alpha\left\{\frac{(1+\rho_{\mathrm{x}}\rho_{\mathrm{y}})+2\sqrt{(1-\rho_{\mathrm{x}}^2)(1-\rho_{\mathrm{y}}^2)}}{2}\right\}\frac{1}{\rho_{\mathrm{y}}}\,. \end{aligned} \tag{3.35}$$

The coefficients for the y polarization can be determined by switching $x \leftrightarrow y$. Each of these coefficients is directly related to the α coefficient for the CME (3.31), but modified by a multiplicative factor [21]. The terms in the braces arise because the Bloch functions of the periodic medium deviate from plane waves. Near a band edge, where $\rho_{\mathrm{i}} \to 0$, the Bloch functions are standing waves, rather than the travelling waves that appear as $\rho_{\mathrm{i}} \to 1$. The "peaks and valleys" in the local intensity of the Bloch function that appear as one moves towards the band gap result in a larger effective nonlinearity there. The dependence on ρ_{i}^{-1} arises because, although the fundamental nonlinearity is proportional to the square of the electric field amplitude, our equations are for quantities a_{i} such that $|a_{\mathrm{i}}|^2$ is the intensity. For fixed intensity, the slower the velocity of a pulse, the larger its electric field amplitude; hence the coefficients increase as $\rho_{\mathrm{i}} \to 0$. Finally, the effect of the nonlinearity on a given pulse is further enhanced as $\rho_{\mathrm{i}} \to 0$ because the pulse is travelling more slowly, and sees the nonlinearity for a longer time as it propagates over a given length. This last effect is accounted for in the equations (3.32) themselves, rather than the multiplicative factors in (3.35).

For small birefringences, in the regime where the CNLSE is a useful approximation, the multiplicative factors (3.35) are roughly equal, so the nonlinear coefficents are in the ratio $\{\alpha^i_{\mathrm{spm}} : \alpha^i_{\mathrm{cpm}} : \alpha^i_{\mathrm{pc}}\} \simeq \{3:2:1\}$. We define an enhanced nonlinear coefficient that accounts both for the multplicative factors in (3.35), and also for the increased time a pulse spends in a nonlinear medium of given length:

$$n_2^{\mathrm{eff}} = \left\{\frac{3-\rho_{\mathrm{x}}^2}{2\rho_{\mathrm{x}}^2}\right\} n_2 = \left\{\frac{2f^2+1}{2\,(f^2-1)}\right\} n_2\,,$$

in terms of which

$$\begin{aligned} \alpha_{\mathrm{spm}} &\simeq \omega'_{\mathrm{x}}\frac{\widetilde{\omega}}{c} n_{2\mathrm{eff}}\,, \\ \{\alpha_{\mathrm{spm}} : \alpha_{\mathrm{cpm}} : \alpha_{\mathrm{pc}}\} &\simeq \{3:2:1\}\,. \end{aligned}$$

As the pulse gets closer to the Bragg wave number, the effect of nonlinearity gets very large. In practice the nonlinear Schrödinger equation has been used

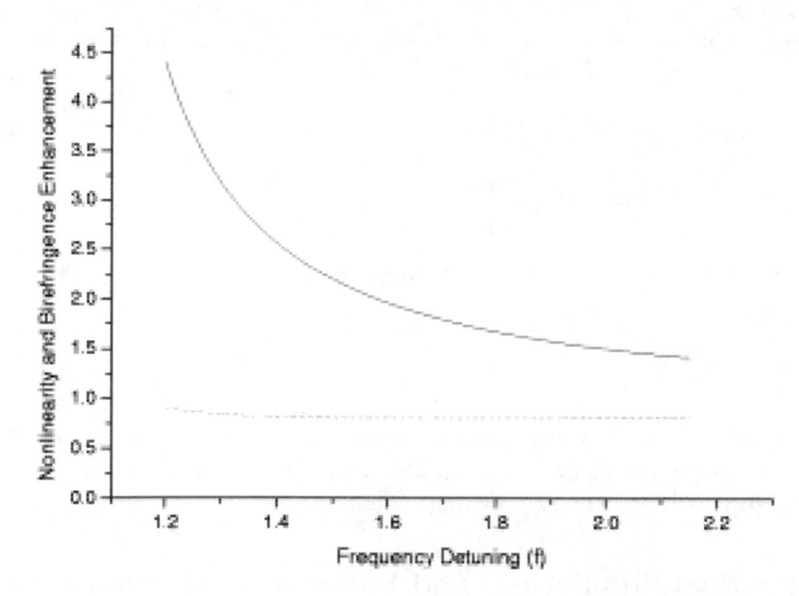

Fig. 3.5. Comparison of the birefringence enhancement (dotted line) and the nonlinear enhancement (solid line) for $M = 0.6$ and $\nu = 0.02$. For these parameters the birefringence is essentially unaffected by the grating until f is almost unity. The nonlinearity is greatly enhanced due to the reduced group velocity and the standing-wave nature of the underlying Bloch functions of the periodic medium near the band edge

to describe pulses travelling at about 50% $c/\overline{n}_{\mathrm{i}}$ [6], where the nonlinearity is enhanced by a factor of about 6. In Fig. 3.5 we plot n_2^{eff}/n_2 and $n_b^{\mathrm{eff}}/n_{\mathrm{b}}$ as a function of detuning, f, for $M = 0.6, \nu = 0.02$. For these parameters, the nonlinearity is enhanced much more than the birefringence as the detuning is decreased.

In the absence of nonlinearity, equations (3.32) describe pulses propagating with a group velocity and group velocity dispersion defined by (3.15). If we include nonlinearity then three effects emerge: α_{spm} governs self phase modulation; α_{cpm} governs cross phase modulation; and α_{pc} governs phase conjugation, or energy exchange. We note, however, that the energy exchange term is not phase matched, so we would not ordinarily expect it to occur. The interpretations of $\{\alpha_{\mathrm{spm}}, \alpha_{\mathrm{cpm}}, \alpha_{\mathrm{pc}}\}$ for the nonlinear Schrödinger equation roughly correspond to those of $\{\alpha, \beta, \gamma\}$ for the nonlinear coupled mode equations, except that the β coefficient in the coupled mode equations governs a very weak energy exchange, while the α_{cpm} term does not.

If we ignore both α_{pc} and the group velocity dispersion, we can define an effective index of refraction for each polarization,

$$n_{\text{NL}}^{x} = \overline{n}_{\text{x}} + n_2^{\text{eff}} \left\{ |a_{\text{x}}|^2 + \frac{2}{3} |a_{\text{y}}|^2 \right\} ,$$
$$n_{\text{NL}}^{y} = \overline{n}_{\text{y}} + n_2^{\text{eff}} \left\{ |a_{\text{y}}|^2 + \frac{2}{3} |a_{\text{x}}|^2 \right\} , \quad (3.36)$$

from which we can define a nonlinear birefringence,

$$n_{\text{NL}}^{\text{b}} = n_{\text{NL}}^{\text{y}} - n_{\text{NL}}^{\text{x}} - n_{\text{b}}$$
$$\simeq \frac{1}{3} n_2^{\text{eff}} \left\{ |a_{\text{y}}|^2 - |a_{\text{x}}|^2 \right\} . \quad (3.37)$$

When we include $\alpha_{\text{pc}}^{\text{i}}$ and group velocity dispersion, the concept of a nonlinear birefringence will still be valid, but the dynamics of the pulse will be complicated, so it is best used as a heuristic guide.

3.3 Numerical Simulations and Experimental Results

In this section we give some experimental results and numerical simulations for birefringent gratings. Except where noted, our simulations represent UV induced optical fiber gratings, with the material parameters given in Table 1, with λ_{vac} defined as the vacuum wavelength that corresponds to the Bragg frequency of the x polarization.

We first discuss a number of polarization effects using the coupled nonlinear Schrödinger equations. These equations have a long history of being used in a variety of situations [12–14], so the nature of their solutions is better understood than that of solutions of the coupled mode equations. Furthermore, the CNLSE involve fewer nonlinear coefficients, since there is no explicit coupling between waves travelling in opposite directions, and thus it is inherently easier to identify the physics in their solutions.

In the CNLSE associated with the grating system, all of the parameters are strongly frequency dependent, which means that frequency detuning gives us access to the different propagation regimes studied in the literature [17]. For light far detuned from the Bragg wave number, the parameters of the equation become similar to the basic material parameters of glass. Because of the birefringence induced by UV grating growth, this puts us in a regime similar to that studied by Menyuk [13], where the energy exchange between polarizations can be ignored. However, by detuning closer to grating we find that the nonlinearity is enhanced more than the birefringence, so we enter a regime where energy exchange is feasible. Finally, very close to the photonic band gap, we enter a regime where the vector soliton described by Ahkmediev et al. [22,23] might be observed. In the following we quantify the three regimes of propagation just discussed. We then present numerical simulations and review experiments performed in one of the regimes.

Although we use the CNLSE to describe the results we expect at given detunings, we use the full nonlinear coupled mode equations to simulate the results of experiments. We do this for three reasons: First, the coupled mode equations provide the better description of what is happening to the light near the grating; second, the successful use of the full CME to describe results that follow from the CNLSE confirms the validity of the approximations that went into the derivation of the CNLSE; third, although both the coupled mode equations and the CNLSE describe infinite systems, it is easier to modify the coupled mode equations to describe the finite, apodized systems studied in experiments. To modify the coupled mode equations to describe a system where light travels from bare fiber into a fiber grating, which is necessary since we inject light from a laser, we allow the index contrast to take on a slowly varying spatial dependence, so that our model for the grating becomes

$$n_i(z) = \overline{n}_\mathrm{i} + \delta n_\mathrm{i} A(z) \cos(2k_0 z) , \tag{3.38}$$

where $A(z)$ is the apodization profile of the grating. When $A(z) = 0$ the grating, in effect, vanishes, and the coupled mode equations describe bare fiber. We note that the only coefficient in the coupled mode equations that depends explicitly on δn_i is the grating strength, κ_i. In the spirit of our perturbation theory, the small corrections that δn_i would impose upon the nonlinear coefficients is ignored. As demonstrated by Sipe et al. [24], we can use the coupled mode equations (3.29) with $\kappa_\mathrm{i} \to \kappa_\mathrm{i}(z) = A(z)(\delta n_\mathrm{i} k_0 / 2\overline{n}_\mathrm{i})$. In our simulations we use a 7.7 cm grating, of which the middle 6 cm have $A(z) = 1.0$. The remaining 0.85 cm on either end of the grating have a Gaussian apodization profile, with a half-width, half-maximum of 0.425 cm. A more complete model for a physical grating would add a slow variation in the background index, $\overline{n}_\mathrm{i}$, itself, that arises as the grating is being written by exposure to UV light. This has been discussed in detail [17,19], but is not directly relevant to the birefringent effects in this chapter, and can in fact be eliminated by a design pattern of irradiation with UV light after the grating is written [19].

Finally, turning to the nonlinear coupled mode equations themselves, we present simulations that demonstrate the all optical AND gate proposed by Lee and Ho [25] and observed by Taverner et al. [7] The dynamics of such a device are very complicated and, although certain heuristics do exist to aid in understanding, one needs to turn to numerical simulations to gain insight into the details of the solutions.

3.3.1 Three Regimes of Propagation

In the optical fiber literature, much of the study of coupled nonlinear Schrödinger equations similar to (3.32) has centered around two effects: soliton propagation [15] and polarization instability [18]. The familiar isotropic soliton exists either when the system birefringence and grating strength mismatch vanishes, or when the light is completely and unambiguously polarized along one of the principal axes of the system. In either case one can consider the first of equations (3.32) with $\alpha_{\text{cpm}}^{\text{x}} = \alpha_{\text{pc}}^{\text{x}} = 0$, so that no coupling exists between the polarizations. It then follows that for the correct pulse shape – specifically, a hyperbolic secant – and peak power, the nonlinearity of the system can balance the group velocity dispersion, and the pulse will propagate without changing its shape. For the positive nonlinearity that describes our system, we need an anomalous group velocity dispersion to excite a soliton. Such a dispersion occurs in the upper band of the dispersion relation. For a given full-width, half-maximum (FWHM) pulse width, T_{FWHM}, an isotropic soliton, polarized completely along the x axis, will form with a peak intensity [15]

$$I_{\text{sol}} = \frac{3.11\,|\omega_{\text{x}}''|}{\alpha_{\text{spm}}\,(\omega_{\text{x}}')^2\,T_{\text{FWHM}}^2}\,. \tag{3.39}$$

Such solitons have been observed both in bare optical fibers [15], where $n\,(z) = \overline{n}_{\text{x}}$ and $\omega_{\text{x}}' = c/\overline{n}_{\text{x}}$, and in fiber gratings [26]. The difference between the two situations is that near the Bragg frequency the grating leads to a group velocity dispersion many orders of magnitude higher than that in the bare fiber. This means that a much higher peak power is required to excite a soliton, but a correspondingly shorter interaction length is required for the soliton to form. In a bare fiber a typical soliton length – the distance over which a soliton forms – is 80 m, while in a fiber grating the length is about 4 cm.

Upon returning to the birefringent situation, we must include the terms involving α_{cpm} and α_{pc}. We have mentioned that the terms with α_{pc} contain a phase factor, $e^{\pm i\Delta z}$, which is related to the mismatch in phase accumulation between the polarizations. These terms are sometimes ignored because the phase factor means that the energy exchange process is not phase-matched, and hence will not build up. However, under the right conditions the nonlinear birefringence (3.37) can cancel the linear birefringence (3.3), and energy exchange can become phase matched. We can write down a critical intensity for the onset of polarization instability, in the case of a CW beam, but before doing so we identify the fast and slow axes of polarization. The terminology *fast* and *slow* arises in the study of instability in the absence of a grating, where the group and phase velocities are roughly equal: $v_{\text{x/y}} = c/\overline{n}_{\text{x/y}}$. The slow axis is the axis with the *higher* index of refraction. Of course, intense light leads to an effective nonlinear contribution to the index of refraction. If we inject intense light polarized near the slow axis, then the nonlinear birefringence (3.37) will augment the effective birefringence (3.23), thus increasing the phase mismatch of the energy exchange. On the other hand, if we inject light polarized near the fast axis, the nonlinear birefringence can

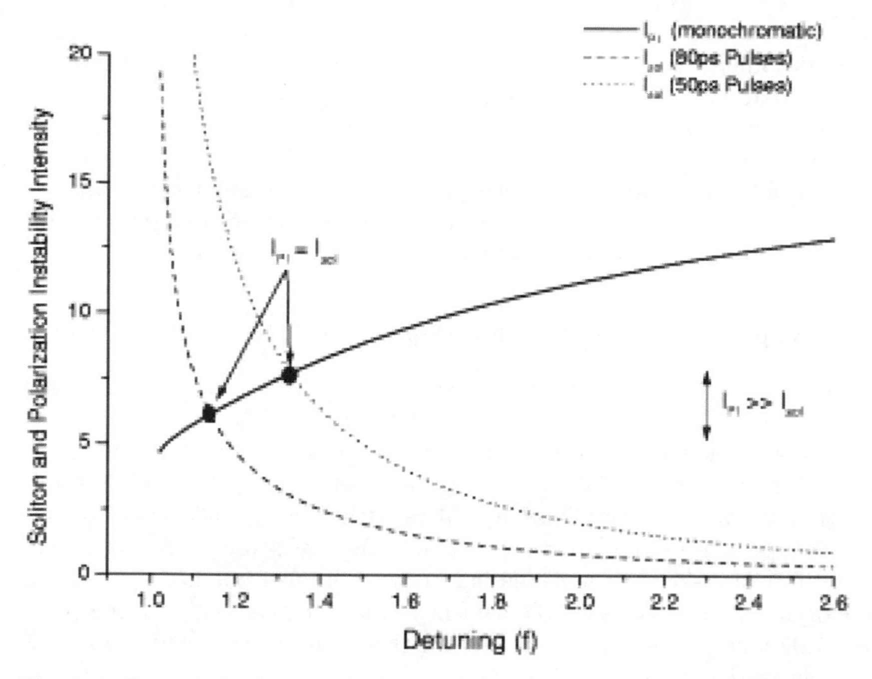

Fig. 3.6. I_{PI} and I_{sol} for $M = 0.6$ and $\nu = 0.02$, and $T_{\mathrm{FWHM}} = 50$, 80 ps. At large detunings $I_{\mathrm{sol}} \ll I_{\mathrm{PI}}$, but eventually the two curves cross. For large detunings, an approximate theory can be used to determine the Stokes vector evolution. For very small detunings, a vector soliton might be observed. In the middle, where $I_{\mathrm{sol}} \simeq I_{\mathrm{PI}}$, the dynamics are complicated, and are not yet fully understood. The I_{PI} curve assumes a CW input. The I_{sol} curve is dependent on the pulse width. Shorter pulse widths increase I_{sol}

cancel the effective birefringence, and energy exchange will occur. For a CW input, the required energy for this cancellation is given by [15]

$$I_{\mathrm{PI}} = \frac{3}{2}\frac{n_{\mathrm{b}}^{\mathrm{eff}}}{n_2^{\mathrm{eff}}}, \tag{3.40}$$

where we use the effective values of n_2 and n_{b}. For a CW input, I_{PI} would phase-match the energy exchange if essentially all of the light were polarized along the fast axis. For higher intensities one can polarize increasingly towards the slow axis and still observe energy exchange.

In Fig. 3.6 we plot I_{sol} and I_{PI} for $M = 0.6$, $\nu = 0.02$ with two different pulse widths: $T_{\mathrm{FWHM}} = 50$ ps, 80 ps. It is evident that the birefringent grating gives us access to three distinct propagation regimes. For large frequency detunings, we find $I_{\mathrm{sol}} \ll I_{\mathrm{PI}}$, so that we can excite pulses that roughly retain their shape upon propagation, and exchange no energy between polarizations; for frequencies close to the gap $I_{\mathrm{sol}} \gg I_{\mathrm{PI}}$; while for frequencies between these extremes we find $I_{\mathrm{sol}} \simeq I_{\mathrm{PI}}$. The formula for I_{sol} is a function of pulse width.

In Fig. 3.6, the I_{sol} curve with $T_{FWHM} = 80\,ps$ intersects the I_{PI} at a much smaller detuning. We could, for even larger pulsewidths, suppress the value of I_{sol} so that $I_{sol} < I_{PI}$ for all detunings where we would expect (3.32) to hold. In the regime where $I_{sol} \gg I_{PI}$ we expect a vector soliton [22] to be observable.

In the following two sections we present numerical simulations and review experimental data that deal with soliton formation and polarization instability. We use the material parameters shown in Table 3.1

3.3.2 Approximate Solution for Polarization Evolution

Ahkmediev et al. [22], seeking vector soliton solutions to equations similar to (3.32), presented a simple theory to describe the evolution of the Stokes parameters for nonlinear pulses whose profile remains constant with propagation. Referring to Fig. 3.6, in the regime where $I_{sol} \ll I_{PI}$ we can apply their theory. We simulate Gaussian pulses with varying initial polarization states, whose parameters are very close to those of solitons. Although the long term behaviour of these Gaussian pulses will involve a complex process of energy radiation and isotropic soliton formation, they will, for the propagation distances considered here, retain their pulse profile while changing their polarization state.

To summarize the theory we write

$$a_x(z,t) = X(z)\,h(t)\,,$$
$$a_y(z,t) = Y(z)\,h(t)\,, \qquad (3.41)$$

where $h(t)$ is the common temporal pulse profile. We define the normalized Stokes parameters [14]

$$S_0 = |X|^2 + |Y|^2$$
$$S_1 = \left(|X|^2 - |Y|^2\right)/S_0\,,$$
$$S_2 = \left(X^*Y + XY^*\right)/S_0\,,$$
$$S_3 = -i\left(X^*Y - XY^*\right)/S_0\,, \qquad (3.42)$$

and the Stokes vector

$$\mathbf{S} = (S_1, S_2, S_3)\,. \qquad (3.43)$$

In the absence of material absorption or other loss mechanisms, both S_0 and $|\mathbf{S}|$ are uniform, so the evolution of the Stokes parameters can be visualized as a path traced on the surface of a sphere. From the CNLSE (3.32) we find that the Stokes parameters satisfy the vector differential equation

$$\frac{d}{dz}\boldsymbol{S} = 2g_1\left[\boldsymbol{e}_1 \times \boldsymbol{S}\right] + 2g_3S_3\left[\boldsymbol{e}_3 \times \boldsymbol{S}\right]\,, \qquad (3.44)$$

where $\mathbf{e}_{1/3}$ are unit vectors pointing in the $S_{1/3}$ directions. The parameters g_1, half the phase-velocity mismatch, and g_3, the nonlinear parameter, are defined as

$$\begin{aligned} g_1 &= \Delta/2\,, \\ g_3 &= \frac{1}{3}\alpha_{spm}\frac{\int f^4(t)\,dt}{\int f^2(t)\,dt}\,. \end{aligned} \tag{3.45}$$

In Figs. 3.7 and 3.8 we compare the final polarization states of pulses simulated with the full nonlinear coupled mode equations to the final polarization states predicted by (3.44). In both figures we use 80 ps pulses with a detuning $f = 2.00$, which puts us well in the region where $I_{sol} \ll I_{\mathrm{PI}}$. In Fig. 3.7 we simulate the propagation of light injected with equal intensity on both axes; we vary the initial phase lag between the polarizations from -90^0 to $+90^0$. In the language of the Stokes parameters this is equivalent to keeping S_1 fixed, but varying S_2 and S_3. In Fig. 3.8 we fix the input phase lag at 0^0, but vary the initial value of S_1 from -0.9 to $+0.9$.

Figures 3.7 and 3.8 show an excellent agreement between the theory (3.44) and the full numerical solution of (3.29). This verifies the existence of the high detuning regime shown in Fig. 3.6. Its existence is confirmed, too, by the fact that in Fig. 3.8 the output S_1 is essentially equal to the input S_1, which means that no energy was exchanged between the pulses. The slight deviation between the theory and the simulations is largely due to the fact that effects

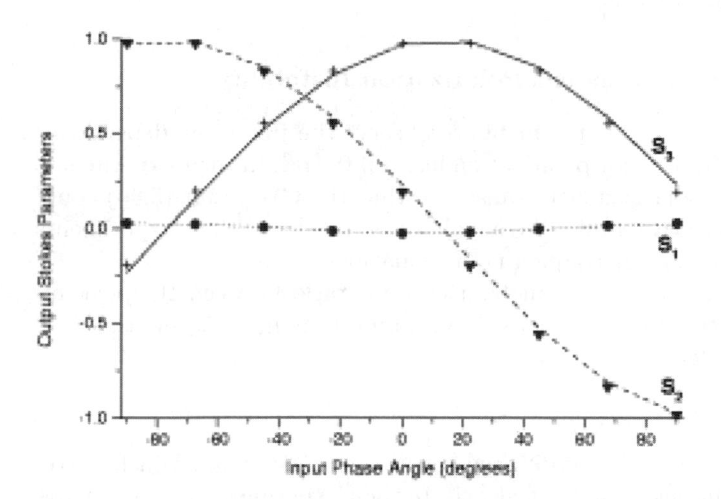

Fig. 3.7. Output values for S_1, S_2 and S_3 when the input pulse has soliton-like parameters, and is detuned to $f = 2.00$ with equal intensities in the x and y polarizations. The control parameter – the phase lag between polarizations – is varied between $\pm 90^0$. The symbols, which represent the approximate theory of Akhmediev et al., match up very well with the results of the full numerical solution of the nonlinear coupled mode equations (lines)

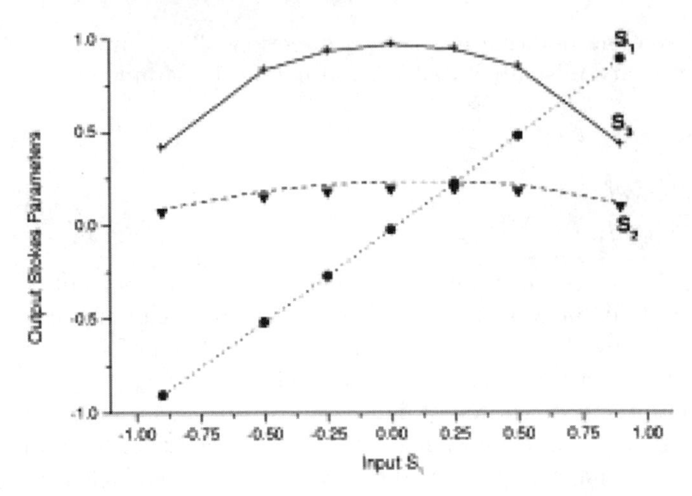

Fig. 3.8. Output values for S_1, S_2 and S_3 when the input pulse has soliton-like parameters, and is detuned to $f = 2.00$ with equal intensities in the x and y polarizations. The control parameter – the input value of S_1 – is varied between ± 0.9. The symbols, which represent the approximate theory of Akhmediev et al., match up very well with the results of the full numerical solution of the nonlinear coupled mode equations (lines)

of the apodization profile, discussed after (3.38), are not accounted for by the approximate theory (3.44).

3.3.3 Frequency Dependent Polarization Instability

Our colleagues performed experiments to observe the frequency dependence of the threshold required for polarization instability [18]. In their experiments they used linearly polarized 80 ps pulses, so that the CW result (3.10) could only act as a qualitative guide. We simulate the results of their experiments by using the full nonlinear coupled mode equations (3.29).

We choose as a control parameter the input ratio between the peak intensity on the slow axis and fast axis. Assuming that $\overline{n}_{\mathrm{y}} > \overline{n}_{\mathrm{x}}$, so that y is the slow axis, we define

$$\sigma = |Y|^2 / |X|^2 .$$

To observe polarization instability we require σ to be small, which corresponds to a large intensity on the fast axis. In their experiments our colleagues fixed $\sigma = 1/10$. They then varied the frequency detuning of the pulse, and searched for the polarization instability threshold. At each frequency detuning they increased the intensity of their input pulse until the output pulse had $\sigma = 1/3$, which was chosen as the criterion for the polarization instability threshold.

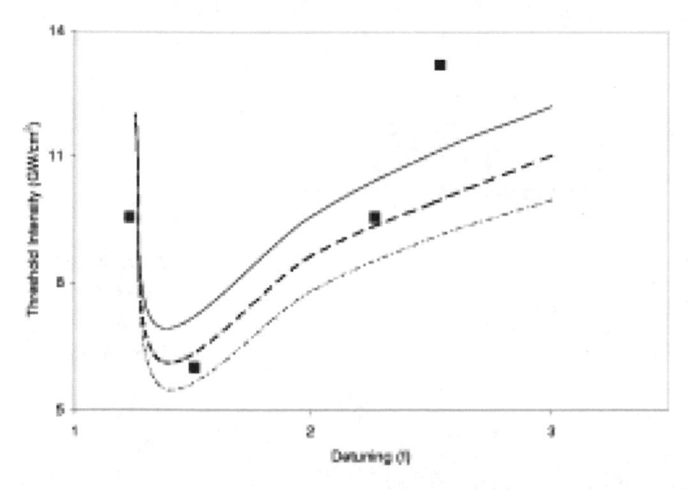

Fig. 3.9. Threshold intensities for polarization instability for experimental (circles) and numerically simulated (line) fiber Bragg gratings with material parameters in Table 1, except as given below. Both simulations and experiments follow the trend of $I_{\rm PI}$ from Fig. 3.6 until, at small detunings, we enter a regime where the CNLSE are no longer a good approximation for light propagating in the grating, and $I_{\rm PI}$ rises. The simulations are performed using three values of n_2: 2.52×10^{-16} (solid), 2.8×10^{-16} (dash) and 3.08×10^{-16} (dash-dot) W/cm^2

In Fig. 3.9 we plot the results of the experiment performed by Slusher et al. [18], along with numerical simulations of the full nonlinear coupled mode equations using the parameters in Table 3.1, but with three values of n_2: 2.52×10^{-16}, 2.8×10^{-16} and 3.08×10^{-16} W/cm^2. Increasing the value of n_2 lowers the threshold for polarization instability, but the behaviour of the frequency dependence is otherwise unchanged. The experimental points agree qualitatively with the numerical simulations. As the detuning approaches unity, the polarization instability threshold intensity gets *larger*, which contradicts the plot of $I_{\rm PI}$ in Fig. 3.6. This is because in the small detuning regime the CNLSE is no longer a good approximation for pulse evolution, so the $I_{\rm PI}$ that emerges from an analysis based on those equations is no longer valid. The CNLSE account neither for the higher-order dispersion at lower detunings, nor for the reflection from the photonic band gap, both of which serve to invalidate the discussion of polarization instability given in Sect. 3.3.2.

3.3.4 Logic Gates using Birefringence

In a 1998 paper Taverner et al. demonstrated the use of a birefringent grating to construct a logical AND gate [7]. In their experiment the input bits were the intensities of light polarized along the x and y directions. The frequency of the light was set such that it was within the photonic bandgaps of both

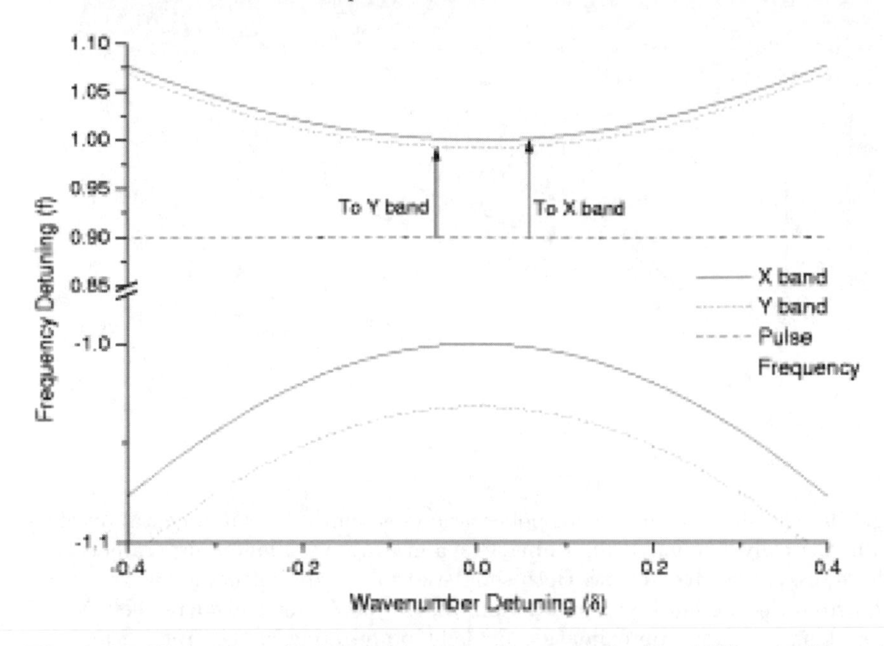

Fig. 3.10. Dispersion relation and incident frequency for the AND gate. The pulse frequency is deeper within the band gap of x-polarized, so a higher intensity is needed to lift that polarization alone out of the gap. However, when both polarizations are injected simulataneously, the process of nonlinear energy exchange can favour the x-polarized light

polarizations; the intensity was chosen such that each polarization, when injected alone, was almost completely reflected. When both polarizations were injected together into the system, their combined intensity allowed some of the light to be transmitted.

To describe such experiments we can no longer rely on the CNLSE, since they do not account for the reflection of the photonic bandgap. Instead we turn to the full coupled mode equations, both for insight and for numerical simulations. The key to understanding the AND gate is the phase modulation incurred both by self- and cross-phase modulation. Referring to Fig. 3.10, we consider the situation where light is injected at a frequency within the photonic bandgaps of both polarizations but, since we are assuming $\overline{n}_{\mathrm{y}} > \overline{n}_{\mathrm{x}}$, deeper within the bandgap of the x polarization. Using the terminology of polarization instability, the y axis is the slow axis so, if any instability occurs, we would expect energy to shift towards the y polarization, at least initially.

Considering only phase modulation for the moment, we recall that it has the effect of adding a nonlinear index of refraction. The Bragg frequency of the light is defined by the index of refraction so that, in the presence of phase

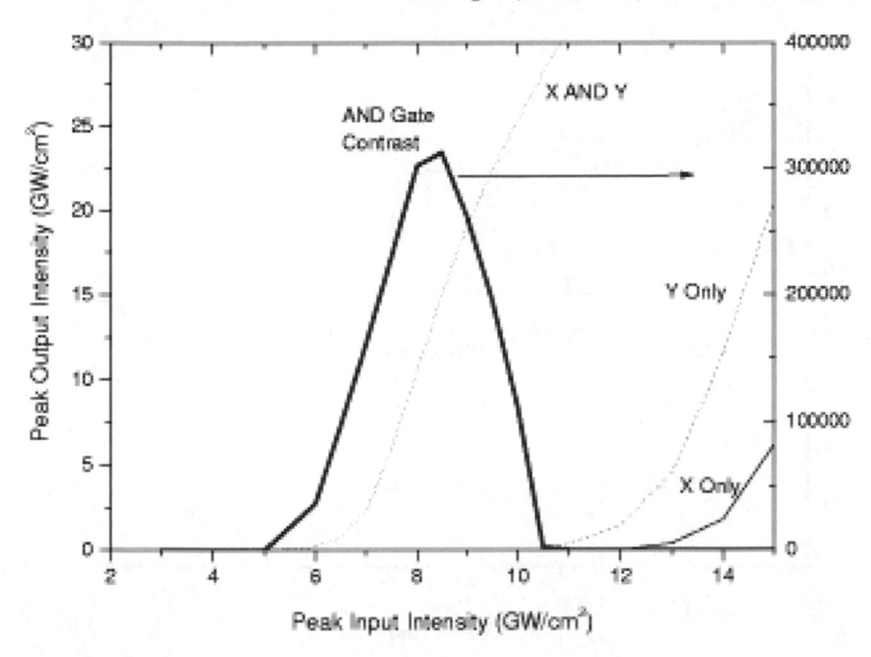

Fig. 3.11. Switching intensities for the AND gate, referenced to the left hand scale. As suspected the x polarization has a higher switching threshold than the y polarization. When both polarization are injected together, they can more easily escape the band gap. The switching contrast ratio, which is referenced to the right hand scale, reaches several orders of magnitude

modulation, we can define a nonlinear Bragg frequency as

$$\omega_{0\mathrm{i}}^{\mathrm{NL}} = \frac{c}{\overline{n}_{\mathrm{i}} + n_{\mathrm{i}}^{\mathrm{NL}}} k_0 \simeq \omega_{0i} \left(1 - n_{\mathrm{i}}^{\mathrm{NL}} / \overline{n}_{\mathrm{i}}\right) . \tag{3.46}$$

We expect the Bragg frequency to shift down with increasing phase modulation. On the other hand, we expect no change in the index *contrast* of the grating, so that the width of the photonic band gap should remain constant. The net effect is that intense light can lift itself out of the photonic bandgap. Returning to Fig. 3.10, we suspect that if we only inject light polarized in the y direction, there should be some input intensity at which an appreciable amount of energy manages to escape the photonic band gap. The same should occur with only the injection of the x-polarized light, but the threshold for switching out of the gap should be higher. To confirm this we performed numerical simulations with the parameters in Table 1, using linearly polarized pulses with $T_{\mathrm{FWHM}} = 250\,\mathrm{ps}$ and a Gaussian profile. Using the left hand scale of Fig. 3.11 we plot the output peak intensity versus input peak intensity for light injected along one of the two polarization axes. We clearly observe the switching behaviour just described. The x-polarized light

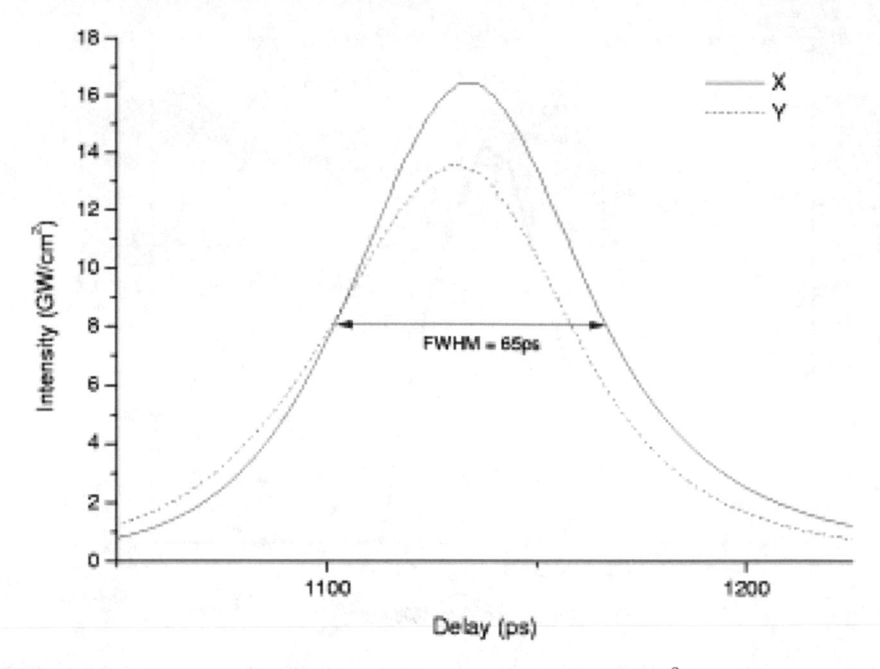

Fig. 3.12. Output pulse for the AND gate with 8.5 GW/cm^2 incident in each polarization. The pulse width has been compressed by a factor of 4. The x-polarized pulse has a higher peak intensity as a consequence of the complicated nonlinear interplay between the polarizations

has a switching threshold at around 13 GW/cm^2 while the y-polarized light has a switching threshold at about 11 GW/cm^2.

The line labelled X AND Y in Fig. 3.11 gives one half the sum of the output intensity along the x and y axes if both polarizations are injected with the same peak intensity. The threshold for switching has now dropped to 6 GW/cm^2. The thick black line in Fig. 3.11, which is referred to the right hand scale, gives the output intensity constrast ratio for the AND gate. We divide the total output intensity by two to account for the fact that we are injecting twice as much energy, and then compare this number to the output intensity detected from the injection of only y-polarized light. In our simulations the AND gate constrast reached as high as 3×10^5. We note, though, that there is a trade off involved: Longer pulses give lower thresholds and higher contrast ratios, at the expense of slower devices.

In Fig. 3.12 we plot the output intensity along the two axes of polarization for the AND gate with 8.5 GW/cm^2 injected on each axis. The output pulses have experienced significant nonlinear compression: The larger output pulse has $T_{\mathrm{FWHM}} = 65$ ps. Notice that the polarization of the larger output pulse lies along the x axis. This is counter-intuitive because we know, from Fig. 3.10, that the frequency is closer to the band edge of y-polarized light, and thus the y polarized light should have an easier time escaping the band gap. Furthermore, were energy exchange to occur, we would expect the y axis (the slow axis) to gain energy, and the x axis to lose it. However, the energy exchange process is oscillatory, and although energy tends toward the slow axis, it sloshes back and forth between the axes for a long time. Thus, to some extent the grating length determines which polarization will be dominant at the output.

3.4 Conclusion

We have discussed the propagation of light in birefringent, nonlinear, periodic media. We have shown that a birefringent grating strongly affects the mismatch between the pulse propagation parameters associated with the orthogonal polarizations of light. By controlling the grating strength associated with each of the two polarizations, one can find a frequency where the effective linear birefringence of the system disappears. The familiar linear, isotropic effects – reduced group velocity and enhanced group velocity dispersion – still occur, and the mismatch between them is also a function of frequency.

In the isotropic situation, the inclusion of nonlinearity leads to phase modulation. In the birefringent system, this isotropic phase modulation is complemented by a cross phase modulation, where a given polarization sees the intensity of the orthogonal polarization. Most importantly, birefringence can lead to an exchange of energy between the polarizations which, if correctly exploited, can be used to enhance the performance of devices. We have shown the frequency dependence of the threshold energy required for energy exchange. We have also discussed the role that energy exchange plays in an all optical AND gate based on a birefringent grating system.

The work presented here has two implications for device applications. First, devices intended to exploit linear birefringence and grating strength mismatch can be designed using the formulae presented in this chapter. Second, if devices are intended to operate in an isotropic system, it is unlikely that the system will be entirely isotropic. In optical fibers it is known that a birefringence is induced by UV grating growth. In other systems, such as polymers or semiconductors, imperfections in fabrication techniques will likely lead to the presence of some birefringence. Because of such stray birefringences, the threshold for polarization instability will naturally be a parameter of key importance for device design.

Although this work is strictly one dimensional, it can be extended to treat dielectric waveguides with TE and TM polarization, or other quasi-one dimensional systems. There the strong grating formalism, which we presented earlier [10], would be the logical starting point. A more exact numerical agreement would require a more sophisticated theory, but we expect the qualititative effects to be well described by the work presented in this chapter.

Acknowledgements. This work was supported by the National Science and Engineering Research Council of Canada. The authors would like to thank Dr. R.E. Slusher and Dr. S. Spälter for helpful discussions, for their enthusiasm, and for their perseverance on the polarization experiments, and F. Nastos for his help with LaTeX.

References

1. H.G. Winful, J.H Marburger, and E. Garmire: Appl. Phys. Lett. **35**, 379 (1979)
2. W. Chen, D.L. Mills: Phys. Rev. Lett. **58**, 160 (1987)
3. C.M. de Sterke, J.E. Sipe: Phys. Rev A **38**, 5149 (1988)
4. C.M. de Sterke, D.G. Salinas, J.E. Sipe: Phys. Rev. E **54**, 1969 (1996)
5. C.M. de Sterke, J.E. Sipe: 'Gap Solitons'. In: *Progress in Optics XXXIII*, ed. by E. Wolf (North-Holland, Amsterdam 1994) pp. 203–260
6. B.J. Eggleton, C.M. de Sterke, R.E. Slusher: J. Opt. Soc. Am. B **16**, 587 (1999)
7. D. Taverner et al.: Opt. Lett. **20**, 246 (1998)
8. T. Erdogan, V. Mizrahi: J. Opt. Soc. Am. B **11**, 2100 (1994)
9. S. Pereira, J.E. Sipe: Opt. Express **3**, 418 (1998)
10. S. Pereira, J.E. Sipe: Phys. Rev. E **62**, 5745 (2000)
11. Y. Silberberg, Y. Barad: Opt. Lett. **20**, 246 (1995)
12. N. Akhmediev, J.M. Soto-Crespo: Phys. Rev. E **49**, 5742 (1994)
13. C.R. Menyuk: Opt. Lett. **12**, 614 (1987)
14. N.N. Akhmediev, A. Ankiewicz: *Solitons, Nonlinear Pulses and Beams* (Chapman and Hall, New York 1997)
15. G.P. Agrawal: *Non-Linear Fiber Optics* (Academic Press, San Diego 1989)
16. S. Pereira, J.E. Sipe: Phys. Rev. E **66**, 026606 (2002)
17. S. Pereira, J.E. Sipe, R.E. Slusher: to be published in J. Opt. Soc. Am. B
18. R.E. Slusher et al.: Opt. Lett. **25**, 749 (2000)
19. S. Pereira, J.E. Sipe, R.E. Slusher, S. Spälter: J. Opt. Soc. Am. B **19**, 1509 (2002)
20. R.W. Boyd, *Nonlinear Optics* (Academic Press, San Diego 1992)
21. H.-M. Keller, S. Pereira, J.E. Sipe: Opt. Comm. **170**, 35 (1999)
22. N. Akhmediev, A. Buryak, J.M. Soto-Crespo: Opt. Comm. **112**, 278 (1994)
23. S.T. Cundiff et al.: Phys. Rev. Lett. **82**, 3988 (1999)
24. J.E. Sipe, L. Poladian, C.M. de Sterke: J. Opt. Soc. Am. A **11**, 1307 (1994)
25. S. Lee, S.T. Ho: Opt. Lett. **18**, 962 (1993)
26. B.J. Eggleton et al., Phys. Rev. Lett. **76**, 1627 (1996)

4 Raman Gap Solitons in Nonlinear Photonic Crystals

H.G. Winful and V.E. Perlin

4.1 Introduction and Brief History

In this chapter we review some of our early work on nonlinear periodic structures and describe how stimulated Raman scattering in these structures makes possible the generation of stationary gap solitons and the operation of distributed feedback Raman lasers.

Wave propagation in periodic structures is characterized by the presence of stopbands in frequency within which the wave vector is imaginary and the amplitude of the wave decays exponentially with distance as a result of Bragg reflection [1]. Near the stopband a linear periodic structure is highly dispersive, exhibiting very large values of both positive and negative group velocity dispersion as well as large group delays. The transmission of the structure is high outside the stopband and periodically attains values of unity at frequencies that correspond to linear resonances of the finite-length structure. These properties of linear periodic structures form the basis for applications that include optical filters, dispersion compensators, multiplexers, and distributed feedback lasers [2].

One of the first applications of periodic structures to nonlinear optics was a suggestion by Armstrong, et al to phase match nonlinear optical interactions through a periodic reversal of the sign of the nonlinear susceptibility [3]. This technique, known as quasi-phase matching, has finally come into its own with the development of periodically-poled lithium niobate [4]. Other ideas for the use of periodic structures in nonlinear optics involve taking advantage of the structural dispersion and the enhanced photon density of states near the edge of the stopband [5]. In all these applications, however, the periodic structure plays a purely passive role and its properties do not change.

In 1979 Winful et al. first described nonlinear periodic structures whose refractive index depends on the local intensity [6,7]. An incident wave that satisfies the Bragg condition at low intensity can tune itself out of the bandgap at high intensity. When this occurs the transmissivity of the structure jumps to a high value and reaches unity for certain intensities (Fig. 4.1). For the same input intensity there can be two or more values for the transmitted intensity and hence the structure exhibits optical bistability [6]. For each of these values of transmitted intensity there is an associated spatial distribution of intensity. In Fig. 4.2 the three distributions of forward intensity in the

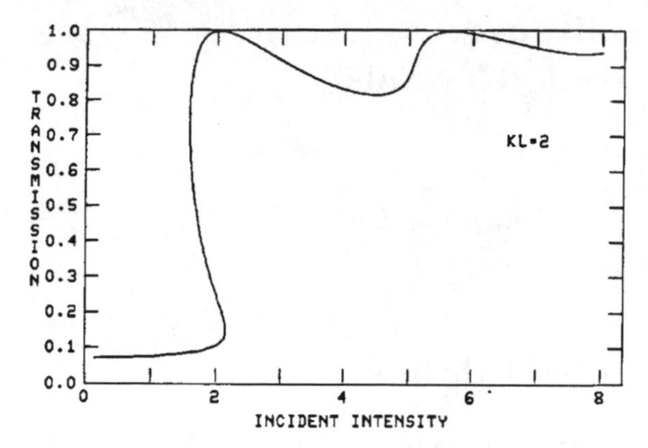

Fig. 4.1. Transmissivity of a nonlinear periodic structure as a function of normalized input intensity for a wave that satisfies the Bragg condition at low intensity. From [7]

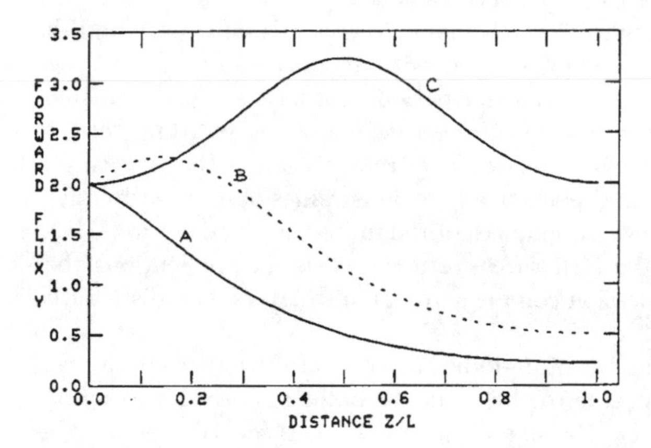

Fig. 4.2. The three distributions of forward intensity that correspond to the same value of normalized input intensity of 2 in Fig. 4.1. The gap soliton is the symmetric localized distribution labeled C. From [7]

periodic structure correspond to the same normalized input intensity of 2 [7]. The distribution labeled A is the exponentially decaying field belonging to the lower branch of the transmission curve of Fig. 4.1. Distribution C, corresponding to the upper branch, is a nonlinear resonance that represents a kind of self-induced transparency. This symmetric, stationary intensity profile at a point of unity transmission is what was originally termed a gap soliton [8]. It is, in fact, a spatial soliton. Similar bistable behavior was also predicted for naturally occurring photonic bandgap structures such as cholesteric liquid crystals [9]. In such crystals the pitch of the helical structure can be tuned by an intense light wave. Almost periodic gratings, such as those with chirp or taper were also shown to exhibit hysteresis and bistability [7]. The possibility of intensity tuning of transient gratings in four-wave mixing processes has also been considered [10]. The result is optical bistability in a dynamic photonic crystal formed by the interacting light waves themselves.

The dynamic behavior of nonlinear periodic structures was first addressed by Winful and Cooperman [11]. They showed through numerical solutions of the time-dependent nonlinear coupled-mode equations that these structures exhibit an instability that can transform an input cw beam into a train of pulses. Such an instability could form the basis of a distributed-feedback optical pulse generator [12,13]. Winful also suggested that the negative group velocity dispersion and nonlinear refraction of periodic structures could make possible the compression of optical pulses and the propagation of solitons [14]. Exact solitary wave solutions of the time-dependent nonlinear coupled mode equations were found by Christodoulides and Joseph [15] and by Aceves and Wabnitz [16]. The stationary gap solitons were shown to be a limiting case of solitary wave solutions of the nonlinear coupled mode equations with velocities that range from zero to c [16]. In certain limits, pulse propagation in nonlinear periodic structures can also be described in terms of a nonlinear Schröedinger equation [17].

Most of the predicted phenomena described above have now been observed in fiber Bragg gratings and other nonlinear photonic crystals [18–23]. One notable exception is the stationary gap soliton. Since the gap soliton is essentially trapped inside the periodic structure, its presence can only be inferred from the observation of hysteresis or the compression of an incident pulse. As a practical matter, the actual launching of a gap soliton is not an easy matter since the structure is highly reflective for even fairly intense light whose wavelength satisfies the Bragg condition.

One approach to the creation of stationary or ultraslow gap solitons is through the mediation of stimulated Raman scattering [24]. In this chapter we show that an intense pump wave whose frequency is far from the Bragg resonance and hence propagates unimpeded can nevertheless excite a gap soliton at a Raman shifted frequency if this frequency coincides with the Bragg frequency. Because the pump is not reflected by the grating, large amounts of power can be coupled in. Once coupled into the structure, the pump supplies gain to a weak signal at the Bragg peak. The pump can transfer a significant fraction of its energy to the Stokes-shifted wave whose intensity can become strong enough to create a nonlinear resonance or gap soliton at this shifted wavelength. We call these excitations Raman gap solitons. Once formed, the Raman gap soliton remains trapped even after the pump pulse has exited the grating. It will decay slowly as a result of coupling to the external world through the ends of the grating. We find the trapped Raman gap soliton bounces back and forth with turning points at locations where the net grating reflectivity is high. It moves with a velocity that is less than 0.1% of the speed of light in the unperturbed medium. This scheme thus provides a means to trap and store light pulses. Stimulated Raman scattering in a nonlinear periodic structure also makes possible distributed feedback Raman lasers with certain unique properties [25]. We discuss such lasers here as well.

4.2 Propagation Equations

Stimulated Raman scattering in a nonlinear photonic bandgap structure can be described by a set of coupled mode equations for the forward pump (A_p), the forward Stokes (A_{s+}), and the backward Stokes (A_{s-}) amplitudes. Here we consider only the first Stokes component since the higher order components lie outside the grating stop band and hence do not experience any reflective feedback. As a result their amplitude will be small compared to the first Stokes component. We allow for pump depletion but ignore the material group velocity dispersion since it is much smaller than the dispersion due to the periodic structure [14]. While the ideas presented here apply to general nonlinear periodic structures, we will use terminology and parameters suitable for fiber Bragg gratings since they are the most likely candidates for experimental verification. The coupled mode equations are as follows [24]

$$\frac{\partial A_p}{\partial z} + \frac{1}{v_p}\frac{\partial A_p}{\partial t} = -\frac{g_p}{2}\left(|A_{s+}|^2 + |A_{s-}|^2\right) A_p + i\gamma_p\left(|A_p|^2 + 2|A_{s+}|^2 + 2|A_{s-}|^2\right) A_p, \tag{4.1}$$

$$\frac{\partial A_{s+}}{\partial z} + \frac{1}{v_s}\frac{\partial A_{s+}}{\partial t} = \frac{g_s}{2}|A_p|^2 A_{s+} + i\kappa A_{s-} + i\Delta\beta A_{s+} + i\gamma_s\left(|A_{s+}|^2 + 2|A_p|^2 + 2|A_{s-}|^2\right) A_{s+}\,, \tag{4.2}$$

$$\frac{\partial A_{s-}}{\partial z} - \frac{1}{v_s}\frac{\partial A_{s-}}{\partial t} = -\frac{g_s}{2}|A_p|^2 A_{s-} - i\kappa A_{s+} - i\Delta\beta A_{s-} - i\gamma_s\left(|A_{s-}|^2 + 2|A_p|^2 + 2|A_{s+}|^2\right) A_{s-}\,. \tag{4.3}$$

Here κ is the grating coupling constant which is proportional to the amplitude of the periodic refractive index modulation. The intensity-dependent refractive index is described by the parameters $\gamma_\mathrm{p} = \pi n_2/\lambda_\mathrm{p} A_\mathrm{eff}$ and $\gamma_\mathrm{p} = \pi n_2/\lambda_\mathrm{s} A_\mathrm{eff}$, with n_2 the nonlinear index coefficient, λ_s and λ_p respectively the pump and Stokes wavelengths, and A_eff the effective area of the guided mode. The gain coefficients g_p and g_s are related to the Raman gain coefficient g_R through $g_\mathrm{s} = g_R/A_\mathrm{eff}$ and $g_\mathrm{p} = (\lambda_\mathrm{s}/\lambda_\mathrm{p})g_\mathrm{s}$. The group velocities of the pump and Stokes waves in the unperturbed medium are v_p and v_s, respectively. The detuning of the Stokes frequency (ω_s) from the Bragg frequency (ω_B) is described by the parameter $\Delta\beta = n(\omega_\mathrm{s} - \omega_\mathrm{B})/c$, where n is the average refractive index of the fiber core. This set of equations includes forward and backward Raman amplification, grating coupling between the Stokes waves, as well as self- and cross-phase modulation. These equations are solved numerically with the boundary condition that at $z = 0$, the pump pulse is a given function of time and the initial condition that the forward and backward Stokes amplitudes evolve from a weak cw input seed whose intensity is about 10^{-8} that of the pump. We use a fourth-order collocation scheme to integrate these equations numerically [26]. The parameter values used in the simulations were chosen as appropriate for a typical fiber Bragg grating. The grating length L varies between 3 cm and 6 cm, the coupling constant is $6\,\mathrm{cm}^{-1}$, the nonlinear index is taken as $3.2 \cdot 10^{-16}\,\mathrm{cm}^2/\mathrm{W}$, the Raman gain coefficient is $0.7 \cdot 10^{-13}\,\mathrm{m/W}$ and the fiber effective area is 50 µm. The Raman shift of silica fiber is $440\,\mathrm{cm}^{-1}$ and the width of the Raman gain ($\sim 100\,\mathrm{cm}^{-1}$) is much larger than the bandwidth of the fiber Bragg grating ($\sim 0.1\,\mathrm{cm}^{-1}$). The input pump pulses are taken as gaussian with pulsewidths τ (FWHM) of about 500 ps.

4.3 Ultraslow and Stationary Raman Gap Solitons

There are a number of distinct operating regimes that depend on the input pump power. For low input power there is no observable output at the Stokes wavelength and the pump propagates through the structure without any distortion or loss. As the pump power is increased, it begins to detune the grating stopband so that the signal can propagate inside the grating. Simultaneously energy begins to be transferred to the Stokes wave through stimulated Raman scattering. There is a Raman threshold beyond which the Stokes power becomes comparable to the pump power. This threshold power is given roughly by [27]

$$P_{th} = \frac{16 A_{eff}}{g_R L} \,. \tag{4.4}$$

For a 6-cm long piece of fiber this threshold power is about 130 kW. As this threshold value is exceeded an intense Stokes pulse is generated. This Stokes pulse also detunes the grating through the intensity-dependent refractive index. It is able to maintain a gap detuning even after the pump pulse has exited the grating. This pulse proceeds to shed energy until it reaches a value of intensity that is just enough to create a gap soliton at the Raman frequency. This Raman gap soliton is trapped inside the grating and executes slow oscillations within the structure as it bounces back and forth between turning points that mark the locations of equivalent lumped mirrors. At each reflection a small amount of energy leaks out. Eventually the magnitude decays to a value such that there is no intensity-dependent contribution to the dynamics. At that point the pulse evolution is linear and the rate of energy loss increases. Figure 4.3 shows the evolution of a slow Raman gap soliton with a velocity only 0.1% the speed of light in the unperturbed medium. This gap soliton consists of forward and backward components of nearly equal amplitude that are coupled together by the grating. The velocity of the soliton is

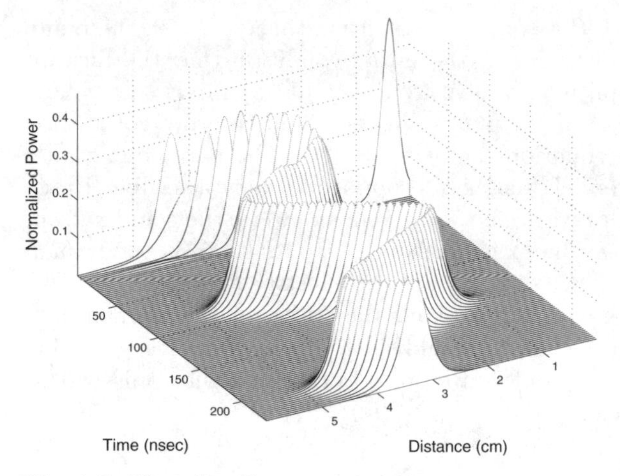

Fig. 4.3. Ultraslow Raman gap soliton

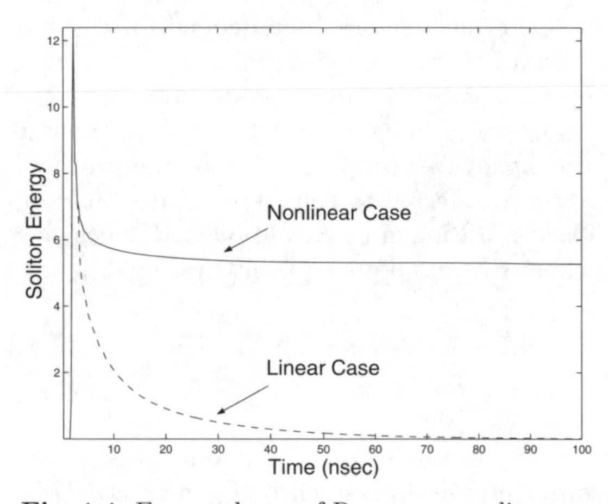

Fig. 4.4. Energy decay of Raman gap soliton compared to linear case

proportional to $\left(|A_{s+}|^2 - |A_{s-}|^2\right)$. The small difference in the amplitudes of the forward and backward components (here about 0.1%) is what causes the gap soliton to move with the correspondingly low velocity. Figure 4.4 shows the energy in a trapped soliton as a function of time compared to the energy in a field of the same initial amplitude but in the absence of nonlinearity (we turn off the nonlinearity and Raman effect after initializing the field). The soliton energy remains constant for times of order microseconds while the energy in the linear field decays rapidly.

The scheme described here makes it possible to trap a light pulse and hold it stationary for a significant length of time. In numerical simulations we have been able to store a pulse for as long as a microsecond before nu-

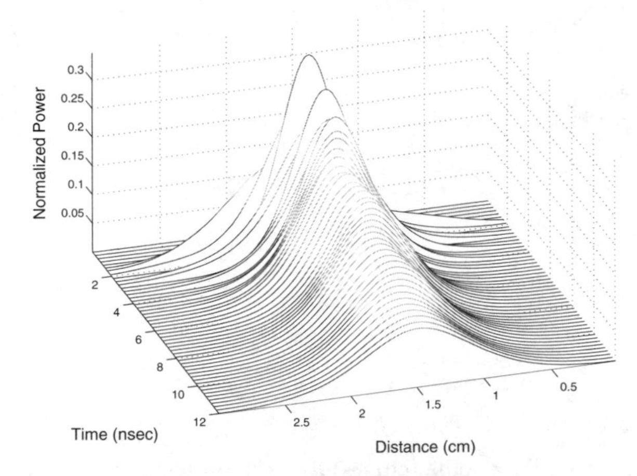

Fig. 4.5. Stationary Raman gap soliton

merical instabilities eventually terminated the program. The trapped soliton can be released by sending in another intense control pulse that momentarily changes the refractive index of the grating [28]. This method provides a controllable way of storing and releasing optical pulses. We note that recently pulses of light have been trapped in an atomic vapor through the use of atomic excitations coupled to the light by Raman transitions [29]. The scheme proposed here uses an all-solid state approach which makes it compatible with fiber-optic communication systems.

Stable, immobile Raman gap solitons have been found for certain parameter values. Figure 4.5 shows such a stable, stationary gap soliton in a grating of length 3 cm. For this soliton the amplitudes of the forward and backward components are equal. The lifetime of this soliton is at least 20 ns, which is forty times the duration of the pump pulse. The steady state Raman gap soliton is accurately described by a sech^2 profile. Figure 4.6 shows the numerically computed intensity profile of the stationary Raman gap soliton, given by the sum of the intensities of the forward and backward Stokes waves. The dotted curve is a sech^2 fit to the data. The excellent agreement confirms our identification of the field profile as a gap soliton. Note that the Raman gap soliton persists after the pump has left the grating. The dynamics of the Raman gap soliton can then be analyzed by setting the pump intensity equal to zero. The remaining equations are exactly the same equations that govern nonlinear pulse propagation and soliton formation in periodic structures [14]. For infinitely long periodic structures analytic soliton solutions exist in the form of sech functions.

Since the Raman gap soliton is trapped in the Bragg grating, the question remains as to the experimental signature of such an excitation. The gap soliton does couple to the outside world for any finite length grating. The energy leakage through the ends of the structure can be used to characterize

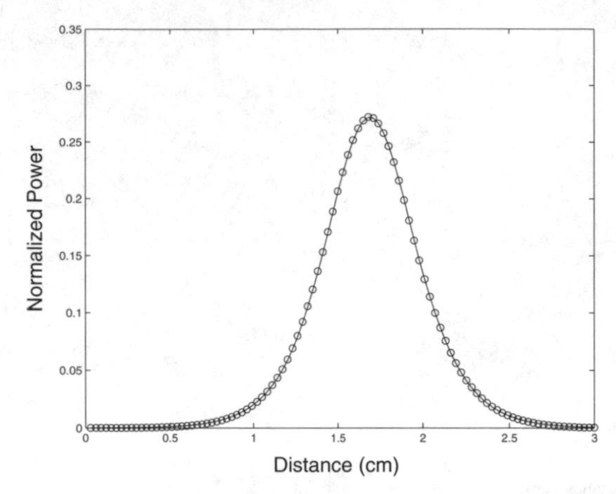

Fig. 4.6. Comparison of computed pulse profile (circles) to sech^2 profile (solid line)

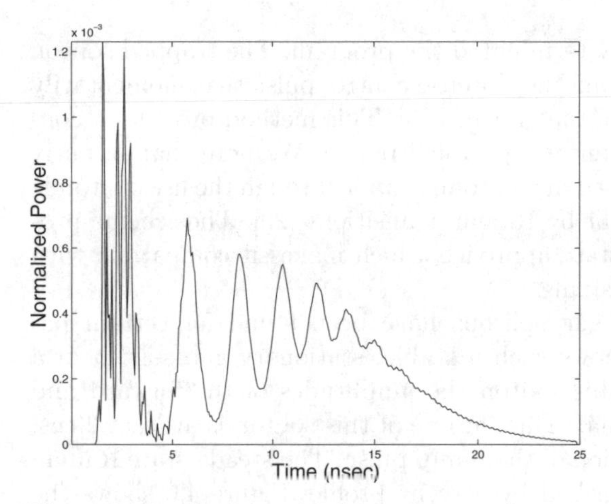

Fig. 4.7. Output intensity for a slow moving gap soliton

the soliton. Figure 4.7 shows the output intensity at the left end of the grating as a function of time for the slow moving Raman gap soliton of Fig. 4.2. After the initial transient, the output oscillates as a result of the periodic motion of the soliton. The energy loss at each reflection leads to a broadening of the soliton which in turn results in higher intensities at the grating edges and consequently a higher rate of energy loss. Here the output intensity (after the transient) peaks at about 12 ns after which point it decays exponentially, consistent with the evolution of a linear resonance.

In the linear case, no significant field amplitude can exist inside the structure for frequencies inside the stop band. This is well known from the theory of distributed-feedback lasers. What makes it possible for the Raman gap soliton to live for long periods inside the bandgap are the following effects: (1) through cross-phase modulation, the pump shifts the bandgap so that the frequency of the weak seed Stokes input, located at the Bragg peak in the absence of the pump, now lies somewhat to the side of this peak. (2) Through stimulated Raman scattering this seed begins to grow to high intensities. Eventually the Stokes pulse becomes intense enough to maintain the bandgap shift through its own intensity-dependent refractive index changes even after the pump has exited the grating. The excitation of the Raman gap soliton thus involves both an active component and a reactive component. The active component transfers energy from the pump wave to the Stokes wave through the usual Raman gain which is related to the imaginary part of the third order susceptibility $\chi^{(3)}$. The reactive component, proportional to the real part of $\chi^{(3)}$ tunes the stopband relative to the Stokes frequency through self- and cross-phase modulation.

In the simulations presented here, the Raman gap soliton was seeded by a weak cw input signal whose intensity is typically 10^{-8} that of the pump. This seed permits the soliton to evolve in a deterministic manner. In the absence of a cw seed the signal would grow from spontaneous Raman noise. We have attempted to simulate this evolution by using random sources in the coupled mode equations. Under these conditions the output is erratic unless the pump is quasi-cw. This is because a short pump does not provide enough gain for the peak of the Raman line to dominate the noise spectrum. On the other hand, with cw pumping, as in the case of the laser configuration described in the next section, the initial noise distribution evolves into a uniform steady state distribution.

4.4 Distributed-Feedback Fiber Raman Laser

The case of cw pumping corresponds to the operation of a Raman laser with distributed feedback [25]. In this case the signal is able to grow from noise to establish a mode determined by the cavity resonances shifted by the intensity-dependent refractive index. The threshold for a DFB Raman laser can be found by solving the coupled mode equations at steady state with fixed pump intensity:

$$\frac{\partial A_{s+}}{\partial z} = \frac{g_s}{2}\left|A_p\right|^2 A_{s+} + i\left(2\gamma_s\left|A_p\right|^2 + \Delta\beta\right) A_{s+} + i\kappa A_{s-} \, , \tag{4.5}$$

$$\frac{\partial A_{s-}}{\partial z} = -\frac{g_s}{2}\left|A_p\right|^2 A_{s-} - i\left(2\gamma_s\left|A_p\right|^2 + \Delta\beta\right) A_{s-} - i\kappa A_{s+} \, . \tag{4.6}$$

Here the nonlinear contributions of the Stokes waves are neglected since the signal intensity is small at the threshold for laser oscillation. The coupled mode equations 4.5 are then linear in the Stokes amplitude and can be solved in the manner of Kogelnik and Shank for the threshold gain and the detuning $\Delta\beta$ [2]. For strong gratings ($\kappa L \gg 1$) we find that the threshold pump power

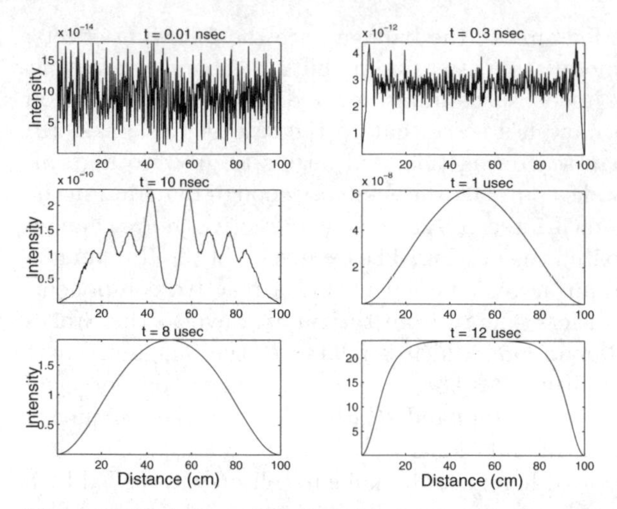

Fig. 4.8. Evolution of Stokes signal intensity from noise in DFB Raman laser

for Raman lasing is [25]

$$P_{th} \simeq \frac{\pi^2 A_{eff}}{g_R \kappa^2 L^3} \,. \tag{4.7}$$

For grating of length 1 m and coupling constant $100\,\mathrm{m}^{-1}$ this yields a threshold power of 0.7 W. The corresponding detuning of the first lasing mode at threshold is

$$\Delta\beta = \pm\kappa \left(1 + \frac{\pi^2}{3\kappa^2 L^2}\right) - 2\gamma_s I_p \,, \tag{4.8}$$

where the pump intensity I_p makes a contribution to the detuning through the nonlinear index. As the signal intensity grows it also contributes to the intensity-dependent detuning and so the lasing frequency shifts down continuously until steady state is reached. The evolution of the Stokes signal distribution from noise is shown in Fig. 4.8. Note the flat nature of the steady-state Stokes field distribution. This is a result of intensity dependent detuning which reduces the effective resonator length to the region where the intensity is approximately constant.

4.5 Conclusion

The interplay of Raman amplification, Kerr nonlinearity, grating dispersion, and distributed feedback leads to a number of interesting and potentially useful phenomena in nonlinear photonic crystals. In this chapter we have shown how stimulated Raman scattering in periodic structures can be used to excite extremely slow moving and even stationary gap solitons. This makes it possible to trap and store light for extended periods of time. The grating feedback also enables the operation of a distributed feedback Raman laser. These phenomena should be observable in fiber Bragg gratings and other nonlinear photonic bandgap structures.

References

1. L. Brillouin: *Wave Propagation in Periodic Structures* (Dover, New York 1953)
2. H. Kogelnik and C.V. Shank: J. Appl. Phys. **43**, 2327 (1972)
3. J.A. Armstrong, N. Bloembergen, J. Ducuing, and P.S. Pershan: Phys. Rev. **127**, 1918 (1962)
4. M.M. Fejer, G.A. Magel, D.H. Jundt, and R.L. Byer: IEEE J. Quantum Electron. **28**, 2631 (1992)
5. N. Bloembergen and A.J. Sievers: Appl. Phys. Lett. **17**, 483 (1970)
6. H.G. Winful, J.H. Marburger, and E. Garmire: Appl. Phys. Lett. **35**, 376 (1979)
7. H.G. Winful: Optical Bistability in Periodic Structures and in Four-Wave Mixing Processes, Ph.D. Dissertation, University of Southern California, Los Angeles (1981)
8. W. Chen and D.L. Mills: Phys. Rev. Lett. **58**, 160 (1987)
9. H.G. Winful: Phys. Rev. Lett. **49**, 1179 (1982)
10. H.G. Winful and J.H. Marburger: Appl. Phys. Lett. **36**, 613 (1980)
11. H.G. Winful and G.D. Cooperman: Appl. Phys. Lett. **40**, 298 (1982)
12. H.G. Winful and G.D. Cooperman: Optical pulse generator, US Patent No. 4 497 535 (Feb. 5, 1985)
13. H.G. Winful, R. Zamir, and S.F. Feldman: Appl. Phys. Lett. **58**, 1001 (1991)
14. H.G. Winful: Appl. Phys. Lett. **46**, 527 (1985)
15. D.N. Christodoulides and R.I. Joseph: Phys. Rev. Lett. **62**, 1746 (1989)
16. A.B. Aceves and S. Wabnitz: Phys. Lett A **141**, 37 (1989)
17. J.E. Sipe and H.G. Winful: Opt. Lett. **13**, 132 (1988)
18. B.J. Eggleton, R.E. Slusher, C.M. de Sterke, P.A. Krug, and J.E. Sipe: Phys. Rev. Lett. **76**, 1627 (1996)
19. B.J. Eggleton, C.M. de Sterke, and R.E. Slusher: J. Opt. Soc. Am. B **14**, 2980 (1997)
20. B.J. Eggleton, C.M. de Sterke, and R.E. Slusher: J. Opt. Soc. Am. B **16**, 587 (1999)
21. C.J. Herbert and S. Malcuit: Opt. Lett. **18**, 1783 (1993)
22. D. Taverner, N.G.R. Broderick, D.J. Richardson, R.I. Laming, and M. Ibsen: Opt. Lett. **23**, 328 (1998)
23. P. Millar, R.M. DeLaRue, T.F. Kraus, J.S. Aitchison, N.G.R. Broderick, and D.J. Richardson: Opt. Lett. **24**, 685 (1999)
24. H.G. Winful and V.E. Perlin: Phys. Rev. Lett. **84**, 3586 (2000)
25. V.E. Perlin and H.G. Winful: IEEE J. Quantum Electron. **37**, 38 (2001)
26. C.M. de Sterke, K.R. Jackson, B.D. Robert: J. Opt. Soc. Am. B **8**, 403 (1991)
27. R.G. Smith: Appl. Opt. **11**, 2489 (1972)
28. V.E. Perlin and H.G. Winful: Phys. Rev. A, to be published
29. D.F. Phillips, A. Fleischhauer, A. Mair, R.L. Walsworth, and M.D. Lukin: Phys. Rev. Lett. **86**, 783 (2001)

5 Self-transparency and Localization in Gratings with Quadratic Nonlinearity

C. Conti and S. Trillo

5.1 Introduction

In noncentrosymmetric media, second-order or $\chi^{(2)}$ nonlinearities dominate the nonlinear response of the material. It is therefore natural to wonder about the possibility to observe nonlinear behavior in periodic media made out of this class of materials, and in particular effects of self-transparency and switching mediated by localized envelopes such as those observed in cubic media (e.g., fibers [1] and AlGaAs [2]) and discussed elsewhere in this book. The first motivation to undertake these studies is that nowadays it is fairly well understood that $\chi^{(2)}$ support spatial (non-diffractive) or temporal (non-dispersive) solitons in *homogeneous* media [3]. It is thus expected that trapping phenomena can be extended to periodic media. The second, more important reason is related to the availability of periodic structures in quadratic materials. A properly tailored periodicity of the effective $\chi^{(2)}$ nonlinearity is commonly employed to achieve phase-matching of frequency conversion process, by means of so called quasi-phase-matching (QPM) techniques [4,5], recently extended in planar (2D) geometry [6,7]. On the other hand also periodic variations of linear properties over submicron scale has been introduced in quadratic materials to achieve Bragg gratings [8–13]

In this chapter we discuss the physics of second-harmonic generation (SHG) in periodic $\chi^{(2)}$ media. We concentrate on one-dimensional periodic structures with a stopband in the through-propagation, or 1D photonic crystals. Our main focus will be on effects of self-transparency and light localization which can be induced at relatively high intensities. Since either linear or nonlinear properties can be changed periodically along the dielectric medium, we will analyse and compare the two cases.

In the former case, an arbitrarily weak electromagnetic field sees a stopband around the Bragg frequency associated with the linear distributed feedback (DFB) mechanism. It well known that Kerr nonlinearities induce bistability [14], limiting [15], and formation of localized solutions hereafter termed linear-gap solitons (LGSs), i.e., solitons originating from a linear stopband [16–20]. We will show that $\chi^{(2)}$ permits a similar behavior [21–37]. Important qualitative differences, however, arise from the fact that the quadratic nonlinearity couples envelopes at different carrier frequencies.

In the case of a periodic nonlinearity, the grating is effective only at high intensity. Therefore, properly engineered structures can exhibit a stopband of nonlinear origin. We will analyze the most simple of these structures, namely SHG occuring in the backward direction (BSHG). Although the proposal of using backward geometries dates back to 1966 [38], and SHG in reflection was observed in 1976 [39], it is only recently that efficient BSHG was reported in periodically-poled $LiNbO_3$ [40,41] and $KTiOPO_4$ (KTP) [42,43], and different theoretical aspects have been tackled [44–54]. Yet, the existence of localized states is against intuition because BSHG, unlike linear DFB structures, lacks a linear reflectivity which ensures exponential damping along soliton tails, where the behaviour is governed by linear terms. Nevertheless, we will explain how a new family of localized states, namely nonlinear-gap solitons (NGSs) arises from higher-order unavoidable nonlinear contributions [55].

As a general remark, the approach which we follow here is based on the standard coupled-mode analysis for weakly nonlinear systems and shallow gratings. We point out, however, that the validity of the results can be widened by employing a different approach based on Bloch functions which are the exact modes of the second-order wave equations with periodic coefficients. The equivalence between the two approaches was established by studies explicitly dedicated to Bragg gratings with quadratic nonlinearity to which the reader is referred to [30,32].

The chapter is organized as follows. In Sect. 5.2 we will discuss the basic features of linear and nonlinear stopbands which are relevant to quadratic materials. In Sect. 5.3 we will describe self-transparency occurring under ideal stationary conditions. Sections 5.4 and 5.5 focus on the existence of moving solitons due to linear and nonlinear stopbands, respectively. The copropagating case which deals with a forbidden band in the momentum space is treated in Sect. 5.6. Finally Sect. 5.7 is aimed at discussing the stability aspects.

5.2 Stopbands Originating from Linear Versus Nonlinear Periodic Properties

For definiteness, let us discuss the origin of stopbands in the two different physical settings. The first situation refers to a Bragg grating induced by means of a periodic perturbation of linear refractive index, which can be always Fourier expanded as $\Delta n(Z) = \sum_{\mathrm{m}} c_{\mathrm{m}} \exp(im\beta_0 Z)$, where $\Lambda = 2\pi/\beta_0$ is the grating period. In this case, the m-th order of the grating perturbation couples efficiently counterpropagating waves with optical frequency around the Bragg harmonics ω_{Bm} such that $k(\omega_{\mathrm{Bm}}) = m\beta_0/2$. The forward (E_+) and backward (E^-) envelopes which constitute the total field $E(Z,T) = [E^+(Z)\exp(ikZ) + E^-(Z)\exp(-ikZ)]\exp[-i(\omega_{\mathrm{Bm}} + \Delta\omega)T]$, where $k = k(\omega_{\mathrm{Bm}} + \Delta\omega)$, obey the coupled-mode problem

$$-i\frac{dE_m^+}{dZ} = \Gamma_m E_m^- e^{-i\Delta\beta_m Z}\,; \quad i\frac{dE_m^-}{dZ} = \Gamma_m^* E_m^+ e^{i\Delta\beta_m Z}\,, \tag{5.1}$$

where $\Delta\beta_{\rm m} = 2k - m\beta_0$ stands for the detuning from Bragg condition, and $\Gamma_{\rm m}$ is the m-th order coupling coefficient (proportional to $c_{\rm m}$). Equations (5.1) are conveniently recast, by introducing the vector $U = (U^+, U^-)^T$ of phase-shifted amplitudes $U^\pm = E^\pm \exp(\pm i\Delta\beta Z)$, in the following standard form

$$\frac{dU}{dz} = AU\,; \quad A = \begin{pmatrix} i\delta_m & i\kappa_m \\ -i\kappa_m & -i\delta_m \end{pmatrix}, \tag{5.2}$$

where $z = Z/Z_0$, $\delta_{\rm m} = \Delta\beta_{\rm m} Z_0/2$, $\kappa_{\rm m} = \Gamma_{\rm m} Z_0$ (henceforth considered real and positive without loss of generality), and Z_0 is a suitable length scale (usually taken to be the device length L or Γ_1^{-1}). The solution of the linear problem (5.2), yields a characteristic equation $\lambda_m^2 = \kappa_m^2 - \delta_m^2$ for the eigenvalues λ_m [i.e., $U \propto \exp(\lambda_{\rm m} z)$] of A. The two roots $\lambda_m^\pm \equiv \pm i K_m$ are imaginary as long as $|\delta_m| > \kappa_m$, $K_m = \sqrt{\delta_m^2 - \kappa_m^2}$ representing a wavevector shift of the linear (oscillating) modes of the structure. Conversely, $\lambda_m^\pm = \pm\sqrt{\kappa_m^2 - \delta_m^2}$ are real as long as $|\delta_m| < \kappa_m$, entailing a forbidden gap of frequencies. Inside this stopband the solutions $U^\pm(Z)$ are of hyperbolic type and describe the optical power dropping down along the DFB structure due to *linear* reflection into the backward mode. One can easily recover the usual dispersion relation $\delta\omega_m = \delta\omega_m(\lambda_m)$ explicitly in wavevector-frequency $(\lambda, \delta\omega_m)$ plane, by expanding at first-order the wavevector around the m-th order Bragg frequency ω_{Bm} as $k = k_{Bm} + dk/d\omega|_{\omega_{Bm}} \Delta\omega_m = m\beta_0/2 + V_m^{-1}\Delta\omega_m$, where V_m is the group-velocity of the host material at frequency ω_{Bm}, and $\Delta\omega_m \equiv \omega - \omega_{Bm}$. Substitution in the expression (characteristic equation) $\delta_m^2 = \kappa_m^2 + K_m^2$ yields, in terms of normalized frequency deviation $\delta\omega_m \equiv \Delta\omega_m Z_0/V_m$, the lower-bound (LB, $\delta\omega_m^-$) and upper-bound (UB, $\delta\omega_m^+$) of the dispersion relation

$$\delta\omega_m(K_m) = \delta\omega_m^\pm = \pm\sqrt{\kappa_m^2 + K_m^2}, \tag{5.3}$$

shown in Fig. 5.1 for the two lowest Bragg resonances $m = 1, 2$. Getting back to real-world units, the dispersion relation is symmetric around the wavevectors π/Λ ($m = 1$) and $2\pi/\Lambda$ ($m = 2$), at which two frequency gaps of widths $2\Gamma_1 V_1$ ($|\Delta\omega_1| < \Gamma_1 V_1$) and $2\Gamma_2 V_2$ ($|\Delta\omega_2| < \Gamma_2 V_2$) open up around Bragg frequencies ω_{B1} and ω_{B2}, respectively. In general $\omega_{B2} \neq 2\omega_{B1}$, unless $k(\omega_{B2}) = 2k(\omega_{B1})$, i.e., perfect phase-matching of SHG is achieved.

With the Kerr effect, one deals with a single carrier frequency and only one of the Bragg resonances (usually $m = 1$) is effective. Conversely nearly phase-matched SHG couples beams at FF and its SH, and a grating made to resonate with the FF beam at $m = 1$ order is nearly resonant with the optical SH at $m = 2$ order. The band structure pertaining to SHG is indeed that of a *doubly-resonant* grating shown in Fig. 5.1. The detuning of the two optical harmonics with respect to their nearest Bragg frequency can be different unless perfect phase-matching is achieved. Importantly, SHG can take place also in a *singly-resonant* grating when the $m = 2$ resonance is negligible because (i) the even harmonics of the linear perturbation are

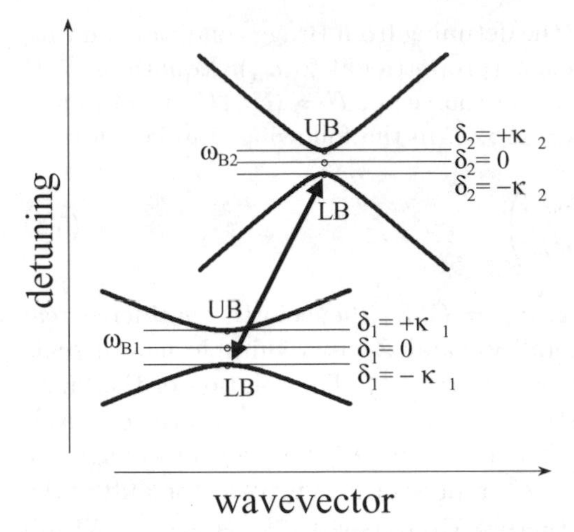

Fig. 5.1. Double stopband structure in an anharmonic Bragg grating. There are two gaps of normalized width κ_1 and κ_2 centered around Bragg resonant frequencies ω_{B1} and ω_{B2} associated with the two lowest order harmonics of the grating. LB and UB indicate the lower- and upper-band edge, respectively. Near band-edge bright gap solitons exist for LB-LB coupling (arrow)

absent as, e.g., in a square-wave corrugation; (ii) birefringence is used to offset wavevector dispersion to achieve phase-matching, making the grating effective only for the FF polarization. Also other configurations are possible such as a singly-resonant grating at SH ($m = 1$ resonance for the SH beam), or mixing involving nearly degenerate or non-degenerate frequencies, which, however, we will not treat here.

A different type of coupling between counterpropagating waves occurs in a structure engineered to phase-match a SH beam propagating in the backward direction. This can be achieved in a QPM structure where the sign of the effective nonlinearity is reversed periodically to form a *nonlinear grating* described in terms of its Fourier expansion $\chi_2(Z) = \hat{\chi}_2 \sum_{m>0} c_\mathrm{m} \exp(\pm im2\pi Z/\Lambda)$, where $\hat{\chi}_2$ characterizes the nonlinearity of the host medium. In bulk media, for field amplitudes normalized so that their $|*|^2$ yields the intensity, $\hat{\chi}_2 = k_0[2Z_0/(n_{\omega_0}^2 n_{2\omega_0})]^{1/2} d_\mathrm{eff}$, with Z_0 vacuum impedance, n linear refractive index, and d_eff effective nonlinear coefficient (measured in pm/V). In the following, we might refer to the case of waveguide geometries as well, where the difference is that $|*|^2$ measures the modal powers, and the expression for the nonlinear coefficients must be straightforwardly extended by introducing the effective depth (planar waveguides) or area (ridge waveguides) of the modes.

Perfect phase-matching is achieved at order $m = \overline{m}$ when the grating wavevector exactly compensates for the bulk nonlinear mismatch $\Delta k = k_2 + 2k_1$. This occurs at FF ω_0 such that $\Lambda = 2\overline{m}\pi/\Delta k(\omega_0)$, where Λ is the period of the grating. In such a QPM structure the FF (E_1) and SH (E_2) detuned components of the total field $E(Z,T) = E_1(Z,T)\exp[ik_1 Z - i(\omega_0 + \Delta\omega)T] + E_2(Z,T)\exp(-ik_2 Z - i2(\omega_0 + \Delta\omega)T)$ obey the following coupled-mode problem [46,48,53].

$$-i\frac{dE_1}{dZ} = \chi_2(Z)E_2E_1^* e^{i\Delta kZ}\,; \quad i\frac{dE_2}{dZ} = \chi_2(Z)E_1^2 e^{-i\Delta kZ}\,, \tag{5.4}$$

where $\Delta k = k_2 + 2k_1 = k(2\omega_0 + 2\Delta\omega) + 2k(\omega_0 + \Delta\omega)$. Retaining only the effective term of the grating and introducing the dimensionless variables $u_2 = [E_2/\sqrt{I_r}]\exp(i\Delta kZ)$, $u_1 = \sqrt{2}E_1/\sqrt{I_r}$, (5.4) can be cast in the following dimensionless form

$$-i\frac{du_1}{dz} = u_2u_1^*\,; \quad i\frac{du_2}{dz} = \delta k u_2 + \frac{u_1^2}{2}\,, \tag{5.5}$$

where $z = Z/Z_{\rm nl}$, $Z_{\rm nl} = (\hat{\chi}_2 c_{\overline{m}}\sqrt{I_r})^{-1}$ being the nonlinear length scale associated with the reference intensity (or power) $I_{\rm r}$, and $\delta k = (\Delta k - 2\overline{m}\pi/\Lambda)Z_{\rm nl}$ is the residual mismatch due to the nonzero detuning $\Delta\omega$ from phase-matching frequency. Following the approach of [56], Eqs. (5.5) can be solved by exploiting the conservation of the Hamiltonian $H = \delta k|u_2|^2 + (u_1^2u_2^* + u_1^{*2}u_2)/2$ and the photon flux (Poynting vector) $P = |u_1|^2/2 - |u_2|^2$. It is easy to obtain the equations for the effective variables $\eta = |u_2|^2$ and $\phi = \phi_2 - 2\phi_1 = \mathrm{Arg}(u_2) - 2\mathrm{Arg}(u_1)$, which play the role of action-angle conjugate variables in the following reduced Hamiltonian system

$$\begin{aligned}
\frac{d\eta}{dz} &= -2(\eta + P)\sqrt{\eta}\sin\phi = \frac{\partial H_r}{\partial\phi}\,,\\
\frac{d\phi}{dz} &= -\delta k + \frac{3\eta + P}{\sqrt{\eta}}\cos\phi = -\frac{\partial H_r}{\partial\eta}\,,\\
H_r &= \delta k\eta + 2(\eta + P)\sqrt{\eta}\cos\phi\,.
\end{aligned} \tag{5.6}$$

From (5.5) decoupled equation for the backward intensity $\eta(z)$ can be obtained from the first of (5.6) by expressing ϕ through the Hamiltonian $H_{\rm r}$, obtaining that η obeys the equation

$$\dot{\eta} = \pm\sqrt{2[E - U(\eta)]}\,, \tag{5.7}$$

where the $\pm$ sign must be chosen accordingly with the initial value of $\sin\phi$ in (5.6). Equation (5.7) is reminiscent of the velocity ($\dot{\eta} = \frac{d\eta}{dz}$, z playing the role of time) of an ideal particle with unitary mass, moving in a one-dimensional potential well $U(\eta)$ with total (kinetic plus potential) energy $E = \frac{1}{2}\dot{\eta}^2 + U(\eta)$. The explicit calculation yields $2[U(\eta) - E] = -4\eta^3 + (\delta k^2 - 8P)\eta^2 - 4\left(P^2 + H_0\delta k\right)\eta + 4H_0^2$, where H_0 is a particular value of the Hamiltonian $H_{\rm r}$ fixed by the initial conditions (i.e., calculated either on input or output section). The spatial evolution of the backward power $\eta(z)$ can be obtained by integrating (5.7). In the relevant case of SHG from an

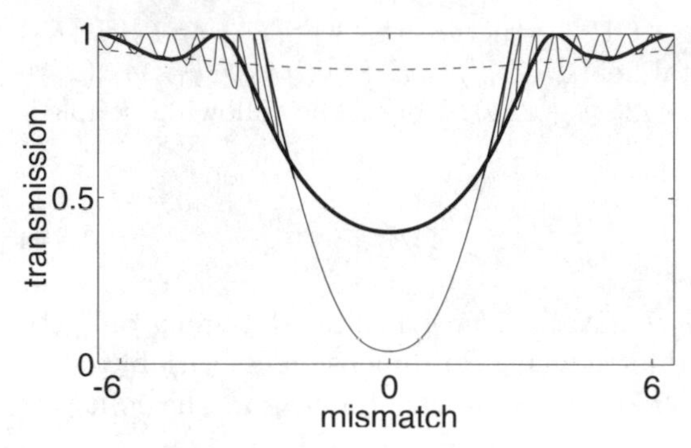

Fig. 5.2. Transmitted power fraction versus residual wavevector mismatch δk in BSHG, in the quasi-linear regime ($z_{\mathrm{L}} = L/Z_{\mathrm{nl}} = 0.5$, dashed curve), intermediate ($z_{\mathrm{L}} = 2$, thick solid curve) and high fluences ($z_{\mathrm{L}} = 10$, thin solid curve)

input beam at FF without injected seed at the grating output $z_L \equiv L/Z_{\mathrm{nl}}$, explicit solutions $\eta(z)$ can be derived from the quadrature formula

$$z - z_L = \pm \int_0^{\eta(z)} \frac{d\eta}{\sqrt{2[E - U(\eta)]}}, \tag{5.8}$$

where $E - U(\eta)$ must be calculated with $H_0 = H_{\mathrm{r}}(\eta = 0) = 0$. The integral in (5.8) can be inverted in terms of Jacobian elliptic functions and their hyperbolic or circular limits [57]. The FF intensity is then obtained as $|u_1(z)|^2 = 2(P + |u_2(z)|^2)$, where $2P = 2P(z_{\mathrm{L}}) = |u_1(z_{\mathrm{L}})|^2$ gives the transmitted (output) power. Let us discuss the physical meaning of these solutions in terms of transmittance. Figure 5.2 shows the transmited fraction of power $\mathcal{T} = |E_1(L)|^2/|E_1(0)|^2$ (the reflectivity is $\mathcal{R} = |E_2(0)|^2/|E_1(0)|^2 = 1 - \mathcal{T}$) for increasing values of z_{L} corresponding to a larger fluences or a longer structures. The device behaves as a *nonlinear reflector*. While in the low intensity limit the structure is transparent, for high fluences the transmission drops inside a stopband centered around phase-matching. A signature of the nonlinear origin of this gap is bistability (i.e., multiple transmission for a fixed mismatch) which appear clearly in the sidelobes of the high fluence transmittance (thin solid curve) in Fig. 5.2.

Inside the stopband the two fields decay monotonically along the device, whereas outside the stopband they exhibit only weak periodic conversion. The transition between the two regimes occurs abruptly, thanks to a phenomenon of separatrix crossing which has no similarity in forward SHG. From the dynamical viewpoint the two regimes correspond to bound motion (transparency) and unbound motion (reflection) in the cubic potential well $U(\eta)$. The bound solutions obtained explicitly from (5.8) read as

$$\eta(z) = \eta_- \mathrm{sn}^2(\tilde{z}|m), \tag{5.9}$$

where $\tilde{z} \equiv \eta_+^{1/2}(z - z_{\rm L})$, $m = \eta_-/\eta_+$ is the modulus of the Jacobian sine sn, and $\eta_\pm = \frac{1}{8}(\delta k^2 - 8P \pm \sqrt{\delta k^2 - 16P})$. For $\delta k = 0$, (5.9) yields the circular limit $\eta = P \tan^2[P^{1/2}(z_{\rm L} - z)]$. From (5.9) we could formulate the device response in terms of input power, which, however, involves cumbersome expression which we do not report here. The physics of the device can be more easily understood by thinking in terms of a given output power P. In this case a critical mismatch exist, $\delta k = \pm\sqrt{16P}$, for which the period of the solution (5.9) diverges. It discriminates between the transparent and reflective regimes. In such critical condition the SH power evolves along a separatrix as $\eta(z) = \eta_s \tanh^2[\delta k(z_{\rm L} - z)/4]$, reaching (in a device long enough) the asymptotic (reflected) value $\eta(0) = \eta_s = [\delta k^2 - 6P + \sqrt{\delta k^2(\delta k^2 - 12P)}]/18 = \delta k^2/16$, which corresponds to a saddle point of the oscillator model (5.6). Viceversa, at fixed mismatch δk, bound periodic evolutions and unbound evolutions with strong nonlinear backreflection occurs below and above the critical power $P_{\rm c} = \delta k^2/16$, respectively. Slight increase of output power P above $P_{\rm c}$ results into large increase of input power and BSHG acts as a limiter with clamped output power $P \simeq P_{\rm c}$ [46,53,55].

To summarize the results of this section, as much as a Bragg reflector is a 1D photonic crystal with photonic bandgap of linear origin, BSHG should be considered as a 1D photonic crystal with stopband of nonlinear origin. From the dynamical viewpoint, the gap of the Bragg reflector is associated with a couple of real eigenvalues of the linear problem. When nonlinearity is added to the problem, the linearization (i.e., arbitrarily weak amplitudes) entails the existence of a saddle point in the origin $U = 0$ with stable and unstable manifold corresponding to the negative and positive eigenvalue, respectively. This allows LGSs to exist in the whole stopband as envelopes which decay exponentially to zero along the manifolds of the origin, and hence have no pedestal (they are described by closed homoclinic orbits). Conversely the nonlinear nature of the stopband in BSHG does not imply that the origin ($u_1 = u_2 = 0$) is a saddle point. In this case, localized NGSs with exponential decay could still be supported, owing to a saddle point $(u_1, u_2) \neq 0$, which represents the finite excitation needed to induce backreflection and thus the stopband itself. In view of this consideration NGSs are envisaged to possess a nonzero pedestal.

5.3 Self-transparency in the Stationary Regime

It is instructive to investigate first the stationary response of the two 1D photonic crystals considered in Sect. 5.2. Although the effect that we describe in this section can hardly be observed without resorting to pulsed sources, the analysis of the stationary regime give us the opportunity to discuss the role of localization in relatively simple terms. When the two structures are excited within their stopband (either of linear or nonlinear origin for the DFB and BSHG, respectively), switching to a self-transparent state requires an effective contribution from next-order nonlinearity to bleach the reflection. In the Bragg grating, the gap is due to the linear effect and hence quadratic $\chi^{(2)}$ terms are the leading order contributions. Conversely, in BSHG, the stopband is induced by $\chi^{(2)}$ backreflection, and hence the next-order contribution arises from cubic terms.

5.3.1 The Models

To model the effect of these terms it is convenient to exploit the Hamiltonian structure of the nonlinear coupled-mode equations

$$\dot{U} = J \frac{\partial}{\partial U^*} H \, , \tag{5.10}$$

where U is the array of modal complex amplitudes, H is the z-invariant Hamiltonion related to the conservation of the dot product between the polarization and the electric field, and $J = \mathrm{diag}(i, -i)$. Under proper conditions to be discussed below the conservation of photon flux (Poynting vector) permits to reduce (5.10) to a one-dimensional, hence integrable, Hamiltonian oscillator where the role of conjugated action and angle variables will be played by the backward propagating intensity $\eta(z)$ and an effective phase $\phi(z)$, which obey the equation (compare with (5.6))

$$\dot{\eta} = \frac{\partial H_r}{\partial \phi} \, , \quad \dot{\phi} = -\frac{\partial H_r}{\partial \eta} \, . \tag{5.11}$$

When SHG takes place in a doubly-resonant Bragg grating, we have a double ($m = 1$ and $m = 2$) resonance described in the linear limit by two sets of (5.2). These pair of equations are coupled through the quadratic terms in the following way (see, e.g., [23])

$$\begin{aligned} \mp i \dot{u}_1^{\pm} &= \delta_1 u_1^{\pm} + \kappa_1 u_1^{\mp} + u_2^{\pm} (u_1^{\pm})^* \, , \\ \mp i \dot{u}_2^{\pm} &= \delta_2 u_2^{\pm} + \kappa_2 u_2^{\mp} + \frac{(u_1^{\pm})^2}{2} \, . \end{aligned} \tag{5.12}$$

Here the dot stand for derivative against $z = Z/Z_0$, $\delta_{\mathrm{m}} = \Delta\beta_{\mathrm{m}} Z_0/2 = (2k_{\mathrm{m}} - m\beta_0)Z_0/2$, $m = 1, 2$, is the normalized detuning from m-th Bragg resonance, and $\kappa_{\mathrm{m}} = \Gamma_{\mathrm{m}} Z_0$ is the normalized coupling coefficient. The real-world total field is $E(Z,T) = \sum_{\mathrm{m}} E_{\mathrm{m}}^{+}(Z,T) \exp[ik_{\mathrm{m}} Z - im\omega_0 T] + E_{\mathrm{m}}^{-}(Z,T) \exp[-ik_{\mathrm{m}} Z - im\omega_0 T]$, and phase shifted variables $u_2^{\pm} \equiv I_{\mathrm{r}}^{-1/2} E_2^{\pm} \exp(\pm i \Delta\beta_2 Z/2)$, $u_1^{\pm} \equiv (\chi_1 I_{\mathrm{r}}/2\chi_2)^{-1/2} E_1^{\pm} \exp(\pm i \Delta\beta_1 Z/2)$, $I_{\mathrm{r}} = (\chi_1 Z_0)^{-2}$ being a reference intensity (or power in waveguide geometry), are conveniently introduced to get rid of rotating exponential terms in (5.12). Note that the nonlinear phase-matching $\delta k = \Delta k Z_0 = \delta_2 - 2\delta_1$ is fixed by the Bragg detunings, and vanishes on exact Bragg resonance. In other words, a perfect doubly-resonant linear grating is actually a triply resonant device, including resonance of the mixing process on phase-matching. Viceversa, nonlinear resonance (phase-matching) can be also obtained off-resonance from Bragg frequencies.

Equations (5.12) are readily cast into the form of (5.10) with $U = (u_1^+, u_1^-, u_2^+, u_2^-)^T$ and $H = \mathrm{Re}[(u_1^+)^2(u_2^+)^* + (u_1^-)^2(u_2^-)^*] + \sum_{j=1,2} \delta_j(|u_j^+|^2 + |u_j^-|^2) + 2\kappa_j \mathrm{Re}[u_j^-(u_j^+)^*]$. As we will show afterwards the dynamics ruled by (5.12) can be still very complex, due to nonintegrability. Yet, the integrable limits where one of the resonances is negligible and the FF components have a leading role, give us the opportunity to show that localization can occur in a regular fashion. Either the case of strong SH Bragg coupling ($|\kappa_2| \gg 1$), or large mismatches ($|\delta k| \sim |\delta_2| \gg 1$) in a DFB singly-resonant at FF ($\kappa_2 = 0$), allows for adiabatic elimination of SH amplitudes [35,36]. Here we consider the latter case because the engineering of the grating is simpler, and does not require significant deviation from a harmonic or square-wave index change. Moreover, as already discussed, a grating made to be Bragg resonant for the FF polarization is likely to be ineffective for the SH polarization, which is obviously different in any phase-matching scheme based on birefringence.

The key to understand LGSs in the singly-resonant grating is the so-called cascading limit of SHG [3], such that the dominant effect is not frequency conversion, but rather the Kerr-like phase-shift of FF beams acquired through repeated (cascaded) up- and down-conversion processes. Adiabatic elimination of SH amplitudes in (5.12) leads, in terms of amplitudes $u_\pm = u_1^\pm/|2\delta_2|$, to the following model

$$\mp i\dot{u}_\pm = \delta_1 u_\pm + \kappa_1 u_\mp + |u_\pm|^2 u_\pm \,, \tag{5.13}$$

which is reminiscent of a Kerr DFB, except for the absence of cross-phase modulation terms. Although, the other case of strong SH Bragg resonance $|\kappa_2| \gg 1$ leads to nonlinear coupling between the forward and backward FF beams, it yields an integrable model which describes localization phenomena similar to those governed by (5.13) which entail only linear coupling. For this reason, we prefer to refer the reader to the original paper [35], and focus more specifically on the model (5.13).

The model for BSHG can be easily obtained from (5.4) by considering the additional self- and crossed-induced phase-shifts owing to cubic nonlinearity as follows

$$\begin{aligned} -i\dot{u}_1 &= u_2 u_1^* + (X|u_2|^2 + S_1|u_1|^2)u_1 \,, \\ i\dot{u}_2 &= \delta k u_2 + \frac{u_1^2}{2} + (X|u_1|^2 + S_2|u_2|^2)u_2 \,, \end{aligned} \tag{5.14}$$

where we have assumed equal cross-phase modulation coefficients in the two equations. In this case we recast the form of (5.10) with $U = (u_1, u_2)^T$ and $H = \mathrm{Re}[u_1 u_2^*] + (S_1|u_1|^4 + S_2|u_2|^4) + X|u_1 u_2|^2 + \delta k|u_2|^2$.

Following the procedure outlined in Sect. 5.2, both (5.13) and (5.14) can be reduced to a 1D oscillator in the form of (5.11). The reduced Hamiltonian for the FF-Bragg resonant case read

$$H_r = 2\delta_1 \eta - s\eta(\eta + P) + 2\sqrt{\eta(\eta + P)} \cos\phi\,, \tag{5.15}$$

where $\eta(z) = |u^-(z)|^2$ and $\phi(z) \equiv \phi_-(z) - \phi_+(z)$ with $\phi_\pm \equiv \mathrm{Arg} u^\pm$, $P = |u^+|^2 - |u^-|^2$, and $s = \mathrm{sign}(\delta_2) = \mathrm{sign}(\delta k)$.

On the other hand, in BSHG, the Kerr terms only modifies the reduced Hamiltonian (5.6) as follows

$$H_r = \Delta\eta + \frac{\sigma}{2}\eta^2 + 2(P + \eta)\sqrt{\eta}\cos\phi\,, \tag{5.16}$$

where $\Delta = \delta k + 2(X + 2S_1)P$, $\sigma = 4X + 4S_1 + S_2$ and the other quantities have the same meaning as in (5.6).

Following the approach outlined in Sect. 5.2, the Hamiltonian oscillator (5.15) or (5.6) permits to obtain explicit expressions by means of quadrature [see (5.8)] for the backward propagating power $\eta(z)$ (and forward $|u_+|^2$ through conservation of photon flux), thus permitting to calculate the relative transmittance. In the following section we will discuss the physics hidden behind these solutions.

5.3.2 In-gap Self-transparency

Let us consider the DFB and BSHG reflectors illuminated by a FF beam with frequency set right in the center of their respective stopband. The transmitted power fraction $\mathcal{T}$ at FF is reported in Fig. 5.3 for a DFB grating [as obtained numerically from (5.12) in good agreement with explicit solutions of (5.13)] and in Fig. 5.4 for the BSHG reflector. Comparing Fig. 5.3a and Fig. 5.4a, it is evident that the two structures behave in opposite way in the linear (i.e., low-intensity) regime. In the low-power limit, the Bragg grating exhbits a strong reflectivity ($\mathcal{T} \sim 0$), while BSHG is totally transparent ($\mathcal{T} \sim 1$). As the input intensity is raised, however, both structures exhibit bistability and become fully transparent at the points labeled T in Fig. 5.3a and Fig. 5.4a. Incidentally note that the transparency points can be reached only through a hysteresis cycle.

The inspection of the field profiles at T-points reveals that transparency is associated in both cases (the similarity between Fig. 5.3b and Fig. 5.4b is evident) with excitation of a strongly localized structure. Loosely speaking, these confined fields constitute the first evidence for the formation of a stationary gap soliton in $\chi^{(2)}$ periodic media. More precisely, the localized fields correspond to a suitable portion of a periodic or quasi-periodic [for (5.12)] solution of the coupled-mode problem which is such to fulfil the boundary value problem in the finite structure. Indeed multiple adjacent peaks can be excited at higher fluences or in longer devices. Nevertheless, the linkage between gap solitons and localization effects shown in Fig. 5.3b and Fig. 5.4b is sufficiently justified, not only physically, but also from dynamical viewpoint

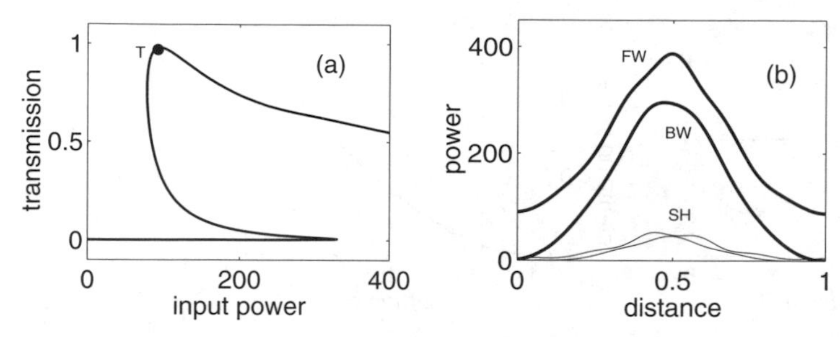

Fig. 5.3. Nonlinear transparency in a nearly singly-resonant ($\kappa_2 = 10^{-3}$) Bragg grating: **a** Transmitted power fraction $\mathcal{T}$ at FF; **b** Profiles of the forward (FW) and backward (BW) power at FF (thick curves), corresponding to the transparency point labeled T in **a**. The thin curves show the intensity profile of the weak second-harmonic components, labeled SH. The results are obtained by means of numerical integration of (5.12) with parameter values are $\delta_1 = \Delta\beta_1 L/2 = 0$, $\delta_2 = \Delta\beta_2 L/2 = 20$ (the normalized mismatch being $\delta k = \Delta k L = \delta_2 - 2\delta_1 = 20$), and $\kappa_1 = \Gamma_1 L = 4$

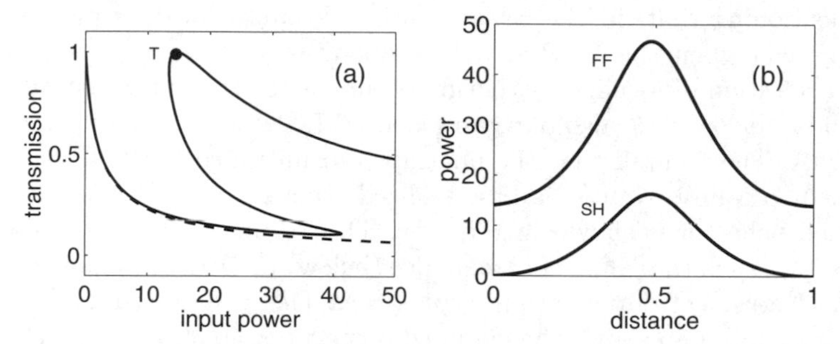

Fig. 5.4. Nonlinear transparency in a phase-matched ($\delta k = 0$) BSHG reflector: **a** Transmitted fraction of intensity with ($S_{1,2} = X = 0.1$, solid line) and without ($S_{1,2} = X = 0$, dashed line) cubic contributions; **b** Profiles $|u_1(z)|^2$ at FF and $|u_2(z)|^2$ at SH, corresponding to the transparency point labeled T in **a**

by the close proximity of the relative orbits in phase space. Even though DFB and BSHG show close similarity at transparency the localized structure will be termed LGSs and NGSs, respectively, to recall their different physical origin. In a DFB, it is the $\chi^{(2)}$ nonlinearity which counteracts linear grating reflectivity, in the simplest case through an induced detuning originating from an effective intensity-dependent index. Conversely, in BSHG it is the $\chi^{(3)}$ nonlinearity which bleaches the reflection induced by $\chi^{(2)}$ at moderate fluences, and the origin of localization phenomena is fully nonlinear. In fact, when Kerr terms are not effective, BSHG behaves as an ideal nonlinear reflector (see dashed line in Fig. 5.4, or more generally as a limiter out of phase-matching [46,53,55]), never becoming self-transparent.

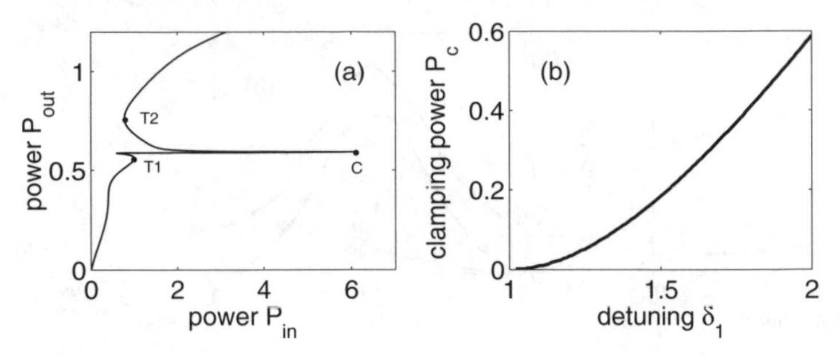

Fig. 5.5. a Transmitted power $P_{\rm out} = |u_1^+(z_{\rm L})|^2$ vs. input power $P_{\rm in} = |u_1^+(0)|^2$ as obtained for a DFB with Bragg resonance at FF from (5.12), with $\kappa_1 = 1$, $\delta_1 = 2$ ($|\delta_1| > \kappa_1$ means to be out-gap), $\delta k = 50$ (effective defocusing nonlinearity), $\kappa_2 = 10^{-3}$, and a device length $z_L = 5$. The transmission is clamped to the critical value $P_{\rm out} = P_{\rm c} \simeq 0.6$. **b** Critical power $P_{\rm c}$ versus the FF detuning δ_1

5.3.3 Out-gap Dynamics in Quadratic DFBs

It is worth mentioning that, in the Bragg grating, localization does play a significant role also when the DFB structure operates out-gap ($|\delta_1| > 1$). While in-gap excitation of localized structure occurs at relatively low output power, we will see below that also out-gap localized LGSs with a significant pedestal do exist. These entail that the out-gap transmittance of the device can be clamped to a finite value, as first realized for a Kerr DFB [15]. An example of such behavior is shown in Fig. 5.5a. This regime occurs only on one side of the gap, such that $s\delta_1 > 0$, i.e., above (below) the Bragg frequency for an effective Kerr effect of defocusing type $s = \delta k/|\delta k| = 1$ (focusing, $s = \delta k/|\delta k| = -1$). Figure 5.5b shows the clamped output power $P_{\rm out} \equiv P = P_c$ as a function of detuning δ_1. The origin of such phenomenon becomes clear by looking at the fields inside the DFB, displayed in Fig. 5.6. The virtual transparent behavior of the DFB up to the point T1 is due to a regime where the fields experience only weak oscillations. Above this point clamping is due to a dramatic sensitivity of the spatial period on slight changes of output power, characteristic of a regime of large oscillations. A localized structure is finally excited at the transparency point T2.

5.3.4 Regularity Versus Disorder in Quadratic DFBs

The features of a $\chi^{(2)}$ Bragg grating discussed in Sects. 5.3.3 and 5.3.2 are reminiscent of a Kerr NLDFB. However, in general this is not so. A Kerr DFB involves the interaction of only two counterpropagating waves, thus resulting into a low-dimensional coupled-mode problem which is fully integrable [14,15]. Viceversa, SHG couples four waves (counterpropagating in pairs) as clear from (5.12). In general this causes a loss of integrability and, as a consequence, the possibility that a quadratic DFB exhibits irregular behavior or

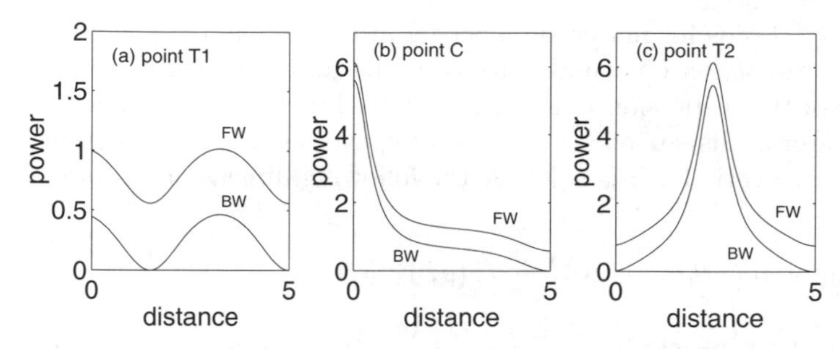

Fig. 5.6. Power evolution inside a quadratic NLDFB, corresponding to the three excitation conditions labeled T1,C,T2 in Fig. 5.5a. BW and FW denote backward $|u_1^-(z)|^2$ and forward $|u_1^+(z)|^2$ powers, respectively. The SH fields (not shown) follow adiabatically the FF fields

more precisely deterministic spatial chaos. In spite of the fact that quadratic DFBs have been numerically studied earlier [58], it is only recently that the physical mechanisms behind the onset of chaos was disclosed [35,36]. For instance, in a singly-resonant grating operating close to phase-matching, the additional degrees of freedom associated with the SH fields become effective and (5.13) cease to be valid. Under such conditions, the DFB shows a transition to chaos somehow driven by the localized structures. These indeed correspond to separatrices of the integrable reduction [e.g., (5.13)], around which chaos develop first, following a well-known mechanism which is generic for Hamiltonian systems [59]. In this case, the manifestation of chaos is an erratic formation of strongly localized structures which, under small changes of operating conditions (e.g., power) causes in turn multistable disordered jumps from transparency to reflection and viceversa. For further details, we refer the interested reader to [35,36], where this mechanism is discussed more deeply. It is worth pointing out that this qualitatively new feature of a DFB is intrinsically related to the increasing number of interacting waves and hence is shared neither by a Kerr ($\chi^{(3)}$) DFB structure, nor by BSHG even when $\chi^{(3)}$ nonlinearity becomes effective.

5.4 Moving Solitons in the Bragg Grating

In the nonstationary case, following a standard approach, the coupled-mode theory can be generalized by including temporal derivatives $\partial^n_{T..T} = \frac{\partial^n}{\partial T^n}$ of the fields arising from Taylor expansion of wavevector around the carrier frequency of the optical beams. In order to have an absolute reference frequency, such expansion is conveniently carried out around FF Bragg frequency, i.e., around the FF and SH carriers ω_{B1}, $2\omega_{\mathrm{B1}}$, respectively. Formally this is obtained, retaining terms up to the second-order, by means of the substitution $\pm i\partial_{\mathrm{Z}} \to \pm i\partial_{\mathrm{Z}} + ik'_m\partial_{\mathrm{T}} - (k''_m/2)\partial^2_{\mathrm{TT}}$ in the cw models discussed in Sect. 5.3. Here $k'_m = V_m^{-1} = dk/d\omega|_{\omega_m}$ is the inverse group-velocity and

$k''_m = d^2k/d\omega^2|_{\omega_m}$ stands for material dispersion, $m = 1, 2$, and $\omega_m = m\omega_{\mathrm{B1}}$. The dispersion terms k''_m are normally neglected because the grating dispersion (curvature of the dispersion relationship plotted in Fig. 5.1) is huge and prevails over material dispersion. Under such hypothesis, the nonstationary behavior of the quadratic DFB is ruled by the following dimensionless model [23–25]

$$\pm i\partial_z a_1^{\pm} + iv_1^{-1}\partial_t a_1^{\pm} + \kappa_1 a_1^{\mp} + a_2^{\pm}(a_1^{\pm})^* = 0\,,$$

$$\pm i\partial_z a_2^{\pm} + iv_2^{-1}\partial_t a_2^{\pm} + \delta k a_2^{\pm} + \kappa_2 a_2^{\mp} + \frac{(a_1^{\pm})^2}{2} = 0\,. \tag{5.17}$$

The link with real-world quantities is as follows. The total electric field is $E(Z,T) = \sum_{m=1,2} E_m^+(Z,T)\exp[ik_mZ - im\omega_{\mathrm{B1}}T] + E_m^-(Z,T)\exp[-ik_mZ - im\omega_{\mathrm{B1}}T]$, where $a_2^{\pm} \equiv I_r^{-1/2}E_2^{\pm}\exp(\pm i\Delta\beta_2 Z/2)$ $a_1^{\pm} \equiv (\chi_1 I_r/2\chi_2)^{-1/2}E_1^{\pm}$ (note that, thanks to the choice of carrier frequency, $\Delta\beta_1 = 0$, and $\Delta\beta_2 = -2\Delta k$), and $I_r = (\Gamma_1/\chi_1)^2$ is a reference intensity (or power in waveguide geometry). The choice $Z_0 \equiv \Gamma_1^{-1}$ yields the normalized distance $z = \Gamma_1 Z$, and time $t = \Gamma_1 V_1 T$, and permits to reduce the number of physical *external* parameters in (5.17) to the minimum number of three: the nonlinear mismatch $\delta k = \Delta k/\Gamma_1$, the velocity ratio $v_2 = V_2/V_1$, and the coupling ratio $\kappa_2 = \Gamma_2/\Gamma_1$ (sticking to the notation of [23,27,28] we introduce $v_1 = \kappa_1 = 1$ with the purpose to write (5.17) in symmetric form), which depends on the choice of material, phase-matching geometry, and grating features. The frequency detunings $\delta\omega$, $2\delta\omega$ from carrier frequencies ω_{B1}, $2\omega_{\mathrm{B1}}$ can be explicitly introduced by the change of variables

$$a_1^{\pm}(z,t) = u_1^{\pm}(z,t)\exp(-i\delta\omega t)\,, \quad a_2^{\pm}(z,t) = u_2^{\pm}(z,t)\exp(-i2\delta\omega t)\,, \tag{5.18}$$

where the new variables $u_{1,2}^{\pm}$ obey the equations

$$\pm i\partial_z u_1^{\pm} + iv_1^{-1}\partial_t u_1^{\pm} + \delta_1 u_1^{\pm} + \kappa_1 u_1^{\mp} + u_2^{\pm}(u_1^{\pm})^* = 0\,,$$

$$\pm i\partial_z a_2^{\pm} + iv_2^{-1}\partial_t u_2^{\pm} + \delta_2 u_2^{\pm} + \kappa_2 u_2^{\mp} + \frac{(u_1^{\pm})^2}{2} = 0\,, \tag{5.19}$$

which are conveniently used to model numerically the nonstationary (pulse) propagation in the grating, and correctly reduces to (5.12) in cw limit ($\partial_t = 0$). Noteworthy, the location of the two frequencies inside the gaps is fixed by the two detunings

$$\delta_1 \equiv \delta\omega = \frac{\Delta\omega}{V_1\Gamma_1}\,, \quad \delta_2 \equiv \frac{2\delta\omega}{v_2} + \delta k = \frac{2\Delta\omega}{V_2\Gamma_1} + \frac{\Delta k}{\Gamma_1}\,. \tag{5.20}$$

which, can be easily interpreted, consistently with the validity of (5.17), as first-order Taylor expansions around $m\omega_{\mathrm{B1}}$ of Bragg mismatches $2k(m\omega_{\mathrm{B1}} + m\Delta\omega) - m\beta_0$, $m = 1, 2$ (see also the discussion in Sect. 5.3). Solitons (LGSs in this case) can be found from (5.19) by searching for localized profiles which travel locked at a common velocity V. Therefore LGS profiles can be found as non-periodic (separatrix) solutions of the ODE system

$$\frac{du_1^{\pm}}{d\zeta} = -i(V/v_1 \mp 1)^{-1} \left[\delta_1 u_1^{\pm} + \kappa_1 u_1^{\mp} + u_2^{\pm}(u_1^{\pm})^*\right] ,$$

$$\frac{du_2^{\pm}}{d\zeta} = i(V/v_2 \mp 1)^{-1} \left[\delta_2 u_2^{\pm} + \kappa_2 u_2^{\mp} + \frac{(u_1^{\pm})^2}{2}\right] , \tag{5.21}$$

which are obtained from (5.19) with the substitution $u_{1,2}^{\pm} = u_{1,2}^{\pm}(\zeta)$, where $\zeta = z - Vt$. The family of LGSs depends, for fixed values of external parameters, on two *internal* parameters: the frequency detuning from Bragg condition $\delta\omega$ and the velocity V ($\Delta\omega = \delta\omega V_1 \Gamma_1$ and $V_{\rm sol} = VV_1$ in real-world units). The internal nature of this parameters has a precise meaning: (i) they have one-to-one correspondence with symmetries (invariants) of (5.17), see [60] for a math-oriented discussion, and Sect. 5.7; (ii) they can change in the presence of perturbations, or whenever the launching conditions do not match a LGS of the two-parameter family. Note that δ_2 can be conveniently used to specify the location of the SH frequency inside the stopband $|\delta_2| < \kappa_2$, though it involves a combination of internal ($\delta\omega$) and external (v_2 and δk) parameters.

To date, basically four different approaches have been pursued to characterize LGS solutions of (5.17–5.19), which can be summarized as follows

- whenever both frequencies $m(\omega_{\rm B1} + \Delta\omega)$, $m = 1, 2$, lie in the proximity of the UB or LB of the respective gap, an envelope function approach has been proved useful to reduce (5.19) to a model formally identical to that governing SHG in a homogeneous medium with second-order dispersion and group-velocity matching [23,24],
- the symmetry holding for the stationary ($V = 0$) case together with suitable constraints on the parameters (playing the role of imposed extra-resonances) allows to construct analytical solutions which fill the gap [27],
- the Kerr limit of parametric mixing (i.e., cascading) permits to describe analytically the whole family of moving LGSs [25,26,29,33,37],
- in contrast with previous three techniques, which imply a reduction of the effective number of equations, LGS profiles can be also contructed numerically by means of efficient routines employing shooting or relaxation methods, which permit to explore the entire parameter space [25].

The first of these approaches was originally used to predict LGSs of true parametric nature (i.e., not reminiscent of Kerr LGSs), by reducing the original coupled-mode problem for the periodic medium to the usual model for SHG in a homogeneous (dispersive) medium [23,24]. Different type of LGSs are found depending on the location of the two frequencies inside the gaps (for instance, bright LGSs require LB-LB coupling for $\kappa_{1,2} > 0$ as sketched in

Fig. 5.1). Since this approach is discussed elsewhere in this book (see Drummond et al.) we will not go into deeper details here. We point out that, in spite of the noticeable simplification achieved by reducing the problem from four to two coupled equations for leading-order envelopes, this approach remains semi-analytical because LGS profiles of the reduced model can be found only numerically except for few solutions which are isolated in parameter space. We will rather discuss here the limits which are amenable of a full analytical description (second and third points listed above), thus allowing for a more thorough discussion of the link with the cw dynamics and other important subtle points in the physics of LGSs.

First, let us consider the Kerr-limit of (5.22). Similarly to the stationary case discussed in Sect. 5.3, different reductions arise in the opposite cases of strong Bragg resonance at SH [25], and cascading in a grating singly-resonant at FF [26,29,33]. For the reason outlined in Sect. 5.3, we will deal with the latter case, for which adiabatic elimination of SH amplitudes in (5.17) yield in terms of amplitudes $a_\pm = a_1^\mp/(2|\delta k|)^{1/2}$

$$\pm i\partial_z a_\pm + i\partial_t a_\pm + a_\mp - s|a_\pm|^2 a_\pm = 0\,, \quad a_2^\pm = -\frac{(a_1^\mp)^2}{2\delta k} = -s a_\pm^2\,, \quad (5.22)$$

where $s = \mathrm{sign}(\delta k)$. LGS solutions of (5.22) can be contructed analytically in the whole domain of existence. Generally, one finds [29]

$$a_\pm = u_\pm(\zeta)\exp(-i\delta\omega t) = A_\pm \eta(\zeta)^{1/2}\exp[-i\delta\omega t + i\beta\zeta + \psi_\pm(\zeta)]\,, \quad (5.23)$$

where $A_\pm$ are velocity dependent amplitudes, $\eta(\zeta)$ is a common intensity profile, $\beta = \beta(\delta\omega, V)$ is the soliton propagation constant (do not confuse it with an independent new internal parameter), and $\psi_\pm$ are soliton chirps. We point out that, though (5.22) differ from those governing the Kerr (e.g., fiber) case for the absence of cross-phase-modulation, LGS solutions of the two models are qualitatively similar. Nevertheless, since the full picture of the Kerr case is still missing in the literature, we discuss here the features of LGSs of (5.23), limiting ourselves to the subluminal case $|V| = |V_{\mathrm{sol}}|/V_1 \le 1$. LGS solutions are summarized in Fig. 5.7, where we report their intensities $|u_\pm(\zeta)|^2 = A_\pm^2\eta(\zeta)$, sampled at different points in the parameter plane $(\delta\omega, V)$. Bright solitons exist within the circle $\delta\omega^2 + V^2 = 1$. It is easy to verify that this is the domain where exponentially damped solutions of the linear problem, say $a_\pm = f_\pm(\zeta)\exp(-\lambda\zeta - i\delta\omega t)$ with $\lambda > 0$, exist or, in other words, where the ODEs for $u_\pm(\zeta)$ linearized around the origin $(0,0)$ have real eigenvalues (compare also with a similar discussion in Sect. 5.3). Therefore it is natural to consider the inner region $\delta\omega^2 + V^2 \le 1$ as a generalized stopband. This means that the frequency gap seen by a soliton moving at velocity V, is $|\delta\omega| < \sqrt{1-V^2}$ ($|\Delta\omega| < V_1\Gamma_1\sqrt{1-V_{\mathrm{sol}}^2/V_1^2}$ in real-world units). Inside the generalized stopband, zero-velocity LGSs have symmetric envelopes, which changes from low-amplitude (LA) to high-amplitude (HA) as the stopband is spanned. Moving LGSs become asymmetric (AS) with

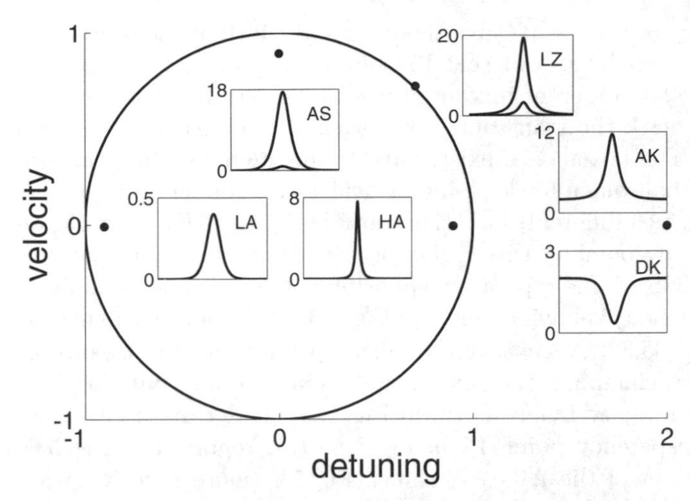

Fig. 5.7. LGS solutions in the detuning-velocity plane $(\delta\omega, V)$ for a singly-resonant $\chi^{(2)}$ grating operating in the Kerr (cascading) limit with effective defocusing nonlinearity, i.e., $s = 1$ $(\delta k > 0)$. The focusing case $(s = -1, \delta k < 0)$ can be obtained by simply reversing the detuning sign. The insets show LGS profiles sampled at the nearest marker (filled circle). Bright LGSs exist within the gap (unit circle). Still $(V = 0)$ LGSs are symmetric solutions $(u_+ = u_-)$ with low-amplitude (LA) and high amplitude (HA) close to LB and UB of the gap, respectively. Moving $(V \neq 0)$, LGSs become asymetric (AS) with $|u_+| > |u_-|$ $(|u_+| < |u_-|)$ when they travel in forward (backward) direction. Over the circle LGSs have Lorentzian profiles (LZ), whereas, any point in the right-half plane outside the stopband support the simultaneous existence of a dark (DK) and an antidark (AK) LGS

the stronger component being that of the direction along which the LGS moves. Close to the stopband edge, LA-LGSs are well described by a nonlinear Schrödinger (NLS) Eq. [26]. In the case shown in Fig. 5.7 this equation has soliton solutions, despite the fact that the grating dispersion is normal, because the effective nonlinearity is of defocusing type. Unlike the Kerr case, however, the location of LA-LGSs changes from the LB edge to the UB edge (and viceversa for HA-LGSs) when the sign of δk is reversed, so to pass from an effective self-defocusing to self-focusing. Since only LGS which are close to the LA solutions are easily excited in practice (see also Sect. 5.7), the $\chi^{(2)}$ environment offers the advantage that LGS can be observed on both edges of the stopband in the same physical DFB system by changing the mismatch.

Right on the circle, the eigenvalues of the linearization around the origin vanish, and LGSs become Lorentzian (LZ) solitons with intensity profile

$$\eta(\zeta) = \frac{c_1^2}{1 + (c_2\zeta)^2}, \tag{5.24}$$

where $c_1 = 2$ and $c_2 = 2/\sqrt{1-V^2}$ give the peak and width of the soliton, respectively. It is worth pointing out that LZ solitons are not new in nonlinear optics, and arise also in different mixing processes [61–64]. Importantly, in a DFB, LZ solitons mark the transition from bright HA-LGSs to LGSs with nonzero pedestal. The latter ones exist, outside the stopband (unit circle), in the right or left half-plane for focusing or defocusing nonlinearity, respectively. Noteworthy, two different dark (DK) and antidark (AK, or bright on pedestal) exist at any point in this region of the $(\delta\omega, V)$ plane. In particular, zero-velocity LGS of this type are responsible for the strong cw limiting (frustrated transmission) behavior shown in Fig. 5.5a. Without discussing the details further (see [35,36]), a separatrix-crossing phenomenon is responsible for the transmission clamping, the separatrices being nothing but the zero-velocity out-gap stationary LGSs shown in Fig. 5.7. The localized envelopes excited at the transparency point T2 in Fig. 5.5a and reported in Fig. 5.6c are indeed reminiscent of the AK-LGS shown Fig. 5.7 (more strictly speaking, those shown in Fig. 5.6c are periodic solutions of (5.13) which are close in phase space to AK-LGS). It is worth pointing out that, in the framework of our definitions, these LGSs with non-zero pedestal are the only solitons existing out-gap. Moreover, they exist on the opposite side of the stopband with respect to LA (easily excitable) bright LGSs. The excitation of moving DK- and AK-LGS poses a series of intriguing riddles which have not been addressed yet.

The solutions of (5.22) show that bright bell-shaped LGSs exist everywhere within the generalized stopband at FF. Things get more involved when the Bragg resonance at SH becomes effective. In this case the two pairs of (5.17) decouple in the low-power limit, and the linearized analysis, straightforwardly extended to deal with (5.17), dictates that exponentially decaying solutions exist indipendently for the FF ($m = 1$) and the SH ($m = 2$) inside the domains

$$\frac{\delta_m^2}{\kappa_m^2} + \frac{V^2}{v_m^2} = 1 \,, \quad m = 1, 2 \,, \tag{5.25}$$

which correctly reduces to the circle $\delta\omega^2 + V^2 = 1$ at FF. At SH, (5.25) yields a generalized stopband of elliptic shape $[(\delta\omega - \delta\omega_0)/a]^2 + V^2/v_2^2 = 1$, which is centered on $\delta\omega = \delta\omega_0 = -v_2\delta k/2$ and has a frequency opening $2a = \kappa_2 v_2$. Based on intuition, one might think that bright LGSs exist only within the region where the two stopbands overlap (see Fig. 5.8), which guarantees the expected decay of both fields along soliton tails. However, bright LGSs do not obey this simple rule. This is exemplified, in the limit of zero-velocity, by those bright analytical solutions which can be found (in the spirit of the second of the approaches listed above) to fulfil the extra-resonance condition $2\delta_1 + \delta_2 + \kappa_2 = 0$ [27]. In particular note that these LGSs exist also in the singly-resonant limit κ_2. Their intensity profiles are shown in Fig. 5.9 for different detunings $\delta_1 = \delta\omega$ inside the FF stopband. Their importance is twofold: (i) they feature the existence of bright LGSs

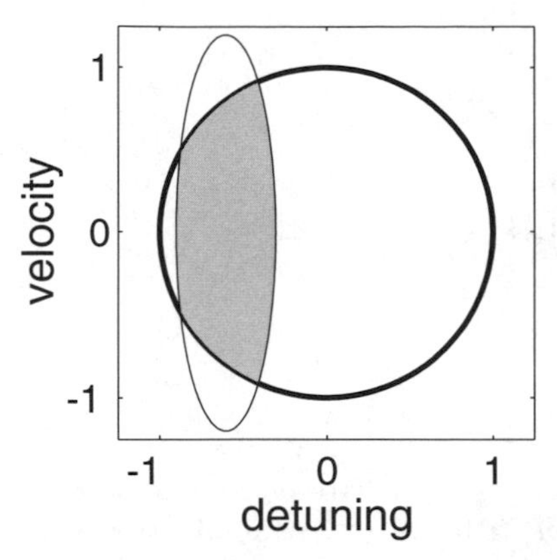

Fig. 5.8. The stopbands at FF (thick line) and SH (thin line) in the $(\delta\omega, V)$ plane. Here $\delta k = 1$, $v_2 = 1.2$, and $\kappa_2 = 0.5$. The center of the elliptic SH gap is the point $(\delta\omega, V) = (-\delta k v_2/2, 0)$. In the shaded region where the gaps overlap, exponentially damped solutions at both FF and SH exist. In general, the larger $|\delta k|$ the smaller the overlap region, whereas the larger κ_2, the larger the overlap region. Increasing velocity ratios v_2 affect the overlap in a more subtle way, because they cause both enlargement and shift of SH gap

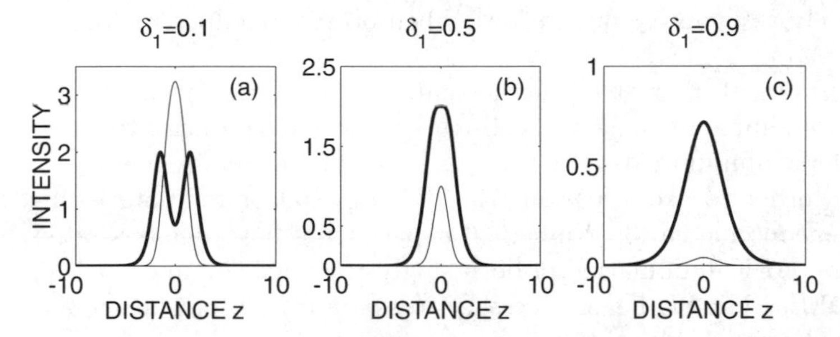

Fig. 5.9. FF (thick line) and SH (thin line) profiles of LGS solutions satisfying the extra-resonance $2\delta_1 + \delta_2 + \kappa_2 = 0$, for different detunings δ_1 spanning the FF gap, and implying, for fixed κ_2, also different detunings $\delta_2 = -(2\delta_1 + \kappa_2)$

which have a two-hump FF intensity profile for $|\delta\omega| < 0.5$); (ii) more importantly LGS solutions exist also when the SH beam is out-gap $|\delta_2| > \kappa_2$, i.e., outside the overlap region in Fig. 5.8, the decay of the SH beam being of nonlinear origin. These two features are both confirmed by searching numerically for LGS solutions in extended domains of the parameter space [25]. In this respect, it was also found in [25] that LGSs do not fill uniformly the overlap region shown in Fig. 5.8. It is important to emphasize that the extra-resonance condition may also be satisfied in the limit case

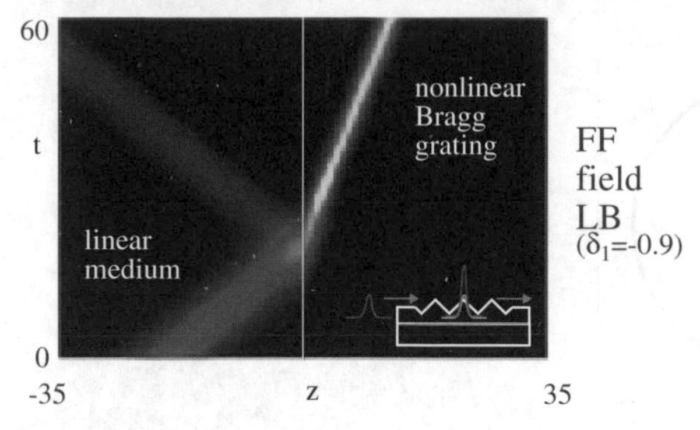

Fig. 5.10. Excitation of a low-amplitude LGS in a singly-resonant grating close to stopband LB ($\delta_1 = -0.9$), with $s = 1$ ($\delta k > 0$). A $\chi^{(2)}$ semi-infinite grating (nonlinear Bragg grating, $z > 0$) is illuminated by a nearly monochromatic pulse at FF impinguing on the structure from a homogeneous medium (linear medium, $z < 0$) with the same average index. Although a consistent portion of energy is Bragg reflected, the "refracted" energy is sufficient to form a LGS slowly traveling at about 40% of the linear group-velocity of the host medium. The distinctive feature of such LGS is a localized SH beam (not shown) which is generated at the interface and travel locked to the FF components

$\kappa_2 = 0$ (the singly resonant configuration), when no stopband exists for the SH fields.

A point of practical interest is whether parametric LGSs can be excited by making use of simple arrangements. In this respect it is crucial to operate with the DFB illuminated by a beam at a single carrier frequency, as usually done in order to excite parametric SHG (spatial or temporal) solitons in homogeneous media [3]. Numerical experiments have demonstrated the effectiveness of FF illumination in both singly-resonant [26] and doubly-resonant [28] DFBs. The SH beam necessary to sustain the propagation of a LGS is generated right inside the nonlinear medium and travel locked to the FF beam. An example of LGS formation in a singly-resonant DFB is shown in Fig. 5.10. Remarkably, also the formation of stationary LGSs have been predicted either from coalescence of two symmetric nearly in-phase LGSs formed by beams impinguing on the DFB from opposite directions [28], or by a radiative damping mechanism which lead to soliton deceleration [33].

5.5 Moving Solitons in a Nonlinear-Gap

Let us consider BSHG in the nonstationary regime. We generalize (5.4) by including time-dependent terms and the intrinsic (material) Kerr effect which we have already shown to be responsible for self-transparency in the stationary regime. In order to have an absolute reference frequency in this case too, we take the fundamental QPM phase-matching frequency ω_0 such that

$\Delta k(\omega_0) = \overline{m}2\pi/\Lambda$. The envelopes $E_1(Z,T)$, $E_2(Z,T)$ at carrier frequency ω_0, $2\omega_0$, respectively, obey the equations

$$L_1 E_1 + \chi_2(Z) E_2 E_1^* e^{i\Delta k Z} + (\widetilde{X}_1 |E_2|^2 + \widetilde{S}_1 |E_1|^2) E_1 = 0\,,$$
$$L_2 E_2 + \chi_2(Z) E_1^2 e^{-i\Delta k Z} + (\widetilde{X}_2 |E_1|^2 + \widetilde{S}_2 |E_2|^2) E_2 = 0\,, \tag{5.26}$$

where $L_1 = iV_1^{-1}\partial_T + i\partial_Z - k_1''\partial_{\mathrm{TT}}^2$, $L_2 = iV_2^{-1}\partial_T - i\partial_Z - k_2''\partial_{\mathrm{TT}}^2$ are linear propagators, and consistently with the approach employed for the DFB, we analyze the dispersionless case ($k_{1,2}'' = 0$).

The QPM grating, besides resulting into an effective reduced $\chi^{(2)}$ coefficient in (5.26), induces a cubic (Kerr-like) phase shift, as shown originally in [65]. Following the perturbative approach of [65], (5.26) can be reduced to the following system with constant coefficients

$$i(\partial_t + \partial_z)u_1 + u_2 u_1^* + (X_1' |u_2|^2 + S_1' |u_1|^2) u_1 = 0\,,$$
$$i(v^{-1}\partial_t - \partial_z)u_2 + \frac{u_1^2}{2} + (X_2' |u_1|^2 + S_2' |u_2|^2) u_2 = 0\,, \tag{5.27}$$

where we made use of the scaling employed in (5.5) and we have additionally introduced the normalized time $t = V_1 T/Z_{\mathrm{nl}}$ and the velocity ratio $v = V_2/V_1$. The normalized cubic coefficients are $S_l' = l(\widetilde{S}_l + \hat{S}_l)\sqrt{I_r}/(2\hat{\chi}_2)$ and $X_l' = (\widetilde{X}_l + \hat{X}_l)\sqrt{I_r}/(l\hat{\chi}_2)$, with $l = 1, 2$, where $\hat{S}_l$, $\hat{X}_l$ account for the grating corrections of the material Kerr coefficients.

Equations (5.27) support localized envelopes, or NGSs, which span the bandwidth of the $\chi^{(2)}$ mixing process (see Fig. 5.2). In order to describe the location inside the stopband, we introduce the frequency detuning $\delta\omega$ ($\Delta\omega = \delta\omega V_1/Z_{\mathrm{nl}}$ in real-world units) from phase-matching frequency, which is responsible for a cw mismatch $\delta k_0 = 2\delta\omega(1+v)/v$, consistently with the first-order expansion of wavevectors employed in (5.27). NGSs will be also characterized by an other internal parameter, namely their normalized velocity V ($V_{\mathrm{sol}} = VV_1$ in real-world units). Therefore we seek for traveling waves of the form

$$u_1 = \sqrt{(1-V)\,(1+V/v)}u(\zeta)\exp\left[i\delta\omega\,(z-t)\right]\,,$$
$$u_2 = (1-V)\,w(\zeta)\exp\left[i2\delta\omega\,(z-t)\right]\,, \tag{5.28}$$

where $\zeta = z - Vt$. We then substitute (5.28) in (5.27). The resulting ODEs can be reduced with lengthy but straightforward algebra to an oscillator of the form (5.11) with the Hamiltonian

$$H_r = \Delta\eta + \sigma\frac{\eta^2}{2} + 2(P+\eta)\sqrt{\eta}\cos\phi\,, \tag{5.29}$$

which is formally identical to the system ruling cw mixing [see (5.11)–(5.16)]. In (5.29) the variables are $\eta = |w|^2$, $\phi = \phi_2 - 2\phi_1 = \mathrm{Arg}(w) - 2\mathrm{Arg}(u)$, the dot stands for $d/d\zeta$. Moreover we set $\sigma = 2X_1 + 2X_2 + S_2 + 4S_1$, where

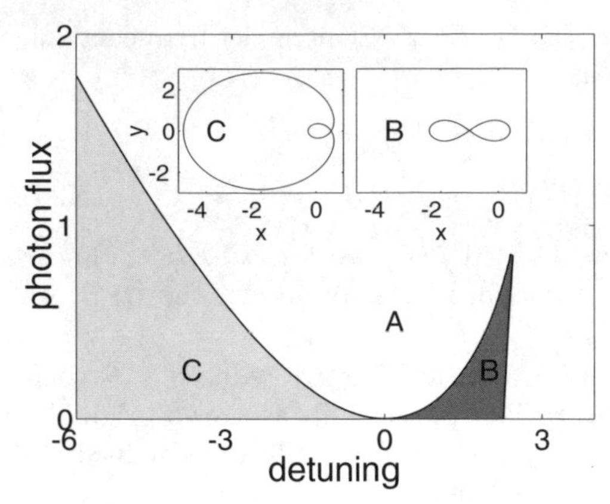

Fig. 5.11. Existence map of NGSs in the parameter plane $(\Delta_\sigma, P_\sigma)$ of effective detuning and photon flux. NGSs correspond to double loop separatrix trajectories displayed in the insets the B and C (corresponding to the two shaded regions B and C, respectively). The coordinates in the phase-plane shown in the insets are $(x, y) = (\eta \cos\phi, \eta \sin\phi)$. No NGSs exist in region A (blank domain of parameter plane)

$X_{1,2} = X'_{1,2}(1-V)$, $S_1 = S'_1(1+V/v)$, and $S_2 = S'_2(1-V)^2/(1+V/v)$ are renormalized Kerr coefficient. The two parameters

$$\Delta = \delta k + 2(X_2 + 2S_1)P \quad P = |u|^2/2 - |w|^2 \tag{5.30}$$

have the meaning of overall detuning (the soliton mismatch $\delta k = \frac{2(1+1/v)\delta\omega}{(1+V/v)}$ modified by the Kerr contributions), and photon flux, respectively. The existence of solitons can then be inferred from the presence of saddle points of the Hamiltonian (5.29) revealed by standard bifurcation analysis. A simple rescaling of the Hamiltonian (5.29) shows that the fixed points depend on the reduced set of two parameters

$$\Delta_\sigma = \sigma[\delta k + 2(X_2 + 2S_1)P], \quad P_\sigma = \sigma^2(|u|^2/2 - |w|^2). \tag{5.31}$$

The regions of existence of NGSs, given by those for which saddle points (representing the asymptotic value of the NGS) exist, can be summarized in the plane $(\Delta_\sigma, P_\sigma)$, as shown in Fig. 5.11.

Their SH intensity profile can be derived explicitly by means of a quadrature integral [see (5.8)], while the FF beam intensity follows from $|u|^2 = 2(P + |w|^2) = 2(P + \eta)$ (see [55] for details). Once $\eta(z)$ is known, also the equations for the phases of the single beams can be integrated. Figure 5.12 shows the intensity and phase profiles of the two qualitatively different NGS (hyperbolic) solutions coexisting in region C (solutions in region B are qualitatively similar), and the Lorentzian (LZ) NGS existing in the point where regions B and C merges. The LZ-NGS intensity is described by (5.24)

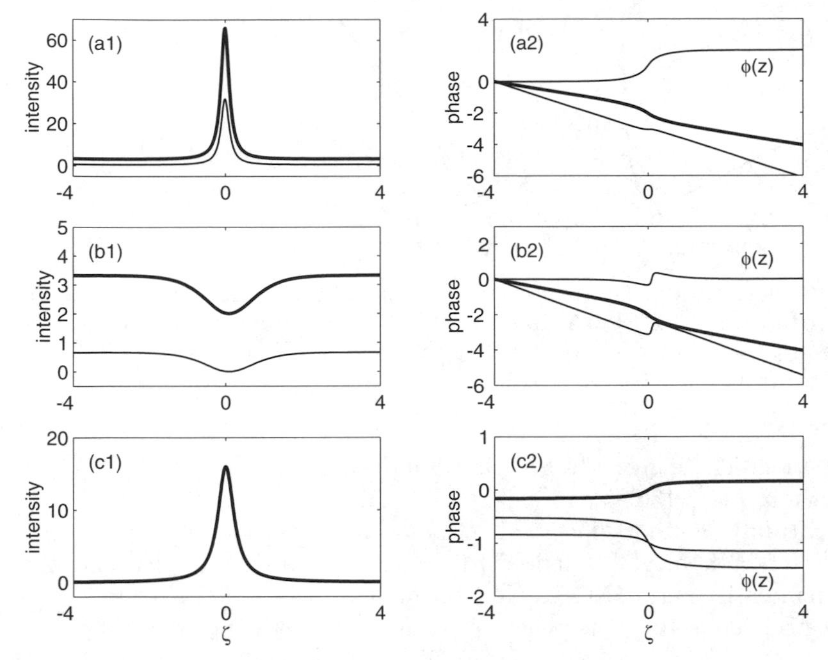

Fig. 5.12. Intensity ($|u(\zeta)|^2$, $|w(\zeta)|^2$) and phase profiles of NGSs calculated with $\sigma = 1$. Thick and thin lines refer to FF and SH beam, respectively. We show also the overall phase $\phi = \phi_2 - 2\phi_1$ (thin line labeled $\phi(z)$). **a1–a2** antidark NGS corresponding to $P = 1$ and $\delta k = -5$; (**b1–b2**) coexisting dark NGS for the same values of parameters; **c1–c2** Lorentzian NGS (intensity $|w(\zeta)|^2 = |u(\zeta)|^2/2$) corresponding to $P = \delta k = 0$

with $c_1 = c_2 = 4/\sigma$, and is accompanied by phase profiles with an asymptotic phase jump $\phi(+\infty) - \phi(-\infty) = \pi$. For NGSs with hyperbolic decay, $\phi(+\infty) - \phi(-\infty) = 2k\pi$ consistently with the fact that they are described by orbits of the Hamiltonian (5.29) which are homoclinic to the saddle points. The physical features of NGSs can be summarized as follows

- NGSs constitute a two-parameter continuous family of solitary envelopes.
- For any value of the internal parameters $\delta\omega$ and V within the region of existence, two types of NGSs coexist: an antidark (bright on pedestal, see Fig. 5.12a) and a dark soliton (see Fig. 5.12b).
- Antidark NGSs represent true localization, considered that the peak intensity is well above the pedestal, at least for $\Delta_\sigma < 0$ (region C) as shown in Fig. 5.13a. The pedestal vanishes right on phase-matching ($\delta k = 0$) for symmetric envelopes ($P = 0$) where the soliton exhibits Lorentzian decay (Fig. 5.12c).
- NGSs are slow waves in the sense that SH and FF beams are locked together and can travel with any velocity $-v < V < 1$, or, in real-world units, in between V_1 in the forward direction and V_2 in the backward

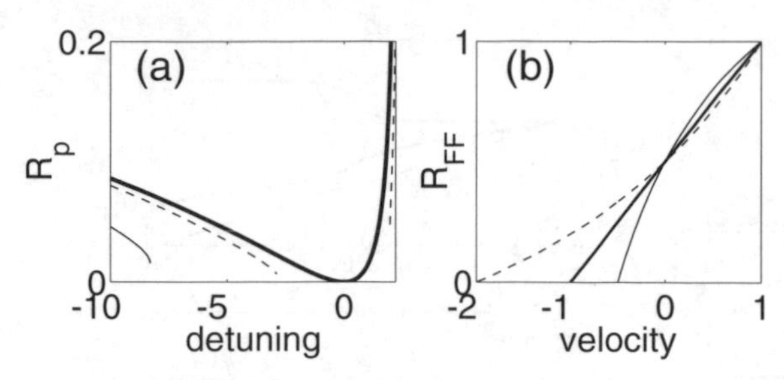

Fig. 5.13. Features of NGS: **a** Pedestal to peak ratio R_p versus detuning Δ for $P = 0$ (thick solid), $P = 0.5$ (dashed), $P = 3$ (thin solid). **b** Fraction of FF intensity $R_\mathrm{FF} = |u_1(0)|^2/(|u_1(0)|^2 + 2|u_2(0)|^2)$ versus velocity V for $P = 0$ and $v = 1$ (thick solid), $v = 0.5$ (thin solid), $v = 2$ (dashed)

direction. The higher the FF (SH) component, the closer to V_1 (V_2), as shown in Fig. 5.13b.

- Importantly antidark localized NGSs do not exist in the limit of pure $\chi^{(2)}$. The role of $\chi^{(3)}$ is indeed to induce transparency by detuning the high intensity part of NGSs, while the pulse is taken together by reflective behavior induced by the pedestal which supports the nonlinear gap.

It is worth emphasizing that, unlike the DFB, here Lorentzian solitons are the only solitons of the bright type. Similarly to what discussed for the DFB, however, they mark a transition in this case, too. In fact they separate two different branches of NGSs sitting on top of an in-phase eigenmode (region C) and an out-of-phase eigenmode (region B), respectively. The two branches have remarkably different properties in terms of existence (see Fig. 5.11) and degree of localization (see Fig. 5.13a). The features of Lorentzian LGSs and NGSs illustrated in this chapter suggest that they might have a universal role in the nonlinear propagation along bandgap materials.

In analogy with Bragg gratings, it is important to be able to observe localization effects with a simple arrangement involving unidirectional illumination at FF. The excitation of a Lorentzian soliton is discussed in [55]. Also NGSs with a non-zero pedestal can be excited as shown in Fig. 5.14. The signatures of soliton behavior are: (i) localization clearly seen only in the presence of the Kerr effect, as shown in Fig. 5.14a; (ii) the SH envelope traveling locked to the FF beam in the (forward) direction, opposite to its natural (backward) direction of propagation. Note, however, that the ascertainment of the latter feature requires to access the fields inside the structure, since the absence of a forward wave at SH together with the boundary condition $u_2(z_\mathrm{L}) = 0$ forces backconversion to the FF field at the output of the grating. In this respect, the physics of NGSs is markedly different from that of LGSs.

Let us briefly comment about links of NGSs with existing results in the literature. In discrete (lattice) systems, solitons which exist in a self-induced

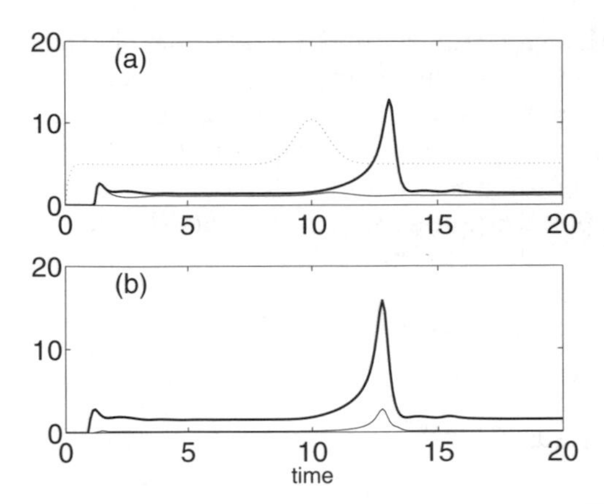

Fig. 5.14. A slow NGS excited by a pulse at FF superimposed on a cw excitation, as obtained numerically from (5.27) with coefficients such that $\sigma = 1$ and the boundary condition $u_2(z_L) = 0$. In **a** we show the temporal profiles of the input (dashed line) and output FF intensity without (thin solid line) and with (thick solid line) the Kerr contribution; **b** SH (thin line) and FF (thick line) intensity profiles in proximity of the output section of the structure ($z = 0.8z_L$)

gap, so called self-induced gap solitons, do exist as well [66–68]. However, these objects are standing kink-like coherent excitations of the lattice, which are therefore reminiscent of the dark solitons found here. No truly localized wave such as bright or antidark waves havc been reported in this case.

In QPM structures, diffraction can be overcome by the action of $\chi^{(2)}$, leading to spatial solitons [65,69–71]. In this respect, also the backward propagating geometry was explicitly studied [71]. In combination with such trapping in the transverse plane, the longitudinal one discussed here, leads to the fascinating possibility of having light localized in multi-dimensions (bullets) slowly traveling along a linearly transparent structure (see also [52] for the pure $\chi^{(2)}$ case).

Moreover, the QPM technique can be extended to $\chi^{(3)}$ media as shown in [72] for frequency conversion in fibers. Periodic structures similar to the $\chi^{(2)}$ grating discussed in this section can be realized by means of stacks of materials with matched linear refractive index but different nonlinear index [73,74], thus making a grating which is effective only at high intensities. It will be worth pursuing the investigation of localization properties of such structures.

5.6 The Copropagating Case: Polarization Resonance Solitons

So far we have discussed localization which involves coupling of counterpropagating modes. Solitons, however, can be formed via SHG in a grating also in the copropagating geometry. In this case, the coupled modes differ by their polarization [75] and localized waves are usually denoted as *polarization resonance* solitons. Although different configurations can be envisaged, the basic physics can be captured in the specific case of two FF polarization components linearly coupled through a grating and nonlinearly coupled through type II (polarization) SHG. The polarization envelopes $E_{1,2}$ at FF and the SH beam E_3 obey the coupled Eqs. [22]

$$
\begin{aligned}
-i\left(\frac{\partial E_1}{\partial Z}+\frac{1}{V_1}\frac{\partial E_1}{\partial T}\right) &= \Gamma E_2+\chi E_3 E_2^* e^{i\Delta k Z}\,,\\
-i\left(\frac{\partial E_2}{\partial Z}+\frac{1}{V_2}\frac{\partial E_2}{\partial T}\right) &= \Gamma E_1+\chi E_3 E_1^* e^{i\Delta k Z}\,,\\
-i\left(\frac{\partial E_3}{\partial Z}+\frac{1}{V_3}\frac{\partial E_3}{\partial T}\right) &= \chi E_1 E_2 e^{-i\Delta k Z}\,,
\end{aligned}
\tag{5.32}
$$

where χ is the effective nonlinear coefficient, and $\Delta k = k_3 - k_2 - k_1$ is the nonlinear mismatch. In such copropagating geometry linear mode coupling occurs efficiently if the grating is resonant with the modal wavevector difference or birefringence $k_2 - k_1$. In (5.32) we have assumed perfect resonance, i.e., $\Delta\beta = (k_2 - k_1) - \beta_0 = 0$. Comparing with the Bragg case for which $\Delta\beta = 2k - \beta_0$, this requires in turn a grating with much longer wavelength Λ since $|k_2 - k_2| \ll 2k$. If we further consider the operating regime of large nonlinear mismatch (cascading), the SH beam can be adiabatically eliminated. In this case, however, conversion and backconversion results into a cross-induced phase-shift and (5.32) reduce to the integrable massive Thirring model [22]

$$
\begin{aligned}
i\frac{\partial u_1}{\partial z}+i\frac{\partial u_1}{\partial t}+u_2-s|u_2|^2u_1 &= 0\,,\\
i\frac{\partial u_2}{\partial z}-i\frac{\partial u_2}{\partial t}+u_1-s|u_1|^2u_2 &= 0\,,
\end{aligned}
\tag{5.33}
$$

where we have exploited the relation between the SH and FF fields $u_3 = -su_1u_2$ derived from the third of (5.32). In (5.33) $z \equiv \Gamma Z$, $t \equiv 2\Gamma\delta V(T - z/V_g)$ is a normalized retarded time in a frame traveling at average group-velocity $V_g = \frac{1}{2}(V_1^{-1}+V_2^{-1})^{-1}$, and $\delta V = |V_1^{-1}-V_2^{-1}|^{-1}$ is the group-velocity difference at FF. Moreover the normalized fields are $u_{1,2} = [\chi/(\Gamma|\delta k|^{1/2})]E_{1,2}$ and $u_3 = (\chi/\Gamma)E_3\exp(i\delta kz)$, where $\delta k \equiv \Delta k/\Gamma$, and $s \equiv \mathrm{sign}(\delta k)$. Noteworthy features of the (5.33) are

- interchanged role of t and z with respect to the counterpropagating case, e.g. compare with (5.22). As a consequence this type of structure exhibits a gap in k-space rather than in frequency space. Moreover, unlike Bragg gratings, no still solitons are possible since zero-velocity solutions of (5.33) travel indeed at average group-velocity V_g,

- integrability by inverse scattering method as a consequence of the absence of self-coupling [22]. The localized solutions of (5.33) are true solitons which possess remarkable feature of stability in their whole domain of existence. In general, however, such properties are not necessarily shared by the original model,
- solitons of (5.33) are qualitatively similar to those of (5.22), i.e., they form a continuous family parametrized by soliton velocity and propagation constant (which is equivalent to the frequency detuning $\delta\omega$ of LGSs in a DFB),
- the relative content of u_1 and u_2 and hence the resulting *polarization angle* of resonance solitons depend on soliton velocity $V_{\rm sol}$ which is bound between the group-velocity of the two modes, i.e., $V_1 < V_{\rm sol} < V_2$. The higher the u_1 (u_2) component the closer $V_{\rm sol}$ to V_1 (V_2).

The last feature is clearly visible in Fig. 5.15, which shows the excitation of two polarization resonance solitons with different velocity.

5.7 Stability of Localized Excitations

The stability of solitons is an essential prerequisite for their observability. Localized waves might be destroyed through two basic diverse mechanisms

- instability of the soliton itself. These include both translational and oscillatory instabilities,
- stability of the background (pedestal) which includes modulational instability (MI) and self-pulsing instability.

As far as the first issue is concerned, one generally wonders about the stability of solitons against weak perturbations (i.e., linear stability). Except for (5.33), all the coupled-mode models discussed here belong to the class of nonintegrable models, for which the stability is not guaranteed a priori. The general problem requires to solve the linear evolution equation obtained for the perturbation by linearizing around the soliton. Instabilities stem from eigenvalues of the linear problem crossing into the right-half of the complex plane. In the case of a generalized (nonintegrable) scalar NLS, such bifurcation is related by means of an exact result (Vakhitov-Kolokolov or Q-theorem [76]) to change of sign of the derivative of soliton invariant (power Q) against the internal parameter β (propagation constant). In this case, the unstable eigenvalue is real (meaning that the perturbation is exponentially amplified) and crosses through the origin (zero eigenvalue). Since the origin is always an eigenvalue of the linear problem associated with the translational symmetry of the original equation, this mechanism is often refered to as *translational* instability. When NLS models possessing solitons with two internal parameters are considered (a new internal parameter arises from an additional model symmetry in turn implying a new invariant), the threshold for translational

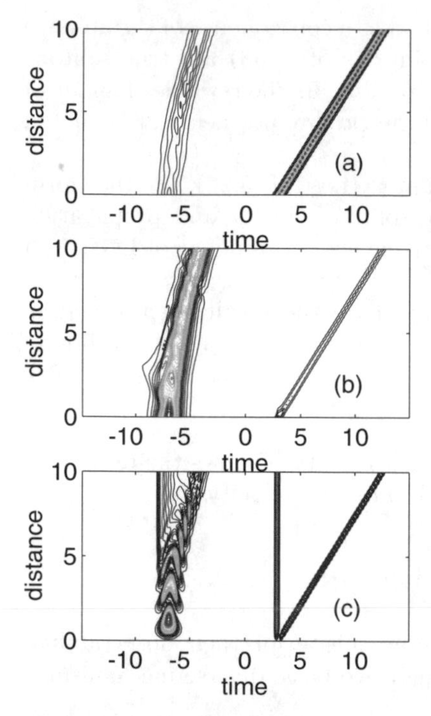

Fig. 5.15. Excitation of polarization resonance solitons supported by coupling of forward polarization modes in a long-wavelength grating. The figure shows the contour plot of the intensity of the three fields in the frame traveling at average group-velocity V_g, as obtained from numerical integration of (5.32): **a** FF field $|E_1|^2$; **b** FF field $|E_2|^2$; **c** SH field $|E_3|^2$. Two solitons with different velocity are excited by launching a FF beam with diverse input polarization $|E_1(0)|^2/|E_2(0)|^2 = 1.2$ (left) and $|E_1(0)|^2/|E_2(0)|^2 = 19$ (right), respectively. They have a suitable delay at $z = 0$ such that their interaction remains negligible. The SH beam is generated inside the grating and travels partly as a linear wave (we assume a SH velocity equal to V_g) and partly locked to the FF soliton components. Here $\delta k = -20$

instabilities can be given again in terms of derivatives of invariants against internal parameters in a form of a vanishing determinant [77]. Gap soliton models differ from NLS ones because of a different linear propagator, and are rather similar to spinorial (Dirac) models, for which stability was preliminarly addressed in other contexts [78]. It is only recently that instabilities of optical bright LGSs have been understood [79,80]. The threshold condition for translational instability of LGSs is described by the criterion

$$\begin{vmatrix} \partial_{\delta\omega} Q & \partial_{\delta\omega} M \\ \partial_V Q & \partial_V M \end{vmatrix} = \frac{\partial Q}{\partial \delta\omega}\frac{\partial M}{\partial V} - \left(\frac{\partial Q}{\partial V}\right)^2 = 0 \,, \tag{5.34}$$

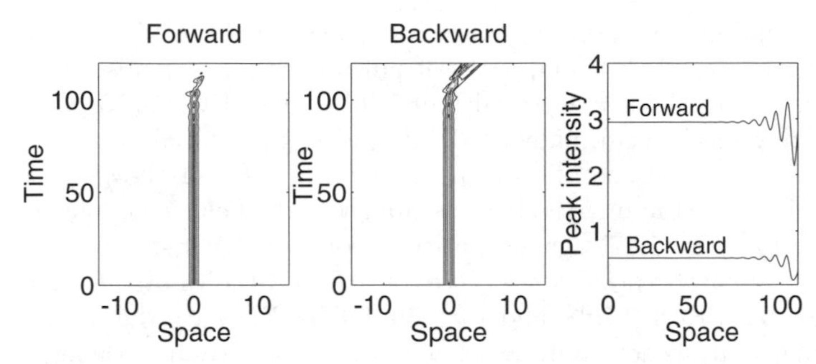

Fig. 5.16. Oscillatory instability of an exact LGS launched in a singly-resonant $\chi^{(2)}$ grating operating in the cascading defocusing ($\delta k > 0$) limit. The figure shows the contour plot of forward and backward FF intensity in the soliton reference frame (space $z - vt$, time t), and the peak power evolution of the two components. Soliton parameters are $\delta\omega \simeq 0.6$ and $V \simeq 0.7$

where $Q = Q(\delta\omega, V)$ and $M = M(\delta\omega, V)$ stand for the dependence of soliton photon flux and linear momentum (related with phase and space-translational invariance of the model, respectively [80]), on the parameters. This instability is also related to singular folding of soliton invariant surface [80] or catastrophe [81].

Importantly, however, a new kind of instability of the *oscillatory* type occurs. This is related to complex eigenvalues crossing into the right half plane, for which no successful and general enough criteria have been developed to date. This mechanism, first found for LGSs in Kerr (e.g., fiber) DFBs [79,80], takes place for quadratic LGSs as well [29,82]. It is worth emphasizing that, though oscillatory instabilities are commonly found in dissipative system, they are rarely displayed by conservative systems like, e.g., the generalized NLS equation. Its occurence was also later recognized for other conservative models [83,84]. Figure 5.16 shows the typical breakup of a $\chi^{(2)}$ LGS described by (5.22). The oscillatory character of the instability is evident.

For LGSs, oscillatory instabilities are found to prevail over translational ones. For instance this is the case for cascading in-gap LGSs shown in Fig. 5.7 [29]. As a rule of thumb, low-amplitude LGSs (LA in Fig. 5.7) are stable whereas high-amplitude LGSs (HA in Fig. 5.7) exhibit oscillatory breakup. Therefore, not only HA-LGSs require more power to be formed, but they can be easily destroyed by the onset of instabilities. A general assessment as to whether LGSs are stable in the general model (5.17) appears much more challenging due to the large dimensionality of the whole (internal and external) parameter space.

The stability of gap solitons (either LGSs or NGSs) with non-zero pedestal was never addressed to date. In this case, however, the stability of the background should be investigated first. Potentially, two mechanisms concur to determine loss of stability. The system may be destabilized through a Hopf bifurcation which induces periodic (at least close to threshold) or disordered (usually much above threshold) temporal self-pulsing. Studies of this mechanism have been reported for a Kerr DFB [85,86] and BSHG [48,53]. The second destabilizing mechanism, namely MI, provides exponential growth of spatially periodic perturbations. The gain bandwidth $g(K)$ of the process can be easily derived by studying the linear stability of the fields components ($u_{1,2}^{\pm}$ in (5.17) and $u_{1,2}$ in (5.27) against perturbations involving spatial sideband pairs $\varepsilon(z,t) = \varepsilon_S(t)\exp(iKz) + \varepsilon_A^*(t)\exp(-iKz)$. Studies of this kind were reported for both Kerr [87,88] and $\chi^{(2)}$ [89] DFBs. We warn the reader, however, that this analysis does not always lead to physical results. Depending on the specific values of the parameters, an infinite bandwith, i.e. $g \neq 0$ for $K \in [0, \infty)$, can be found, which is inconsistent with the coupled-mode approach itself. In this case, the analysis should be carried out by including dispersion $k_{1,2}''$ (see, e.g. [90]), at the price of a higher complexity. Moreover, in finite gratings, the stability problem should be correctly tackled as a boundary value problem [50,90].

Finally we mention that the (global) stability of the soliton solutions against inclusion of material dispersion k'' should be assessed. This analysis, reported so far only for Kerr DFB [91], seems more important for structures with stopband of nonlinear origin such as BSHG, where the grating does not introduce linear dispersion, and the material dispersion remains the dominant mechanism.

5.8 Conclusions and Further Developments

To summarize, we have shown that nonlinear effects in periodic media with quadratic response offer the opportunity to localize the electromagnetic radiation and induce intriguing effects of self-transparency. We have shown that solitons play a key role in the nonlinear dynamical behavior of several physically different structures. Both linear and nonlinear feedback, copropagating and counterpropagating geometries support traveling-wave localized envelopes. The basic technology needed to implement efficient $\chi^{(2)}$ gratings is available and our auspice is to see these effect demonstrated in the near future. Yet, we believe what we have presented here represents only a modest foretaste of what the richness of $\chi^{(2)}$ mixing phenomena joint to the grating technology can offer in the near future. A lot of theoretical efforts will be necessary to generalize the concepts presented in this chapter to other structures and to a higher number of dimensions. We hope to have reached our main goal of convincing the reader that this is a young and rapidly evolving field where the best has to come yet, and will deserve the coordinated efforts of both theoreticians and experimentalists.

Acknowledgment. We are indebted to our collaborators Gaetano Assanto, Alex Buryak, and Alfredo De Rossi, who took part in the work summarized in this chapter. We also thank Yuri Kivshar for enlightning discussions.

References

1. B.J. Eggleton, R.E. Slusher, C.M. de Sterke, P.A. Krug, J.E. Sipe: Phys. Rev. Lett. **76**, 1627 (1996)
2. P. Millar, R.M. De La Rue, T.F. Krauss, J.S. Aitchison, N.G.R. Broderick, D.J. Richardson: Opt. Lett. **24**, 685 (1999)
3. for a review see: G.I. Stegeman, D.J. Hagan, L. Torner: Opt. Quantum Electron. **28**, 1691 (1996); Y.S. Kivshar: Opt. Quantum Electron. **30**, 571 (1998); A.V. Buryak, P. Di Trapani, D. Skryabin, S. Trillo: "Optical solitons due to quadratic nonlinearities: from basic physics to futuristic applications" Phys. Rep., in press
4. S. Somekh, A. Yariv: Appl. Phys. Lett. **21**, 140 (1972)
5. M.M. Fejer, G.A. Magel, D.H. Jundt, R.L. Byer: IEEE J. Quantum Electron. **QE-28**, 2631 (1992)
6. V. Berger: Phys. Rev. Lett. **81**, 4136 (1998)
7. N.G.R. Broderick, G.W. Ross, H.L. Offerhaus, D.J. Richardson, D.C. Hanna: Phys. Rev. Lett. **84**, 4345 (2000)
8. J. Söchtig: Electron. Lett. **24**, 845 (1988)
9. W.P. Risk, S.D. Lau: Opt. Lett. (1993) **18**, 272 (1999)
10. J. Söchtig, R. Grob, I. Baumann, W. Sohler, H. Shütz, R. Widmer: Electron. Lett. **31**, 551 (1995)
11. Z. Weissman, A. Hardy, M. Katz, M. Oron, D. Eger: Opt. Lett. **20**, 674 (1995)
12. C. Becker, A. Greiner, Th. Oesselke, A. Pape, W. Sohler, H. Suche: Opt. Lett. **23**, 1195 (1998)
13. B. Wu, P.L. Chu, H. Hu, Z. Xiong: IEEE J. Quantum Electron. **35**, 1369 (1999)
14. H.G. Winful, J.H. Marburger, E. Garmire: Appl. Phys. Lett. **35**, 379 (1979)
15. A. Mecozzi, S. Trillo, S. Wabnitz: Opt. Lett. **12**, 1008 (1987)
16. W. Chen, D. L. Mills: Phys. Rev. Lett. **58**, 160 (1987)
17. C.M. de Sterke, J. E. Sipe: Phys. Rev. A **38**, 5149 (1988); *ibid* **39**, 5163 (1989); *ibid* **42**, 550 (1990)
18. A.B. Aceves, S. Wabnitz: Phys. Lett. A **141**, 37 (1989)
19. D.N. Christodoulides, R.I. Joseph: Phys. Rev. Lett. **62**, 1746 (1989)
20. J. Feng, F.K. Kneubül: J. Quantum Electron. **QE-29**, 590 (1993)
21. Y.S. Kivshar: Phys. Rev. E **51**, 1613 (1995)
22. S. Trillo: Opt. Lett. **21**, 1732 (1996)
23. C. Conti, S. Trillo, G. Assanto: Phys. Rev. Lett. **78**, 2341 (1997); Opt. Lett. **22**, 445 (1997)
24. H. He, P.D. Drummond: Phys. Rev. Lett. **78**, 4311 (1997); Phys. Rev. E **58**, 5025 (1998)
25. T. Peschel, U. Peschel, F. Lederer, B.A. Malomed: Phys. Rev. E **55**, 4730 (1997)
26. C. Conti, G. Assanto, S. Trillo: Opt. Lett. **22**, 1350 (1997)
27. C. Conti, S. Trillo, G. Assanto: Phys. Rev. E **57**, R1251 (1998)
28. C. Conti, S. Trillo, G. Assanto: Opt. Lett. **23**, 334 (1998); C. Conti, G. Assanto, S. Trillo: Electron. Lett. **34**, 689 (1998)
29. C. Conti, A. De Rossi, S. Trillo: Opt. Lett. **23**, 1265 (1998)

30. C. Conti, G. Assanto, S. Trillo: Opt. Exp. **3**, 389 (1998)
31. W.C.K. Mak, B. A. Malomed, P.L. Chu: Phys. Rev. E **58**, 6708 (1998)
32. A. Arraf, C.M. de Sterke: Phys. Rev. E **58**, 7951 (1998)
33. C. Conti, G. Assanto, S. Trillo: Phys. Rev. E **59**, 2467 (1999)
34. T. Iizuka, Y.S. Kivshar: Phys. Rev. E **59**, 7148 (1999)
35. A.V. Buryak, I. Towers, S. Trillo: Phys. Lett. A **267**, 319 (2000)
36. S. Trillo, C. Conti, G. Assanto, A.V. Buryak: Chaos **10**, 590 (2000)
37. T. Iizuka, C.M. de Sterke: Phys. Rev. E **62**, 4246 (2000)
38. S.E. Harris: Appl. Phys. Lett. **9**, 114 (1966)
39. J.P. Van Der Ziel, L.M. Ilegems: Appl. Phys. Lett. **28**, 437 (1976)
40. J.U. Kang, Y.J. Ding, W.K. Burns, J.S. Mellinger: Opt. Lett. **22**, 862 (1997)
41. X. Gu, R.Y. Korotkov, Y.J. Ding, J.U. Kang, J.B. Khurgin: J. Opt. Soc. Am. B **15**, 1561 (1998)
42. X. Gu, M. Makarov, Y.J. Ding, J.B. Khurgin, W.P. Risk: Opt. Lett. **24**, 127 (1999)
43. X. Mu, I.B. Zotova, Y.J. Ding, W.P. Risk: Opt. Commun. **181**, 153-159 (2000)
44. P.S.J. Russell: IEEE J. Quantum Electron. **27**, 830 (1991)
45. Y.J. Ding, S.J. Lee, J.B. Khurgin: Phys. Rev. Lett. **75**, 429 (1995)
46. M. Matsumoto, K. Tanaka: J. Quantum. Electron. **31**, 700 (1995)
47. Y.J. Ding, J.B. Khurgin: Opt. Lett. **21**, 1445 (1996); IEEE J. Quantum Electron. **32**, 1574 (1996)
48. G. D'Alessandro, P.S.J. Russell, A.A. Wheeler: Phys. Rev. A **55**, 3211 (1997)
49. G.D. Landry, T. Maldonado: Opt. Lett. **22**, 1400 (1997); Applied Optics **37**, 7809 (1998); IEEE J. Lightwave Tech. **17**, 316 (1999)
50. P.M. Lushnikov, P. Lodahl, M. Saffman: Opt. Lett. **23**, 1650 (1998)
51. Y.J. Ding, J.U. Kang, J.B. Khurgin: IEEE J. Quantum Electron. **34**, 966 (1998)
52. A. Picozzi, M. Haelterman: Opt. Lett. **23**, 1808 (1998); Phys. Rev. E **59**, 3749 (1999)
53. C. Conti, G. Assanto, S. Trillo: Opt. Lett. **25**, 1134 (1999)
54. K. Gallo, P. Baldi, M. De Micheli, D.B. Ostrowsky, G. Assanto: Opt. Lett. **25**, 966 (2000)
55. C. Conti, S. Trillo, G. Assanto: Phys. Rev. Lett. **85**, 2502 (2000)
56. G. Cappellini, S. Trillo, S. Wabnitz, R. Chisari: Opt. Lett. **17**, 637 (1992)
57. P. Byrd, M. Friedman: *Handbook of elliptic integrals for engineers and physicists* (Springer, Berlin Heidelberg 1971)
58. M. Picciau, G. Leo, G. Assanto: J. Opt. Soc. Am. B **13**, 661 (1996)
59. A.J. Lichtenberg and M.A. Lieberman: *Regular and Stochastic Motion* (Springer, Berlin Heidelberg New York 1983)
60. M. Grillakis, J. Shatah, and W. Strauss: J. Funct. Anal. **74**, 160 (1987); **94**, 308 (1990)
61. F.V. Marchevskii, V.L. Strizhevskii, V.P. Feshchenko: Sov. J. Quantum Electron. **14**, 192 (1984)
62. T.V. Makhviladze, M.E. Sarychev: Sov. Phys. JETP **44**, 471 (1975)
63. A.P. Sukhorukov: *Nonlinear Wave Interactions in Optics and Radiophysics* (Nauka, Moscow 1988) (in Russian)
64. J.M. Soto-Crespo, N.N. Akhmediev, V.V. Afanasjev: Opt. Commun. **118**, 587 (1995)
65. C.B. Clausen, O. Bang, Y.S. Kivshar: Phys. Rev. Lett. **78**, 4749 (1997); O. Bang, C.B. Clausen, P.L. Christiansen, L. Torner: Opt. Lett. **24**, 1413 (1999)

66. B. Denardo, B. Galvin, A. Greenfield, A. Larraza, S. Putterman, W. Wright: Phys. Rev. Lett. **68**, 1730 (1992)
67. Y.S. Kivshar: Phys. Rev. Lett. **70**, 3055 (1993)
68. G. Huang, Z. Jia: Phys. Rev. B **51**, 613 (1995)
69. C.B. Clausen, Y.S. Kivshar, O. Bang, P.L. Christiansen: Phys. Rev. Lett. **83**, 4740 (1999)
70. B. Bourliaguet, V. Couderc, A. Barthelemy, G.W. Ross, P.G.R. Smith, D.C. Hanna, C. De Angelis: Opt. Lett. **24**, 1410 (1999)
71. K.Y. Kolossovski, A.V. Buryak, and R.A. Sammut: Opt. Lett. **24**, 835 (1998)
72. S.G. Murdoch, R. Leonhardt, J.D. Harvey, T.A.B. Kennedy: J. Opt. Soc. Am. B **14**, 1816 (1997)
73. Q. Li, C.T. Chen, K.M. Ho, C.M. Soukoulis: Phys. Rev. B **53**, 15577 (1996)
74. L. Brzozowski, E.H. Sargent: IEEE J. Quantum Electron. **36**, 550 (2000)
75. S. Wabnitz: Opt. Lett. **14**, 1071 (1989)
76. M.G. Vakhitov, A.A. Kolokolov: Radiophys. Quantum Electron. **16**, 783 (1973)
77. A.V. Buryak, Y.S. Kivshar, S. Trillo: Phys. Rev. Lett. **77**, 5210 (1996)
78. Bogolubsky: Phys. Lett. 73A, 87 (1979); A. Alvarez, B. Carreras: Phys. Lett. **86A**, 327 (1981); W.A. Strauss, L. Vazques: Phys. Rev. D **34**, 641 (1986); A. Alvarez, M. Soler: Phys. Rev. D **34**, 644 (1986)
79. V.I. Barashenkov, D.E. Pelinovsky, E.V. Zemlyanaya: Phys. Rev. Lett. **80**, 5117 (1998); V.I. Barashenkov, E.V. Zemlyanaya: Computer Phys. Commun. **126**, 23 (2000)
80. A. De Rossi, C. Conti, S. Trillo: Phys. Rev. Lett. **81**, 85 (1998)
81. R. Gilmore: *Catastrophe theory for scientists and engineers* (Dover, New York 1993)
82. J. Schöllmann, R. Scheibenzuber, A.S. Kovalev, A.P. Mayer, A.A. Maradunin: Phys. Rev. E **59**, 4618 (1999)
83. D. Mihalache, D. Mazilu, L. Torner: Phys. Rev. Lett. **81**, 4353 (1998)
84. M. Johansson, Y.K. Kivshar: Phys. Rev. Lett. **82**, 85 (1999)
85. H.G. Winful, G.D. Cooperman: Appl. Phys. Lett. **40**, 298 (1982)
86. H.G. Winful, R. Zamir, S.F. Feldman: Appl. Phys. Lett. **58**, 1001 (1991)
87. A.B. Aceves, S. Wabnitz, C. De Angelis: Opt. Lett. **17**, 1566 (1992)
88. C.M. de Sterke: J. Opt. Soc. Am. B **15**, 2660 (1998)
89. H. He, A. Arraf, C.M. de Sterke, P.D. Drummond, B.A. Malomed: Phys. Rev. E **59**, 6064 (1999)
90. M. Yu, C.J. McKinstrie, G.P. Agrawal: J. Opt. Soc. Am. B **15**, 607 (1998)
91. A.R. Champneys, B.A. Malomed, M.J. Friedman: Phys. Rev. Lett. **80**, 4169 (1998)

6 Photonic Band Edge Effects in Finite Structures and Applications to $\chi^{(2)}$ Interactions

G. D'Aguanno, M. Centini, J.W. Haus, M. Scalora, C. Sibilia,
M.J. Bloemer, C.M. Bowden, and M. Bertolotti

6.1 Introduction

An intense investigation of electromagnetic wave propagation phenomena at optical frequencies in periodic structures has occurred over the past two decades. Usually referred to as photonic band gap (PBG) crystals [1], the essential property of these structures is the existence of allowed and forbidden frequency bands and gaps, in analogy to energy bands and gaps of semiconductors. An early paper by Armstrong and co-workers [2] and a second by Bloembergen and Sievers [3] outlined the advantage of periodic dielectric structures to achieve phase matching. The further advantage of band edge enhancement and group velocity changes had not been worked out at that time.

Many applications have been envisioned in one-dimensional systems, which usually consist of multilayer, dielectric stacks or patterned waveguides. The proposed devices include a nonlinear optical limiter [4] and an optical diode [5]; a photonic band edge laser [6]; a true-time delay line for delaying ultra-short optical pulses [7]; a high-gain optical parametric amplifier for nonlinear frequency conversion [8]. More recently, new applications have been proposed with transparent metal-dielectric layers [9] and all-optical switching [10]. Some demonstrations of the potential applications of these structures in higher dimensional systems have been highlighted recently with the realization of photonic crystal fibers [11], and in the microwave regime with the development of a PBG structure for applications to antenna substrates [12].

Second harmonic generation (SHG) in PBG structures has been experimentally observed under different circumstances. As an example, SHG was observed in a centrosymmetric, crystalline lattice of dielectric spheres [13]; in a semiconductor microcavity [14]; and near the band edge of a ZnS/SrF multilayer stack [15]. From a theoretical point of view, the study of nonlinear optical interactions in PBG structures has been undertaken mainly in regard to soliton-like pulses (often referred to as gap-solitons) in cubic $\chi^{(3)}$ [16] and quadratic $\chi^{(2)}$ media [17]. One of the most intriguing aspects related to the $\chi^{(2)}$ response of PBG crystals that is of interest for a number of applications is the possibility of significantly increasing the conversion efficiency of nonlinear processes. We cite the enhancement of SHG as the simplest and well-known parametric nonlinear process, although in general our discussion is valid even for more complicated multi-wave mixing processes. The processes that we discuss owe their increased efficiency to the simultaneous availability of: (a) exact phase matching conditions; and (b) enhancement of the fields with frequencies tuned to transmission resonances near the photonic band edge. These unusual conditions were first reported in [8], where it was numerically demonstrated that by pumping a 20-period, GaAs/AlAs multilayer stack approximately 10 microns in length with a 3-micron wavelength pump pulse one could generate short, picosecond SH pulses at a wavelength of 1.5 microns, with power levels enhanced by two to three orders of magnitude with respect to the output from an equivalent length of an exactly and ideally phase-matched bulk GaAs (or AlAs) material.

Recently, we theoretically investigated the possibility of nonlinear intensity-dependent phase shifts in a structure similar to that of [8]. In our research we developed devices with a nonlinear phase shift of order π in a structure approximately 10 microns in length, with threshold intensities of only a few tens of MW/cm^2 [18]. Also recently, a theoretical description based on the introduction of an effective medium describing the PBG structure as a whole was used [19] in order to explain the results of [8] in terms of appropriate phase matching conditions.

Although it is generally agreed that phase matching conditions and strong field localization are responsible for the enhancement of nonlinear interactions near the band edge, here we set out to review how these effects specifically influence nonlinear field dynamics in structures of finite length, i.e., with the introduction of entry and exit interfaces. We note that the conditions that we discuss in this paper are those where the spatial extent of incident pulses may exceed the spatial extent of a typical structure by several orders of magnitude, as in [8]. The circumstances that arise in this case are not the same as those that are typically considered [16,17], where the structure may be much longer compared to the spatial extent of the pulse, and we will discuss the respective roles of phase matching conditions and high field localizations available near the band edge. These roles will become clear as the formalism is developed, and will help us identify the reasons behind the remarkable predictions of enhancement of SHG [8,19] and recently reported cascading processes [18].

6.2 The PBG Structure as an "Open Cavity": Density of Modes (DOM) and Effective Dispersion Relation

Our analysis begins by considering a one-dimensional, finite, $N-$period structure consisting of pairs of alternating layers of high and low linear refractive indices. The layer thicknesses are a and b respectively; $L = a + b$ is the length of the elementary cell (see inset of Fig. 6.1a), and for N periods the length of

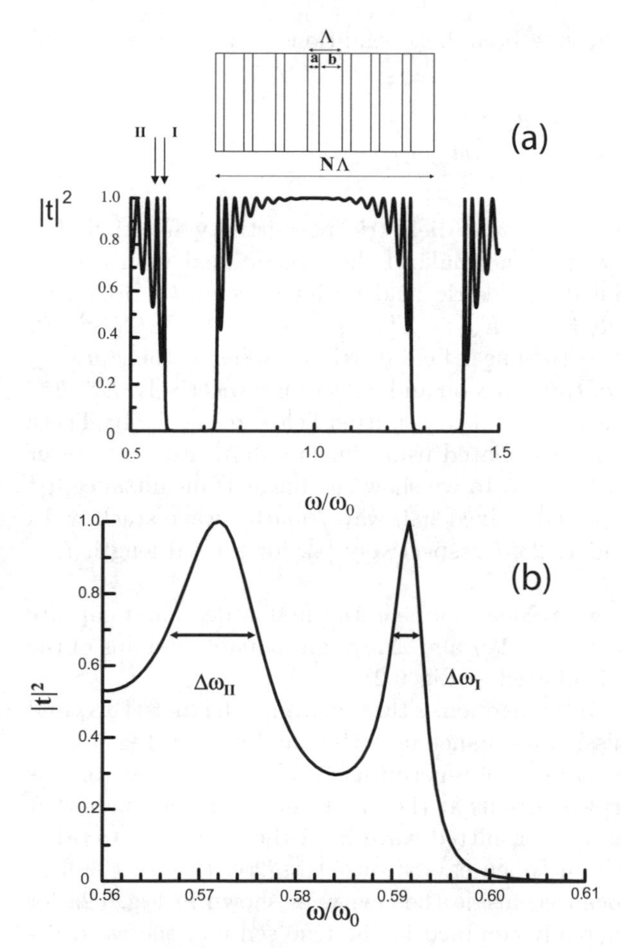

Fig. 6.1. **a** Transmission vs normalized frequency for a 20-period structure, half/quarter- wave stack. The indices of refraction are $n_a = 1$ and $n_b = 1.42857$. The layers have thicknesses $a = \lambda_0/(4n_a)$ and $b = \lambda_0/(2n_b)$, $\lambda_0 = 1\,\mu\text{m}$, $\omega_0 = 2\pi c/\lambda_0$. *Inset:* schematic representation of the structure. **b** The labels 'I' and 'II' identify the first and the second transmission resonance, respectively, near the first order band-gap, with respective bandwidths $\Delta\omega_I$ and $\Delta\omega_{II}$

the structure is $L = N\lambda$. We assume the structure is surrounded by air. Under the monochromatic plane-wave approximation, the Helmholtz equation for the evolution of the electric field in a *lossless* PBG structure is:

$$\frac{d^2\Phi_\omega}{dz^2} + \frac{\omega^2}{c^2}\epsilon_\omega(z)\Phi_\omega = 0\,, \tag{6.1}$$

which is subject to the following boundary conditions at the input ($z = 0$) and output ($z = L$) surfaces:

$$1 + r_\omega = \Phi_\omega(0)\,, \quad t_\omega = \Phi_\omega(L)\,,$$
$$i\frac{\omega}{c}(1 - r_\omega) = \frac{d\Phi_\omega(0)}{dz}\,, \quad i\frac{\omega}{c}t_\omega = \frac{d\Phi_\omega(L)}{dz}\,. \tag{6.2}$$

$\varepsilon_\omega(z)$ is the spatially dependent, real dielectric permittivity function. For simplicity, in (6.1)–(6.2) we have normalized the electric field with respect to the amplitude of the incident electric field by introducing the following dimensionless quantities: $\Phi_\omega(z) = E_\omega(z)/E^I_\omega$. Also, $t_\omega = (E^t_\omega/E^I_\omega)e^{i(\omega/c)L}$, $r_\omega = E^\omega_r/E^\omega_I$, where $\Phi_\omega(z)$ is the linear field distribution inside the stack, t_ω and r_ω are the coefficients of transmission and reflection respectively. E^ω_I, E^ω_r, and E^ω_t are the incident, reflected and transmitted fields, respectively. These quantities can be numerically calculated using the standard matrix transfer technique. As an example, in Fig. 6.1a we show the linear transmittance $|t|^2$ versus frequency for a 20-period, mixed half-wave/quarter-wave stack, with alternating indices of 1 and 1.42857 respectively [8], for a total length $L =$ 12 μm.

In Fig. 6.1b, the first two resonances near the first order band-gap are highlighted; their bandwidths are $\Delta\omega_I$ and $\Delta\omega_{II}$. The square modulus of the field distribution $|\Phi_\omega(z)|^2$ is plotted in Fig. 6.2.

The curves correspond to the frequency that identifies the first (I-2a) and the second (II-2b) transmission resonance near the band edge in Fig. 6.1.

Equatios (6.1–(6.2) describe a non-hermitian problem because (6.1) is supplemented by boundary conditions at the input and output surfaces that give rise to a reflected and a transmitted wave from the structure. In other words, we are dealing with an "open cavity" problem [20]. Even if the field appears to become well localized inside the cavity, as shown in Fig. 6.2a for example, the field is never really confined in the true sense of the word because the field enters and eventually exits the structure. Thus, while the field distribution in Fig. 6.2a strongly resembles a bound-state function, a time-domain analysis of the problem shows that the wave gives rise to a highly localized, metastable state [21], that slowly leaks out through the input and output surfaces.

As an example, let us consider the temporal dynamics of an incident Gaussian pulse of unitary amplitude. The carrier frequency is chosen to correspond to the first transmission resonance peak near the first order band gap. We take the pulse width to be $\tau \gg \frac{1}{\Delta\omega_I}$, i.e., the bandwidth of the input pulse is much smaller compared to resonance bandwidth. This is equivalent to restating the condition that the spatial extent of the pulse is much larger compared to structure length. In the system whose transmission is represented by Figs. 6.1, this means incident pulse duration should be larger than one picosecond in duration; in our simulations we choose $\tau = 2$ ps. In Fig. 6.3a–c we summarize the dynamics of the pulse. In Fig. 6.3a, we show the pulse before and after the scattering event. We note the extent of the

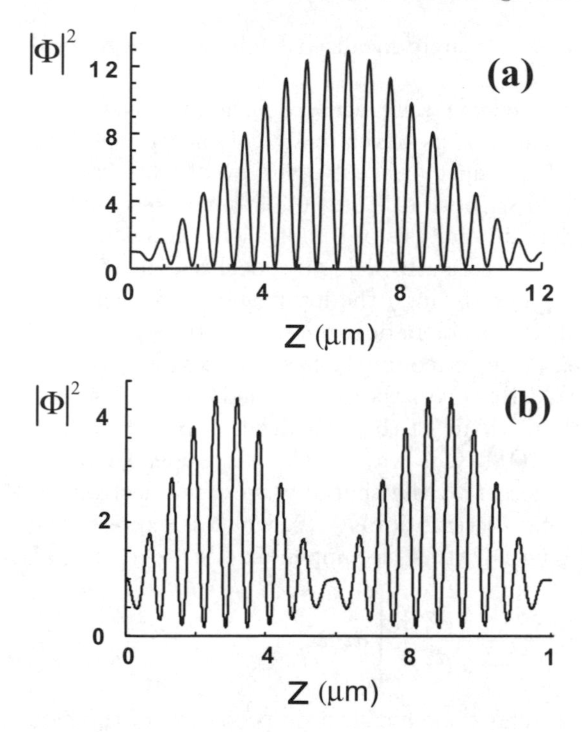

Fig. 6.2. Square modulus of the field distribution corresponding to the frequency that identifies the first transmission resonance **a** labeled I in Fig. 6.1, and the second transmission resonance **b** labeled II in Fig. 6.1) away from the band edge

pulse with respect to structure length. In Fig. 6.3b, the center of the pulse has reached the structure. At this instant we identify what we earlier referred to as a localized metastable state. Examination of the figure shows that in this quasi-monochromatic regime, the field distribution inside the stack is very similar to the field distribution shown in Fig. 6.2a, which was calculated using the stationary Helmholtz equation and its related boundary conditions (6.1)–(6.2). The inset of Fig. 6.3b shows the pulse in its entirety, as it crosses the PBG structure. The large spike visible in the inset corresponds to the field profile of the figure. Finally, in Fig. 6.3c we compare the localization of the 2 ps pulse with the localization of a 200 fs pulse. In this latter case, the spatial extent of the pulse is of the same order as structure length, and as a result the field appears to become far less localized inside the structure. So, the ability to resolve individual resonances places strict conditions that are necessary in order to first excite and then observe the localized metastable states that we show in Figs. 6.2a and 6.3b. In short, then, in studying the physics of the photonic band edge [4–9,18,19,22], we set out to explore the peculiar but remarkable properties of those particular metastable states that transiently manifest themselves as strongly localized fields, with large local field intensities.

Even if the field appears to become well localized inside the cavity, as shown in Fig. 6.3 for example, the field is never really confined in the true sense of the word because the field must first enter and eventually exit the structure. Thus, while the field distribution in Fig. 6.3 strongly resembles a bound-state function, a time-domain analysis of the problem shows that the wave first enters the structure, gives rise to a highly localized, metastable state [21], and then slowly leaks out through the input and output surfaces as shown in Fig. 6.3. As a result, the problem does not admit true eigenstates, intended in the traditional sense, i.e., a complete basis of expansion in the space between $z = 0$ and $z = L$. However, the problem admits metastable, or quasi-stationary states that by virtue of their localization properties can be associated with an effective DOM. One way to define a proper DOM for structures of finite length is to resort to the spatially averaged electromagnetic energy density. For the field distribution $\Phi_\omega(z)$, which is expressed in dimensionless units, we define the DOM as (see appendix A):

$$\rho_E^{(\omega)} \equiv \frac{1}{2Lc}\int_0^L \left[\epsilon_\omega(z)\left|\Phi_\omega\right|^2 + \frac{c^2}{\omega^2}\left|\frac{d\Phi_\omega}{dz}\right|^2\right] dz\,. \tag{6.3}$$

In general, the DOM should reflect the localization properties of the field inside the structure. Our definition in (6.3) does so in that it establishes a clear link between large field localization and large values of the DOM. We decompose the field in the form: $\Phi_\omega(z) = |\Phi_\omega(z)|e^{i\theta(z)}$. It follows that $|\Phi_\omega|^2\frac{d\theta}{dz}$ is a conserved quantity admitted by (6.1). Integrating (6.3) by parts, and using the boundary conditions (6.2), the expression for the DOM takes on the following, more suggestive form:

$$\rho_\omega = \frac{1}{Lc}\int_0^L \epsilon_\omega(z)\left|\Phi_\omega\right|^2 dz - \frac{1}{L\omega}\mathrm{Im}(r_\omega)\,. \tag{6.4}$$

There is an alternative definition of the DOM, as discussed in [22], namely $\rho_\omega = \frac{1}{L}\frac{d\varphi_t}{d\omega}$, where φ_t is the phase of the transmission function, defined as $t_\omega = |t_\omega|e^{i\varphi_t}$. In Fig. 6.4, the DOM is shown for the 20-period structure of Fig. 6.1; it is clearly largest at each transmission resonance, where $|t|^2 = 1$, and it reaches an absolute maximum at the transmission peak nearest to the band edge. There, the corresponding field becomes well localized over the entire structure – see Fig. 6.2. The solid line corresponds to the DOM as calculated by the definition given in (6.3), and the dotted line corresponds to the DOM as calculated using the definition. The two different definitions yield the same results in the pass-band, and show only a modest quantitative difference inside the band gap (see inset of Fig. 6.4). Although quantitative differences are slight, our definition given in (6.3) is much more appealing from a physical and a conceptual point of view because it establishes a clear

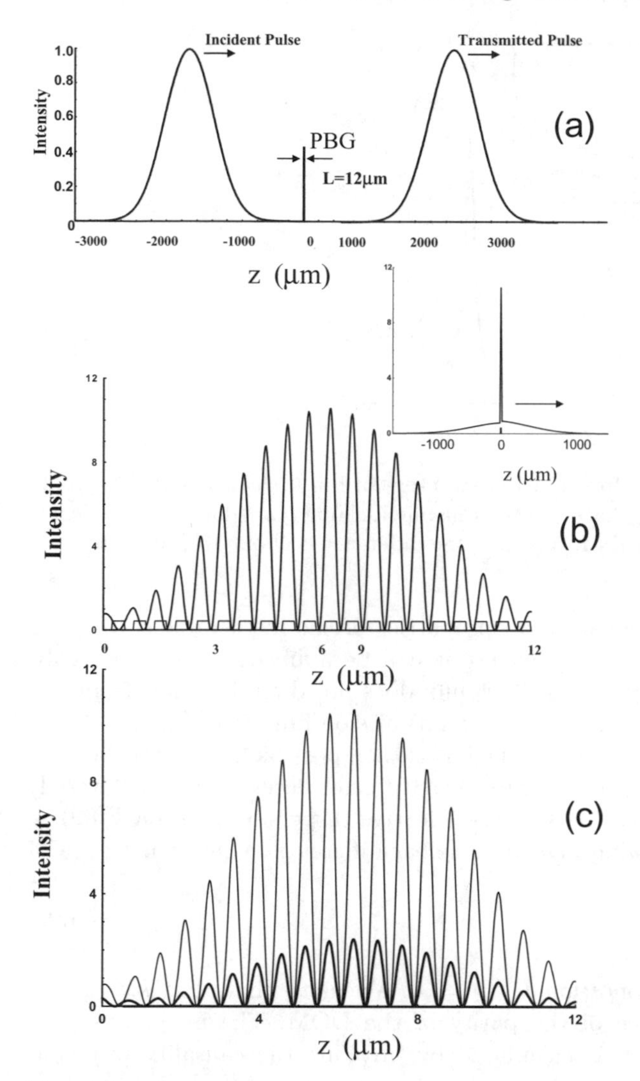

Fig. 6.3. a 2-ps incident Gaussian pulse and the structure. Note the spatial extent of the pulse with respect to structure length is to scale. **b** Internal field profile when the peak of the pulse reaches the structure. *Inset*: the pulse is shown in its entirety. The large spike visible in the inset corresponds to the field profile of the figure. Note that most of the pulse is located outside the structure. **c** Pulse has exited the structure. **d** Metastable states for 2-ps (Fig. 6.3b) (thin line) and for the 200-fs pulses (thick line). Note that the longer pulse is much more localized than the shorter pulse due to the different frequency bandwidth

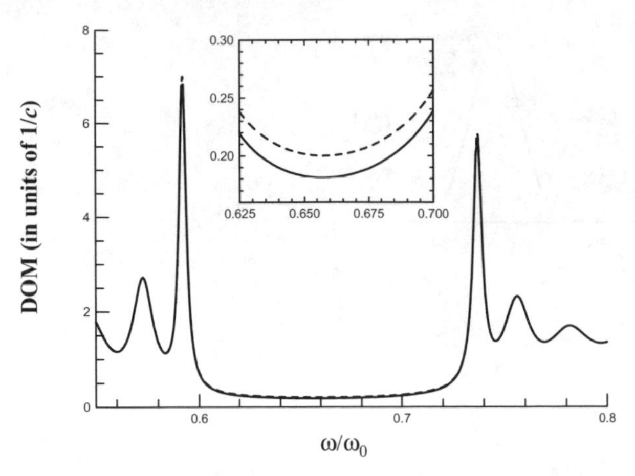

Fig. 6.4. Density of modes for the PBG structure of Fig. 6.1 calculated using (6.3) (solid line) and using the phase of the transmission function (dotted line). *Inset*: inside of the gap is magnified. In this case, the difference is about 10% at center-gap

direct link between the localization properties of the metastable states and the concept of the DOM, and because it can be applied much more easily in a three-dimensional geometry. This link does not directly follow from the definition given in [22] in terms of the transmission function.

Once the DOM ρ_ω has been defined, it is then possible to find an effective dispersion relation for a generic, finite structure, not necessarily multilayered, in the usual way. The real part of the effective dispersion relation $k_r(\omega)$ is the solution of the following first order linear differential equation:

$$\frac{dk_r(\omega)}{d\omega} = \rho_\omega \tag{6.5}$$

supplemented by the condition $k_r(0) = 0$. We note that that $k_r(-\omega) = -k_r(\omega)$ as a consequence of the parity of the DOM. The imaginary part of the dispersion relation is calculated by invoking the causality principle through the Kramers-Kronig relations:

$$k_i(\omega) = -\frac{2}{\pi} P \int_0^\infty \frac{\Omega k_r(\Omega)}{\Omega^2 - \omega^2} d\Omega \tag{6.6}$$

thus obtaining the effective dispersion relation for the finite PBG structure:

$$\widetilde{k}(\omega) = k_r(\omega) + i k_r(\omega) \,. \tag{6.7}$$

Following [19], we can also introduce an effective index as:

$$\widetilde{k}(\omega) = \frac{\omega}{c} \left[n_r(\omega) + i n_r(\omega) \right] \,. \tag{6.8}$$

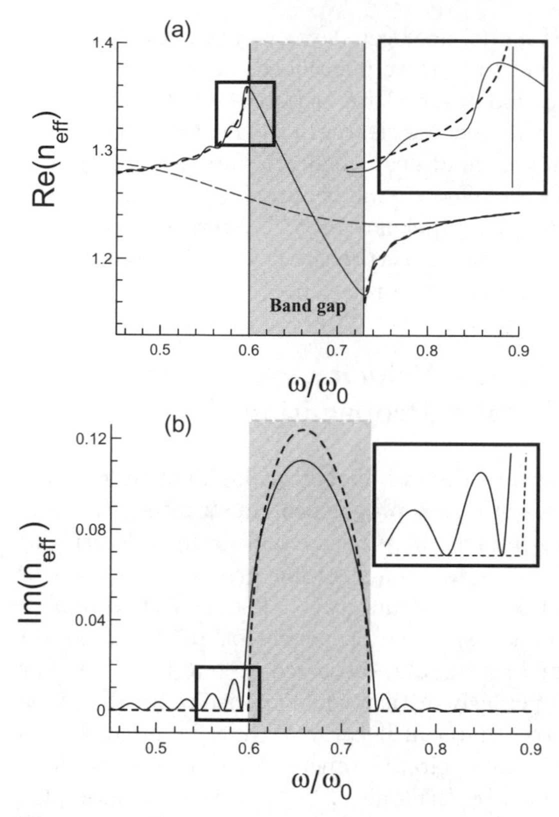

Fig. 6.5. **a** Real part of the effective index for a 2-(long dashes) and 20-period (solid curve) structure; dispersion relation for the infinite structure (short dashes). In the inset the band edge is magnified. **b** Imaginary part of the effective index, same as **a**, without the 2-period curve

Both real and imaginary parts of the effective index of refraction are plotted in Fig. 6.5a–b for a 2-and a 20-period structure, and compared with the dispersion relation of the infinite structure. We note that the real part of the index displays anomalous dispersion inside the gap. The imaginary component is small and oscillatory in the pass-bands; it attains its maximum at the center of each gap, where the transmission is a minimum, and it is identically zero at each transmission resonance. The figures show that the effective dispersive properties are modified by the number of periods (geometrical dispersion), and rapidly converge to the infinite-structure band edge by increasing N. However, we emphasize that the dispersion that we find displays clear oscillatory behavior outside the gap, unlike the infinite structure, since in the quasi-monochromatic regime that we are studying individual resonances can always be resolved.

Although our effective index model has been developed principally in connection with finite, periodic structures, as we mentioned earlier its validity is general because it holds for any kind of structure, periodic or not. The advantages that this approach afford will become clearer in the next sections, when we apply the model to the study of nonlinear quadratic interactions. Suffice it to say for the time being that the effective index approach can simplify the problem of calculating the conversion efficiency for $\chi^{(2)}$ interactions in PBG structures, because in the effective index picture the fields "see" an effective bulk material with a well-defined dispersion relation.

6.3 Group Velocity, Energy Velocity, and Superluminal Pulse Propagation

Recent experimental studies of one-dimensional have highlighted particularly interesting linear properties of pulses propagating in structures of finite length. Examples are the measurement of superluminal group velocities at mid-gap frequencies [23], and the measurement of low group velocities near the band edge of a semiconductor heterostructure [7]. These effects originate with the remarkable but peculiar dispersive properties of finite multi-layer stacks which we have highlighted above, and discussed at length in [19]. The concept of group velocity is particularly critical when applied to an absorbing (or gain), homogeneous dielectric material because V_g may be greater than c, and it can even be negative in some circumstances. These topics were discussed at length in a seminal book by Brillouin [24], in 1970 by Loudon [25], and by Garret and McCumber [26]. The physical relevance of V_g regarding pulse propagation with superluminal or negative group velocities was experimentally studied by Chu and Wang [27], who measured the transmission time of a laser pulse tuned at a GaP:N resonance. Recently, Peatross et al. [28] have theoretically shown that in the context of an absorbing, homogenous material, the group velocity may still be meaningful even for broadband pulses, and when V_g is superluminal or negative. In [28], the group velocity is related to pulse arrival time via the time expectation integral over the Poynting vector.

In the cited references [24–28] the work deals with pulse propagation in absorbing, homogenous dielectric material. More recently, Wang et al. [29] used gain assisted linear dispersion to demonstrate superluminal light propagation in atomic cesium: the group velocity of a laser pulse under conditions of anomalous dispersion in the presence of gain can exceed c as a result of classical interference between the different frequency components. Here, we discuss the case in which a 1-D PBG structure displays anomalous effective index behavior *not as a result of gain or absorption, but as a result of scattering.* More importantly, in our case the presence of entry and exit interfaces play a crucial role in determining the definition of the group velocity and its relationship with the energy velocity. In contrast, boundary conditions in the sense of an entry and an exit interface can play no role in the determination of either group or energy velocity in inhomogeneous, periodic structures of infinite length. Our simple and straight forward analysis shows that there are significant conceptual, qualitative, and quantitative differences between energy and group velocities in finite structures in contrast to the case of infinite structures. In fact, for a periodic, infinite structure a unique dispersion relation exists between K_β (the Bloch vector) and ω. The group velocity. $V_g^{(\omega)}$ is defined as $V_g^{(\omega)} = 1/[dK_\beta/d\omega]$, and it can be demonstrated that $V_g^{(\omega)} = V_E^{(\omega)}$ [30].

In order to discuss the case of 1-d, finite PBG structures, we consider *plane monochromatic waves*, and Eqs. ([1,2]). The averaged energy velocity, which measures energy flow across the sample, is defined as the ratio of the spatial average of the Poynting vector to the spatial average of the energy density within the same volume [30], which in our 1-d geometry is given by:

$$V_{\mathrm{E}}^{(\omega)} = \frac{\frac{1}{L}\mathrm{Re}\left[\frac{ic^2}{\omega}\int_0^L \Phi_\omega \frac{d\Phi_\omega^*}{dz}dz\right]}{\frac{1}{2L}\int_0^L \left[\epsilon_\omega(z)\left|\Phi_\omega\right|^2 + \frac{c^2}{\omega^2}\left|\frac{d\Phi_\omega}{dz}\right|^2\right]dz} . \tag{6.9}$$

Integrating by parts, and using the boundary conditions (6.2), the energy velocity $V_E^{(\omega)}$ takes a form that involves both the transmittance $|t_\omega|^2$ and the imaginary part of the reflectivity r_ω of the stack:

$$V_{\mathrm{E}}^{(\omega)} = \frac{c\left|t_\omega\right|^2}{\frac{1}{L}\int_0^L \epsilon_\omega(z)\left|\Phi_\omega\right|^2 dz - \frac{c}{L\omega}\mathrm{Im}(r_\omega)} . \tag{6.10}$$

The second term in the denominator indicates that the energy is generally not equally shared by the electric and magnetic components, and becomes identical only at each peak of transmittance, where r_ω vanishes. In Fig. 6.6 we depict $V_E^{(\omega)}$ for a 20 and 100-period stack, and compare with the $V_E^{(\omega)}$ (for infinite structures, $V_g^{(\omega)}$ coincides with $V_E^{(\omega)}$) of an infinite structure made with the same elementary cell. We find that the energy velocities of the infinite and finite structures do not converge to one another by increasing the number of periods. This is due to the fact that in the monochromatic approximation it is always possible to resolve each transmission resonance, and hence its curvature, even if more periods are added. In this regime, incident pulses are propagating through the structure tuned at one transmission resonance, for example, with their bandwidth significantly smaller compared to resonance bandwidth. Put another way, the spatial extent of the pulse is orders magnitude greater then the physical length of the structure. As a result, the pulse samples all internal and external interfaces simultaneously, it is delayed and in the end completely transmitted, with minimal distortion or scattering losses [7]. Therefore, the interaction should more properly be referred to as a scattering event.

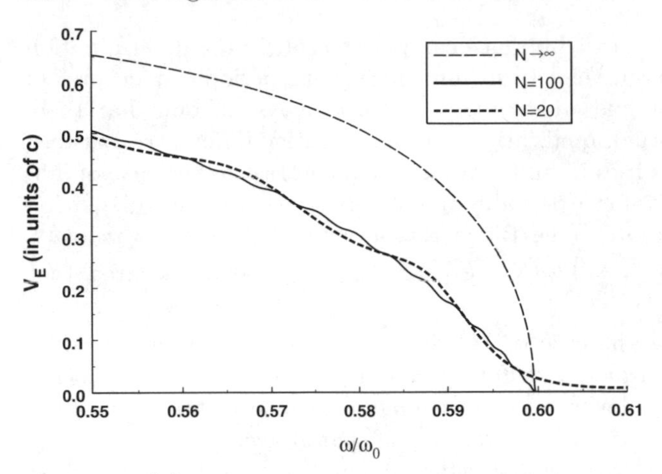

Fig. 6.6. $V_{\mathrm{E}}^{(\omega)}$ for a 20-period (short dashes), 100-period (solid line), and infinite structure (long dashes) vs frequency. Increasing the number of periods, the energy velocity does not converge to the results obtained using the dispersion of the infinite structure. The elementary cell is composed of a combination of half-wave/quarter-wave layers. The indices of refraction are $n_a = 1$ and $n_b = 1.42857$; the respective thicknesses are $a = \lambda_0/(4n_a)$ and $b = \lambda_0/(2n_b)$, with $\lambda_0 = 1\,\mu\mathrm{m}$, $\omega_0 = 2\pi c/\lambda_0$

To better clarify this situation, we consider the case of a short pulse incident on a structure several pulse widths in length. If the spatial extent of the pulse is so short that it traverses the structure without simultaneously sampling both entry and exit interfaces, then we may expect that the discontinuity at the entrance and exit interfaces will not significantly affect the dynamics [31]. This is shown in Fig. 6.7, where we plot $V_E^{(\omega)}$ as calculated in the quasi-monochromatic limit via (6.3), and also as numerically calculated for an incident gaussian pulse 150 fs in duration tuned in the passband of a 100-period structure. The energy (group) velocity of the infinite structure is also shown in the figure. The comparison between the length of the structure and pulse width is made in the inset. As the inset shows, the structure is several pulse widths in length. As a result, the energy velocity of the pulse propagating *inside* this structure approximates better the energy velocity of the infinite structure (Bloch's velocity). The energy velocity of the pulse does not show the same sharp cut off near the band edge we observe for both the infinite and the 100-period structures because the pulse is ultrashort, and even if the carrier frequency is tuned inside the gap a good fraction of the energy is still trasmitted. We find the same degree of convergence only if pulse width and structure length are significantly increased simultaneously, so that the pulse can better resolve the frequencies near the band edge, but still fits well inside the structure, as in the inset. In Fig. 6.8, the total momentum and energy inside the 100-period structure depicted in Fig. 6.7 are shown as a function of time. The total electromagnetic momentum for the pulse inside

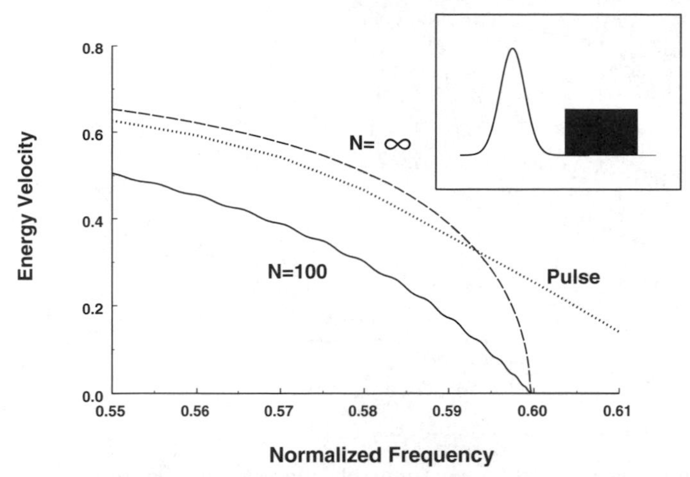

Fig. 6.7. $V_{\mathrm{E}}^{(\omega)}$ vs frequency for 150fs incident pulse (solid line), and monochromatic regime (dotted line). The monochromatic wave regime is also obtained using pulses at least several tens of picoseconds in duration. The dashed line corresponds to Bloch's velocity for the infinite structure. The structure is similar to that outlined in Fig. 6.1, but contains 100 periods. Inset: the 100-period structure is approximately 60 microns in length, while the spatial extent of the FWHM of the pulse is approximately 20 microns

the structure can be written as follows:

$$g(t) = \frac{1}{c^2} S(t) = \frac{1}{c^2} \int_0^L E_\omega(z,t) \times H_\omega(z,t) dz \,. \tag{6.11}$$

The total energy is given by:

$$U(t) = \int_0^L \left[\epsilon_\omega(z) \left| E_\omega(z,t) \right|^2 + \frac{c^2}{\omega^2} \left| B_\omega(z,t) \right|^2 \right] dz \,. \tag{6.12}$$

As the pulse enters the structure, there is a rapid rise in both energy and momentum, which settle to constant values once the whole pulse travels inside the structure. When the pulse is totally inside, both group and energy velocities are equal and given by Bloch's velocity. However, even if the structure is long, it is nevertheless finite, and so the pulse must eventually exit, leading to a reduction of energy and a reversal in sign in the total momentum. Once the momentum becomes negative, we track the first pulse reflected from the exit interface. In fact the momentum undergoes several sign reversals, until all the energy has left the structure. From Fig. 6.3 it should be evident that the energy velocity can equal Bloch's velocity *only after the entire pulse has entered and remains inside the structure*, while in general the time-averaged energy velocity will be different.

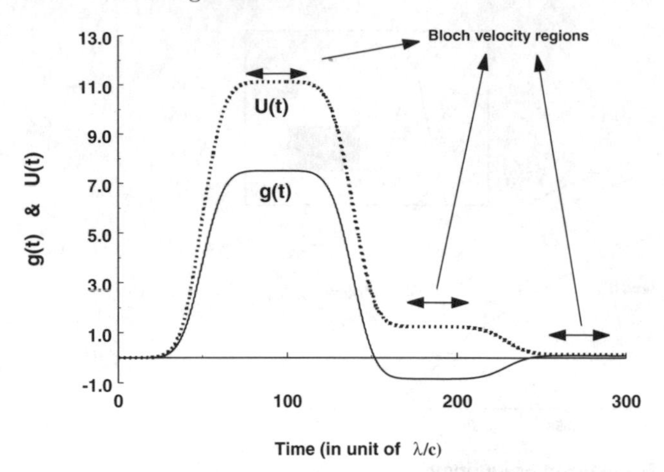

Fig. 6.8. Total energy (dashed) and momentum (solid) inside the structure for the pulse shown in the inset of Fig. 6.2. Both momentum and energy increase as the pulse traverses the entry boundary. The energy velocity for the entire process, i.e., the ratios of the areas under the curves, which is a measure of energy flow through the structure in both directions, cannot be the same as Bloch's velocity, which monitors only the velocity of the transmitted pulse. Multiple reflections inside the stack leads to ringing in reflection and transmission from the structure

With these considerations in mind, *we define the tunneling time* in a quasi-monochromatic regime, consistent with our approach, as:

$$\tau_\omega = \frac{1}{c}\int_0^L \epsilon_\omega(z)\,|\Phi_\omega|^2\,dz - \frac{1}{\omega}\mathrm{Im}(r_\omega)\,. \tag{6.13}$$

This definition of tunneling time, derived by imposing boundary conditions on our finite structure and directly suggested by (6.4) and (6.10), is the electromagnetic analogue of Smith's "dwell time" [32], which addresses electron wave packet tunneling times through a potential barrier. According to Smith, a quantum particle spends a mean time proportional to $\int_0^L |\Psi(z)|^2\,dz$ in the region of space between 0 and L, which is just the probability of finding the particle within the same region of space. Following Bohm [21], we use the concept of electromagnetic energy density, instead of the quantum probability density, to define the tunneling time (see appendix A): Equation (6.13) states that the time the field spends inside the structure is proportional to the energy density integrated over the volume. The term $-Im(r_\omega)/\omega$ represents the difference in energy between the electric and magnetic components, and it has no counterpart in the quantum case. Equation (6.13) thus establishes a clear link between large delay times and field localization, as experimentally verified for pulse propagation near the band edge [7]. One may also define a group velocity associated with the delay of the transmitted pulse as $V_g^{(\omega)} \equiv L/\tau_\omega$. In the eyes of an observer, this definition of group velocity is an extremely useful and powerful concept. However, we remind the reader that we are not considering propagation in a uniform medium, where a true group velocity can be defined. Our system consists of a pulse whose spatial extent can be orders of magnitudes larger compared to the length of the structure, which is therefore entirely contained within the pulse [7]. As a consequence, the dynamics can only properly be described as a scattering event, with an associated tunneling time.

Once a convenient group velocity has been defined in the manner indicated, (6.13) can finally be recast in the following simple form:

$$V_E^{(\omega)} = |t_\omega|^2 \, V_g^{(\omega)} . \tag{6.14}$$

Equation (6.14) is a strikingly simple new result that makes it clear that for finite structures the tunneling velocity V_g and the energy velocity V_E are the same only at each transmission resonance, and can be very different from each other, especially at frequencies inside the gap. The implications of (6.14) are even more profound and far reaching if we consider superluminal tunneling behavior. We begin with the assertion that the energy velocity can never take on values greater than c, namely $V_E^{(\omega)} \leq c$. From (6.14) it immediately follows that the tunneling velocity satisfies $V_g^{(\omega)} \leq c/\left|t_\omega\right|^2$. That is, *the simple requirement that the energy velocity should be subluminal does not prevent superluminal tunneling times. In fact, this inequality places an unambiguous upper limit on the tunneling velocity that can be achieved without violating the requirement that the energy velocity remain subluminal.* Based on these simple considerations, statements regarding superluminal pulse propagation should always be qualified by the energy velocity and transmittance. In Fig. 6.9 we plot $V_g^{(\omega)}$ and $V_E^{(\omega)}$ versus frequency for the 20-period structure of Fig. 6.1. Inside the gap, the group velocity becomes superluminal, while the energy velocity always remains causal. In this case, minimum transmittance can be as low as 1 part in 10^5. We note that the maximum superluminal group velocity is approximately 5.5 times the speed of light in vacuum (see Fig. 6.10), far below the upper limit imposed by the condition that the energy velocity should remain subluminal.

In Fig. 6.5 we compare the tunneling velocity calculated using (6.6), and the tunneling velocity calculated using the phase time. The group velocity corresponds to the inverse of the DOM, which has already been shown in Fig. 6.4. While in the pass band the two methods yield similar results, our method gives slightly higher estimates (10%, as in Fig. 6.4) for the maximum superluminal velocity compared to the method of the phase time. This difference corresponds to a time delay of the order of 1 fs, which is small but measurable [25]. The integration of the equations of motion in the time domain yield results consistent with our predictions, namely a group velocity of approximately 5.5 c for pulses tuned inside the gap.

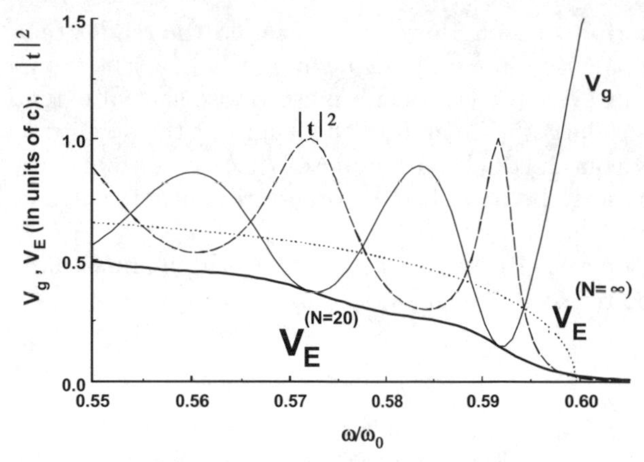

Fig. 6.9. $V_g^{(\omega)}$ (thin solid line), $V_{\mathrm{E}}^{(\omega)}$ (thick solid line), $|t|^2$ (dashes) for the 20-period structure described in the caption of Fig. 6.1. In the gap, the group velocity becomes superluminal. At resonance, $V_g^{(\omega)} = V_{\mathrm{E}}^{(\omega)}$, and $V_g^{(\omega)}$ is a minimum. The group velocity for the infinite structure is also depicted (dotted line) for comparison

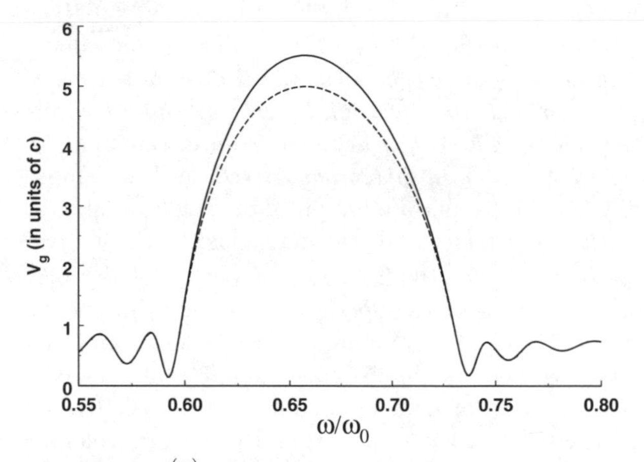

Fig. 6.10. $V_g^{(\omega)}$ as calculated using the definition of tunneling times given in (6.4) (solid line) and as calculated using the definition of phase time (dashed line)

6.4 $\chi^{(2)}$ Interactions and the Effective Medium Approach

The equations that describe the interaction of a fundamental and a SH field in a PBG structure have been discussed in [8], in a regime referred to the Slowly Varying Envelope Approximation in Time, or SVEAT. Operating under the assumption that field envelopes are allowed to vary rapidly in space but not in time, i.e., the temporal evolution of the field envelope proceeds slowly with respect to the optical cycle, it can be shown that the equations of motion reduce to [8]:

$$\varepsilon_{\Omega}\frac{\partial}{\partial\tau}E_{\Omega}(\xi,\tau)=\frac{i}{4\pi\Omega}\frac{\partial^2}{\partial\xi^2}E_{\Omega}(\xi,\tau)\qquad(6.15)$$
$$-\frac{\partial}{\partial\xi}E_{\Omega}(\xi,\tau)+i\pi\left[\varepsilon_{\Omega}-1\right]\Omega E_{\Omega}+i8\pi^2\Omega\chi^{(2)}E_{2\Omega}E_{\Omega}^{*}$$

and

$$\varepsilon_{2\Omega}\frac{\partial}{\partial\tau}E_{2\Omega}(\xi,\tau)=\frac{i}{4\pi 2\Omega}\frac{\partial^2}{\partial\xi^2}E_{2\Omega}(\xi,\tau)\qquad(6.16)$$
$$-\frac{\partial}{\partial\xi}E_{2\Omega}(\xi,\tau)+i\pi\left[\varepsilon_{2\Omega}-1\right]2\Omega E_{2\Omega}+i8\pi^2 2\Omega\chi^{(2)}E_{\Omega}^{2}\,.$$

In (6.16), $\Omega=\omega/\omega_0, \xi=z/\lambda_0$, and $\tau=ct/\lambda_0$. The spatial coordinate z has been conveniently scaled in units of a reference wavelength λ_0, and $\omega_0=2\pi/\lambda_0$; the time is then expressed in units of the corresponding optical period. It is evident that the large refractive index contrast preempts a more general slowly varying envelope approximation, or SVEA. However, if the discussion is restricted to pulses whose duration far exceeds a few optical cycles, then (6.16,6.17) are more than adequate.

Alternative options to the numerical integration of (6.16,6.17) also exist in the form of more approximate solutions. One approach is based on the hypothesis that the modulation of the linear dielectric permittivity function is small with respect to the averaged dielectric permittivity of the structure: $\varepsilon_{\omega}=\overline{\varepsilon}_{\omega}+\Delta\varepsilon_{\omega}(z)$, where $|\Delta\varepsilon_{\omega}(z)|\ll\overline{\varepsilon}_{\omega}$ (shallow grating limit). Consequently, the fields are expanded in terms of forward and backward propagating modes of the equivalent linear bulk medium. As a result, mode amplitudes are slowly modulated by the combined action of nonlinear coupling and the weak longitudinal variation of the refractive index [33]. The second approach is based on the expansion of the fields using Bloch states of the corresponding infinite linear structure. The grating may be of arbitrary depth, and the expansion coefficients are slowly modulated by the action of the quadratic coupling [34]. In the shallow grating limit, this method yields coupled mode equations identical to those that result using the first approach.

In this work, we propose an alternative approach based on our effective medium model that it is useful when strong localization effects and the appearance of metastable states of the kind discussed above come into play. The method that we propose can also be applied to intrinsically disordered layered structures, where translational invariance cannot be invoked, and analytical solutions would be almost impossible to find. We write the coupled mode equations for the nonlinear quadratic interactions in a finite PBG structure of length L as if the interaction were taking place in a bulk material of the same length L, but with an effective dispersion relation given by (6.8). In the case of two monochromatic waves at fundamental frequency (FF) ω and second harmonic frequency (SH) 2ω, each tuned at a peak of transmittance where the imaginary part of the effective dispersion relation (6.12) is zero [19], the coupled mode equations written for the effective medium are:

$$\frac{d}{dz}A_\omega = i\frac{\omega}{n^r_{\text{eff}}(\omega)c}d_{\text{eff}}A_{2\omega}A^*_\omega e^{i\Delta k_{\text{eff}}z} \tag{6.17}$$

and

$$\frac{d}{dz}A_{2\omega} = i\frac{\omega}{n^r_{\text{eff}}(2\omega)c}d_{\text{eff}}A^2_\omega e^{-i\Delta k_{\text{eff}}z}, \tag{6.18}$$

where $\Delta k_{\text{eff}}(\omega) = k_r(2\omega) - 2k_r$ is the effective phase mismatch calculated using the real part of the effective dispersion relation (6.12); n^r_{eff} is the real part of the effective index; and d_{eff} is the effective nonlinear coupling coefficient defined as:

$$d_{\text{eff}} = \frac{1}{L}\int_0^L d^{(2)}(z)|\Phi_\omega(z)|^2|\Phi_{2\omega}(z)|dz, \tag{6.19}$$

where $d^{(2)}(z)$ is the quadratic coupling function. It is overlapped with the square modulus of the FF and the modulus of SH. The value of the effective nonlinear coupling coefficient strictly depends on the localization properties of the FF field. From (6.4), at the peak of transmittance $\rho_\omega = \frac{1}{Lc}\int_0^L \epsilon_\omega(z)\,|\Phi_\omega|^2 \times dz$. As a consequence, d_{eff} is enhanced by a factor proportional to the DOM when the FF field is localized inside the nonlinear layers

$$d_{\text{eff}} \propto \rho_\omega d^{(2)}_{\text{layer}}, \tag{6.20}$$

where is $d^{(2)}_{\text{layer}}$ is the actual second order susceptibility of the nonlinear layer.

Now that we have developed all the major components of the model, we mention its fundamental limitations. Coupled mode (6.17) cannot give detailed information about the actual nonlinear dynamics of the fields inside the structure. This is evident if we recall that our approach to the problem consists in substituting the finite PBG structure with an equivalent length of bulk material and an effective dispersion relation given by (6.12). In other words, we cannot expect that this model will yield both reflected and transmitted components. This can only be achieved if we resort to the numerical integration of (6.16), as was done in [8]. However, the solutions of (6.17) yield remarkably accurate energy conversion efficiencies when compared to the conversion efficiencies calculated by integrating (6.16). By establishing the validity of the model with a direct comparison with the numerical integrations, conversion efficiencies can then be calculated very efficiently by resorting to the coupled mode (6.17). As an example, for second harmonic generation under the non-depleted pump approximation, we obtain the expression for the conversion efficiency h in analogy with the conversion efficiency calculated for bulk materials:

$$\eta = \frac{8\pi^2 d^2_{\text{eff}}L^2\widetilde{I}_p}{[n^r_{\text{eff}}(\omega)]^2 n^r_{\text{eff}}(2\omega)\varepsilon_0 c\lambda^2}\sin c^2\left(\frac{\Delta k_{\text{eff}}L}{2}\right), \tag{6.21}$$

where $\widetilde{I}_p = \frac{1}{2}\varepsilon_0 c n^r_{\rm eff}(\omega)|A_\omega|^2$ is the scaled input pump intensity. Equation (6.17) clarifies the role played by the linear FF metastable state inside the structure in the enhancement of the effective nonlinearity $d_{\rm eff}$. When the FF metastable state overlaps well the nonlinear layers, we expect an enhancement of the conversion efficiency approximation proportional to ρ_ω^2:

$$\eta \approx \left[\rho_\omega d^{(2)}_{\rm layer} L\right]^2 \widetilde{I}_p \sin c^2 \left(\frac{\Delta k_{\rm eff} L}{2}\right) . \tag{6.22}$$

In order to compare with an equivalent length of quadratic bulk material we can define, from (6.19), a figure of merit (FM) as:

$$\eta = L^2 \left[\frac{d_{\rm eff}}{d^{(2)}_{\rm layer}}\right]^2 \sin c^2 \left(\frac{L}{2L_c}\right) , \tag{6.23}$$

where $L_c = (1/\Delta k_{\rm eff})$ is either one coherence length in the case of bulk material, or the effective coherence length calculated via the effective index $n_{\rm eff}$ in the case of a PBG structure.

6.5 Examples: Blue and Green Light Generation

In order to highlight the potential of 1-D PBG structures for nonlinear frequency conversion processes, we discuss a new type of blue laser that could operate in the 400 nm range. The device is composed of 30-periods alternating SiO_2/AlN layers, to form a quarter-wave/ half-wave stack similar to that described in the previous section. With a reference wavelength of 0.52 microns, the total length of the structure is approximately 6 microns. The AlN layers are assumed to have a nonlinear coefficient of about 10 pm/V. The FF beam is assumed to be incident at an angle of 300 degrees with respect to the surface of the structure. In Fig. 6.11, we show the linear transmittance of this structure. The FF frequency, at 800 nm, is tuned to the first transmission resonance near the first order band gap, where the corresponding linear mode is well localized, and a high density of modes (DOM) is achieved. The SH frequency is tuned to 400 nm, which corresponds to the second resonance peak near the second order gap, as indicated on Fig. 6.11. We note that we are using experimentally available data for both materials, and that aligning the resonances as prescribed can be done by varying the thickness of the layers, i.e., by adjusting the geometrical dispersion of the structure [19]. The structure is exactly phase-matched in the sense of the effective index [19]. Through (6.23) we calculate the SH conversion efficiency as a function of the input FF intensity, and show the results in Fig. 6.12. We find a conversion efficiency of approximately 15% in a single pass through the device, with input power levels of 40 GW/cm^2. Although this intensity may seem high, we stress that the structure is only 30 periods long, and only 6 microns length.

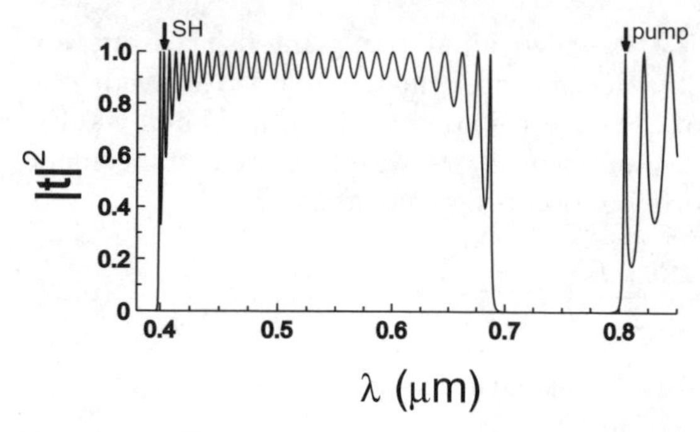

Fig. 6.11. Transmittance for 30-period PBG structure composed of alternating layers of SiO_2/AlN, quarter- wave/ half-wave stack respect to a reference length of 0.53 microns. The FF field (pump) and SH field (generated) are phase matched

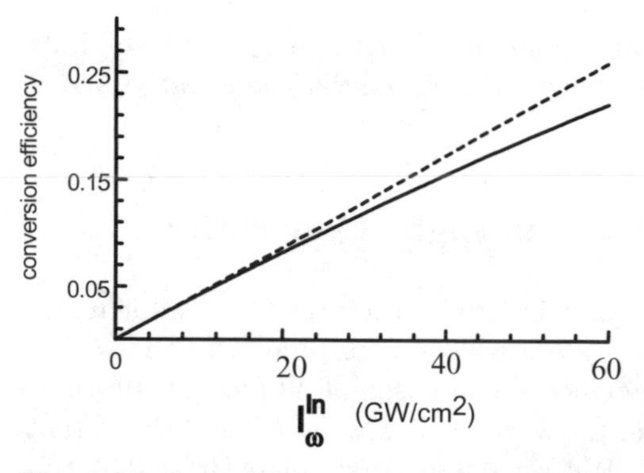

Fig. 6.12. Second harmonic conversion efficiency vs the input pump intensity under undepleted pump approximation (dotted line) and in the case of pump depletion (continuous line) 30-periods alternating SiO_2/AlN. A conversion efficiency of approximately 15% is reached for input pump intensity of 40 GW/cm^2

To test the validity of the effective medium model we perform a full numerical integration of the nonlinear coupled wave equations as discussed in [8]. We plot the results in Fig. 6.13, where we show the total energy converted from the FF to the SH field as a function of time. The conversion efficiency obtained numerically is approximately equal to what the effective medium model predicts. The conversion efficiency could be dramatically improved, and thresholds lowered, by increasing the number of periods. To conclude this section, we point out that a semiconductor multilayer stack composed of AlGaAs/AlAs has been fabricated and tested according to the criteria that exploit the simultaneous availability of high density of modes and phase matching conditions. The results are in good agreement with the effective index predictions and will soon be published [35].

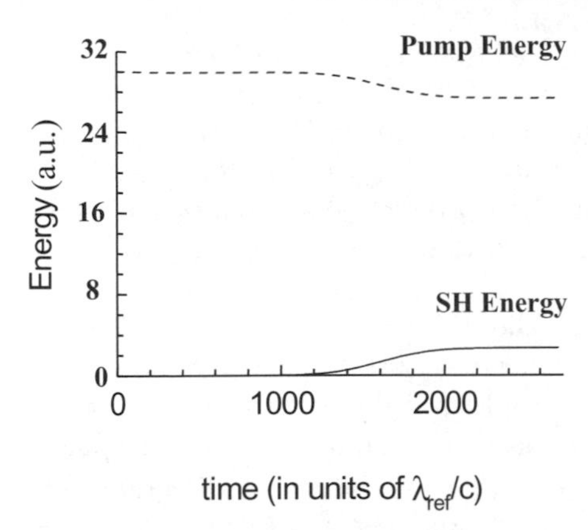

Fig. 6.13. Energy converted from the pump to the SH field as function of the time. λ_{ref} corresponds to a wavelength of 1 µm. We pump with a 2-ps pulse incident from the left, with 40 GW/cm^2 of peak input intensity. The conversion efficiency, given by the amount of energy converted to SH divided the total amount of initial energy in the FF, is approximately $(4/30) \approx 13\%$, in good agreement with that predicted by the effective medium model

6.6 Frequency Down-Conversion

In this section we extend the analysis we began with the introduction of an effective dispersion relation for the finite structure by examining in some detail another important aspect of nonlinear frequency conversion, that is down-conversion processes. Using the effective index approach, we first derive simple phase matching conditions for three-wave coupling, and then test them by numerically integrating the equations of motion in the time domain. We specifically set out to explore the possibility of generating near- and far-infrared radiation through a three-wave mixing process in micron-sized PBG structures because there is a conspicuous absence of small-volume, reliable, coherent radiation sources in that range. The amplification of sum-frequency generation was recently reported [36] for pumps that were tuned near the band edge of a one-dimensional structure. In [36] the authors show that the increase in the density of modes near the band edge can lead to the generation of 400 nm radiation at a rate that was approximately one order of magnitude higher compared to the case where the pumps where not tuned near the band edge. However, no particular care was exercised in that case to make sure that phase matching conditions that we describe in [19] were satisfied. Therefore, we predict that if the scheme we propose for phase matching is implemented in a one-dimensional geometry, enhancements will be much larger.

As we have seen, the typical 1-D structure that we consider consists of alternating layers of high and low index materials. We assume that two pump fields, for example, at ω_1 ($\lambda = 1$ micron) and ω_2 ($\lambda = 1.5$ microns), are incident normally on the structure, and that their interaction mediated by a $\chi^{(2)}$ process generates a down-converted signal at a third frequency ω_3 ($\lambda = 3$ microns). Initially there is no ω_3 signal. The specific wavelengths that we use here are not crucial, and are only used for illustrative purposes. It will become clear later, as we also showed previously [19], that both material and geometrical dispersion can be combined to produce exact phase matching conditions for a variety of situations, a fact that highlight the flexibility of these devices.

The approach that we pursue is as follows: first, we design a multilayer stack, and determine its linear phase properties using the effective index formulation. We seek appropriate tuning of all the fields with respect to their respective band edges in order to simultaneously optimize and combine phase matching conditions with high DOM [19]. Then, we integrate the nonlinear, coupled equations of motion in the time domain. For clarity and completeness, we briefly outline both the effective index approach and the integration method that we use.

6.7 Effective Index Method

We now review a method based on a formulation of the effective dispersion relation that is derived from the phase properties of the transmission function [22] that allows us to evaluate the effective refractive index for a multilayered structure. We note that this method yelds results that are nearly identical to the more general approach highlighted in the previous sections. Using the matrix transfer method [19,37], we define a general transmission function for any structure as follows:

$$t = x + iy = \sqrt{T}e^{i\varphi_t}\,, \tag{6.24}$$

where $\sqrt{T}$ is the transmission amplitude, $\varphi_t = \tan^{-1}(y/x) \pm m\pi$ is the total phase accumulated as light traverses the medium, and m is an integer number. In analogy with the propagation in a homogeneous medium, we can express the total phase associated with the transmitted field as

$$\varphi_t = k(\omega)D = (\omega/c)n_{\mathrm{eff}}(\omega)D\,, \tag{6.25}$$

where $k(\omega)$ is the effective wave vector, and n_{eff} is the effective refractive index that we attribute to the layered structure whose physical length is D. As described in [19], we obtain the following expression of the effective index of refraction:

$$\widetilde{n}_{\mathrm{eff}} = \frac{c}{\omega D}\left[\varphi_t - \frac{i}{2}\ln(x^2+y^2)\right]\,. \tag{6.26}$$

Equation (6.26) suggests that at resonance, where $T = x^2 + y^2 = 1$, the imaginary part of the index is identically zero. We can also define the effective index as the ratio between the speed of light in vacuum and the effective phase velocity of the wave in the medium. We have $k(w) = \frac{\omega}{c}\widetilde{n}_{\rm eff}(\omega)$. This is the effective dispersion relation of the finite structure. For periodic structures, the phase matching conditions are automatically fulfilled if the fields are tuned at the right resonance peaks of the transmission spectrum. Using the formalism introduced in [22], the expression for the effective index for the N-periods finite structure can be recast as follow [19]:

$$\widetilde{n}_{\rm eff} = \frac{c}{\omega {\rm Na}} \left\{ \tan^{-1}\left[z \tan(N\beta)\cot(\beta)\right] + Int\left[\frac{N\beta}{\pi} + \frac{1}{2}\right]\pi \right\}, \qquad (6.27)$$

where β is the Bloch's phase for an infinite structure having the same identical unit cell as the finite structure in question. Equation (6.27) contains additional information regarding the location of the resonances where phase-matching for a three wave mixing (TWM) process can occur. For illustration purposes, we consider a 20-period, mixed half-wave/eighth-wave periodic structure. We choose this arrangement because it allows easy tuning of all the fields near their respective band edges, thus allowing us to simultaneously access a high density of modes for all fields. We note that this arrangement is not unique, in that higher or lower order band edges can be combined to yield the phase matching conditions, within the context of the effective index approach outlined in [19]. We assume that two pump fields are tuned at $\omega_3 = 3\omega$, $\omega_2 = 2\omega$; the interaction via a $\chi^{(2)}$ process then generates a down-converted signal at a third frequency $\omega_1 = \omega$ ($\lambda = 3$ microns). First, we tune the field of frequency ω_1 at the first resonance near the first order band edge; this insures a high density of modes. We have [19]:

$$\beta_1 = \frac{\pi}{N}(N-1). \qquad (6.28)$$

Then, we tune the field ω_2 at the first resonance near the second order band edge, once again securing a high density of modes; we have:

$$\beta_2 = \frac{\pi}{N}(2N-1). \qquad (6.29)$$

We now impose the phase-matching condition for the TWM process, namely:

$$K_3(\omega_3) - K_2(\omega_2) - K_1(\omega_1) = 0. \qquad (6.30)$$

Substituting (6.28) and (6.29) into (6.30), we obtain

$$\beta_3 = \frac{\pi}{N}(3N-2), \qquad (6.31)$$

which is the value of the Bloch's phase that will correspond to the field at frequency ω_3. This means that phase-matching conditions will be satisfied for this structure if the thickness of the layers are combined with material dispersion such that the first pump field (3ω) is tuned to the second resonance

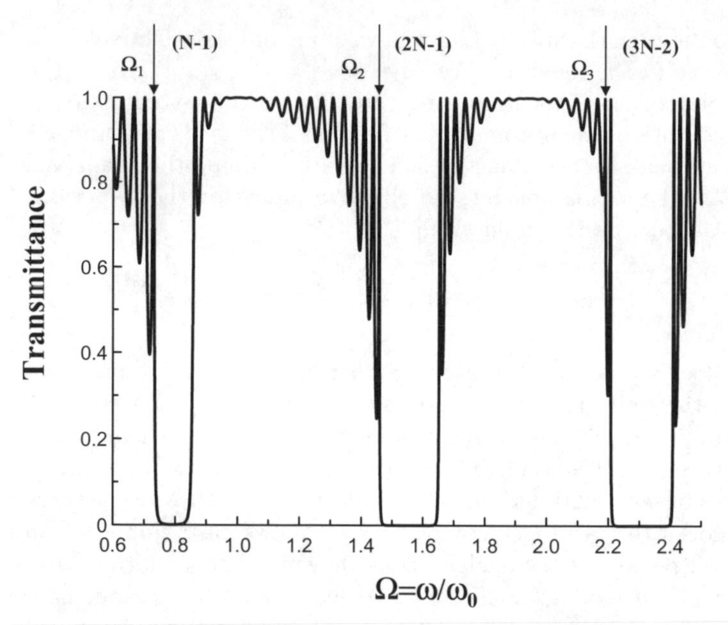

Fig. 6.14. Transmittance vs normalized frequency for a 20 period mixed half-wave/eigth-wave structure. We use $n_1 = 1$ for the low index material at all frequencies. We have introduced a small amount of dispersion, typical of semiconductors, for example, such that: $n_2(\omega_3) = 1.502$, $n_2(\omega_2) = 1.475$, and $n_2(\omega_1) = 1.4285714$. The fields are tuned near their respective band edges as follows: ω_1 at the (N-1)th resonance, ω_2 at the (2N-1)th resonance, and ω_2 is tuned to the (3N-2)th resonance, to satisfy exact phase matching conditions for the three-wave mixing process

away from the low frequency band edge of the third order gap (3N-2). Here $3N$ identifies the gap order, while -2 fixes the frequency 2 resonances away from the band edge. The second pump and the generated fields are then tuned to the first resonance near their respective band edges. In Fig. 6.14 we depict the transmission spectrum of the structure we are considering, and also point out the relative tuning of all the fields. The caption contains the structure's data. We emphasize that in order to be able to tune the fields as specified we have used both the natural dispersion of the material and the geometrical dispersion of the structure. Thus, the dispersion induced by the geometry of the structure, for a particular layer thickness, can be used to compensate the natural dispersion of material as shown in the Fig. 6.14.

6.8 Equations of Motion and Pulse Propagation Formalism

The crystal is composed of 40 dielectric layers and the index of refraction alternates between a high and a low value. For simplicity, we assume normal dispersion, with $n_2(\omega) = 1.4285$, $n_2(2\omega) = 1.475$, $n_2(\omega) = 1.502$, and $n_1 = 1$ (air) at all frequencies. We assume we are operating in a region where losses can be neglected, and the nonlinear material is distributed in the high index layers only, with $\chi^{(2)} \approx 100\,\text{pm/V}$. We note that this particular choice of the active layer location is not crucial, and has been chosen because at the low frequency band edges the fields localize in the high index layers. Therefore, one might envision a scheme for tuning the pumps at the high frequency band edge, which would require relocation of the nonlinear material to those layers. From a computational point of view, one may choose any configuration.

For a reference wavelength λ_0, corresponding to a frequency ω_0, the layers have thicknesses $a = \frac{\lambda_0}{8n_1}$ and $b = \frac{\lambda_0}{2n_2}$, respectively. A range of frequencies is reflected, as shown in Fig. 6.14, where we plot the transmittance as a function of the scaled frequency $\Omega = \omega/\omega_0$, where $\omega_0 = 2\pi/\lambda_0$. We now describe in more details the process of arriving at coupled nonlinear equations of motion that are then integrated in the time domain. We begin by writing Maxwell's equation for the total electric field, in Gaussian units:

$$\frac{\partial^2}{\partial z^2}E(z,t) - \frac{n^2(z)}{c^2}\frac{\partial^2}{\partial t^2}E(z,t) = \frac{4\pi}{c^2}\frac{\partial^2}{\partial t^2}P_{NL}(z,t)\,. \tag{6.32}$$

Here, P_{NL} is the total nonlinear polarization. Without loss of generality, the fields can arbitrarily (and conveniently) be decomposed as follows:

$$E(z,t) = \sum_{j=1}^{j=3} \mathcal{E}_j(z,t)e^{i(k_j z - w_j t)} + c.c. \tag{6.33}$$

and

$$P(z,t) = \sum_{j=1}^{j=3} \mathcal{P}_j(z,t)e^{i(k_j z - w_j t)} + c.c. \tag{6.34}$$

This decomposition highlights the fields' angular frequencies. Then we assume that $\omega_1 = \omega$, $\omega_2 = 2\omega$, $\omega_3 = \omega_1 + \omega_2$ with $k_1 \equiv k = \frac{\omega}{c}$, $k_2 = 2\omega/c$, and $k_3 = 3\omega/c$. The nonlinear polarization takes the following form:

$$\begin{aligned} P_{\mathrm{NL}}(z,t) = \chi^{(2)}E^2(z,t) = 2\chi^{(2)} \Big[&\{\mathcal{E}_1^*\mathcal{E}_2 + \mathcal{E}_2^*\mathcal{E}_3\}\, e^{i(kz-wt)} \\ &+ \left\{\mathcal{E}_1^*\mathcal{E}_3 + \frac{1}{2}\mathcal{E}_1^2\right\} e^{i(2kz-2wt)} + \mathcal{E}_1\mathcal{E}_2 e^{i(3kz-3wt)} \Big] + c.c.\,, \end{aligned} \tag{6.35}$$

where $\mathcal{E}_1$, $\mathcal{E}_2$ and $\mathcal{E}_3$ are the fields at the frequencies ω_1, ω_2, and ω_3, respectively. While we can assume initial left- or right-propagating pump pulses (2ω and 3ω), the generated signal at frequency ω is initially zero everywhere. The direction of propagation of the spontaneously generated field, and the exact nature of the quasi-standing wave inside the structure, are dynamically determined by the nature of the initial and boundary conditions, pump frequency tuning with respect to the band edge, and the distribution of nonlinear dipoles inside the structure. We are also considering propagation in the presence of large index discontinuities, and so we retain all second-order spatial derivatives. However, we assume that pulse envelopes have a duration that is always much greater than the optical cycle, thus allowing the application of the SVEAT only [5–8,38]. The equations of motion for the fields can then be derived by substituting (6.33)–6.35) into (6.32), and by applying the SVEAT. The choice of wave vectors that we highlighted above corresponds to an initial condition consistent with pump fields initially propagating in free space, located away from the structure. Any phase modulation effects that ensue from propagation, i.e., multiple reflections and nonlinear interactions, are fully accounted for in the dynamics of the field envelopes, which maintain their general form. The inclusion of all second-order spatial derivatives in the equations of motion means that reflections are accounted for to all orders, without any approximations. Details on the propagation method can be found in [5–8,38]. Therefore, assuming that pulses never become so short as to violate SVEAT (usually this means a few tens of optical cycles, if propagation distances are on the order of pulse width, as in our case. Neglecting the second-order temporal derivative leads to pulse spreading if the pulse is allowed to propagate over long enough distances), neglecting all but the lowest order temporal contributions to the dynamics, and using the nonlinear polarization expansions of (6.35), we have:

$$n_{2\Omega}^2(\xi)\frac{\partial}{\partial\tau}\mathcal{E}_{2\Omega}(\xi,\tau) = \frac{i}{8\pi\Omega}\frac{\partial^2}{\partial\xi^2}\mathcal{E}_{2\Omega}(\xi,\tau) - \frac{\partial}{\partial\xi}\mathcal{E}_{2\Omega}(\xi,\tau)$$
$$+i\pi\left[n_{2\Omega}^2(\xi)-1\right]2\Omega\mathcal{E}_{2\Omega} + i16\pi^2\Omega\chi^{(2)}(\xi)\left\{\mathcal{E}_{\Omega}^*\mathcal{E}_{3\Omega}+\frac{1}{2}\mathcal{E}_{\Omega}^2\right\} \quad (6.36)$$

and

$$n_{3\Omega}^2(\xi)\frac{\partial}{\partial\tau}\mathcal{E}_{3\Omega}(\xi,\tau) = \frac{i}{12\pi\Omega}\frac{\partial^2}{\partial\xi^2}\mathcal{E}_{3\Omega}(\xi,\tau) - \frac{\partial}{\partial\xi}\mathcal{E}_{3\Omega}(\xi,\tau)$$
$$+i\pi\left[n_{3\Omega}^2(\xi)-1\right]3\Omega\mathcal{E}_{2\Omega} + i24\pi^2\Omega\chi^{(2)}(\xi)\mathcal{E}_{\Omega}\mathcal{E}_{2\Omega} \quad (6.37)$$

and

$$n_{\Omega}^2(\xi)\frac{\partial}{\partial\tau}\mathcal{E}_{\Omega}(\xi,\tau) = \frac{i}{4\pi\Omega}\frac{\partial^2}{\partial\xi^2}\mathcal{E}_{\Omega}(\xi,\tau) - \frac{\partial}{\partial\xi}\mathcal{E}_{\Omega}(\xi,\tau)$$
$$+i\pi[n_{\Omega}^2(\xi)-1]\Omega\mathcal{E}_{2\Omega} + i8\pi^2\Omega\chi^{(2)}(\xi)\left\{\mathcal{E}_{\Omega}^*\mathcal{E}_{2\Omega}+\mathcal{E}_{2\Omega}^*\mathcal{E}_{3\Omega}\right\}. \quad (6.38)$$

From (6.35)–(6.38) it is apparent that since we are considering harmonics of a fundamental field, there will be multiple contributions to the nonlinear source terms which might otherwise be ignored in the usual rotating wave approximation. In (6.38)-(6.39), $\Omega = \omega/\omega_0$, $\xi = z/\lambda_0$, and $\tau = \mathrm{ct}/\lambda_0$. The spatial coordinate z has been conveniently scaled in units of λ_0; the time is then expressed in units of the corresponding optical period. As we will see

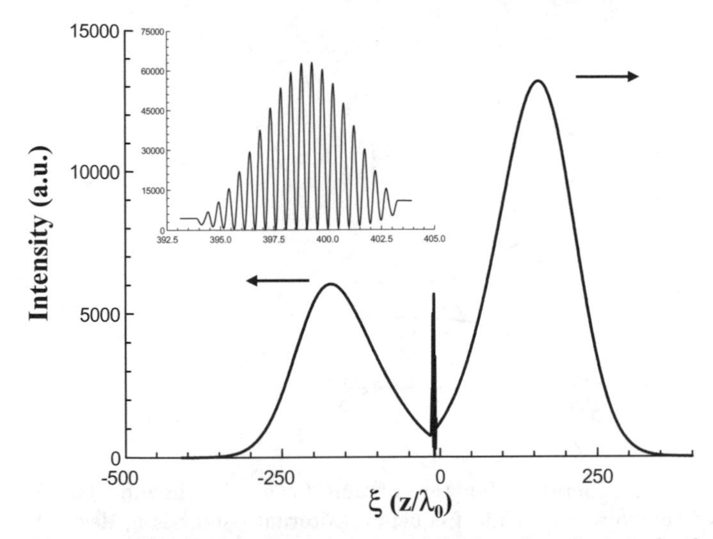

Fig. 6.15. Snapshot of the generated field at $\lambda = 3\,\mu\text{m}$ in both forward and backward directions. The field leaves the structure as the incident pulses exit to the right, and the interaction comes to an end. Inset: field intensity profile inside the structure at the instant the peaks of the incident pulses reach the structure

below, forward and backward field generation can occur. Equations (6.36)–(6.38) are then integrated using a modified beam propagation method based on the split-step algorithm [5–8,38].

In Fig. 6.15 we depict the generated field at frequency ω. The figure shows that pulses are generated in both the forward and backward directions. The inset shows the field distribution inside the structure at the instant the peak of the incident pulse(s) has reached the center of the structure, and it strongly resembles the linear field profile calculated at the same frequency using the matrix transfer method in the linear case. This follows from the fact that we are dealing with deep gratings, where index discontinuities can be of order unity. As a result, no noticeable band shifts occur [19]. In Fig. 6.16 we plot the total energy as a function of time for all the fields, as the incident pulse traverses the structure. We define the total energy as:

$$W_j(\tau) = \frac{1}{4\pi} \int_{-\infty}^{\infty} n_j^2(\xi) \left|\mathcal{E}_j(\xi,\tau)\right|^2 d\xi \,. \tag{6.39}$$

We note that both the fundamental and second harmonic fields experience gain at the expense of the third harmonic signal; that the third harmonic pump pulse becomes significantly depleted during a single pump pass; and that the total energy remains conserved, as expected. We stress the fact that with a $\chi^{(2)}$ of 100 pm/V and input pump field intensities of order of 100 MW/cm^2 we are able to reach the pump depletion regime with a structure approximately 10 microns (for $\lambda_0 = 1$ micron) in length. Increasing structure

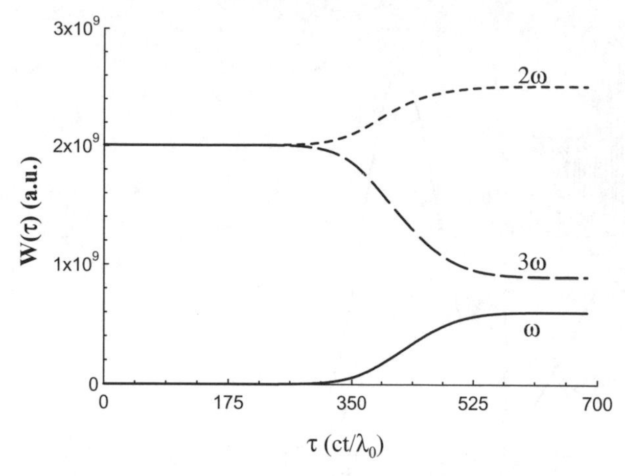

Fig. 6.16. Electromagnetic energy as function of time for all the three fields. We note: **a** depletion of the third harmonic frequency (alternate short-long dashes); **b** amplification of the second harmonic frequency (dashes); and **c** generation of the fundamental signal (solid line)

length can further decrease pump thresholds as a result of a corresponding increase of the density of modes for each of the pump fields.

In Fig. 6.17 we compare the energy output of our structure with that of a structure with the same length of an exactly phase matched, dispersionless bulk medium. We also compare with the energy output of an unmatched bulk material of the same length, and whose refractive indices at ω_1, ω_2, and ω_3 are taken to be $n(\omega_1) = 1.4285714$, $n(\omega_2) = 1.475$, $n(\omega_3) = 1.502$, i.e., no phase matching conditions of any kind are introduced. The conversion efficiency obtained with the 20 period structure is approximately two orders of magnitude larger compared to the conversion efficiency of the unmatched bulk, and approximately 10 times better than the conversion efficiency of *an exactly phase-matched bulk* whose index of refraction is chosen to be $n = 1.475$ at all frequencies, thanks to the simultaneous availability of phase matching conditions and field localization inside the structure. These results are obtained using incident pulses whose duration in approximately 160 optical cycles, which corresponds to approximately 1 ps. Our calculations also show that the conversion efficiency drops significantly if the phase matching conditions shown in Fig. 6.14 are not satisfied. The amount of dispersion we introduced in our sample in order to achieve phase matching conditions is typical of the dispersion present in ordinary semiconductor materials (5–10%), as our Fig. 6.12 shows.

The dashed curve in Fig. 6.17 is calculated using slightly different pump tuning conditions that result in another possible phase-matching scheme, and highlights the flexibility of the system. Referring to Fig. 6.14, tuning the third and second harmonic fields to the (3N-3) and (2N-2) resonances causes the

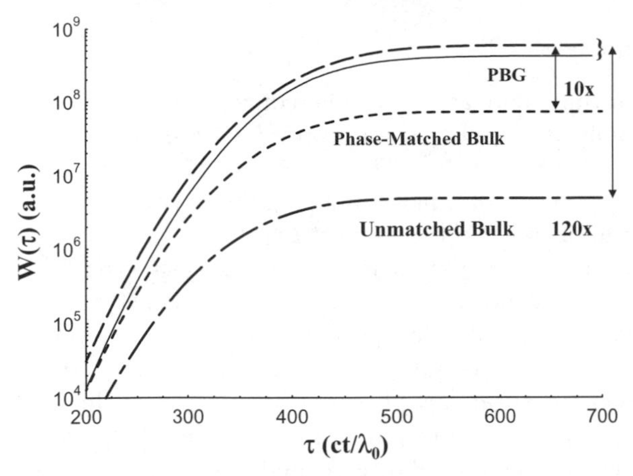

Fig. 6.17. Electromagnetic energy as function of time for the fields generated from the PBG structure depicted in Fig. 6.1 (solid line), and an exactly phase matched bulk of the same length (short dashes) with a refractive index $n = 1.475$. We observe that the conversion efficiency is enhanced by approximately one order of magnitude. We also show the energy attainable with the same length of unprocessed, unmatched bulk material, i.e., the refractive indices for the fields are those specified for the high index layers of Fig. 6.1 (lower curve, with alternating long-short dashes). The higher dashed curve represents the energy output by tuning the fields to the $(3N - 3)$, $(2N - 2)$ and $(N - 1)$ resonances. In this case, an unusual set of circumstances, i.e., simultaneously phase matched fundamental-second harmonic interaction, and three wave mixing process, cause slightly better conversion efficiency, even if the density of modes is smaller for the pump fields as depicted in Fig. 6.15. In both cases the PBG performs approximately 100 times better compared to the unmatched bulk

phase matching condition for generation at the fundamental frequency to occur at the (N-1) resonance. The figure suggests that while the density of modes is smaller for both pump fields, the conversion efficiency increases due to the following, highly unusual set of circumstances: the conditions are right for a double phase-matched interaction. The first is between the fundamental and second harmonic fields, while the second is satisfied for all the fields, according to (6.28). Clearly, these conditions take on a special meaning for up-conversion processes, simultaneous phase-matched second and third harmonic generation in particular. These conditions were first discussed by Akhmanov [37] for three-wave mixing in a phase matched bulk material. However, to our knowledge the ability to achieve simultaneous phase matching conditions for all fields is impossible in bulk materials. The use of geometrical dispersion in the structure to compensate for material dispersion has essentially solved a long-standing problem that can lead to a new class of efficient, compact, nonlinear frequency conversion devices.

6.9 Conclusions

In summary, we have shown that there are nontrivial conceptual, qualitative and quantitative differences between energy and group velocities in structures of finite length, as exemplified in our (6.18). These considerations have naturally lead us to develop the concept of a new tunneling time, which we call the electromagnetic analogue of Smith's tunneling time, that can be useful to understand the limits and meaning of what is referred to as superluminal pulse propagation under general conditions. The only requirements for the validity of our theory are that the scattering potential should be real, and the bandwidth of the incident pulses should be much narrower than a typical resonance bandwidth near the band edge, e.g., [5]. We note that the last requirement is also necessary for the definition of a phase time [20]. However, the tunneling time predicted by our (6.6) is formally and conceptually not the same as the phase tunneling time, by a measure that depends on the interplay between electric and magnetic components. These differences may be accentuated depending on the circumstances, i.e., structure length, frequency, boundary conditions.

We have developed a useful effective index model that can be used to describe the effective dispersive properties of finite structures and it that can be useful also for the study of nonlinear interactions in finite PBG structures [19]. In particular, we have analyzed the properties of nonlinear quadratic interactions near the photonic band edge by studying the linear properties of finite PBG structures and by introducing the concept of an effective medium. We arrive at nonlinear coupled-mode equations that are formally similar to the equations for quadratic interactions in bulk materials, scaled by the appropriate coupling coefficients that define the multilayer stack. These equations predict energy conversion rates that are in agreement both with those predicted by an integration of the coupled wave equations in the time domain and with the experimental results.

We have shown a new method to achieve efficient nonlinear infrared parametric generation in one-dimensional PBG structures. We predict conversion efficiencies well in excess of two orders of magnitude compared to bulk materials of the same length, and approximately one order of magnitude compared to an exactly phase matched bulk medium. We take advantage of the simultaneous availability of high field localization inside a multilayer stack when tuned near band edge resonances, and the engineering of exact phase matching conditions brought about by a combination of material and geometrical dispersion. In our view, these unusual circumstances make PBG structures the best candidates for micron-sized nonlinear frequency converters based on quadratic nonlinearities.

Appendix

The definition of DOM given in our (6.3) can be justified in analogy with the quantum mechanical properties of massive particles. The probability that a quantum particle can be found with a momentum between $\hbar k$ and $\hbar(k+\delta k)$ is given by

$$dP(k) = \Psi^*(k)\Psi(k)dk\,, \tag{6.40}$$

where $\Psi(k)$ is the wave function of the quantum particle in the momentum representation. Mapping (6.40) into the frequency domain can be accomplished with knowledge of the dispersion relation. That is, if $k = k(\omega)$, the mapping $P(k) \longrightarrow p(\omega)$ and $\Psi(k) \longrightarrow \psi(k)$ leads to:

$$dp(\omega) = \psi^*(\omega)\psi(\omega)\rho_\omega d\omega\,, \tag{6.41}$$

where $\rho_\omega = \frac{dk}{d\omega}$ is the DOM. From (6.41) we formally obtain the expression for the DOM in the following form:

$$\widetilde{\rho}_\omega = \frac{\frac{dp(\omega)}{d\omega}}{c|\psi|^2}\,, \tag{6.42}$$

where $\widetilde{\rho}_\omega$ is the dimensionless DOM expressed in units of 1/c: $\rho_\omega = \frac{1}{c}\widetilde{\rho}_\omega$. Note that $dp(\omega)$ is only proportional, and not equal, to the probability that a quantum particle can be found with an energy between $\hbar\omega$ and $\hbar(\omega+\delta\omega)$ and because in general the mapping of the functions from k-space into the frequency domain via the dispersion relation does not correspond to a unitary transformation.

Following Bohm [21], in the case of classical electromagnetic fields, i.e., when many quanta of light are excited in a coherent state and the classical limit is approached [40], we may resort to the electromagnetic energy density which in this case plays a role analogous to that played by the probability density in the case of massive, quantum particles. In fact, in the classical limit, the electromagnetic energy density U_ω is proportional to the mean number of photons in the range from ω to $(\omega+\delta\omega)$ [21]. Keeping these considerations in mind, we express the DOM for the finite 1-D PBG structure as the spatially averaged electromagnetic energy density, which in our case plays a role analogous $dp(\omega)/d\omega$, as in (6.42). The normalization factor found in (6.42), $c|\psi|^2$, is then replaced by the energy density of the incident field $|E_\omega^I|^2$. We therefore write the dimensionless DOM as:

$$\widetilde{\rho} \equiv \frac{\frac{1}{2L}\int_0^L \left[\epsilon_\omega(z)\left|E_\omega\right|^2 + \frac{c^2}{\omega^2}\left|\frac{dE_\omega}{dz}\right|^2\right]dz}{|E_\omega^I|^2}\,. \tag{6.43}$$

Introducing the dimensionless field distribution inside the PBG structure as $\Phi(z) = E_\omega(z)/E^I_\omega$, we obtain:

$$\widetilde{\rho} \equiv \frac{1}{2L}\int_0^L \left[\epsilon_\omega(z)\,|\Phi_\omega|^2 + \frac{c^2}{\omega^2}\left|\frac{d\Phi_\omega}{dz}\right|^2\right] dz\,, \tag{6.44}$$

from which our definition in (6.3) in the text follows.

Acknowledgments. Two of us (G.D. and M.C.) are grateful to the US Army European Research Office for financial support. JWH was supported by NSF grant ECS-9630068. This work was also supported by the OPEN Esprit project ANLM. We also thank Neset Akozbek and Omar El Gwary for helpful discussions relating to this work.

References

1. P. Yeh, *Optical Waves in Layered Media John* (Wiley & Sons, New York Chichester 1988); E. Yablonovitch: Phys. Rev. Lett. **58**, 2059 (1987); S. John: Phys. Rev. Lett. **58**, 2486 (1987); J. Maddox: Nature **348** 481 (1990); E. Yablonovitch and K.M. Leung: Nature **351**, 278 (1991); J. Martorell and N.M. Lawandy: Phys. Rev. Lett. **65**, 1877 (1990); *Development and Applications of Materials Exhibiting Photonic Band Gaps*, edited by C.M. Bowden, J.P. Dowling, and H.O. Everitt, special issue of J. Opt. Soc. Am. B **10**, 279 (1993); J. Mod. Opt. **41**, 171 (1994), special issue on photonic band gaps, edited by G. Kurizki and J.W. Haus; J.D. Joannopoulos, R.D. Mead, and J.N. Winn, *Photonic Crystals* (Princeton University Press, Princeton 1994)
2. J.A. Armstrong, N. Bloembergen, J. Ducuing, and P.S. Pershan: Phys. Rev. **127**, 1918 (1962)
3. N. Bloembergen and A.J. Sievers: Appl. Phys. Rev. Lett. **17**, 483 (1970)
4. M. Scalora, J.P. Dowling, C.M. Bowden, and M.J. Bloemer: Phys. Rev. Lett. **73**, 1368 (1994)
5. M. Scalora, J.P. Dowling, M.J. Bloemer, and C.M. Bowden: J. Appl. Phys. **76**, 2023 (1994)
6. J.P. Dowling, M. Scalora, M.J. Bloemer, and C.M. Bowden: J. Appl. Phys. **75**, 1896 (1994)
7. M. Scalora, R.J. Flynn, S.B. Reinhardt, R.L. Fork, M.D. Tocci, M.J. Bloemer, C.M. Bowden, H.S. Ledbetter, J.M. Bendickson, J.P. Dowling, and R.P. Leavitt: Phys. Rev. E **54**, 1078 (1996)
8. M. Scalora, M.J. Bloemer, A.S. Manka, J.P. Dowling, C.M. Bowden, R. Viswanathan, and J.W. Haus: Phys. Rev. A **56**, 3166 (1997)
9. M. Scalora, M.J. Bloemer, A.S Pethel, J.P. Dowling, C.M. Bowden, and A.S. Manka: J. of Appl. Phys. **83**, 2377 (1998)
10. G. D'Aguanno, E. Angelillo, C. Sibilia, M. Scalora, and M. Bertolotti: J. Opt. Soc. Am. B **17**, 1188 (2000)
11. J.C. Knight, T.A. Birkes, P.S. Russell, and J.P. De Sandro: J. Opt. Soc. Am. A **15**, 748 (1998)

12. E. Yablonovitch: Microwave Journal **Vl**, 66 (1999)
13. J. Martorell, R. Vilaseca, and R. Corbalan: Appl. Phys. Lett. **70**, 702 (1997); J. Martorell, R. Vilaseca, and R. Corbalan: Phys. Rev. A **55**, 4520 (1997); J. Martorell, R. Vilaseca, and R. Corbalan: J. Opt. Soc. Am. B **15**, 2581 (1998)
14. C. Simonneau, J.P. Debray, J.C. Harmand, P. Vidakovic, D.J. Lovering, and J.A. Levenson: Opt. Lett. **23**, 1775 (1997)
15. A.V. Balakin, D. Boucher, V.A. Bushev, N.I. Koroteev, B.I. Mantsyzov, P. Masselin, I.A. Ozheredov, and A.P. Shurinov: Opt. Lett. **24**, 793 (1999)
16. W. Chen and D.L. Mills: Phys. Rev. Lett. **58**, 160 (1987); J.E. Sipe and H.G. Winful: Opt. Lett. **13**, 132 (1988); S. John and N. Akozbek: Phys. Rev. Lett. **71**, 1168 (1993); C.M. de Sterke, D.G. Salinas and J.E. Sipe: Phys. Rev. E **54**, 1969 (1996); B.J. Eggleton, R.E. Slusher, C.M de Sterke, P.A. Krug, and J.E. Sipe: Phys. Rev. Lett. **76**, 160 (1996)
17. C. Conti, S. Trillo, and G. Assanto: Phys. Rev. Lett. **78**, 2341 (1997)
18. G. D'Aguanno, M. Centini, C. Sibilia, M. Bertolotti, M.Scalora, M. Bloemer, and C.M. Bowden: Opt. Lett. **24**, 1663 (1999)
19. M. Centini, C. Sibilia, M. Scalora, G. D'Aguanno, M. Bertolotti, M. Bloemer, C.M. Bowden, and I. Nefedov: Phys. Rev. E **60**, 4891 (1999)
20. A.E. Siegman: Phys. Rev. A **39**, 1253 (1989); P.T. Leung, W.M. Suen, C.P. Sun, and K. Young: Phys. Rev. E **57**, 6101 (1998)
21. D. Bohm *Quantum Theory* (Dover Publications, 1989) pp. 283–295; ibid. pp. 9–10; ibid. pp. 97–98
22. J.M. Bendickson, J.P Dowling, and M. Scalora: Phys. Rev E **53**, 4107 (1996); C. Sibilia, I. Nefedov, M. Bertolotti; and M. Scalora: J. Opt. Soc. Am. B 1947 (1998)
23. A.M. Steinberg, P.G. Kwiat, and R.T. Chiao: Phys. Rev. Lett. **71**, 708 (1993)
24. L. Brillouin, *Wave Propagation and Group Velocity* (Academic Press, New York 1960)
25. R. Loudon: J. Phys. A **3**, 283 (1970)
26. C.G.B. Garret and D.E. McCumber: Phys. Rev. A **1**, 305 (1970)
27. S. Chu and S. Wong: Phys. Rev. Lett. **48**, 738 (1982)
28. J. Peatross, S.A. Glasgow, and M. Ware: Phys. Rev. Lett. **84**, 2370 (2000)
29. L.J. Wang, A. Kuzmich, and A. Dogariu: Nature **406**, 277 (2000)
30. A. Yariv and P. Yeh, *Optical Waves in Crystals* (Wiley & Sons, New York Chichester 1984)
31. P.S.J. Russell: J. Mod. Optics **38**, 1599 (1991); C.M. De Sterke and J.E. Sipe: J. Opt. Soc. Am B **6**, 1722 (1989)
32. For a review article on the definition of tunneling times see: E.H. Hauge and J.A. Stoevneng: Rev. Mod. Phys **61**, 917 (1989); see also R.T. Chiao and A.M. Steinberg, *Progress in Optics*, edited by E. Wolf, Vol. XXXVII, (1997) p. 345
33. J.W. Haus, R. Viswanathan, M. Scalora, A.G. Kalocsai, J.D. Cole, and J. Theimer: Phys. Rev. A. **57**, 2120 (1998) (and references therein)
34. C.M. de Sterke and J.E. Sipe: Phys. Rev. A **38**, 5149 (1988); A. Arraf and C.M. de Sterke: Phys. Rev. E **58**, 7951 (1998) (and references therein)
35. Y. Dumeige et al., to appear in Applied Physics Lett.
36. V. Balakin, V.A. Bushuev, B.I. Mantsyzov, P Masselin, I.A. Ozheredov, and A.P. Shkurinov: JETP Letters **70**, 725 (1999)
37. M. Born and E. Wolf, *Principles of Optics*, 5th Edn., (Oxford, Pergamon Press)
38. M. Scalora and M.E. Crenshaw: Opt. Commun. **108**, 191 (1994)

39. S.A. Akhmanov and R.V. Khokhlovnce, (Gordon and Breach, New York 1972). Translated from the original Russian edition: Problemy Nelineinoi Optiki, Academy of Sciences of the USSR, 1964
40. R. Loudon, *The Quantum Theory Of Light*, 2nd Ed., (Oxford University Press, 1984)

7 Theory of Parametric Photonic Crystals

P.D. Drummond and H. He

7.1 Introduction

A photonic crystal is defined by its band-gap, but it is possible to have more than one bandgap. In this chapter, we focus on cases of two bandgaps where one has twice the frequency of the other. This allows coupling to occur between the bandgaps, provided the dielectric itself has a quadratic or parametric nonlinearity. Since Bloembergen's time, it has been known that a modulated quadratic nonlinearity can enhance the prospects for phase-matching in frequency-doubling. Here, we consider phase-matching obtained through a modulated linear refractive index, giving the advantage that shortened interaction lengths are possible. In the following sections the equations for electromagnetic propagation in a grating structure with a parametric nonlinearity are derived using Maxwell's equation as well as a coupled mode Hamiltonian analysis.

Parametric solitons [1], or simultaneous solitary wave ('simulton' [2]) solutions are proved to exist in photonic crystals both by direct numerical integration, and by using the effective mass approximation (EMA) to transform the equations to the coupled equations describing a nonlinear parametric waveguide [3]. Exact one dimensional numerical solutions in agreement with the EMA solutions are also given. Direct numerical simulations show that the solutions have similar types of stability properties to the bulk case [4], providing the carrier waves are tuned to the two Bragg resonances – and the pulses have a width in frequency space less than the band-gap. These equations describe a physically accessible localized nonlinear wave that is stable in up to $(3+1)$ dimensions. Possible applications include photonic logic and switching devices [5].

Solitons have a generic character in nonlinear, dispersive wave equations, as non-spreading solutions in which the dispersion and nonlinearity effects cancel each other. As such, these entities are basic to nonlinear wave equations. Most solitons observed in optical systems come from the interplay of the cubic nonlinear refractive index and linear dispersion. They are described by the nonlinear Schröedinger equation, and are stable in one space dimension, but unstable in higher dimensions. A different type of soliton is possible in a nonlinear parametric photonic crystal, where large effective dispersion occurs near the two resonant band-gaps. The advantage of the nonlinear parametric bandgap medium compared to a cubic, or nonlinear refractive index medium is that solitons can occur with lower input powers and in higher space dimensions than conventional nonlinear Schröedinger equation gap solitons.

The physics behind this resides in the simple fact that the nonlinear phase shift is much larger at low intensities for parametric nonlinear materials, since it scales as E^2, not E^3. The price that is paid for these nice properties is that the underlying equations generically are non-integrable. Fortunately, there are special cases which can be integrated. Arguments using topological properties or Sobolev inequalities can also be used, and have powerful consequences. In all cases, computer simulation is relatively straightforward.

Thus, parametric solitons certainly have a strong fundamental appeal. As interesting as they are, it would be useful to also have potential applications, which are provided by the relatively strong interactions of these solitons. This means that it might be feasible to use parametric solitons for a number of photonics applications, including spatial guiding, regeneration and logic switching. A number of these ideas have already been experimentally demonstrated. Perhaps the most remarkable is the experimental demonstration of an ultra-fast AND-gate with 100 fs resolution, in accord with theoretical predictions [6]. This photonic logic gate, which utilizes soliton formation dynamics at a nonlinear interface, is the fastest ever demonstrated. Other experimentally observed parametric solitons in quadratic media include both continuous wave spatial solitons [7], and spatio-temporal solitons [6], also in accordance with theoretical predictions [8–22].

There are a number of material requirements for solitons, especially that of group-velocity matching, and the necessity of having dispersions of identical sign in both the signal and its harmonic [4]. Using bulk nonlinear crystals is also not feasible in many real-world photonics applications where compactness and ease of integration is prized above all. This is also a problem for other, more straightforward device applications – like simple frequency doublers, or parametric amplifiers that could be used for wide-band signal regeneration.

One route to miniaturization is the idea of using photonic crystals or Bragg gratings to greatly increase the effective dispersion, and thereby reduce the interaction length. The strong dispersion of photonic crystals or Bragg gratings has been confirmed experimentally [23], which makes them an ideal candidate for the formation of parametric simultons with short interaction distances. Using a Bragg grating also helps to solve other problems that occur with conventional pulsed frequency conversion. Material group-velocity matching is no longer essential with band-gaps, since the group velocity is strongly modified by the band-gap. In addition, it is always possible to choose branches of the dispersion relation that give anomalous dispersion at both wavelengths, thus creating an ideal spatio-temporal soliton environment [3,24,25]. This could lead to 'simultons' with short formation distances that are stable in higher dimensions.

7.2 The Parametric Gap Equations

In this section, one-dimensional parametric gap equations are derived from the Maxwell equations for an isotropic medium. This treatment omits higher dimensions and tensor properties. These can be be added later as necessary, but in essence the important physics is already seen in the simpler case treated here.

7.2.1 One-dimensional Maxwell Equations

The Maxwell equation describing the propagation of a linear polarized electric field $\boldsymbol{E}$ and displacement field $\boldsymbol{D}$ in one space dimension can be written as,

$$\frac{\partial^2 \boldsymbol{E}}{\partial z^2} = \mu_0 \frac{\partial^2 \boldsymbol{D}}{\partial t^2} \tag{7.1}$$

where μ_0 is the vacuum permeability and $\boldsymbol{D}$ and $\boldsymbol{E}$ are perpendicular to the propagation direction z. The displacement field $\boldsymbol{D}$ is defined as usual:

$$\boldsymbol{D} = \epsilon_0 \boldsymbol{E} + \boldsymbol{P} . \tag{7.2}$$

Since we shall later use the displacement field as the canonical variable, it is most natural to expand the electric field in a power series in the displacement field, using inverse permittivity tensors as expansion coefficients [26]. This is also a very useful approach in birefringent or spatially-varying media, as it is the displacement field that obeys the transversality condition of $\nabla \cdot \boldsymbol{D} = 0$, not the electric field. The most general expansion of this type is [28]:

$$\epsilon_0 E_i(t, \boldsymbol{x}) = \sum_n \int d^n \boldsymbol{t} \eta^{(n)}_{i,i_1 \ldots i_n}(\boldsymbol{t}, \boldsymbol{x}) D_{i_1}(t - t_1, \boldsymbol{x}) \ldots D_{i_n}(t - t_n, \boldsymbol{x}) . \tag{7.3}$$

For simplicity, we suppose that $\eta^{(1)}$ is rotationally symmetric. Given a quasi-monochromatic electric field in a second harmonic generation process, the solutions to the Maxwell's equation for frequencies near ω_1 and $\omega_2 = 2\omega_1$ can be written as:

$$\boldsymbol{D} = \sum_{j=1,2} \boldsymbol{\mathcal{D}}^{(j)} + c.c. = \sum_{j=1,2} \sum_{\pm} \boldsymbol{e}^{(j)} A_{j\pm}(z,t) e^{\pm ij\bar{k}z - i\omega_j t} + c.c. , \tag{7.4}$$

where $\boldsymbol{e}^{(j)}$ are the polarizations, the sign $p = \pm$ represents right or left propagation, and $j\bar{k}$ is a reference wave-vector used to describe the corresponding carrier field, which will be chosen to correspond to the grating periodicity. The actual wave-number of the complex propagating fields $\boldsymbol{\mathcal{D}}^{(j)}$ can differ from $j\bar{k}$, since the envelope function can vary in space; thus, $\bar{k}$ is simply chosen as being close to the relevant wave-number, in the case of detunings. The inverse permittivity at each wavelength is treated as a distinct term, in order to allow for material dispersion effects over the large frequency difference that separates the two carrier frequencies. Thus, we introduce:

$$\eta_j(z) = \eta^{(1)}(\omega_j, z) = \int dt \eta^{(1)}(t, z) e^{-i\omega_j t} . \tag{7.5}$$

This can be related to the usual electric permittivity $\varepsilon(z, \omega_j)$ and refractive index $n(z, \omega_j)$ by the relationships:

$$\varepsilon(z, \omega_j) = [1 + \chi^{(1)}(z, \omega_j)]\varepsilon_0 = n^2(z, \omega_j)\varepsilon_0 = 1/\eta_j(z) . \tag{7.6}$$

The nonlinear coefficient is related to the Bloembergen expansion coefficient by $\eta^{(2)} = -\varepsilon_0 \chi^{(2)} \eta_1^2 \eta_2$. We also assume that material dispersion in the vicinity of either carrier frequency is small enough to be neglected in comparison to the induced dispersion of the Bragg grating, along with any higher order nonlinear effects.

7.2.2 Bragg Grating Structure

The spatial variation of the refractive index is chosen to corresponding to a Bragg grating structure with:

$$\eta_j(z) = \overline{\eta}_j \left[1 + \Delta_j(z)\right] , \tag{7.7}$$

so that $\varepsilon(z, \omega_j) \approx \overline{\varepsilon}_j \left[1 - \Delta_j(z)\right]$. Thus, the refractive index at each frequency is a periodic function with period d, and the reference or 'mean' refractive index at each carrier frequency is defined as $\bar{n}_j = 1/\sqrt{\varepsilon_0 \overline{\eta}_j}$. We then choose the reference wavenumber in the carrier so that $\bar{k} = \pi/d$. The notation indicates that dispersion may result in different refractive indices at the two different carrier frequencies of the fundamental and second-harmonic fields. Thus, the depth of spatial modulation present at frequency ω_1 is possibly different to the spatial modulation present at the second-harmonic frequency ω_2. In general, we will consider $\Delta_j(z)$ to be small. Larger modulation depths would require a more complicated analysis, without changing the general physical conclusions. Thus, each carrier wave at these distinct frequencies will experience a modulated refractive index. However, the resonant properties of Bragg gratings means that each will only interact strongly with a Fourier component having half the respective carrier wavelength. Expanding in a Fourier series, results in a Fourier series for $\Delta_j(z)$:

$$\Delta_j(z) = \sum_l \Delta_{jl} \exp(2il\bar{k}z) + cc. \tag{7.8}$$

In general, the Δ_{jl} are complex coefficients, and $\bar{k} = \pi/d = 2\pi\bar{n}_1/\lambda_1$, where λ_1 is the free-space wave-length of the fundamental field. More complicated types of grating can be treated, but this is sufficient to treat the gap soliton.

For reference purposes later, we note that if Δ_{jl} is real and positive, the refractive index has a $4\cos(2l\bar{k}z)$ modulation, with a minimum in the refractive index occurring at the origin.

The simplest grating would just be a super-position of two sinusoidal waves – only terms with Δ_{j1} and Δ_{j2} in (7.8) would then occur. It is possible, for example, that the two coefficients might have opposite signs with the two modulations out of phase with each other. For this reason, the relative phase of the complex coefficients in the Fourier expansion needs to be taken into account. It is highly desirable that bandgaps occur at both the carrier wavelengths, in order to optimize the nonlinear coupling between the waves. This is because the resonant modes of the linear Maxwell equations near a bandgap are quasi-standing waves – and these will only couple strongly to other standing waves.

A more practical modulation of refractive index would have square-wave shape. This could occur by using laser interference patterns in a saturated photo-sensitive material, thus giving rise to higher order harmonics. This type of grating has simultaneous bandgaps at higher harmonics of the optical carrier frequency.

7.2.3 Grating Equations

To solve the Maxwell equations in the presence of the nonlinear response and grating terms, we substitute (7.8) into the Maxwell equation (7.1):

$$\mu_0 \frac{\partial^2}{\partial t^2} \boldsymbol{\mathcal{D}}^{(1)} = \frac{\partial^2}{\partial z^2} \left[\eta_1(z) \boldsymbol{\mathcal{D}}^{(1)} + 2\eta^{(2)} : \boldsymbol{\mathcal{D}}^{(1)*} \boldsymbol{\mathcal{D}}^{(2)} \right] \tag{7.9}$$

$$\mu_0 \frac{\partial^2}{\partial t^2} \boldsymbol{\mathcal{D}}^{(2)} = \frac{\partial^2}{\partial z^2} \left[\eta_2(z) \boldsymbol{\mathcal{D}}^{(2)} + \eta^{(2)} : \boldsymbol{\mathcal{D}}^{(1)} \boldsymbol{\mathcal{D}}^{(1)} \right] . \tag{7.10}$$

Assuming that $\mathcal{A}_j$ are slowly evolving, we neglect group velocity dispersion terms involving $\partial^2 A_j/\partial t^2$ and $\partial^2 A_j/\partial z^2$ in the slowly-varying envelope and rotating-wave approximations. Define $k_j = k(\omega_j) = \omega_j \sqrt{\mu_0/\overline{\eta}_j} = \omega_j/v_j$ as the linear wave-number, and assume $\delta k_j = k_j - j\bar{k} \ll j\bar{k}$. Thus, any terms involving δk_j^2 or first order differentiation and δk_j can be neglected, for they are much smaller than these terms involving only δk_j or first order differentiation, which we call first order terms. Hence we have, after re-arranging terms and neglecting higher-order spatial Fourier components:

$$\begin{aligned}
i[\frac{1}{v_1}\frac{\partial}{\partial t} + \frac{\partial}{\partial z} + \delta k_1]\mathcal{A}_{1+} &= \kappa_1 \mathcal{A}_{1-} + \chi_D \mathcal{A}_{1+}^* \mathcal{A}_{2+} \\
i[\frac{1}{v_1}\frac{\partial}{\partial t} - \frac{\partial}{\partial z} + \delta k_1]\mathcal{A}_{1-} &= \kappa_1^* \mathcal{A}_{1+} + \chi_D \mathcal{A}_{1-}^* \mathcal{A}_{2-} \\
i[\frac{1}{v_2}\frac{\partial}{\partial t} + \frac{\partial}{\partial z} + \delta k_2]\mathcal{A}_{2+} &= \kappa_2 \mathcal{A}_{2-} + \chi_D \mathcal{A}_{1+}^2 \\
i[\frac{1}{v_2}\frac{\partial}{\partial t} - \frac{\partial}{\partial z} + \delta k_2]\mathcal{A}_{2-} &= \kappa_2^* \mathcal{A}_{2+} + \chi_D \mathcal{A}_{1-}^2 \, ,
\end{aligned} \tag{7.11}$$

where the coupling coefficients are given by:

$$\chi_D = k_1 \varepsilon_1 \boldsymbol{e}^{(1)*} \eta^{(2)} : \boldsymbol{e}^{(1)} \boldsymbol{e}^{(2)} = -k_1 \varepsilon_0 \boldsymbol{e}^{(1)*} \chi^{(2)} : \boldsymbol{e}^{(1)} \boldsymbol{e}^{(2)} / (\varepsilon_1 \varepsilon_2) \,, \quad (7.12)$$

$$\kappa_j = j \bar{k} \Delta_{jj} / 2 \,. \quad (7.13)$$

To simplify the equations, we can always choose the phases of $\boldsymbol{e}_j$ so that χ_D is real. We neglect group velocity dispersion (GVD) of the medium, as this is usually much smaller than the gap dispersion. However, we have included the difference in group velocity between the two carriers, as this is not always negligible.

7.2.4 One-dimensional Dispersion Relation

Without the grating structure, the dispersion relation (frequency ω versus wave-number k) would be a continuous straight line in the vicinity of the gap frequency. Introducing a grating structure opens a gap at the edge of the Brillouin zone for each of the carrier frequencies. Inside each gap, light is completely Bragg-reflected, resulting in strong dispersion near the critical gap frequencies. The eigenmodes of the Maxwell equations in the vicinity of the gap are also modified. Instead of the usual plane-waves, the eigenmodes become modulated quasi-standing waves, with a pure standing wave being achieved exactly at the wave-number for resonance. In this case there are two possible standing wave solutions with different spatial phases (i.e., $\sin(\bar{k}z)$ and $\cos(\bar{k}z)$ solutions). These are familiar in electronic bandgap theory, and have the usual property that one has an eigenfrequency above, and the other below the gap center frequency. Propagation of a free field with a frequency in the gap region is, of course, prohibited. However, in the presence of the nonlinear medium, it is possible that propagation can occur, due to nonlinear phase-shifts.

The dispersion relation of the one-dimensional Maxwell equations in the slowly-varying envelope approximation can be obtained by studying the linear part of the gap parametric equations, (7.11). Neglecting the nonlinear terms, we have the following linear coupled equations,

$$\begin{aligned} i[\frac{1}{v_j}\frac{\partial}{\partial t} + \frac{\partial}{\partial z} + \delta k_j]\mathcal{A}_{j+} &= \kappa_j \mathcal{A}_{j-} \\ i[\frac{1}{v_j}\frac{\partial}{\partial t} - \frac{\partial}{\partial z} + \delta k_j]\mathcal{A}_{j-} &= \kappa_j^* \mathcal{A}_{j+} \,. \end{aligned} \quad (7.14)$$

Following standard techniques, we introduce a vector for the right and left propagating fields:

$$\boldsymbol{\mathcal{A}}_j(z,t) = \begin{bmatrix} \mathcal{A}_{j+} \\ \mathcal{A}_{j-} \end{bmatrix} \,. \quad (7.15)$$

Inserting the ansatz,

$$\boldsymbol{\mathcal{A}}_j(z,t) = \boldsymbol{f}_j(Q)e^{i(Qz-\Omega_j t)}\,, \quad j=1,2\,, \tag{7.16}$$

into the linear equation, one obtains the following algebraic equations,

$$\begin{bmatrix} \Omega_j/v_j - Q - \delta k_j \kappa_j \\ -\kappa_j^* \Omega_j/v_j + Q + \delta k_j \end{bmatrix} \begin{bmatrix} f_{j+} \\ f_{j-} \end{bmatrix} = 0\,. \tag{7.17}$$

Solving the above equation for Q, we have two eigenvalues corresponding to $s = \pm 1$,

$$\Omega_j^{(s)}(Q) = v_j(s\sqrt{Q^2 + |\kappa_j^2|} - \delta k_j)\,, \quad j = 1,2\,. \tag{7.18}$$

If $\delta k_j = 0$, this equation becomes the dispersion relationship found in conventional band gap systems. The width of each band-gap in the dispersion relation is then given by

$$\Delta\Omega_j = \Omega_j^+(0) - \Omega_j^-(0) = 2v_j|\kappa_j|\,. \tag{7.19}$$

Substituting the above solutions, (7.18), into the linear equation, we obtain two sets of normalized eigenvectors, corresponding to linear propagation above and below the bandgap:

$$\boldsymbol{f}_j^{(s)}(Q) = \frac{\left[\kappa_j, s\sqrt{|\kappa_j^2| + Q^2} - Q\right]^T}{\sqrt{2\left(|\kappa_j^2| + Q^2 - sQ\sqrt{|\kappa_j^2| + Q^2}\right)}}\,, \tag{7.20}$$

where the sign $s = -1$ corresponds to the lower branch and $s = 1$, to the upper branch.

The physical meaning of the s parameter is clearest in the case of $Q = 0$, which is in the center of the bandgap in k-space. Suppose, for simplicity, that the $j-th$ refractive index has a local minimum at $z = 0$. This corresponds to a $\cos(2\bar{k}jz)$ modulation of the inverse permittivity, so that $\kappa_j > 0$. In general, we can always choose the origin so that this is true for at least one of the carrier frequencies, although it might not be true for both. In this case the upper branch ($s = 1$) has a symmetric form also, with a $\cos(\bar{k}jz)$ modulation. The lower branch has an antisymmetric $\sin(\bar{k}jz)$ mode function. We can understand this physically if we argue that a lower energy – and hence a lower frequency – is obtained when the maximum field intensity in space corresponds to the maximum refractive index in space, which means the maximum dielectric polarization, thus reducing the energy via the $-d \cdot E$ interaction. In the next section, we see that this is justified by the Hamiltonian theory of the dielectric + radiation system.

The dispersion relation, (7.18) is depicted in Fig. 7.1, for one of the bandgaps. Because of the gap, linear propagation is not allowed if the frequency

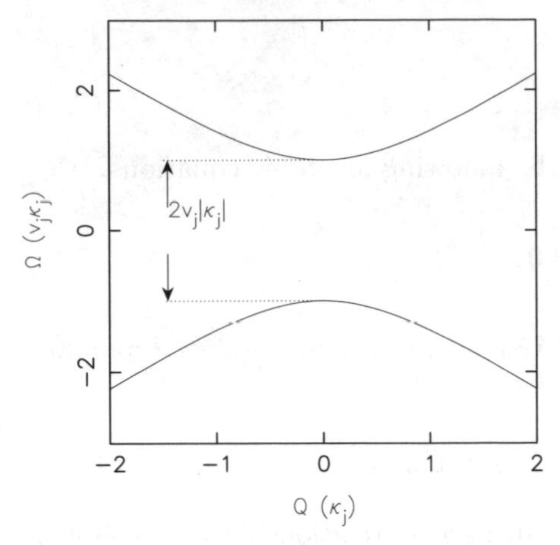

Fig. 7.1. Dispersion relation for light with wave-numbers around $k_j = j\pi/d$ where d is the period of the grating. The width of the gap is given as: $\Delta\Omega = \Omega^+ - \Omega^- = 2v_j|\kappa_j|$. For simplicity, Ω is plotted with a unit of $v_j\kappa_j$ while Q is plotted with a unit of κ_j

shift from the gap center is small, i.e., $|\Omega| < v_j|\kappa_j|$. This results in strong dispersion, so that

$$\frac{d^2\Omega_{js}}{dQ^2} = \frac{sv_j|\kappa_j^2|}{(Q^2 + |\kappa_j^2|)^{3/2}} \,. \tag{7.21}$$

The dispersion ($\omega'' = d^2\omega/dk^2$) of typical nonlinear optical media is of the order of $10^{-1}\mathrm{m}^2/s$ at a wavelength of 1 µm. In the case of lithium niobate, the corresponding refractive index is $\bar{n} \simeq 2.5$. Assuming 0.2% refractive index modulation, so that $\Delta_{jj} = 0.002$, we find that κ_j is of the order of $10^4\mathrm{m}^{-1}$. This indicates a maximum band-gap dispersion of $v_j/|\kappa_j| \simeq 10^4\mathrm{m}^2/s$. Such a strong dispersion gives an advantage in reducing soliton formation length significantly and has been recently confirmed experimentally [23]. The large gap-induced dispersion provides a justification for our neglect of material group-velocity dispersion (GVD) effects in these equations.

7.3 Hamiltonian Method

We will now use a Hamiltonian method to describe propagation near the center of the band-gap, thus giving an approximate pair of coupled second order equations. This theory also permits us to describe waveguide mode-structures and their coupling in a very simple way. We note that the Hamiltonian for a nonlinear medium is most readily written using the displacement field as a canonical variable, as pointed out by Hillery and Mlodinow [26]. From dielectric theory, the complete Hamiltonian [27,28] can be written as:

$$H = H_0 + H_{\mathrm{int}} \,, \tag{7.22}$$

where the linear and nonlinear terms are, respectively,

$$H_0 = \int \sum_j \left(\eta(\boldsymbol{x}, \omega_j) \left| \boldsymbol{\mathcal{D}}^{(j)} \right|^2 \boldsymbol{\mathcal{D}}^{(j)*} + \frac{1}{\mu_0} \left| \boldsymbol{\mathcal{B}}^{(j)} \right|^2 \right) d^3\boldsymbol{x}$$

$$H_{\text{int}} = \frac{1}{3} \int \boldsymbol{D} \cdot \eta^{(2)} : \boldsymbol{D}\boldsymbol{D} d^3\boldsymbol{x} \,. \tag{7.23}$$

Here $\boldsymbol{x} = (\boldsymbol{r}, z)$, while $\boldsymbol{r} = (x, y)$ are the transverse coordinates and z is the longitudinal coordinate. We also define [26], $\boldsymbol{D} = \sum_{j=1,2}(\boldsymbol{\mathcal{D}}^{(j)} + cc.)$ and $\boldsymbol{B} = \sum_{j=1,2}(\boldsymbol{\mathcal{B}}^{(j)} + cc.)$ as previously. In the previous section, the electric field is expressed as a pair of anti-propagating waves based on the coupled mode theory. Here, we start from the general form of these fields in a three dimensional medium.

7.3.1 Linear Part of the Hamiltonian and Mode Expansion

Although we only intend to get classical equations, it is convenient to follow the normalization of the standard canonical procedure [27,29,30], which can be used both classically and quantum mechanically. We introduce the dual potential $\overrightarrow{\Lambda}$ [26], defined so that $\boldsymbol{\mathcal{D}} = \nabla \times \overrightarrow{\Lambda}\Lambda$ and $\boldsymbol{\mathcal{B}} = \mu\partial_t \overrightarrow{\Lambda}$, which is useful for obtaining a nonlinear Hamiltonian theory. We express $\overrightarrow{\Lambda}$ in terms of mode-functions $\overrightarrow{\Lambda}^{(j\boldsymbol{n})}$ normalized over a length L. Periodic boundary conditions are imposed at $x_i = 0, x_i = L$, where $x_i = x, y, z$, and we later take $L \to \infty$. The mode expansion is taken to be:

$$\overrightarrow{\Lambda}(t, \boldsymbol{x}) = \frac{1}{\sqrt{L}} \sum_{j\boldsymbol{n}} a_{j\boldsymbol{n}}(t) \overrightarrow{\Lambda}^{(j\boldsymbol{n})}(\boldsymbol{x}) \,. \tag{7.24}$$

Here $a_{j\boldsymbol{n}} \propto e^{-i\omega_{j\boldsymbol{n}}t}$ and $\boldsymbol{n}$ refers to all mode indices. We choose $\overrightarrow{\Lambda}^{(j\boldsymbol{n})}$ to be transverse.

Hence, we can write:

$$\boldsymbol{\mathcal{D}} = \frac{1}{\sqrt{L}} \sum_{j\boldsymbol{n}} \nabla \times \overrightarrow{\Lambda}^{(j\boldsymbol{n})} a_{j\boldsymbol{n}} \,,$$

$$\boldsymbol{\mathcal{B}} = \frac{-i}{sqrtL} \sum_{j\boldsymbol{n}} \mu_0 \omega_{j\boldsymbol{n}} \overrightarrow{\Lambda}^{(j\boldsymbol{n})} a_{j\boldsymbol{n}} \,. \tag{7.25}$$

For evaluating mode functions, we consider just the linear Maxwell equations. Substituting the above expansion into the Maxwell equations gives a wave-equation, and hence an eigenvalue equation for mode-functions $\overrightarrow{\Lambda}^{(j\boldsymbol{n})}$ with eigenvalues $\omega_{j\boldsymbol{n}}$, as follows:

$$\nabla \times \left(\eta_j(\boldsymbol{x}) \nabla \times \overrightarrow{\Lambda}^{(j\boldsymbol{n})} \right) = \mu\omega_{j\boldsymbol{n}}^2 \overrightarrow{\Lambda}^{(j\boldsymbol{n})} \,. \tag{7.26}$$

Expanding $\mathcal{D}$ and $\mathcal{B}$ in terms of modes that satisfy the above equation in the linear part of the Hamiltonian gives the result:

$$H_0 = \sum_{jn} \hbar\omega_{jn} a^*_{jn} a_{jn} \,. \tag{7.27}$$

In order to use this result, we need to develop approximate expressions for the mode eigenvalues and eigenfunctions, in the typical case of weakly guided waves in one and two dimensions – as well as a full three-dimensional bulk crystal layered structure. In a weakly guided waveguide, we assume that the inverse permittivity is given to first order by:

$$\eta(\boldsymbol{x},\omega) \approx \bar{\eta}(\omega)[1 - \Delta_{jr}(\boldsymbol{r}) - \Delta_j(z)] \,. \tag{7.28}$$

Next, suppose that $\boldsymbol{n} = (\boldsymbol{m}, n)$, and each spatial mode is approximately factorizable into the form:

$$\overrightarrow{\Lambda}^{(jn)}(\boldsymbol{r}, z) = \boldsymbol{u}^{(jm)}(\boldsymbol{r})\Lambda_{jn}(z) \,, \tag{7.29}$$

where the direction of the $\overrightarrow{\Lambda}^{(jn)}(\boldsymbol{x}, z)$ totally depends on $\boldsymbol{u}^{(jm)}(\boldsymbol{r})$, and $\nabla\Lambda_{jn} \cdot \boldsymbol{u}^{(jm)} \approx 0$. Since $\nabla \cdot \overrightarrow{\Lambda}^{(jn)} = 0$, we have $\nabla \cdot \left(\Lambda_{jn}\boldsymbol{u}^{(jm)}\right) = 0$, which gives $\nabla \cdot \boldsymbol{u}^{(jm)} \approx 0$. This allows us to spilt up the mode frequency into longitudinal and transverse contributions, so that:

$$\omega_{jn} = \omega_{jn} + \Delta\omega_{jm} \,, \tag{7.30}$$

where ω_{jn} is the total longitudinal eigenvalue, as obtained in the previous sections, while $\Delta\omega_{jm}$ is the transverse part.

7.3.2 Transverse Modes

It is possible to solve the equation for $\boldsymbol{u}^{(jm)}$, using the standard techniques for weakly guided waves at the j-th carrier frequency. The transverse modes can also be normalized so that:

$$\int \boldsymbol{u}^{(jm)*} \cdot \boldsymbol{u}^{(jm')} d^2\boldsymbol{r} = \delta_{\boldsymbol{m},\boldsymbol{m}'} \,. \tag{7.31}$$

These modes take different forms, depending on the specific type of waveguide:

- *One dimension*: In the one dimensional case (for example, in a single mode fiber), higher order modes are usually neglected, so:

  $$\boldsymbol{u}^{(jm)} = \boldsymbol{u}^{(j)}(\boldsymbol{r})$$

  where $\boldsymbol{u}^{(j=0)}$ is the zero-th order transverse mode at the j-th carrier frequency.

- *Two dimensions*: In a planar waveguide case (two dimensional), $\boldsymbol{u}^{(jm)}$ can be written in the form:

  $$\boldsymbol{u}^{(jm)} = \boldsymbol{u}^{(j)}(y)e^{ik_{jm}x}/\sqrt{L}$$

  (assuming the waveguide is confined along the y direction and L is the transverse normalization distance), where $\boldsymbol{u}^{(j=0)}$ is the zero-th order transverse mode and k_{jm} is a scalar.
- *Three dimensions*: In a bulk crystal (three dimensional), $\boldsymbol{u}^{(jm)}$ can be written in the form:

  $$\boldsymbol{u}^{(jm)} = \boldsymbol{u}^{(j)}e^{i\boldsymbol{k}_{jm}} \cdot \boldsymbol{r}/L\,.$$

  Here $\boldsymbol{u}^{(j)}$ is the polarization direction, and L^2 is the transverse normalization area.

7.3.3 Longitudinal Modes

We next consider the longitudinal mode equation. We assume a waveguide which is longitudinally modulated in a similar manner as discussed previously. Expanding Λ_{jn} into a Fourier series, we look for modes that are described by a momentum factor Q_n such that:

$$\Lambda_{jn}(z) = \lambda_{jn} \sum_l C_{jn}(k_{nl}) \exp(ik_{nl}z)\,, \tag{7.32}$$

where $k_{nl} = Q_n + l\bar{k}$ and l is an integer.

Solving the longitudinal eigenvalue equation yields the exact values of C_{jn} and the dispersion relationship between $\tilde{\omega}_{jn}$ and Q_n. Any two longitudinal modes of the dual potential must satisfy the very general – and extremely useful and simple – orthonormality requirement [27,29,30]:

$$\int \left(\Lambda_{jn}^* \Lambda_{jn'}\right) dz = \frac{L\hbar\delta_{nn'}}{2\mu_0\omega_{jn}}\,, \tag{7.33}$$

which determines the value of λ_{jn}. We are interested in waves near the Bragg condition, for if the wave vector is too far away from the Bragg condition, the medium acts just like a homogeneous medium. We therefore assume $Q_n \ll \bar{k}$ and $\bar{\eta}\bar{k}_j^2 \approx \mu_0\tilde{\omega}_{jn}^2$.

If we substitute $\omega_{jn} = \omega_j + \Omega_{jn}$ into the above equation, we obtain the same dispersion relationship obtained in the previous section, (7.18), so that:

$$\Omega_{jn} = \Omega_j^{(s)}(Q_n) = v_j(s\sqrt{|\kappa_j|^2 + {Q_n}^2} - \delta k_j)\,, \tag{7.34}$$

where $\delta k_j = k(\omega_j) - j\bar{k}$ as before.

The total eigenvalue can be written as a sum of carrier frequency, transverse frequency and gap frequency:

$$\omega_{jn} = \omega_{jm}^{(s)}(Q_n) = \omega_j + \Delta\omega_{jm} + \Omega_j^{(s)}(Q_n)\,, \tag{7.35}$$

and we can write Λ_{jn} and C_{jn} as,

$$\Lambda_{jn}(z) = \Lambda_j^{(s)}(Q_n, z) = \lambda_j^{(s)}(Q_n) u_{j\Lambda}^{(s)}(Q_n, z) e^{iQ_n z}$$
$$C_{jn}(\pm\bar{k} + Q_n) = C_{j\pm}^{(s)}(Q_n) , \quad (7.36)$$

where the rapidly varying part of the Bragg grating mode function is given by:

$$u_{j\Lambda[D]}^{(s)}(Q, z) = C_{j+}^{(s)}(Q) e^{ij\bar{k}z} + [-] C_{j-}^{(s)}(Q) e^{-ij\bar{k}z} . \quad (7.37)$$

Here $u_{j\Lambda}$ is the dual potential mode function, while u_{jD} is the displacement (or electric) field mode function in a slowly varying envelope approximation. From now on, we shall omit the mode index n on Q_n, since these modes become infinitely closely spaced in the limit of large quantization volume, where L is large.

The coefficients $C_{j+}^{(s)}(Q)$ and $C_{j-}^{(s)}(Q)$ are normalized such that

$$\left(C_{j+}^{(s)}(Q)\right)^2 + \left(C_{j-}^{(s)}(Q)\right)^2 = 1 .$$

Using these conditions, we derive the explicit forms in first order approximation:

$$C_{j\pm}^{(s)}(Q) = \pm f_{j\pm}^{(s)}(Q) , \quad (7.38)$$

where $\boldsymbol{f}_j = (f_{j+}, f_{j-})$ is the the same as in (7.20), in the section on the one-dimensional equation. There is an additional sign correction in the above equation, since the expansion given here is for the dual potential Λ, rather than in terms of the electric or displacement field. Once this is taken into account, the general symmetry properties of the longitudinal modes correspond exactly to those in the one-dimensional case.

The value of $\lambda_j^{(s)}(Q)$ can now be obtained from substituting $\Lambda_j^{(s)}(Q, z)$ into the normalization condition [27,29,30],

$$\frac{1}{L} \int \left[(\Lambda_j^{(s)})^* \Lambda_j^{(s)} \right] dz = \frac{\hbar}{2\mu_0 \omega_{jm}^{(s)}(Q)} . \quad (7.39)$$

Substituting (7.36) into the above equation, we find:

$$\lambda_j^{(s)}(Q) \approx \sqrt{\frac{\hbar}{2\mu_0 \omega_{jm}^{(s)}(Q)}} , \quad (7.40)$$

For later use, we need an approximate expression suitable for evaluating nonlinear interaction terms in the Hamiltonian. We therefore also evaluate this expansion coefficient at the gap center, giving the result:

$$\lambda_j^{(s)}(0) \approx \lambda_j = \sqrt{\frac{\hbar v_j \bar{\epsilon}_j}{2k_j}} \left(1 + O(\Delta)\right) . \quad (7.41)$$

In order to understand the physical properties of these solutions, we recall that these longitudinal mode functions are essentially identical in symmetry to those obtained in the one-dimensional case. Thus, if κ_j is real and positive, the higher energy displacement field solution for $Q = 0$ (which is labeled as $s = 1$), has a $\cos(\bar{k}jz)$ spatial dependence, and is therefore completely symmetric about $z = 0$. This can be understood physically by noticing that the linear Hamiltonian is given by:

$$H_0 = \int \sum_j \left(\eta(\boldsymbol{x}) |\boldsymbol{\mathcal{D}}^{(j)}|^2 + \frac{1}{\mu_0} |\boldsymbol{\mathcal{B}}^{(j)}|^2 \right) d^3\boldsymbol{x} \,. \tag{7.42}$$

In the case that κ_j is real and positive, the inverse dielectric permittivity has a positive $\cos(2\bar{k}jz)$ modulation term, which *increases* the energy of the symmetric $\cos(\bar{k}jz)$ field mode, with $s = 1$; since $\cos^2(\bar{k}jz) = [1 + \cos(2\bar{k}jz)]/2$, while it *reduces* the energy of the anti-symmetric mode with $s = -1$, since $\sin^2(\bar{k}jz) = [1 - \cos(2\bar{k}jz)]/2$.

These energy changes agree precisely with the frequency changes of the mode frequencies worked out from the solutions to the one-dimensional Maxwell equations. While this is as expected, it provides an additional confirmation of the correctness of the the dielectric-radiation Hamiltonian that is used here. This difference in energy is, of course, the physical origin of the band-gap in the dispersion relations.

7.4 The Effective Mass Approximation

We are mostly interested in photon properties near the center of the bandgap region in momentum space, where Q is small, so we further assume here that $Q/\kappa \ll 1$. This expansion is not essential to the problem – we can still write down the Hamiltonian without it – but it greatly simplifies the final equations that are obtained. We first expand the mode frequency around the carrier frequency ω_j. This gives the same dispersion relationship, (7.18), which was obtained from the coupled mode theory. We then expand the resulting expression in a Taylor series up to second order in $Q/|\kappa_j|$:

$$\Omega_j^{(s)}(Q) = \Omega_j^{(s)}(0) + \frac{s\hbar Q^2}{2m_j} \,, \tag{7.43}$$

where the effective mass of the j-th carrier is:

$$m_j = \hbar|\kappa_j|/v_j \,, \tag{7.44}$$

and v_j is the group velocity. The frequency at the band-gap edge is:

$$\Omega_j^{(s)}(0) = (s|\kappa_j| - \delta k_j)v_j \,. \tag{7.45}$$

This is, in fact, the well-known effective-mass approximation (EMA) in solid state physics – although more precisely sm_j is called the effective mass, with opposite signs below and above the band-gap. It should be noticed here that the main effect of material group-velocity terms is to slightly change the curvature of the effective-mass parabola.

The total eigenvalue, $\omega_{j\boldsymbol{n}}$, of the mode equation is the sum of the longitudinal eigenvalue $\omega_j + \Omega_{jn}$, and the transverse eigenvalue, $\Delta\omega_{j\boldsymbol{m}}$. From this relationship we have (near the j-th carrier),

$$\omega_{j\boldsymbol{n}} = \omega_{j\boldsymbol{m}}^{(s)}(Q_n) = \omega_{j\boldsymbol{m}}^{(s)} + \frac{s\hbar Q_n^2}{2m_j}\,, \tag{7.46}$$

where the eigenvalue at the band-gap edge is:

$$\omega_{j\boldsymbol{m}}^{(s)} = \omega_j + \Omega_j^{(s)}(0) + \Delta\omega_{j\boldsymbol{m}}\,. \tag{7.47}$$

It is convenient to work in the coordinate representation. The Hamiltonian can therefore be expressed in terms of the polariton field operators. We introduce an effective dimensionality, $D = 1, 2, 3$. To simplify the analysis, we assume that where there are discrete transverse modes – as in a fiber – only the lowest order mode ($\boldsymbol{m} = (0, 0)$) needs to be considered. In this case, we define $\omega_j^{(s)} = \omega_{j0}^{(s)}$. An envelope for the excitation in the dielectric – which physically is really the polariton density field in D dimension(s) – is defined as:

$$\Psi_j^{(s)}(\boldsymbol{x}) = L^{-D/2} \sum_{\boldsymbol{k}} a_{j\boldsymbol{k}}^{(s)} e^{i\boldsymbol{k}} \cdot \boldsymbol{x}\,, \tag{7.48}$$

where $\boldsymbol{k} = (0, 0, Q)$ in one dimension, $\boldsymbol{k} = (k_{jm}, 0, Q)$ in two dimensions and $\boldsymbol{k} = (\boldsymbol{k}_{jm}, Q)$ in three dimensions.

Substituting the above expression into the linear part of the Hamiltonian, we have:

$$H_0 \approx \sum_{j,s} \hbar \int \left(\frac{s\hbar|\partial_z \Psi_j^{(s)}|^2}{2m_j} + \frac{\hbar|\nabla_\perp^D \Psi_j^{(s)}|^2}{2m_{j\perp}} + \omega_j^{(s)}|\Psi_j^{(s)}|^2 \right) d^D\boldsymbol{x}\,, \tag{7.49}$$

where $\nabla_\perp^D$ is the transverse part of the operator in D dimensions, and $m_{j\perp} = \hbar k_j / v_j$ is the effective transverse mass. The longitudinal and transverse effective masses can have quite different values, especially if the Bragg dispersion is large (i.e., $|\kappa_j|$ is small). Thus, given the parameters quoted in Sect. 7.2.3, the effective masses have the following orders of magnitude: $m_j \sim 10^{-38}$kg, $m_{j\perp} \sim 10^{-35}$kg.

7.4.1 Nonlinear Part of the Hamiltonian

We will only keep the leading terms, when we use the EMA to expand $\mathcal{D}$ for the nonlinear part of the Hamiltonian. Substituting the mode expression into the nonlinear part of the Hamiltonian results in:

$$H_{\text{int}} \approx -\frac{\hbar}{2} \sum_{\boldsymbol{s}} \int \chi(\boldsymbol{s})(\Psi_2^{(s_2)})^* \Psi_1^{(s_1)} \Psi_1^{(s_1')} d^D\boldsymbol{x} + h.c. \tag{7.50}$$

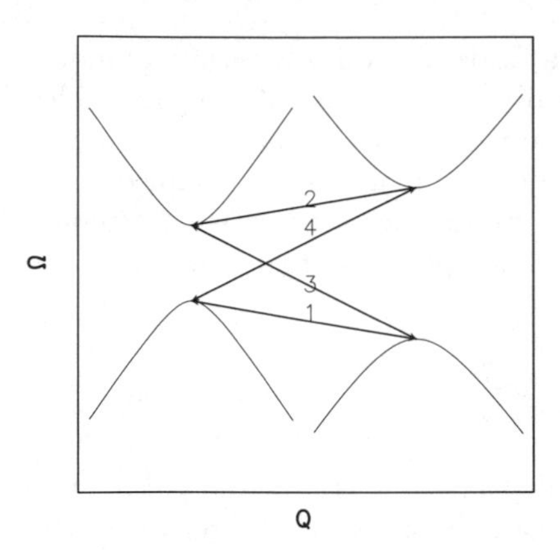

Fig. 7.2. Possible nonlinear couplings between gaps. Coupling 1 and 2 generate bright simultons while coupling 3 and 4 generate dark simultons

We can always assume κ_1 to be real and positive, by shifting the location of the coordinate origin – since only the relative phase between the two gratings is important. The nonlinear coupling then simplifies to:

$$\chi(\boldsymbol{s}) = \frac{i\varepsilon\chi_D v_1}{4}\sqrt{\frac{\hbar k_2 v_2}{\bar{\epsilon}_2 A_c}}(\mathrm{sgn}(\kappa_2) + s_1 s_1' s_2)\,, \tag{7.51}$$

where A_c is an effective mode 'area'. We only consider cases with non-vanishing coupling, and $s_1 = s_1'$, although in a three-wave mixing process, it is possible that $s_1 \neq s_1'$. Thus, the nonlinear part of the Hamiltonian, (7.50), vanishes if the total coupling between gaps is antisymmetric. This condition limits the number of possible couplings. Therefore, these possible couplings between gaps are (1) coupling between lower branches; (2) coupling between upper branches; (3) coupling between upper and lower branches, as illustrated in Fig. 7.2.

Next, suppose that κ_2 is also real, but can have either sign. This would be the case if the overall grating structure was symmetric relative to the origin. We can investigate the possible cases of modes having a nonzero coupling at $Q = 0$, by simply considering whether $(\mathrm{sgn}(\kappa_2) + s_2)$ is zero or not. Thus, it is clear that only the second harmonic mode is restricted in any way, and in this case it is necessary that s_2 has the same sign as κ_2. However, this is precisely the condition for having a symmetric modal solution, from the structure of the eigenfunctions. In summary, while the parity of the fundamental harmonic can have either sign, the parity of the second harmonic must be symmetric. We can understand the physics of this in a very straightforward way. The nonlinear coupling involves the square of the fundamental field, multiplied by the second harmonic. The square of the fundamental field mode is always symmetric whether the mode itself is symmetric or anti-symmetric. This can only give rise to a finite nonlinear coupling if the second-harmonic mode is also symmetric about $z = 0$.

The Hamiltonian approach therefore affords a physically intuitive understanding of the coupling processes. Not only does the use of the gap modes eliminate linear cross-couplings, but it also introduces a powerful symmetry principle in the limit of $Q = 0$; the second harmonic that is coupled must have the same type of symmetry as the product of the two sub-harmonic modes. Because of this, the use of gap modes permits great simplifications even in this nonlinear problem.

7.5 Simulton Solutions

We can now apply the known topological properties [19] of the parametric soliton equations to the bandgap case. From the above Hamiltonian, in the one-dimensional case, we derive classical equations for two coupled waves with symmetries s_1 and s_2:

$$\begin{aligned} \frac{\partial \psi_1^{s_1}}{\partial t} &= i\frac{\hbar}{2m_1^{s_1}}\frac{\partial^2 \psi_1^{s_1}}{\partial z^2} - i\omega_j^{s_1}\psi_1^{s_1} + i\chi(\boldsymbol{s})\psi_2^{s_1}\psi_1^{\dagger s_1} \\ \frac{\partial \psi_2^{s_2}}{\partial t} &= i\frac{\hbar}{2m_2^{s_2}}\frac{\partial^2 \psi_2^{s_2}}{\partial z^2} - i\omega_j^{s_2}\psi_2^{s_2} + i\frac{\chi(\boldsymbol{s})}{2}\psi_1^{2s_1} \,. \end{aligned} \tag{7.52}$$

Transforming the new coupled equations by $\tilde{\psi}_j = \psi_j e^{ij\omega_1^{s_1}t}$, a pair of equations is obtained that is identical to the usual description of a nonlinear dispersive parametric waveguide, with dispersions $\omega_j'' = \hbar/m_j^{s_j}$, and an effective phase mismatch of:

$$\beta = (2\omega_1^{s_1} - \omega_2^{s_2}) = (\delta k_2 + 2s_1|\kappa_1| - s_2|\kappa_2|)v \,. \tag{7.53}$$

From previously known results [19], (7.52) support dark (i.e., topological) solitary waves if $s_1 = -s_2$. Also, bright type solitary waves can occur, if the dispersions have identical sign, i.e., $s_1 = s_2 = s = sgn(\kappa_2)$. We will focus on this case in what follows.

Soliton type solutions to (7.52) can be written as [19]:

$$\begin{aligned} \psi_1 &= \pm|q|\sqrt{\left|\frac{\kappa_1}{\kappa_2}\right|}V_1(z/z_0)e^{i(q-\omega_1)t}/\chi(\boldsymbol{s}) \,, \\ \psi_2 &= s_1|q|V_2(z/z_0)e^{2i(q-\omega_1)t}/\chi(\boldsymbol{s}) \,. \end{aligned} \tag{7.54}$$

where q is an arbitrary parameter describing the (inverse) soliton time-scale, and the corresponding length scale is $z_c = \sqrt{|v/(2q\kappa_1)|}$.

Assuming the condition $Q \ll \kappa_j$, we can expand the mode function $u_{jQ}^s(z)$ into a Taylor's series up to first order about $Q = 0$. Hence, the electric field at $t = 0$ can be expressed as:

$$\mathcal{E}_j(z) = \sum_s \sqrt{\frac{\hbar}{\omega_j}^s 2\bar{n}_j^2\epsilon_0 A}\left(\psi_j^s(z)u_{j0}^s(z) - i\frac{\partial \psi_j^s(z)}{\partial z}\delta u_{j0}^s(z)\right) \,. \tag{7.55}$$

Substituting (7.54) into the above result gives:

$$\mathcal{E}_1 = \pm a_1 \left(V_1 \begin{bmatrix} \mathrm{sgn}(\kappa_1) \\ s \end{bmatrix} - \frac{i}{2|\kappa_1|} \frac{dV_1}{dz} \begin{bmatrix} \mathrm{sgn}(\kappa_1) s \\ -1 \end{bmatrix} \right)$$

$$\mathcal{E}_2 = a_2 \left(s V_2 \begin{bmatrix} 1 \\ 1 \end{bmatrix} - \frac{i}{2|\kappa_2|} \frac{dV_2}{dz} \begin{bmatrix} 1 \\ -1 \end{bmatrix} \right) . \tag{7.56}$$

Here $a_1 = \sqrt{|\kappa_1/2\kappa_2|}|q|e^{i(q-\omega_1)t}/(c\varepsilon\chi_D/\bar{n})$ and $a_2 = |q|e^{2i(q-\omega_1)t}/(c\chi_E/\bar{n})$.

It has been proved that there is a family of one parameter solitary wave solutions of (7.52) [19]. The parameter ρ in this case is:

$$\rho = \left| \frac{\kappa_2}{\kappa_1} \right| \left| 2 - \frac{\beta}{q} \right| . \tag{7.57}$$

Provided that the effective mass approximation is valid, which means that $\kappa_i z_o \gg 1$, each of these known simultaneous solitary waves ('simultons') generates a corresponding bandgap soliton, which has an additional phase modulation when compared with the usual solitons.

For example, in cases with $\rho = 1$, the corresponding solutions can be worked out for the case $\kappa_j > 0$, $s_1 = s_2 = s_3 = 1$, which involves coupling between lower branches. The solutions are:

$$V_1(z) = \frac{3}{\sqrt{2}} \mathrm{sech}^2 \left(\frac{z}{2z_0} \right)$$

$$V_2(z) = \frac{3}{2} \mathrm{sech}^2 \left(\frac{z}{2z_0} \right) . \tag{7.58}$$

Cases that involve cross couplings between the upper branch and the lower branch of different gaps result in dark simulton solutions. These are not available analytically, and therefore must be calculated numerically – as is also necessary for all cases with $\rho \neq 1$.

An experimentally relevant point is that the solutions given already are completely stationary in the laboratory frame. This creates an unexpected problem: how can they be introduced into the bandgap material? In fact, this is easily solved. If the gap structure is fabricated with $m_2 = 2m_1$, there is a symmetry in the equations which allows for moving solutions with an identical form to the stationary ones. Thus, they can be generated at the boundary, and then move into the bulk medium.

The dispersive parametric equations we obtain also support higher dimensional soliton solutions in $(2+1)$ and $(3+1)$ dimensions. These correspond to striped or layered bandgap structures respectively, and are described by adding a transverse Laplacian to each of the earlier propagation equations. Solutions of this type do not appear to exist in conventional ($\chi^{(3)}$) gap solitons. Our Hamiltonian mapping from band-gap to dispersive equations, therefore proves that the parametric bandgap environment is able to support higher-dimensional solitons. Thus, it is possible to obtain parametric gap solitons in up to three spatial dimensions.

In order to demonstrate stability, we have numerically solved the original bandgap Eqs. (7.11), using the effective mass approximation result as an initial condition, in $(1+1)$ and $(2+1)$ dimensions. To provide a suitable perturbation, the (small) imaginary part of the input solution was omitted. In both cases, this perturbation is quickly radiated, and a stable solitary wave is obtained. Our results in Fig. 7.2 show $(2+1)$ dimensional gap simulton propagation, with the complete set of four coupled partial differential equations. After a small initial oscillation, the simulton reaches a steady state, proving the stability of the soliton in higher dimensions.

These characteristics of fast interaction times, low pulse energy and stability in higher dimensions make the gap parametric system an ideal soliton environment for both fundamental physics and applications of solitons.

As a numerical example, we will show briefly that the pulse energy of a one dimensional gap simulton is possibly of the order of pJ. We consider a waveguide made of $LiNbO_3$ and a laser whose free-space wavelength is $\lambda_1 = 1.06\,\mu m$ and whose pulse width is 10 ps. The pulse energy depends on the soliton volume through the volume factors in front of the integral, while the exact dimensionless soliton envelope is included in the integral. This depends on both the dimensionality and the factor γ in the equations themselves, and usually has to be evaluated numerically. We will choose the special case of $\gamma = 1$, where the pulse envelopes have an analytic form.

We use the following typical values: $\chi^{(2)} = 11.9\,\mathrm{pm/V}$ [39] and the average refractive index of the waveguide, $\bar{n} = 2.5$. We assume the refractive index modulation to be 0.2% of $\bar{n}$, at each wavelength, so that $\Delta_{jj} \simeq 0.002$. This gives coupling parameters of $\kappa_1 \simeq 0.001\bar{k} = 0.001 \times 2\pi\bar{n}/\lambda_1 = 1.5 \times 10^4 \mathrm{m}^{-1}$, and $\kappa_2 \simeq 2 \times \kappa_1$. For a 10 ps pulse, we have $z_c \simeq 10^{-11} c/\bar{n} \simeq 1.2\,\mathrm{mm}$, where c is the speed of light in free space. This in turn gives the soliton period or re-shaping time, $q^{-1} \simeq 350\,\mathrm{ps}$. The amplitude of a simulton is therefore of the order $q\bar{n}/(\chi_E c) \simeq 1.2 \times 10^6 \mathrm{Vm}^{-1}$. Using the above values, we find that $W(z_c) \simeq 75\,\mathrm{Jm}^{-3}$. Assuming $\rho = 1$ and $\sigma = 2$, we find the the combined pulse energy is around 40 pJ for an effective waveguide area of $25(\mu\mathrm{m})^2$.

This pulse energy is many orders of magnitude lower than the usual values for the corresponding $\chi^{(3)}$ gap solitons. The energy density requirements are greatly reduced if κ and/or $\chi^{(2)}$ are increased, at fixed z_c. Thus, further reductions in pulse energy can be readily obtained if larger values of κ are used, together with shorter pulse-lengths.

7.5.1 Higher Dimensional Solutions

The dispersive parametric equations we obtain also support higher dimensional soliton solutions [34,19] in $(2+1)$ and $(3+1)$ dimensions. These correspond to striped or layered bandgap structures respectively. Equation (7.11), the one-dimensional coupled Maxwell equation, can be extended to two- or three dimensional structures in the paraxial approximation, by adding a transverse Laplacian to each of the earlier propagation equations. The equations describing the propagation of higher dimensional gap simultons are therefore written as,

$$
\begin{aligned}
i\left[\frac{1}{v_1}\frac{\partial}{\partial t}+\frac{\partial}{\partial z}\right]\mathcal{A}_{1+}+\frac{1}{2\bar{k}}\nabla^2\mathcal{A}_{1+}+\delta k_1\mathcal{A}_{1+} &= \kappa_1\mathcal{A}_{1-}+\chi_E\mathcal{A}_{1+}^*\mathcal{A}_{2+}\\
i\left[\frac{1}{v_1}\frac{\partial}{\partial t}-\frac{\partial}{\partial z}\right]\mathcal{A}_{1-}+\frac{1}{2\bar{k}}\nabla^2\mathcal{A}_{1-}+\delta k_1\mathcal{A}_{1-} &= \kappa_1\mathcal{A}_{1+}+\chi_E\mathcal{A}_{1-}^*\mathcal{A}_{2-}\\
i\left[\frac{1}{v_2}\frac{\partial}{\partial t}+\frac{\partial}{\partial z}\right]\mathcal{A}_{2+}+\frac{1}{4\bar{k}}\nabla^2\mathcal{A}_{2+}+\delta k_2\mathcal{A}_{2+} &= \kappa_2\mathcal{A}_{2-}+\chi_E\mathcal{A}_{1+}^2\\
i\left[\frac{1}{v_2}\frac{\partial}{\partial t}-\frac{\partial}{\partial z}\right]\mathcal{A}_{2-}+\frac{1}{4\bar{k}}\nabla^2\mathcal{A}_{2-}+\delta k_2\mathcal{A}_{2-} &= \kappa_2\mathcal{A}_{2+}+\chi_E\mathcal{A}_{1-}^2\,,
\end{aligned}
\qquad (7.59)
$$

where $\nabla^2=\partial^2/\partial x^2$ for $(2+1)$ dimensions and $\nabla^2=\partial^2/\partial x^2+\partial^2/\partial y^2$ for $(3+1)$ dimensions.

Using the Hamiltonian method to obtain equations in the EMA for this case, once again results in much simpler equations. We can transform the above equation approximately into the reduced form, except with transverse Laplacian terms. This equation has also been analyzed previously, and it has been proved it supports stable simultons in both two- and three dimensions. This indicates that the parametric bandgap environment is also able to support higher-dimensional solitons. Thus, it is possible to obtain parametric gap solitons in up to three spatial dimensions. If $s_2=1$, solutions are cylindrically symmetric and can be obtained exactly using numerical techniques [19]. Such a condition is not necessary satisfied and non-symmetric solutions can be obtained approximately via variational method [4].

An unusual property of the higher-dimensional gap parametric solitons is that they provide an example of a nonlinear, three-dimensional self-confined object. These can even appear stationary in the laboratory frame. Of course, this raises the practical question of how an object of this type could be generated with external laser fields. Apart from inserting a gain medium into the Bragg grating, it is likely that a slightly detuned, and therefore moving soliton would be more practical – since it could then be coupled through a spatial boundary of the nonlinear volume grating. Another practical consideration is the question of losses, which are neglected here. These are likely to be very significant for slowly moving gap solitons, due to the long interaction times with a possibly lossy environment.

As we have calculated previously for the one dimensional gap simultons, the pulse energy of higher-dimensional gap simultons can be estimated similarly. Assuming the same nonlinear material and pulse width, we find that the energy density scaling coefficients are $W(z_c)\simeq 75\mathrm{Jm}^{-3}$ as before. We also find that $r_c\simeq 4\times10^{-5}$m. Assuming cylindrically symmetric solutions, we find that the pulse energy is around 2 nJ for two dimensional case ($\gamma=1$ and width of waveguide $\simeq 5\,\mu$m) and around 55 nJ for the three dimensional case ($\gamma=3$). A larger value of ρ was chosen in the three-dimensional case, as this gives an improved stability. Note that exact phase-matching implies that $\rho = 4$, and this is also stable. In these cases, the dimensionless integrals were carried out numerically, using the shooting technique [19] to obtain the pulse envelopes. Another possibility is to use a variational method [4], which allows the integrals to be evaluated analytically to a good approximation.

The total energy for $D > 1$ depends strongly on the radial parameter r_c, which scales as $z_c\sqrt{\kappa}$ at fixed wavelength. This means that the aspect ratio of the pulse changes as κ increases, which increases the ratio of radius to length – changing the soliton from an elongated 'cigar' at small κ, to a more spheroidal shape at large κ. Despite this, it is still favorable to increase κ, if a lower pulse energy is required at a given pulse length z_c. To reduce the pulse length and pulse energy simultaneously, it is most favorable to fix the product of $z_c\sqrt{\kappa}$ – which determines the radius – while reducing z_c and increasing κ, until the EMA limit of $\kappa z_c \simeq 1$ is nearly reached. Note that a given value of the product $z_c\sqrt{\kappa}$ determines both the corresponding energy density factor *and* the radius scale in higher dimensions; these are modified, of course, by the solutions to the corresponding dimensionless equations.

7.5.2 Stability

The important question of the stability of the parametric band-gap simultons is investigated here by numerically solving the original bandgap Eq. (7.11), using the effective mass approximation solution from the Hamiltonian method as an initial condition. The numerical simulation is based on the implicit central-difference (split-step) Fourier-transform scheme [37]. While the simplified equations are known to be stable, these are not exact equations for the gap-soliton problem. Thus, there are possible additional instabilities that may arise from invalidation of the EMA, group velocity mismatch effects, or other internal properties of the band-gap simultons. This investigation does not address the issue of whether the paraxial and slowly-varying envelope approximations are themselves always applicable here.

To provide a suitable perturbation, the (small) imaginary part of the input solution was usually omitted. If the imaginary part is included, a steady propagation of gap simultons is observed when the simulton is stable, and even an unstable solution can survive in a metastable fashion for a relatively long time. By omitting this, the initial condition is observed to either evolve towards a stable wave, or to rapidly decay. Unless stated otherwise, the inputs used were obtained from the simplified EMA analysis, which in most cases gives an excellent approximate starting profile. An exception to this was the test of stability at small z_c, where we cannot expect the EMA to be even approximately valid. In these cases, we used the exact one-dimensional initial solutions described in the previous section, with the imaginary part included.

For simplicity, (7.11) is treated as if it were a dimensionless equation in all of our simulations. The nonlinear coefficient χ_E is usually taken as 1 and the phase mis-match of the fundamental harmonic is taken as 0. In one dimensional cases when $\rho = 1$, the known analytical form, is used. Otherwise, numerical solutions transformed from solutions in [19] are used. The transverse lattice size is normally 1024 for one dimension, 64×64 for two dimensions and $40 \times 40 \times 40$ for three dimensions. The propagation step size is chosen such that the local error is less than one percent, by comparing results at two different time-steps.

7.5.3 The EMA and Stability

The condition that the EMA is valid can be understood as $z_c\kappa_j \gg 1$. In order to investigate possible instabilities when the EMA is invalid, we reduce the value of z_c while fixing the values of κ_j. We find that gap simultons become unstable when $z_c\kappa_j \approx 1$. In order to verify the existence of the instability, an exact numerical solution of the full coupled gap equations (including the imaginary part of the solution) was also used. Unstable propagation was observed for all cases when $z_c\kappa_j < 1$.

These results show that the EMA, which was introduced here as a useful approximation to simplify the equations, also in a sense delineates the physical region where stable solutions can be expected to occur. If the fields have frequency components which extend well outside the band-gap region, the EMA is of course invalid. In addition to this, the corresponding solitons rapidly become unstable, even if all the calculations are carried out without appealing to the EMA. Thus, the useful region of this approximation also appears to correspond to the region of most physical interest for soliton formation, which provides a justification for the use of this method.

This does not exclude the possibility that some exotic solutions exist when the EMA is invalid. For example, it is possible that there could be solutions which do not occur in the EMA limit, due to symmetry considerations outlined previously – but which are stable in some transition region where there are frequency components near the edges of the band-gap. We have not investigated novel solutions of this type.

7.5.4 Material Group Velocity Mismatch

In a real experiment, the first harmonic and the second harmonic usually have different material group velocity. Does such a group velocity mismatch introduce a new instability?

We have performed a series of numerical simulations in a one-dimensional band-gap environment to answer this question. We first fixed the material group velocity of the second harmonic and then adjusted the material group velocity of the first harmonic so that the ratio (v_1/v_2) varied from 0.5 to 2 with a step size of 0.25. Stable propagation was observed for all cases even under extreme circumstances, such as $v_1/v_2 = 0.5, 2$, although oscillations did occur when the departure of the ratio away from one was more than 0.5. Similar results have also been obtained on varying the velocity of the second harmonic while fixing that of the first harmonic. This result is particularly encouraging for experiments since gap simultons can form within a wide range of material group velocity mis-match, thus avoiding the difficult task of matching group velocities.

7.5.5 Higher Dimensional Stability

Using the EMA mapping relationship, we have obtained higher dimensional band-gap simultons from known higher dimensional conventional parametric simultons. Although non-symmetric solutions (in different spatial directions) are possible, we here only show solutions obtained under circumstances where the simplified dimensionless EMA equations are rotationally symmetric. Under these conditions, appropriate initial conditions are obtained by solving the simplified equations exactly via the numerical shooting technique [19], based on the EMA equations.

In the case of the propagation equations that are applicable in the EMA limit, there are a number of results available on stability in two and three dimensions [35,36,4,34]. In particular, it is known that no self-focusing collapse is possible. Stable propagation is expected from a Lyapunov [35,36] analysis, for all cases with $\delta = 1$, $\sigma = 2$, and $\beta < 0$, which corresponds to $\gamma > 4$ in our dimensionless notation. Here we only consider couplings near the upper band-gap (i.e., $s_1 = s_2 = 1$). A variational and numerical treatment [4] indicates an even wider stability region is possible, although not extending as far as $\rho = 0$ in any case.

After obtaining appropriate initial estimated solitons, they are propagated numerically using the four coupled partial differential equations, as before.

Results in Fig. 7.3 show a $(3 + 1)$ dimensional gap simulton propagation, with the complete set of equations, and parameters corresponding to $\rho = 3$, $\delta = 1$, $\sigma = 2$. No assumption of radial symmetry was used in solving these equations, which were treated on a full four-dimensional space-time lattice. The gap simulton reaches a steady state after a small initial oscillation. This indicates that we can have stable (3+1) dimensional gap simultons, at least with $\rho = 3$. While the general stability in this case is poorly understood as yet, we note that the reduced propagation equations (without a band-gap) are known [35,36] to have absolute stability for the case of perfect phase-matching and above ($\rho \geq 4$). We conjecture that there is stability for (at least) $\rho \geq 3$, given an appropriate initial pulse energy.

7.6 Conclusions

Using a coupled mode theory, the classical band-gap equations describe a nonlinear parametric waveguide containing a Bragg grating. These equations are difficult to analyze for simultaneous solitary wave solutions or 'simultons', due to their nonlinearity. We therefore developed a Hamiltonian theory which

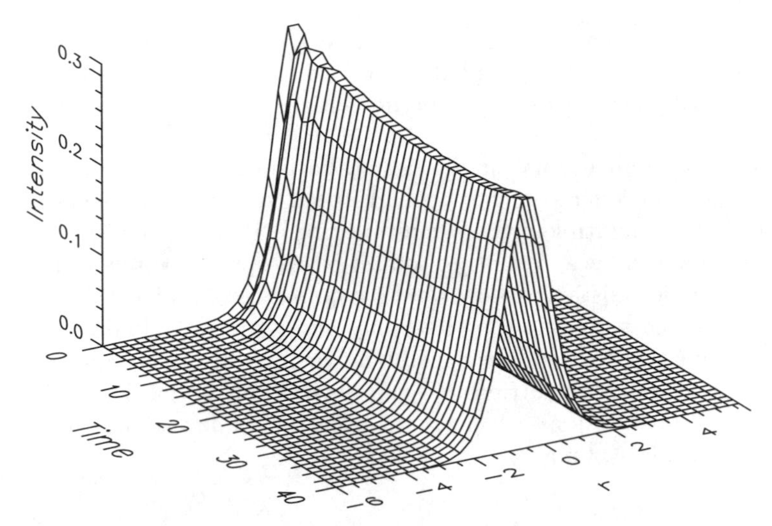

Fig. 7.3. Stable propagation of a (3 + 1) dimensional gap simulton. Only the fundamental harmonic is shown. Initial conditions were cylindrically symmetric in the reduced (EMA) coordinate system. The dimensionless parameters used for simplicity were: $q = 1/8$, $-\kappa_2 = 2\kappa_1 = 8$, $z_c = 1$, $\rho = 3$, $v_1 = v_2 = 1$, $\delta k_1 = 0$, $\delta k_2 = 1/16$ and $\chi_D = 1$

treats one, two and three dimensional propagation of eigenmodes of the linear equations, instead of plane waves. Using the effective mass approximation, we obtained a pair of coupled equations which are formally identical to the coupled equations describing a conventional dispersive parametric medium. The solutions found this way are stable, multidimensional solitons – provided the pulse itself is tuned to the upper edge of the bandgap, and satisfies the restriction that $\kappa z_0 > 1$. This physically means just that the Fourier components of the pulse are themselves mostly within the bandgap.

A mapping relationship between the solutions of the approximate coupled equations and the solutions of the classical band-gap helps to reduce the number of phase space dimensions, and makes analytical solutions possible in special cases. The approximations used in obtaining the mapping relationship are well justified since the EMA solutions agree with exact numerical solutions. Direct numerical simulation of the complete classical band-gap equation show steady propagation of simultons, using the EMA solutions as the initial condition, provided the pulse band-width is small compared to the width of the band-gap in frequency space.

In summary, a parametric bandgap waveguide can provide both large dispersion and large nonlinearity. Perturbations may arise which could limit the lifetime of the solitons, owing to processes omitted in the original equations. These include material losses (absorption), as well as Raman scattering, four-wave mixing, and any diffractive effects which are omitted from the paraxial approximation. Nevertheless, it is clearly physically interesting that at least a quasi-stable type of solitary wave can be generated in one, two or three dimensions. The price that is paid is the use of equations defined in a higher-dimensional phase-space, which do not satisfy classical integrability requirements.

The two- and three dimensional simultons may have potential applications to all-optical signal processing, including high-speed switching, frequency-shifting, pulse-shaping, multiplexing, de-multiplexing and signal replication. There are possible advantages over other soliton-based optical switches, due to the short interaction distances, low power requirements, and novel stability properties of the gap simulton. Response times are only limited by the electronic response of the nonlinear medium, which typically occurs over femtosecond time-scales. Parametric band-gap devices can be fabricated in compact sizes, with fast response times, and using power levels similar to those in communication systems.

References

1. Y.N. Karamzin and A.P. Sukhorukov: Moscow University Physics Bulletin **20**, 339 (1974); Y.N. Karamzin and A.P. Sukhorukov: Zhurnal-Eksperimental'noi-i-Teoreticheskoi-Fiziki. **68**, 834 (1975)
2. B.J. Herman, P.D. Drummond, J.H. Eberly, and B. Sobolewska: Phys. Rev. A **30**, 1910 (1984)
3. H. He and P.D. Drummond: Phys. Rev. Letts. **78**, 4311 (1997); H. He and P.D. Drummond: Physical Review **E58**, 5025 (1998)
4. B.A. Malomed et al.: Phys. Rev. E **56**, 4725 (1997)
5. P.D. Drummond, K.V. Kheruntsyan, H. He: J. Opt. B **1**, 387 (1999)
6. X. Liu, L.J. Qian, and F.W. Wise: Phys. Rev. Lett. **82**, 4631 (1999); X. Liu, L. Beckwitt, and F.W. Wise: Phys. Rev. E. **61**, 4722 (2000)
7. W.E. Torruellas et al.: Phys. Rev. Lett. **74**, 5036 (1995)
8. M.J. Werner and P.D. Drummond: J. O. S. A. B. **12**, 2390 (1993); M.J. Werner and P.D. Drummond: J. Opt. Soc. Am. B **10**, 2390 (1993)
9. R. Schiek: J. Opt. Soc. Am. B **10**, 1848 (1993)
10. M.J. Werner and P.D. Drummond: Optics Lett. **19**, 613 (1994)
11. M.A. Karpierz and M. Sypek: Optics Comm. **110**, 75 (1994)
12. C.R. Menyuk, R. Schiek, and L. Torner: J. Opt. Soc. Am. B **11**, 2434 (1994)
13. K. Hayata and M. Koshiba: Phys. Rev. A **50**, 675 (1994)
14. A.V. Buryak and Y.S. Kivshar: Optics Lett. **19**, 1612 (1994)
15. L. Torner, C.R. Menyuk, and G.I. Stegeman: J. Opt. Soc. Am. B **12**, 889 (1995)
16. L. Torner: Optics Comm. **114**, 136 (1995)
17. L. Torner, C.R. Menyuk, and G.I. Stegeman: Optics Lett. **19**, 1615 (1994)
18. A.V. Buryak and Y.S. Kivshar: Phys. Rev. A **51**, R41 (1995)
19. H. He, M. Werner, and P.D. Drummond: Phys. Rev. E **54**, 896 (1996)
20. A.V. Buryak and Y.S. Kivshar: Optics Lett. **20**, 834 (1995)
21. A.V. Buryak, Y.S. Kivshar, and V.V. Steblina: Phys. Rev. A **52**, 1670 (1995)
22. A.V. Buryak and Y.S. Kivshar: Physics Letters A **197**, 407 (1995)
23. B.J. Eggleton et al.: Electron. Lett. **32**, 1610 (1996); B.J. Eggleton et al.: Phys. Rev. Lett. **76**, 1627 (1996); C.M. de Sterke and J.E. Sipe: Progress in Optics **XXXIII**, 203 (1994)

24. T. Peschel, U. Peschel, F. Lederer, and B.A. Malomed: Phys. Rev. E **55**, 4730 (1997)
25. C. Conti, S. Trillo, and G. Assanto: Phys. Rev. Lett. **12**, 2341 (1997)
26. M. Hillery and L.D. Mlodinow: Phys. Rev. A **30**, 1860 (1984)
27. P.D. Drummond and S.J. Carter: J. Opt. Soc. Am. B **4**, 1565 (1987)
28. P.D. Drummond: Phys. Rev. A **42**, 6845 (1990)
29. M.G. Raymer, P.D. Drummond, and S. Carter: Opt. Lett. **16**, 1189 (1991)
30. P.D. Drummond: Advances in Chemical Physics **85**, 379 (1994)
31. K.J. Ebeling, *Integrated opto-electronics*, English edn. (Springer, Berlin Heidelberg New York, 1992)
32. A. Yariv and P. Yeh, *Optical waves in crystals* (Wiley & Sons, New York Chichester 1984), Chap. 6
33. P.D. Drummond and H. He: Phys. Rev. A **56**, R1107 (1997)
34. K. Hayata and M. Koshiba: Phys. Rev. Lett. **71**, 3275 (1993)
35. A.A. Kanashov and A. Rubenchik: Physica D **4**, 122 (1981)
36. L. Bergé, V.K. Mezentsev, J.J. Rasmussen, and J. Wyller, Phys. Rev. A **52**, R28 (1995)
37. P.D. Drummond: Comp. Phys. Comm. **29**, 211 (1983)
38. D.E. Pelinovsky, A.V. Buryak, and Y.S. Kivshar: Phys. Rev. Lett. **75**, 591 (1995)
39. G.D. Boyd et al.: Appl. Phys. Lett. **5**, 234 (1964)

Part II

Nonlinear Fiber Grating Experiments

8 Nonlinear Propagation in Fiber Gratings

B.J. Eggleton and R.E. Slusher

8.1 Introduction

Exploration of nonlinear phenomena in photonic crystals began with simple one-dimensional photonic crystals in waveguides and glass fibers. Bragg gratings in silica fibers have been developed for many linear applications, for example for optical filters in wavelength division multiplexed optical communication systems. These high quality fiber Bragg gratings have been used to observe nonlinear propagation phenomena for the first time in photonic crystals [1,2]. Low linear and nonlinear losses along with high optical damage thresholds make these gratings ideal for studying nonlinear propagation effects. Nonlinear phenomena in these periodic dielectric structures include many of the phenomena in uniform dielectric media [3] including fiber solitons [4] and polarization instabilities [5]. Fiber Bragg gratings can be fabricated as shown in Fig. 8.1 with great precision and control over many parameters including the spatial dependence of the amplitude and phase of the grating.

Bragg gratings in the core of optical fibers constitute one-dimensional photonic bandgap structures since the fiber core mode constrains the optical field transverse to the fiber axis and the light propagating in the single core mode encounters a strongly reflecting photonic bandgap in the wavelength region near the Bragg wavelength

$$\lambda_B = 2 n_o \Lambda \,, \tag{8.1}$$

where λ_B is the Bragg wavelength, n_o is the average linear index of refraction in the core and Λ is the period of the index grating in the fiber. The gap width is

$$\Delta\lambda = \lambda_B \Delta n / n \,, \tag{8.2}$$

where Δn is the amplitude of the index modulation in the grating.

Nonlinear propagation of light pulses in these grating structures exhibit a beautiful array of new phenomena including Bragg grating solitons [6], modulational instability [7], enhanced polarization instabilities [8] and pulse compression [9,10]. The dispersion for wavelengths near the edges of the photonic bandgap is typically nearly six orders of magnitude larger than for propagation in the bare fiber. This large dispersion along with nonlinear

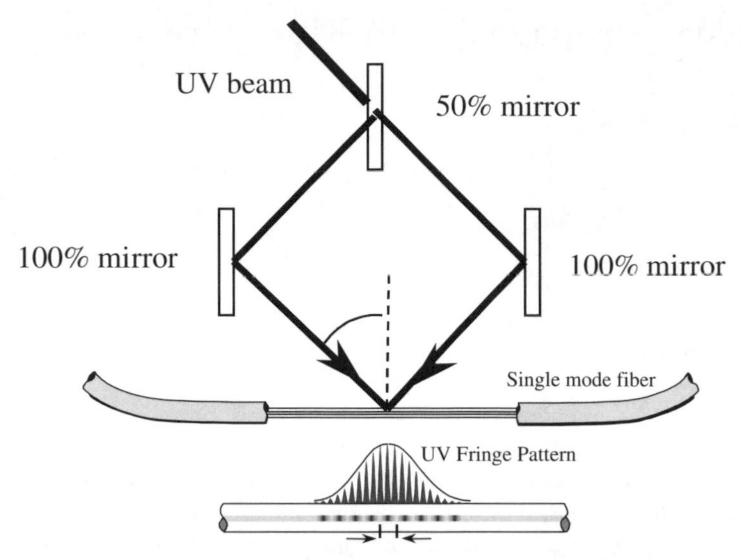

Fig. 8.1. Fiber Bragg grating fabrication. An ultraviolet light source is split into two beams that combine to form an interference pattern in the core of the fiber. This periodic ultraviolet light pattern induces periodic changes in the refractive index of the germanium doped fiber core through creation of germano-oxide defect centers. These refractive index changes form a permanent Bragg grating in the fiber core. The period of the grating can be varied by changing the angle of the interfering beams

changes in the refractive index caused by intense picosecond pulses results in soliton formation lengths that are only centimeters, compared with fiber lengths over a hundred meters required for soliton formation in bare fiber cores. Birefringence near the photonic bandgap can also be much larger than that in the bare fiber. This large birefringence leads to a rich class of nonlinear polarization pulse propagation phenomena [11].

The ability to tailor the amplitude of the periodic index profile, the phase of the grating period, the grating birefringent properties and the amplitude of the average index in the grating region allow us to engineer grating properties that control both linear and nonlinear propagation phenomena. For example, by slowly tapering the grating index amplitude at the beginning and end of the grating, a process called apodization [12–14], we can control the reflection of light at wavelengths near the edges of the photonic bangap. One can also vary the period of the grating along its length and introduce phase or amplitude "defects" in the grating. These variations have not been studied in detail at present but should lead to many interesting new nonlinear phenomena.

In this chapter we will first review the linear properties of fiber Bragg gratings in Sect. 8.2. Nonlinear propagation will be introduced in Sect. 8.3 using an effective nonlinear Schröedinger equation (NLSE) model [6] that approximates the results obtained using the more accurate but much more computationally challenging coupled mode equations. This NLSE is shown to be a good model for Bragg solitons that propagate with wavelength components just outside the bandgap. The Bragg soliton regime is expanded to the case of arbitrary input polarization in Sect. 8.3.4. In Sect. 8.4 we describe the experimental apparatus used to explore nonlinear pulse propagation. The experimental results are then compared with our models in Sect. 8.5. Conclusions and future directions for these experiments are discussed in Sect. 8.6.

8.2 Linear Properties of Fiber Bragg Gratings

8.2.1 Fiber Bragg Gratings

We consider a one-dimensional photonic crystal, such as a fiber Bragg grating, located in the core of an optical fiber, whose axis is in the z-direction. The linear refractive index is given by

$$n(z) = n_o + \delta n(z) + \Delta n(z) cos(2\pi z/\Lambda + \Phi(z)) , \quad (8.3)$$

where n_o is the refractive index of the fiber core, $\delta n(z)$ is the change in the average refractive index, $\Delta n(z)$ is the refractive index modulation depth, Λ is the grating period, and $\Phi(z)$ is the phase that allows for grating period variations. For typical gratings in optical fiber, $n_0 = 1.45$. Both $\Delta n(z)$ and $\delta n(z)$ can be as large as 0.01 in hydrogen loaded fibers [15], and the period is typically of the order of $\Lambda = 0.3\,\mu\text{m}$ for light wavelengths near one micron. All of the gratings referred to throughout this chapter have index gratings that are formed using the ultraviolet photosensitivity in germanosilicate glass [15].

If the wavelength of the incident light lies within the photonic bandgap (see 8.2), the light field envelopes in the grating are evanescent and a major fraction of the light is reflected for low intensity light pulses. For wavelengths outside the photonic band gap, the fields are propagating waves and most of the light is transmitted. In a grating with a uniform amplitude index modulation, spectral side-lobes are present in the transmission spectrum for wavelengths just outside of the photonic bandgap. Spectral components of the pulse at wavelengths near these side-lobes experience severe dispersion because the group velocity dispersion (GVD) [1] varies strongly near the side-lobe ripples. These strong disperive effects can lead to severe pulse envelope distortion. The side-lobes result from a mismatch in the effective index of the grating and the surrounding medium (i.e. bare fiber). The side-lobes can be removed by fabricating the grating so that the index modulation decreases smoothly at each end while the average index is kept constant, a process known as apodization [14]. A schematic diagram of an apodized grating is shown in Fig. 8.2.

Gratings are often fabricated in photosensitive fiber using a phase mask technique instead of the method shown in Fig. 8.1. This phase mask scanning

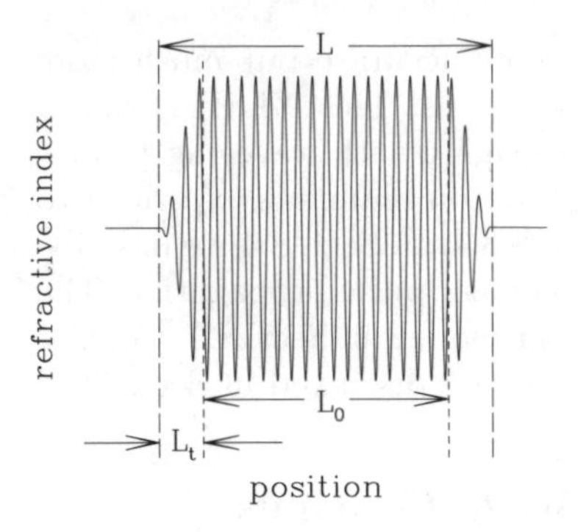

Fig. 8.2. Schematic diagram of an apodized fiber grating refractive index as a function of grating length. The grating has a total length L, a uniform section L_o and tapered end sections, each of length L_t

technique [12,16,17] uses an ultraviolet beam, with a wavelength in the range from 242 to 248 nm, that is scanned along the length of a phase mask etched in a flat glass plate with a period required for the grating to be transfered into the fiber core. The phase mask plate is positioned just above the fiber so that the transmitted UV light forms the required diffraction pattern in the fiber core. Gratings as long as one meter have been fabricated using this method.

Fabrication of apodized gratings involves a two-step procedure [12]. First, an ultraviolet beam is scanned along the length of the phase mask. The intensity of the ultraviolet beam is varied as the beam is scanned along the axis of the fiber creating the modulated component of the grating. Second, the unmodulated component of the grating is then imprinted in the same manner without the phase mask present. This two-step technique ensures that the average refractive index of the grating remains approximately constant along the length of the grating. This procedure minimizes both the reflection and dispersive side-lobes near the photonic bandgap edge.

We model the pulsed optical field using an electric field of the form,

$$E(z,t) = [E_+(z,t)e^{ik_B z} + E_-(z,t)e^{-ik_B z}]e^{i\omega_B t} + c.c.\,, \tag{8.4}$$

where $k_B = \pi/\Lambda$ and $\omega_B = \pi c/(n\Lambda)$ are the wavenumber and the frequency, respectively, at the Bragg condition and E_+ and E_- are slowly varying field envelopes of the forward and backward propagating waves. Pulse propagation through gratings can now be modelled by substituting this form of the field into Maxwell's equations in order to obtain the nonlinear coupled mode equations (NLCME) [18],

$$i\frac{\partial E_+}{\partial z} + i\frac{1}{V}\frac{\partial E_+}{\partial t} + \kappa(z)E_- + \Gamma_S|E_+|^2E_+ + 2\Gamma_X|E_-|^2E_+ = 0 \tag{8.5}$$

$$i\frac{\partial E_-}{\partial z} + i\frac{1}{V}\frac{\partial E_-}{\partial t} + \kappa(z)E_+ + \Gamma_S|E_-|^2E_- + 2\Gamma_X|E_+|^2E_- = 0\,, \tag{8.6}$$

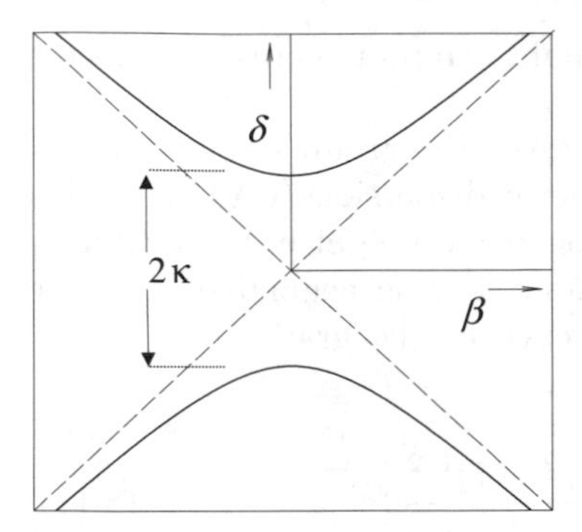

Fig. 8.3. Photonic bandgap diagram shows the dispersion relation for a uniform Bragg grating. The vertical axis is the detuning parameter and the horizontal axis is the propagation constant. The dashed lines correspond to a uniform medium (no grating)

where $V = c/n$ is the phase velocity of light in the fiber, c is the velocity of light and

$$\kappa(z) = \frac{\pi \Delta n(z) \eta}{\lambda_B} \tag{8.7}$$

is the coupling coefficient, that can vary along the length of the grating, and $\eta \approx 0.8$ is the fraction of the total energy in the fiber core. Γ_S and Γ_X are the nonlinear coupling coefficients that model self and cross phase modulation respectively. We will describe the nonlinear processes in Sect. 8.3.

Linear pulse propagation is modeled by setting the nonlinear terms in (8.5) and (8.6) to zero. We consider the dispersion relation $\omega(k)$, which is equivalent to the relationship between the frequency detuning parameter

$$\delta = \frac{1}{V}(\omega - \omega_B) \tag{8.8}$$

and the propagation constant $\beta = k - k_B$, where ω and k are the optical frequency and the propagation wavenumber, respectively. For an infinite uniform grating the dispersion relation can be obtained from the linearized NLCME by insertion of a plane-wave solution of the form

$$E_{\pm} = C_{\pm} \exp\left[i(\beta z - V\delta t)\right] , \tag{8.9}$$

where $C_{\pm}$ are constants. This substitution yields the dispersion relation

$$\delta = \pm\sqrt{\kappa^2 + \beta^2} , \tag{8.10}$$

which is illustrated in Fig. 8.3, for both a uniform medium (dashed lines) and a periodic medium (solid lines).

Figure 8.3 shows that for frequencies in the range $-\kappa \leq \delta \leq \kappa$, defining the photonic band gap, no propagating solutions are allowed; this corresponds to the regions of high reflectivity. Low intensity light can propagate only outside this region in the pass band.

The slope of the dispersion curve $d\delta/d\beta$ gives the group velocity, while the curvature, $d^2\delta/d\beta^2$ gives the group velocity dispersion (GVD). In the absence of the grating, light propagates at the speed of light in the medium. We ignore the background material dispersion, as it is negligible over the bandwidth under consideration. In the presence of the grating the group velocity

$$v_g = \frac{d\omega}{dk} = \frac{V d\delta}{d\beta} = \pm V\sqrt{1 - \left(\frac{\kappa}{\delta}\right)^2} = \frac{V(f^2-1)^{1/2}}{f}, \tag{8.11}$$

where f is the detuning from the edge of the gap in unites of half-gaps from the gap center. Note that the group velocity approaches zero at the band edge and asymptotically approaches the speed of light in the medium far from the Bragg resonance. The reduction in the group velocity can be explained in terms of the multiple Fresnel reflections that occur at each of the individual grating rulings, resulting in additional path length. This extreme variation in the group velocity over a relatively small range of wavelengths (roughly equal to the bandwidth of the grating) leads to very strong GVD. Indeed, it has been shown that the grating GVD can be up to six orders of magnitude larger than that of standard telecommunication fiber [19].

On the long-wavelength side of the band gap ($\delta < -\kappa$), the grating GVD is positive and can be used to compensate for the negative dispersion effects which are caused by propagation in optical fiber at communication wavelengths [19]. On the short-wavelength side of the photonic band gap ($\delta > -\kappa$) the grating GVD is negative, and solitons are supported [1]. At frequencies close to the edge of the photonic band gap the grating also exhibits significant higher-order dispersion that can have a significant effect on light propagation [20,21]. The effect of the various orders of grating dispersion can be obtained by expanding the propagation constant $\beta(\omega)$ in a Taylor series [21,22]

$$\beta(\omega) = \beta_0 + \beta_1(\omega-\omega_0) + \frac{1}{2}\beta_2(\omega-\omega_0)^2 + \frac{1}{6}\beta_3(\omega-\omega_0)^3 + \ldots, \tag{8.12}$$

where β_n is the nth derivative of β with respect to ω. The GVD inside the grating can now be obtained from the second term in the Taylor series expansion of the propagation constant [20–22]

$$\beta_2 = -\left(\frac{1}{V}\right)^2 \frac{1}{\delta} \frac{(\kappa/\delta)^2}{(1-(\kappa/\delta)^2)^{3/2}}. \tag{8.13}$$

The strong variation of GVD near the photonic bandgap is shown in Fig. 8.4.

This dispersion strongly broadens the optical pulses we study experimentally as pulse wavelength approaches the gap edge. An important parameter

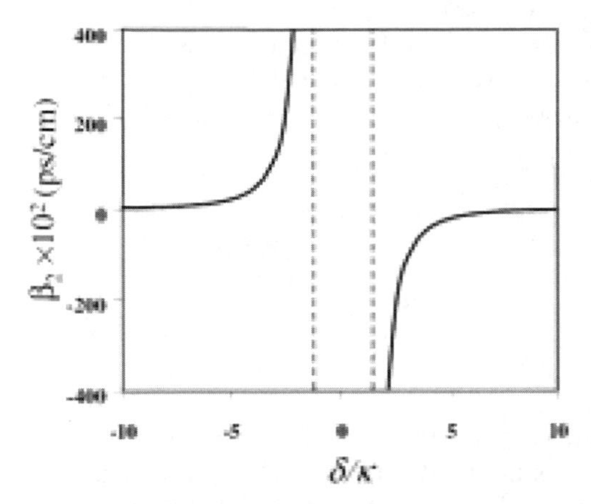

Fig. 8.4. Group velocity dispersion near the bandgap (solid lines). The dashed lines indicate the edges of the photonic bandgap

that characterizes the dispersion in the grating is the dispersion length

$$L_D = \frac{{\tau_0}^2}{s^2 \beta_2} , \tag{8.14}$$

where τ_0 is the full width at half maximum (FWHM) pulse width and s^2 is a numerical factor that depends on pulse shape, e.g. $s = 2.77$ for a Gaussian pulse and s=3.11 for a hyperbolic secant pulse. A pulse will broaden by a factor of roughly two after propagation through a dispersion length. The third order dispersion is also important for pulse propagation, adding asymmetry to the pulse shapes. This asymmetry is evident when the GVD is not constant over the spectral bandwidth of a short pulse.

Consider experimental optical pulse parameters of $\kappa = 10\,\mathrm{cm}^{-1}$, a pulse with duration $\tau_0 = 10\,\mathrm{ps}$, a detuning of $\delta = 15\,\mathrm{cm}^{-1}$, corresponding to $v_g = 0.75\,\mathrm{V}$. The dispersion length for these parameters is $L_D \sim 0.5\,\mathrm{cm}$ so that a pulse with this detuning from the gap will experience major broadening while propagating through a 6 cm long grating. It is impressive to compare this dispersion length with standard communications optical fiber in which the dispersion length for a 10 ps pulse at a wavelength of 1.5 µm is 4 km, i.e. the grating dispersion for this typical example is nearly a factor of 10^6 larger than in standard fiber!

8.2.2 Polarization Properties of Fiber Bragg Gratings

Fiber gratings written by holographic exposure of the Ge/Si fiber core to UV radiation result in index changes in the fiber core that have much larger birefringence than the residual birefringence in a typical silica fiber core without a grating. An incident pulse with an arbitrary polarization with respect to the principle polarization axes, X and Y, of the fiber grating will experience significant change in the polarization state while propagating through a 6 cm long grating. The beat length, corresponding to a rotation of the polarization vector by π, is given by

$$L_P = \frac{\lambda}{2\Delta n_{\rm br}}\,, \tag{8.15}$$

where

$$\Delta n_{\rm br} = n_X - n_Y \tag{8.16}$$

is the difference in the fast (X) and slow (Y) axis index, typically of the order of 10^{-5}, or a few percent of the index modulation in the Bragg grating. This means that the Bragg wavelength is different for the fast and slow axis by a few percent of the gap width.

It is also found experimentally that the magnitude of the index modulation is different along the fast and slow axis. The ratio of the modulation index difference to the bireferingence

$$M = \frac{(\delta n_X - \delta n_Y)}{\Delta n_{\rm br}} \tag{8.17}$$

is near 0.6 for the experiments described in this chapter. The net result is that the fast and slow gap edges are nearly at the same wavelength at the short wavelength side of the gap and displaced in wavelength by a few percent of the gap at the long wavelength edge. This displacement can be seen in the measured transmission of a 6 cm grating shown in Fig. 8.5.

The increase in the effective interaction length due to the multiple reflections along the grating results in an enhancement of the birefringence [11]

$$F_{\rm br} = \frac{(f - M)}{(f^2 - 1)^{1/2}}\,. \tag{8.18}$$

This analytical expression is derived for a single Fourier component of the pulse frequency spectrum and assumes that the birefringence is smaller than the index modulation and that the detuning is sufficiently far from the gap that the group velocity dispersion mismatch is small, i.e. $f^2 \gg (1 + \Delta n_{\rm br}/\delta n_X)$. These limits are satisfied for the experiments described in this chapter.

The enhanced birefringence is shown as a function of detuning in Fig. 8.6 for $M = 0.6$. Note that there is only a small enhancement in birefringence for this M value and even a reduction in the effective birefringence over a large range of detunings. The enhancement is always larger than unity for symmetric gap position case where $M = 1$. This small change in the effective birefringence is important for nonlinear effects, for example, polarization instabilities, described in the next section.

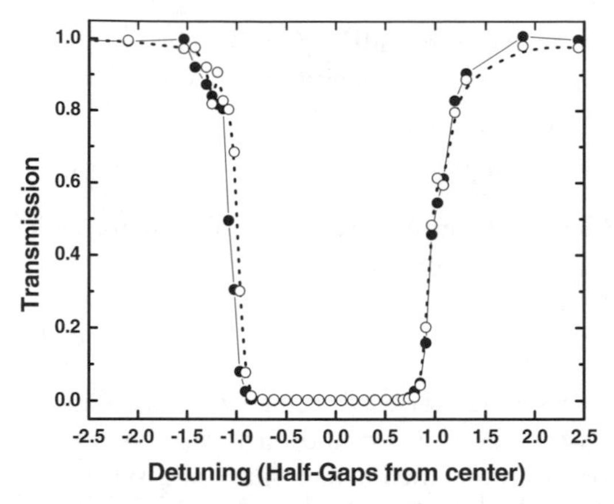

Fig. 8.5. Transmission of a 6 cm long fiber grating along the fast (solid circles) and slow (open circles) axes. The measurement is done with a CW laser in order to eliminate the loss in resolution caused by the broad pulse spectral components. The spectrum is obtained by stretching the fiber in order to change the grating period

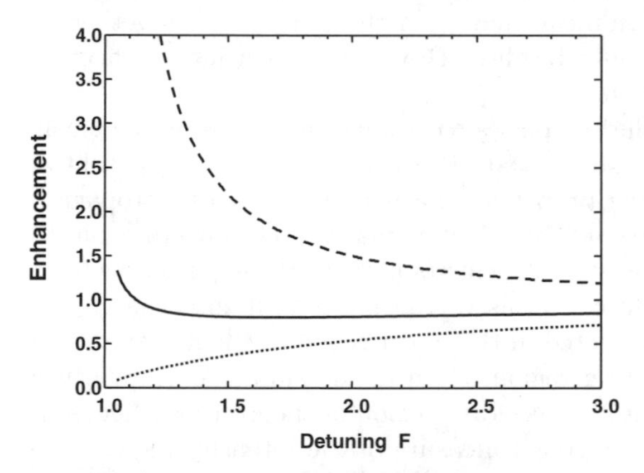

Fig. 8.6. Birefringence, nonlinearity and polarization instability threshold intensity enhancement as a function of detuning from the short wavelength edge of the gap. The solid curve is the birefringent enhancement for $M = 0.6$ (see (8.18)). The dashed curve is the enhancement in the nonlinearity as described by (8.23). The dotted curve is the polarization threshold enhancement factor that results from the combined effects of birefringence and nonlinearity

8.3 Nonlinear Properties of Fiber Bragg Gratings

8.3.1 Nonlinear Coupled Mode equations

Nonlinear pulse propagation in Bragg gratings can be described using a set of nonlinear coupled mode equations (NLCMEs, (8.5) and (8.6)). These equations are valid only for relatively weak gratings, where $\Delta n \ll n_o$, and for detunings no more than a few times the gap width, $f < 5$. The nonlinear response of the fiber grating is described by an intensity dependent refractive index

$$n = n_o + n_{\mathrm{NL}} = n_o + n_2 I\,, \tag{8.19}$$

where $I = P/A_{\mathrm{eff}}$ is the peak optical pulse intensity. The nonlinear terms in (8.5) and (8.6) are related to the nonlinear index by

$$\Gamma_S = \frac{\Gamma_X}{2} = \frac{2\pi n_2}{\lambda Z}\,, \tag{8.20}$$

where Γ_S is self phase modulation by the forward going wave with itself, Γ_X is cross phase modulation between the forward and backward waves and Z is the vacuum impedance. For isotropic media like silica the self-phase modulation is half of the cross-phase modualtion.

In this chapter we will describe the nonlinear pulse propagation for pulses whose wavelengths are near, but outside the photonic bandgap. These pulses propagate through the grating with only small reflection losses in the linear regime. In the nonlinear regime, anomalous dispersion on the short wavelength side of the gap and nonlinear change in the index work together to form Bragg solitons. We will describe these Bragg solitons using an effective nonlinear Schröedinger equation.

The more general case including pulses with wavelengths inside the linear photonic bandgap has been solved analytically [23,24]. These gap soliton solutions have many interesting properties, for example they can propagate at any velocity including zero velocity. Unfortunately they are difficult to explore experimentally because of inefficient coupling to the gap solitons from outside the grating. As the wavelengths of the pulse shift into the gap, a large fraction of the pulse is reflected in the linear regime, reducing the pulse intensity coupled into the grating region. Even in the nonlinear regime it is difficult to couple the pulse into the uniform region of the grating where gap solitons are expected to form. A rough measure of the intensity required to form a gap soliton in the uniform grating region follows from the nonlinear shift in the gap,

$$\Delta\lambda_B = 2 n_2 I \Lambda\,. \tag{8.21}$$

This nonlinear shift should be of the order of the width of the pulse wavelength spread in order to make the pulse propagate in the linear gap spectral region. This condition allows formation of a gap soliton when the input pulse wavelengths are just inside the short wavelength edge of the gap, i.e. the gap is nonlinearly shifted so that the pulse propagates without reflection from the grating. This argument makes it clear that relatively long pulse lengths with their associated narrow range of wavelengths will require lower intensities for gap soliton formation. The gap soliton regime is described in Chap. 9 where relatively long nanosecond pulse widths are used to excite gap solitons in fiber gratings and semiconductor waveguides.

8.3.2 Bragg Solitons

Although the pulse propagation is best described by a complete analysis of the NLCMEs, a very useful, simpler description of the nonlinear propagation can be made if the grating quadratic dispersion is nearly constant over the spectral bandwidth of the pulse (i.e. $c\kappa\tau \gg 1$) and the intensity of light is sufficiently low so that $n_2 I \ll \Delta n$. This regime is of interest for a wide range of experiments. In this case the NLCMEs reduce to much simpler Nonlinear Schröedinger Equations (NLSE) of the form [6]

$$i\frac{\partial E}{\partial z} - \frac{\beta_2}{2}\frac{\partial^2 E}{\partial \tau^2} + \bar{\Gamma}|E|^2 E = 0\,, \tag{8.22}$$

where the effect of the grating has modified the nonlinear coupling, now expressed as an effective nonlinear coupling coefficient

$$\bar{\Gamma} = \frac{(3 - v^2)}{2v}\Gamma \tag{8.23}$$

and an effective group velocity dispersion

$$\beta_2 = -\frac{1}{V^2}\frac{1}{\kappa\gamma^3 v^3}\,, \tag{8.24}$$

where Γ is the nonlinear self phase modulation coefficient and v is the propagation velocity in units of V.

The derivation of (8.23) and (8.24) is based on a multiple scales approach and has been presented elsewhere (see [6] for a complete derivation). We note that, close to the edge of the photonic bandgap.the effective nonlinearity inside the grating (8.23) is enhanced with respect to that outside of the grating. This enhancement is attributed to both the standing wave field inside the grating and the reduced group velocity of light inside the grating associated with multiple reflections from the grating structure.

Equation 8.22 describes nonlinear pulse propagation inside the Bragg grating, and ignores the problem of coupling light into the grating. In order to include this coupling effect we need to relate the incoming field, i.e. the pulse incident upon the grating, to the field just inside the grating at which point (8.22) applies. If we assume that the grating is suitably apodized, then the reflectivity outside of the photonic band gap is negligible, and we can show to a good approximation, that the total field intensity inside the grating is enhanced by a factor of V/v_g, so that $I_{\text{in}} = (V/v_g)I_{\text{out}}$ [6]. Combining this enhancement with that described by (8.23) results in enhancements of the total effective nonlinearity by as much as a factor of three to five in typical fiber grating experiments [6].

The approximate description given by the NLSE is a very valuable tool in understanding nonlinear pulse propagation through Bragg gratings for cases where the wavelengths of the pulse are near but outside the gap. This approach highlights the physics and leads to results that are in good agreement with both NLCME and experimental results described in Sect. 8.5. A very interesting consequence of this equation is a new kind of soliton, referred to as a Bragg soliton [6], which relies on the GVD provided by the Bragg grating rather than on material dispersion. The fundamental Bragg soliton, in the NLSE limit, can be written as

$$\mathcal{E} = \sqrt{\left|\frac{\beta_2}{\bar{\Gamma}}\right|}\frac{1}{\tau_0}\,\mathrm{sech}\left(\frac{t}{\tau_0}\right) . \tag{8.25}$$

We can now apply the standard NLSE results for a soliton in uniform media, and by relating the fields outside and inside the grating we find that the peak intensity required for launching a fundamental, N=1, Bragg soliton is

$$I = \frac{3.107}{V^2}\frac{\lambda}{\pi w^2(3-v^2)\kappa\gamma^3 v n_2} , \tag{8.26}$$

where w is the FWHM of the incoming pulse. The soliton period, the length scale over which the soliton changes shape, is

$$z_0 = \frac{0.322\pi w_e^2}{2|\beta_2|} = 0.161\pi w_e^2 V^2 \kappa\gamma^3 v^3 , \tag{8.27}$$

where ω_e is the width of the emerging soliton. This periodicity is obeyed by all higher order solitons, $N > 1$. In fact, during the evolution of a higher order soliton over one soliton-period, it first contracts to a fraction of it's initial width, splits into distinct pulses, and then merges again, recovering the original shape at the end of the soliton period. The initial compression is often referred to as soliton-effect compression and forms the basis of a well known pulse compression scheme [38]. Soliton-effect pulse compression in fiber Bragg gratings is an interesting technique for applications [38].

Consider a typical pulse used in the experiments with a pulse width of 80 ps (FWHM) and a wavelength of 1.053 µm. The grating has a coupling constant $\kappa = 700\,\mathrm{m}^{-1}$ and the detuning is set so that the effective velocity is $v = 0.5$. The fundamental soliton is then expected to have a external peak intensity of 11 GW/cm^2. The soliton period, assuming $w = w_e$, is $z_0 = 1.8\,\mathrm{cm}$. These values are very nearly obtained in the experiments. Note that the soliton nearly, but not completely, evolves into a full soliton in a 6 cm grating length.

Some of major differences between Bragg solitons and conventional solitons are as follows:

1. Bragg solitons can in principle travel at any velocity between 0 and V. In experiments, velocities as low as 0.5 V have been reported [6].
2. The GVD is caused by the presence of the grating and not by the inherent material dispersion of glass. The magnitude of the grating GVD can be up to five to six orders of magnitude larger than the material dispersion of glass. Thus the soliton period, which is proportional to L_D, is correspondingly smaller.

3. The magnitude of the effective nonlinear coefficient in the NLSE is considerably larger than that in conventional fiber and increasses rapidly as the pusle frequency approaches the edge of the photonic band gap.
4. Bragg solitons represent solutions to the NLCME which are a set of non-integrable equations. These solitary wave solutions are known to reduce to the NLSE solutions given by (8.22) in the appropriate limit [6]. Since these solitary waves are solutions to the full NLCME, they include all orders of dispersion and their existence is not restricted to the regimes with a low ratio of β_3/β_2.

8.3.3 Modulational Instability

At input intensities well above the threshold for fundamental soliton formation, nonlinear propagation is expected to lead to the formation of a pulse train described by a modulational instability [39]. The interest in modulational instability for the fiber grating is the ability to control the propagation parameters near the edge of the photonic bandgap and thereby strongly influence the period of the modulational instability pulse train. We can easily use the effective nonlinear Schröedinger equation approach described in Sect. 8.3.2 to describe the modulational instability for the fiber grating case. If the dispersion is in the anomalous regime at the short wavelength edge of the photonic bandgap, the NLSE predicts an instability over a continuous frequency range on either side of a continuous wave pump frequency. The sideband with the highest growth rate will dominate the instabilty. The dominate instability occurs at a sideband detuning of [39]

$$\pm\Delta_{mi} = \sqrt{\frac{4\pi}{\lambda} n_2 I \frac{\Delta^3}{\kappa^2}\left(1+\frac{\kappa^2}{2\Delta^2}\right)\sqrt{1-\frac{\kappa^2}{\Delta^2}}} \tag{8.28}$$

and the growth rate for this detuning is

$$g_m = \bar{\Gamma} I = \frac{2\pi}{\lambda} n_2 I \frac{(1+\kappa^2/(2\Delta^2))}{(1-\kappa^2/\Delta^2)}\,, \tag{8.29}$$

where I is the intensity of the incoming pump field and $\Delta = \delta/c$ is the detuning parameter in the same units as κ. Both the most unstable detuning and the growth rate depend strongly on the detuning. In a uniform fiber the growth rate depends only on the pump intensity since the effective nonlinear constant in that case is independent of the sideband spacing. We can generate a pulse train with an easily tunable repetition rate by simply varying the detuning parameter Δ, either by varying the pump frequency or by straining the grating along its axis.

8.3.4 Nonlinear Polarization Effects

A set of two NLSEs is required to describe the pulse evolution in the general case where the input is not linearly polarized and aligned along one of the principle axes of the grating. These two equations take the form [11]

$$i\frac{\partial E_X}{\partial z} - \frac{\beta_2}{2}\frac{\partial^2 E_X}{\partial \tau^2} + \left(\bar{\Gamma}_{\rm spm}|E_X|^2 + \bar{\Gamma}_{\rm cpm}|E_Y|^2\right) E_X + \bar{\Gamma}_{\rm pc}|E_Y|^2 {E_X}^* e^{i\mu t} = 0 \qquad (8.30)$$

and

$$i\frac{\partial E_Y}{\partial z} - \frac{\beta_2}{2}\frac{\partial^2 E_Y}{\partial \tau^2} + \left(\bar{\Gamma}_{\rm spm}|E_Y|^2 + \bar{\Gamma}_{\rm cpm}|E_X|^2\right) E_Y + \bar{\Gamma}_{\rm pc}|E_X|^2 {E_Y}^* e^{-i\mu t} = 0\,, \qquad (8.31)$$

where E_X and E_Y are the electric field components along the fast and slow principle axes respectively and $\bar{\Gamma}_{\rm spm}$ now refers to the effective self-phase modulation by the principle axes fields on themselves, $\bar{\Gamma}_{\rm cpm}$ is the effective cross-phase modulation between the polarization components and $\bar{\Gamma}_{\rm pc}$ is the nonlinear coupling between the phase conjugated components of the field. The ratio of the nonlinear coefficients $\bar{\Gamma}_{\rm spm} : \bar{\Gamma}_{\rm cpm} : \bar{\Gamma}_{\rm pc}$ is $3 : 2 : 1$. The important new interaction in these equations is the energy coupling term, the last term in both equations. This term describes an exchange of energy between the two polarizations that can lead to new nonlinear effects. The time scale of the energy coupling is given by the exponential factor in the energy coupling term and is determined by the phase velocity mismatch introduced by birefringence

$$\mu = 2n_{\rm br}^{\rm eff}\,\omega\,, \qquad (8.32)$$

where the effective birefringence is

$$n_{\rm br}^{\rm eff} = \Delta n_{\rm br}\left(\frac{1}{\nu}\left[\sqrt{(f+\nu)^2 - (1+M\nu)^2} - \sqrt{f^2-1}\right]\right) \qquad (8.33)$$

for wavelengths on the short wavelength side of the bandgap.

A polarization instability is expected when the nonlinear change in index becomes comparable to the birefringence, $\Delta n_{\rm NL} = n_2 I \approx \Delta n_{\rm br}$. For a CW signal the instability occurs at a threshold value of

$$I_t = \frac{1.5\Delta n_{\rm br}}{n_2}\,. \qquad (8.34)$$

Both the birefringence and the effective nonlinearity vary as the detuning decreases toward the short wavelength edge of the grating as shown in Fig. 8.6. The nonlinear Schröedinger equation model gives the effective nonlinearity (8.23) that increases rapidly as the detuning approaches the gap. Since the nonlinearity increases much faster than the birefringence the polarization instability threshold intensity decreases as the detuning decreases as shown by the dotted line in Fig. 8.6.

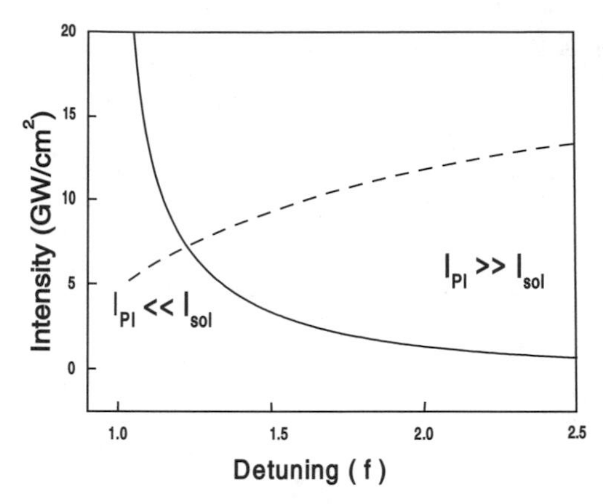

Fig. 8.7. Intensity thresholds for soliton formation (I_{sol}, solid line) and the polarization instability (I_{PI}, dashed line)

At large detuning the threshold for soliton formation is less than the polarization instability threshold. As the detuning decreases, there is a crossover to a region near the gap edge where the threshold for polarization instability is smaller than that for forming a soliton, as shown in Fig. 8.7. This crossover region results in a rich variety of nonlinear propagation effects for fiber Bragg gratings.

8.4 Experimental Apparatus

We explored nonlinear pulse propagation in well-characterized fiber gratings using a stable pulsed laser source. The intensities required to achieve the soliton and polarization instability thresholds are near $10\,\text{GW/cm}^2$. These high intensities, required because of the small nonlinear coefficient of $2.6 \times 10^{-16}\,\text{cm}^2/W$ in silica glass, are below, but near the threshold for optical damage in fiber gratings.

The experimental setup is shown in Fig. 8.8. The light source was a mode-locked Nd:YLF laser generating pulses at a wavelength of 1053.2 nm. High intensities were achieved by Q-switching the mode-locked laser at a repetition rate of 500 Hz. Under these operating conditions the laser emits an intense mode-locked train of several hundred pulses every 2 ms. Each pulse in the train has a duration of approximately 80 ps. An electro-optic pulse selector was used to select one pulse in each train in order to provide a unique intensity pulse and avoid thermal effects in the grating. More details are given in the paper by Eggleton et al. [26].

The pulse incident upon the grating is close to transform limited, i.e. its spectral width is approximately equal to the inverse of the pulse width. This is a necessary condition for interpreting the experimental results. The very high intensities in the Q-switched laser cavity can lead to self-phase modulation

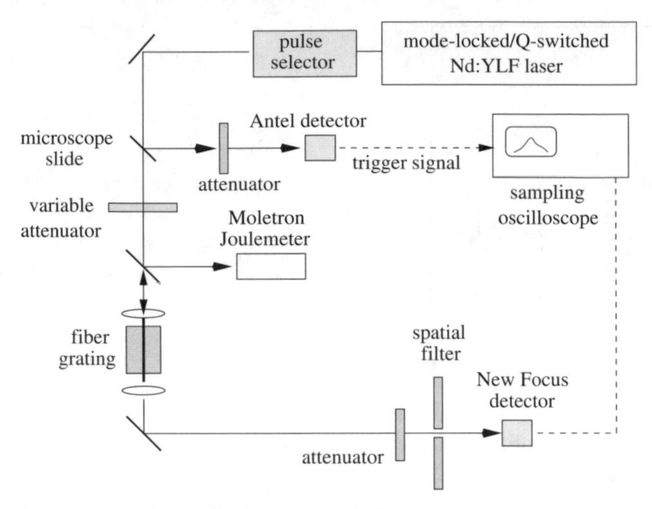

Fig. 8.8. Schematic diagram of the experimental apparatus

in the YLF crystal and the associated spectral broadening [37,38]. This is particularly true near the peak of the Q-switched train where the nonlinear phase shift can exceed 2π radians. By choosing a pulse near the leading edge of the Q-switched train we can ensure high peak intensity pulses and minimize the "prechirp" on the pulse. In the experiments described here the nonlinear phase shift occurring in the laser cavity was inferred to be only approximately $\Delta\Phi = 0.13\pi$ at the peak of the mode-locked pulse. This nonlinear phase is much smaller than the nonlinear phase shifts incurred in our high-intensity experiments by propagating through the grating and does not strongly alter the observed phenomena.

The initial length of bare fiber with no grating was minimized to a length less than 5 mm. This eliminates any significant spectral broadening of the pulse before it reaches the grating. The light transmitted through the grating was detected using a fast photodiode with a response time of 9 ps. The net time resolution, including trigger jitter and oscilloscope resolution was approximately 20 ps.

We fabricated a 75 mm long unchirped, apodized fiber grating in photosensitive fiber using the phase mask scanning technique [12,17,16]. The fiber was measured to have an effective mode area of roughly 20 μm^2. The grating fabrication process consisted of two steps: First, an ultraviolet beam, generated by an excimer-pumped frequency-doubled dye laser, was scanned along the length of a phase mask. The intensity of the ultraviolet beam was varied as the beam was scanned along the length of the phase mask, creating the modulated AC component of the grating. Then the unmodulated DC component of the grating was imprinted in the same manner without the phase mask present [12]. This "post-processing" of the grating ensures that the average refractive index of the grating remains approximately constant

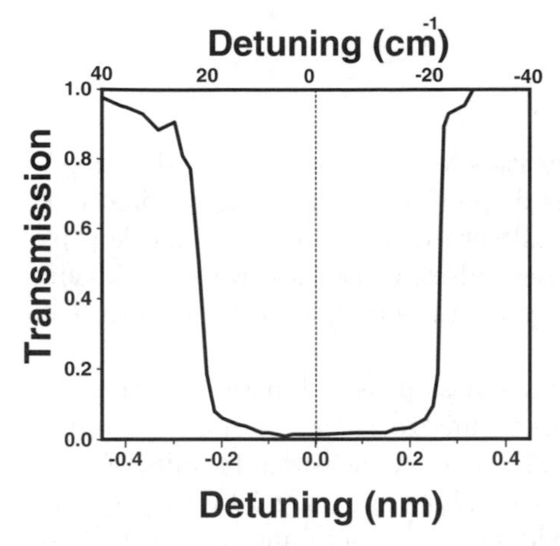

Fig. 8.9. Photonic bandgap spectrum of the fiber grating used in experiments

along the length of the grating. Both apodization and the unifrom average index are required to reduce side-lobes in the reflection spectrum [12,13,40]. The apodization profile was a raised cosine function as shown in Fig. 8.2, which is typical of profiles used in designing gratings for WDM lightwave applications [40].

We obtained a high-resolution reflection spectrum of the fiber grating, centered at 1.051 µm, by strain tuning it with respect to the fixed wavelength of the mode-locked, Q-switched Nd:YLF laser; the relation between the strain and the wavelength or frequency shift is well known [41]. The resolution was limited by the precision mechanical translation stage which was used to strain tune the grating, and by temperature variations. These effects give a combined effective resolution of roughly $50\,\mathrm{m}^{-1}$. The measured reflection spectrum is shown in Fig. 8.9. Note the steep edges in the spectrum indicating the short and long wavelength sides of the photonic bandgap. Such steep edges are characteristic for apodized gratings, and differ from those of uniform gratings, where the reflectivity decreases gradually when tuning away from the photonic band gap edge. The grating bandwidth is approximately 0.17 nm, corresponding to a coupling strength of $\kappa = 700\,\mathrm{m}^{-1}$ in the uniform section of the grating.

We note that the analysis of our high-intensity measurements in Sect. 8.5.2 requires us to know accurately the position of the high-frequency edge of the photonic band gap. However, we cannot deduce this quantity from a direct measurement of the spectrum, because at the photonic band gap edge the spectrum does not exhibit any particular features. Instead, the position of the edge was deduced from the low-intensity pulse propagation measurements discussed in Sect. 8.5.1.

8.5 Experimental Results

8.5.1 Linear Regime

In the first set of experiments the intensities were kept well below the threshold for nonlinear phenomena. Here the dispersive effects of the grating dominate. The results of these experiments were used to deduce some key parameters of the grating and the incoming pulses. Once these were established they were used, unchanged, in the analysis of the high-intensity experiments described in Sect. 8.5.2.

We measured the low-intensity transmission as a function of time for 13 different values of the strain, corresponding to 13 values of the detuning. Note that the experimental detunings $\Delta = \delta/c$ are measured in units of m^{-1}. Seven of the experimental pulse shapes are shown in Fig. 8.10 for a sequence of detunings. Three key trends are evident as the detuning decreases: first, the delay increases; second, the total transmitted energy decreases; and third, the broadening of transmitted pulse increases. The group velocity increases gradually from zero at the band gap edge, to V sufficiently far from the gap due to multiple scattering from the perioodic grating. This explains the trend that the delay increases with decreasing detuning. The strong frequency dependence of the group velocity implies that the group velocity dispersion is large. This dispersion is largest close to the edge of the photonic band gap and it leads to an increase in broadening with decreasing detuning. Note that for the three traces with the largest detunings, and thus the smallest delays, the pulse shape is essentially the same, indicating that for these detunings the dispersion is negligible over the length of the grating. Finally, for frequencies inside the photonic band gap the reflectivity essentially reaches unity for the

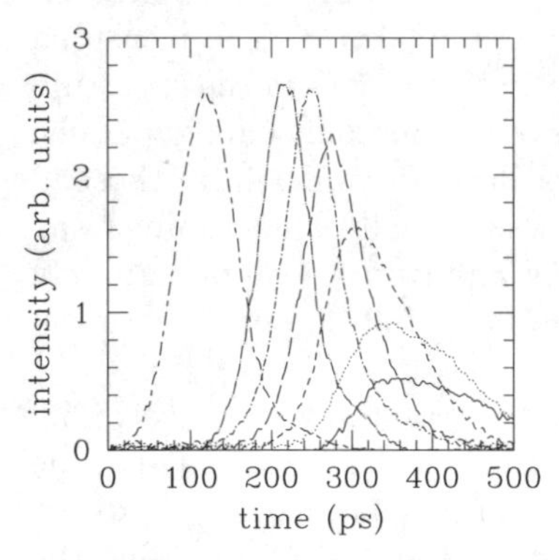

Fig. 8.10. Pulse intensity as a function of time after propagation through a fiber grating at low intensities (linear regime). Seven different values of detuning from the short wavelength grating edge are shown (solid curve, $820\,\mathrm{m}^{-1}$; dotted curve, $850\,\mathrm{m}^{-1}$; short-dashed curve, $910\,\mathrm{m}^{-1}$; long-dashed curve, $960\,\mathrm{m}^{-1}$; short-dashed dotted curve, $1020\,\mathrm{m}^{-1}$; long-dashed-dotted curve, $1170\,\mathrm{m}^{-1}$; and long-short-dashed curve, $3610\,\mathrm{m}^{-1}$)

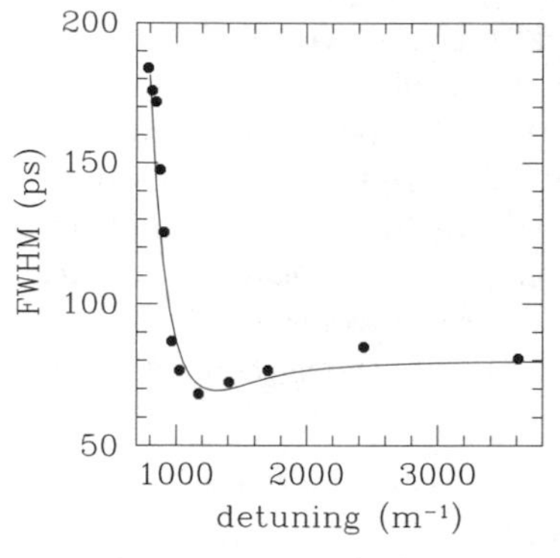

Fig. 8.11. Pulse widths (FWHM) of the optical pulse transmitted through the grating as a function of the detuning from the photonic bandgap edge. Experimental values are shown as the dots and numerical simulations are shown as a solid line

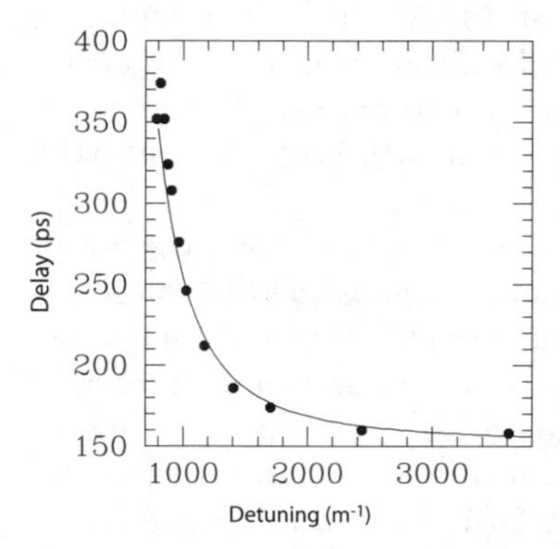

Fig. 8.12. Delay of the transmitted optical pulse as a function of the detuning from the photonic bandgap edge. Experimental values are shown as dots and numerical simulations are shown as a solid line

strong gratings we are considering here. Even though for apodized gratings the out-of-gap reflectivity is reduced compared to that in uniform gratings, it is not eliminated. For small detunings, a fraction of the pulse frequencies will be reflected even for very sharp spectral features at the bandgap edge. This leads to a reduction in the transmitted energy, strongly pronounced for the small detunings.

From the measurements in Fig. 8.10 we extracted the full-width at half-maximum (FWHM) of the transmitted pulse, as well as the relative arrival times, indicating the pulse delay. The results are indicated by the dots in Figs 8.11 and 8.12. The pulse widths were deconvolved to account for the 20 ps response time of the detection system.

From the result at the largest detuning in Fig. 8.11 we conclude that the FWHM of the incoming pulses was 80 ps, since at these large detunings the pulse is not affected by the grating, and thus the width of the pulse is unchanged upon propagation. We determine the photonic bandgap edge wavelength from the data in Fig. 8.12. From this edge position we calibrate the detuning of the optical pulses from the gap edge. Finally, from the depth and the position of the minimum in Fig. 8.11 we deduce that the incoming pulses have a self-phase modulation of 0.13π radians at their peak. Using these values (8.5) and (8.6) were solved numerically. The results of these simulations are indicated by the lines in Figs. 8.11 and 8.12. Clearly the agreement is very good.

8.5.2 Fundamental Soliton Regime

The experimental results in the region where fundamental solitons are predicted for detuning from the short wavelength edge of the gap are described in detail in this section. The pulse intensities for soliton formation depend on detuning and typically require intensities near $10\,\mathrm{GW/cm^2}$ for detuning of the order of the gap width. If the fundamental soliton forms at an intensity of $10\,\mathrm{GW/cm^2}$, the second order, $N = 2$, soliton is found near $25\,\mathrm{GW/cm^2}$. We will show that the experiments agree very well with both the NLCMEs and the simpler NLSE models.

After the low-intensity measurements (Sect. 8.5.1) we increased the power of the incident pulses by removing an attenuating neutral density filter between the laser and the fiber. The small self-phase modulation of the incoming pulses is thus unchanged. We performed six sets of measurements. In each of these we kept the pulse energy constant, and varied the fiber strain, corresponding to the detuning [41]. An example of such a series of measurements, taken at an estimated input pulse energy of $0.20\,\mu\mathrm{J}$ is shown in Fig. 8.13.

These nonlinear results should be compared with its linear counterpart Fig. 8.10. The detuning of $3612\,\mathrm{m^{-1}}$ is sufficiently large that its shape essentially duplicates that of the input pulse. Both figures show that the average velocity of the pulse increases with detuning, as expected. However, in contrast to linear data in Fig. 8.10, the nonlinear pulse shapes Fig. 8.13 also show significant compression at intermediate values of the detuning. At a detuning of $1053\,\mathrm{m^{-1}}$, for example, the width of the transmitted pulse is 33 ps, or 26 ps after deconvolution, corresponding to a compression ratio of over 3. This compression, associated with the formation of solitons [4], is known to occur in uniform fibers, but over length scales of hundreds of meters instead of centimeters in our experiments.

A striking example of slow Bragg soliton propagation where the incident pulse shape is nearly preserved is shown in Fig. 8.14. The experimental conditions are similar to those in Fig. 8.13 except that the pulse energy of $0.66\,\mu\mathrm{J}$ is higher. The two traces in Fig. 8.14 correspond to detunings far from the edge of the photonic band gap (pulse on the left of the figure) and close to the gap edge (delayed soliton pulse on the right). Note that the pulses have roughly the same width and peak intensity, yet the delay is 310 ps. Assuming

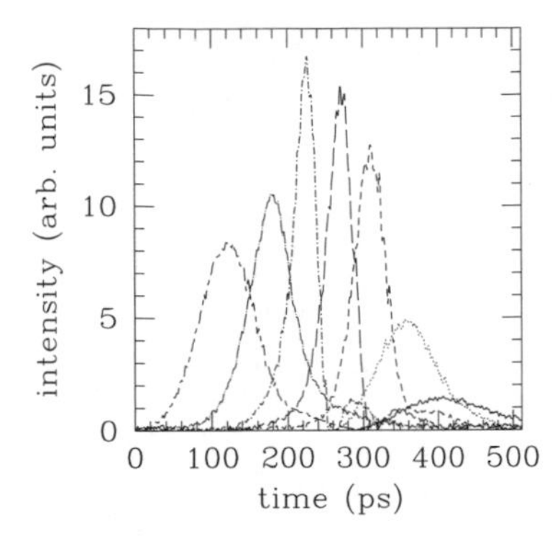

Fig. 8.13. Pulse intensity as a function of time after propagation through a fiber grating at a peak input intensity of 11 GW/cm^2. Seven different values of detuning from the short wavelength grating edge are shown (solid curve, 729 m^{-1}; dotted curve, 788 m^{-1}; short-dashed curve, 847 m^{-1}; long-dashed curve, 935 m^{-1}; short-dashed dotted curve, 1053 m^{-1}; long-dashed-dotted curve, 1406 m^{-1}; and long-short-dashed curve, 3612 m^{-1})

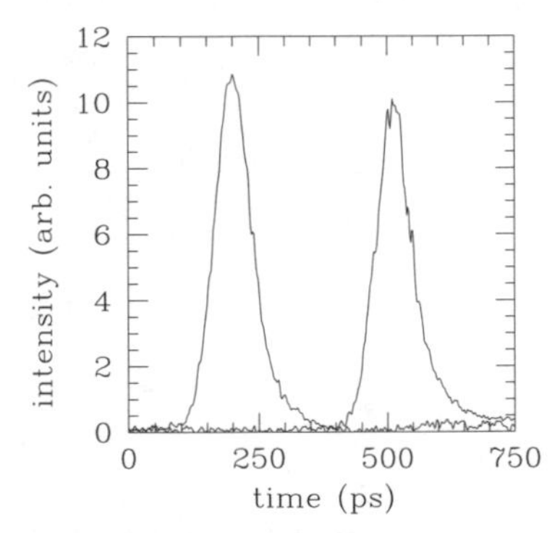

Fig. 8.14. Pulse intensity as a function of time after propagation through the grating at a pulse energy of 0.66 µJ, for a detuning close to the gap edge (pulse peak near 510 ps) and far from the gap edge (pulse peak near 200 ps). The pulse tuned near the gap edge is nearly a soliton with a velocity of 0.5 V

adiabatic adjustment of the pulse velocity in the apodized grating region, this corresponds to a velocity in the uniform section of the fiber of 0.5 V. The slowly propagating pulse in Fig 8.14 is a fundamental Bragg soliton. Numerical simulations show that it propagates long distances without a change in shape. Our limited experimental grating length only allows for propagation over a few formation lengths (see 8.27).

Since the pulse delays are only weakly intensity-dependent, we analyse the nonlinear reults by considering only the pulse widths. The dots in Figs. 8.15 and 8.16 show the measured widths of the transmitted pulses versus detuning, for estimated in-fiber pulse energies of 0.06 µJ and 0.14 µJ, respectively. The

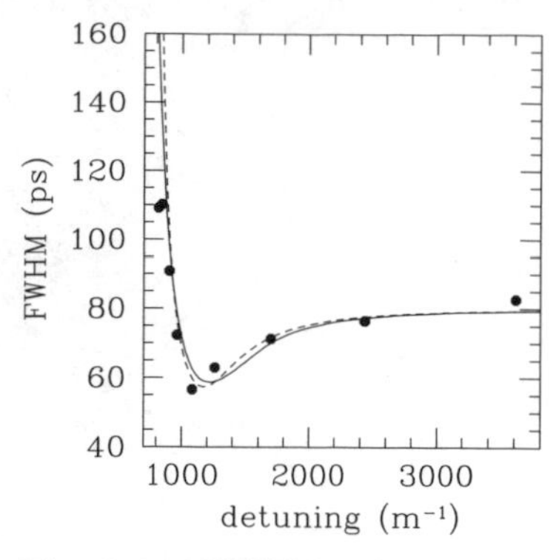

Fig. 8.15. FWHM pulse widths for transmitted pulses as a function of detuning at a pulse energy 0f 0.06 µJ. Experimental results are indicated by the solid circles. Numerical results are shown solving the NLCMEs (solid curve) and the NLSE (dashed curve). Both sets of numerical results assume a peak intensity of 3 GW/cm^2 for the incoming pulse

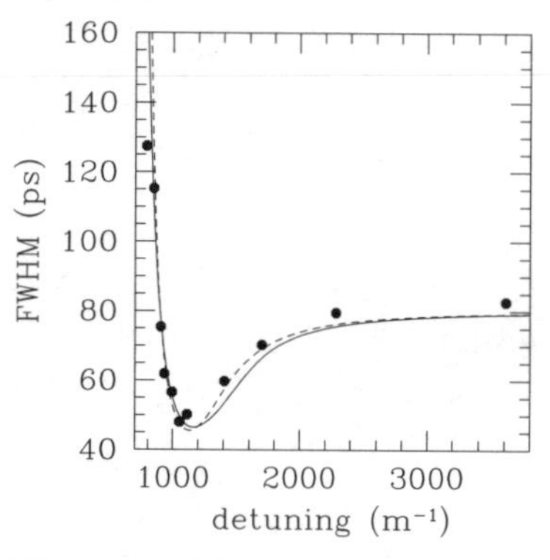

Fig. 8.16. FWHM pulse widths for transmitted pulses as a function of detuning at a pulse energy of 0.14 µJ. Experimental results are indicated by the solid circles. Numerical results are shown solving the NLCMEs (solid curve) and the NLSE (dashed curve). Both sets of numerical results assume a peak intensity of 6 GW/cm^2 for the incoming pulse

solid lines are results obtained by numerically solving the NLCME (8.5) and (8.6) with the parameters determined in Sect. 8.5.1, and peak intensities of 3 GW/cm^2 and 6 GW/cm^2, respectively. The dashed lines are results obtained by numerically solving the nonlinear Schröedinger equation, (8.22), treating the tapered parts of the grating as discussed in [6]. Notice that the agreement between experiment and both sets of simulations is excellent. It is important to note that the pulse widths obtained at detunings of roughly 1000 m^{-1} are substantially shorter than those obtained in the limit, especially at the higher energy (compare Figs. 8.16 and 8.11). This is due to nonlinear soliton

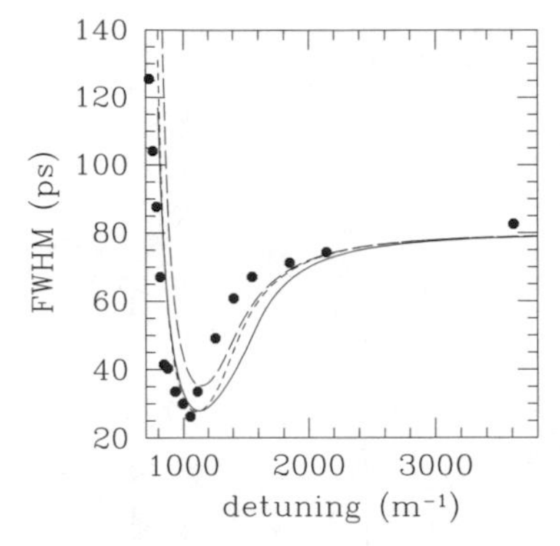

Fig. 8.17. FWHM pulse widths for transmitted pulses as a function of detuning at a pulse energy of 0.2 µJ. Experimental results are indicated by the solid circles. Numerical results are shown solving the NLCMEs (solid curve) and the NLSE (dashed curve). Both sets of numerical results assume a peak intensity of 11 GW/cm^2 for the incoming pulse

compression [3]. At smaller values of the detuning the pulses actually begin to broaden because the grating dispersion begins to outweigh nonlinear effects.

Figure 8.17 is similar to Figs. 8.15 and 8.16 except that the estimated pulse energy is 0.20 µJ. In this case the simulations, indicated by the solid line [for (8.5) and (8.6)], and the short-dashed line [for (8.22)], were performed for peak intensities of 11 GW/cm^2. The long-dashed curve shows results of numerical solutions to the nonlinear Schröedinger equation, but without the field enhancement factor, v^{-1}, that accounts for the difference of the field inside and outside the grating. The inclusion of this factor, which is most prominent at small detunings where v is small, is clearly required to get the correct estimate at small detunings, and to obtain a good estimate for the minimum pulse width.

The final results in the fundamental soliton regime are shown in Fig. 8.18, for which the estimated pulse energy is 0.36 µJ. In the simulations in Fig. 8.18 the peak pulse power was taken to be 13 GW/cm^2. Here the trends from the previous figures continue, particularly the increasingly strong compression at detunings around 1000 m^{-1}. We note that the deduced peak power for this set of measurements is somewhat uncertain since the deconvolved pulse widths are close to the experimental resolution of 20 ps.

Our experimental results are in good agreement with theory, based both on the NLCME and the NLSE. The NLSE treatment is useful since it allows us to interpret our results in terms of fundamental and higher order solitons. The agreement between experiments and theory is particularly good in the fundamental soliton regime. At higher intensities we find that the incident pulse breaks into two pulses as expected for the higher order $N = 2$ soliton regime at intensities above 25 GW/cm^2. These high intensity experiments are difficult to analyze since the temporal structure is finer than our experimental

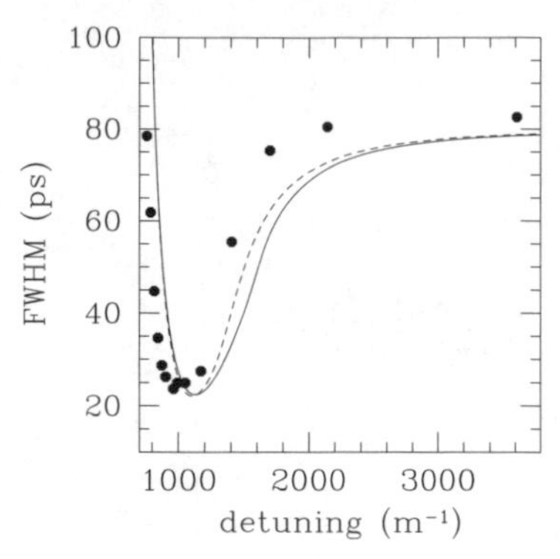

Fig. 8.18. FWHM pulse widths for transmitted pulses as a function of detuning at a pulse energy of 0.36 µJ. Experimental results are indicated by the solid circles. Numerical results are shown solving the NLCMEs (solid curve) and the NLSE (dashed curve). Both sets of numerical results assume a peak intensity of 13 GW/cm^2 for the incoming pulse

resolution of approximately 20 ps. The extreme pulse narrowing also means that the cubic grating dispersion cannot be neglected.

8.5.3 Modulational Instability Experiments

Modulational instabilites in uniform fibers were first seen experimentally in uniform fibers by observing pulse breakup of 100 ps optical pulses [48]. In our fiber grating experiments, at input intensities above those required for the fundamental soliton, we clearly see a modulational instability as shown in Fig. 8.19. These results were obtained for an input pulse intensity of 25 GW/cm^2. The grating is detuned just above the short wavelenght gap edge where the transmission is approximately unity. We found that we could tune the period of the modulational instability by nearly a factor of three from 16

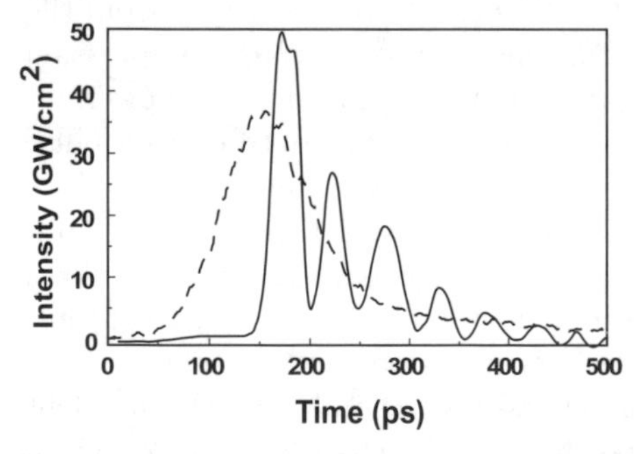

Fig. 8.19. Transmitted intensity through a 6 cm fiber grating at an input intensity of 25 GW/cm^2 at a detuning far from the gap (dotted curve) and near the edge of the gap (solid curve)

to 40 GHz by simply straining the fiber in order to change the detuning (see 8.28). These modulational instability frequencies agree well with both simulations using the coupled mode equations and with the effective NLSE. The modulational instability was better developed at small detunings, consistent with the predicted increase in growth rate at small detuning in (8.28).

8.5.4 Nonlinear Polarization Dependent Propagation Experiments

We now consider the more general nonlinear propagation case where the polarization does not lie along a principle polarization axis or is not linearly polarized. The two major new effects in this case are polarization instabilities and new types of solitons that propagate at various detunings of the pulse from the gap edge.

As described in Sect 8.3.4, a polarization instability is expected when the nonlinear change in the index becomes comparable to the linear birefringence. Above this threshold intensity the principle axes are no longer well defined since the net birefringence depends on the pulse intensity and polarization orientation. For pulses with polarizations initially along the fast polarization axis, we find that in the nonlinear intensity regime a major portion of the pulse energy is transferred to the slow axis as the pulse propagates through a length of fiber grating comparable to the polarization beat period. Polarizations initially aligned near the slow axis are stable and very nearly retain their polarization angle. Similar polarization instability effects have been measured in bare fiber [49]. In the fiber grating case the thresholds for polarization instabilites are much different in magnitude and vary strongly with detuning.

As shown in Fig. 8.6 the birefringence remains nearly constant in our experiments as the detuning decreases near the short wavelength edge of the gap while the nonlinearity increases dramatically. A significant drop in the polarization instability is expected, as shown for a single component in the pulse frequency spectrum in Fig. 8.6. A complete analysis of the pulsed case requires the solution of the NLCMEs, however the polarization threshold effects and their qualitative threshold intensities are close to the CW predictions. Our experimental measurements [8] of the instability threshold intensity and the corresponding simulated numerical results are shown in Fig. 8.20. As expected the threshold intensity decreases with decreasing detuning until reflection from the grating at very small detunings begins to increase the incident intensity required for the instability.

Pulse shapes for the linear and nonlinear regime are shown in Fig. 8.21. Note that in the linear regime 95% of the pulse energy is along the fast axis. At the detuning and intensity for the data in Fig. 8.21b, we are above the instability threshold and it is clear that a major fraction of the energy transfers during the pulse from the fast to the slow axis. One can also see a pulsating

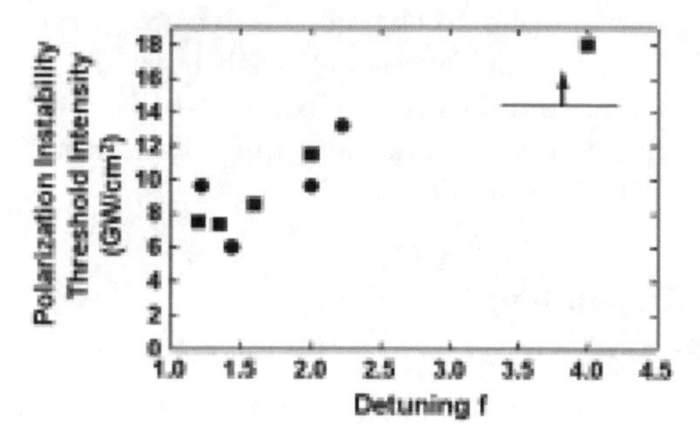

Fig. 8.20. Threshold intensities for the polarization instability for experimental (filled circles) and numerically simulated (filled squares) fiber Bragg gratings 7.7 cm in length with a birefringence of 3.5×10^{-6} and an index modulation of 8×10^{-5}. The detuning is $f = 1.4$

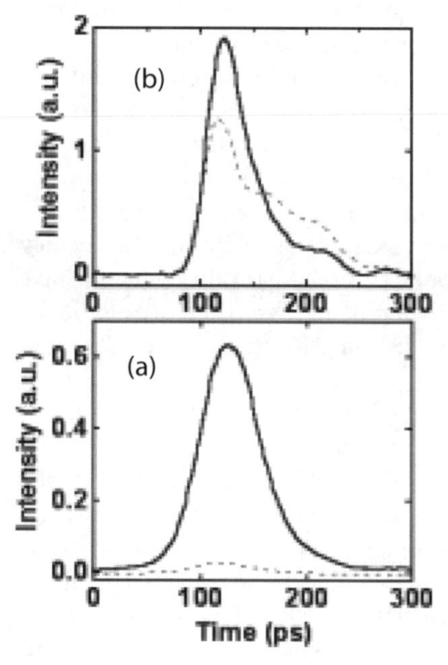

Fig. 8.21. Experimental pulse shapes are shown for the linear (**a**) and nonlinear (**b**) regimes for the fast (solid curves) and the slow (dashed curves) axes. The solid and dashed curves in **a** are for the linear regime at 0.5 GW/cm^2 and the solid and dashed curves in **b** are for the nonlinear regime at 10 GW/cm^2. The detuning is $f = 1.4$

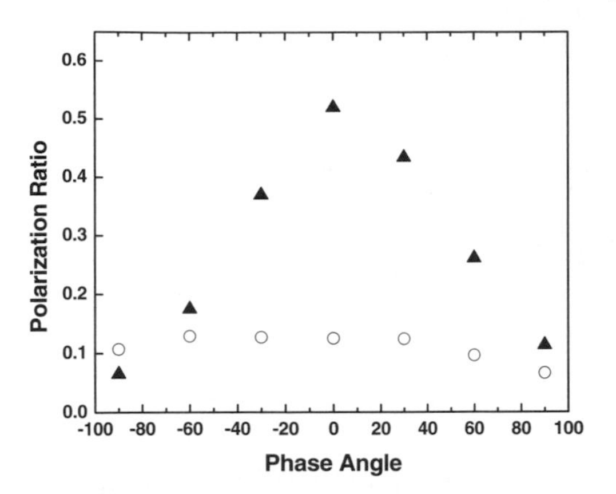

Fig. 8.22. Polarization intensity ratio (slow-intensity/fast-intensity) as a function of phase between the initial fast and slow axis polarization components. The open circles correspond to the linear case where the incident intensity is only 0.2 GW/cm^2. The ratio of intensities along the slow and the fast axis is set at approximately 0.1. In the nonlinear regime (solid triangles) at intensities near 8 GW/cm^2 the polarization instability results in a transfer of energy to the slow axis and an increase in the polarization ratio that depends strongly on the incident phase relation. The detuning is approximately $f = 2$ for these measurements

exchange of energy between the two polarizations as the pulse evolves in time that is often observed in the nonlinear regime at small detunings ($f < 2$).

Another interesting nonlinear polarization effect is the dependence of pulse evolution on the relative phase, $\Delta\phi = \phi_X - \phi_Y$, of the initial polarizations along the fast and slow axes. Note that the polarization instability can be thought of as the nonlinear cancellation in the phase mismatch terms (see (8.30) and (8.31)). Evolution of the instability is expected depend on the initial relative phases of the two polarizations or the degree of linear polarization. This effect is demonstrated by the results shown in Fig. 8.22 and analyzed in more detail in reference [11]. The grating length is approximately half of a birefringent beat period for the experiments described here. This length nearly maximizes the polarization instability effects and their strong dependence on initial phase lag. The experimental results in Fig. 8.22 are in qualitative agreement with numerical simulation results [11].

A complete description of the polarization state of the pulse during its nonlinear evolution in the grating is given by the Stokes parameters [51],

$$S_0 = {E_X}^2 + {E_Y}^2 \,, \tag{8.35}$$

$$S_1 = {E_X}^2 - {E_Y}^2 \,, \tag{8.36}$$

$$S_2 = 2E_X E_Y \cos(\Delta\phi) \,, \tag{8.37}$$

$$S_3 = 2E_X E_Y \sin(\Delta\phi) \,, \tag{8.38}$$

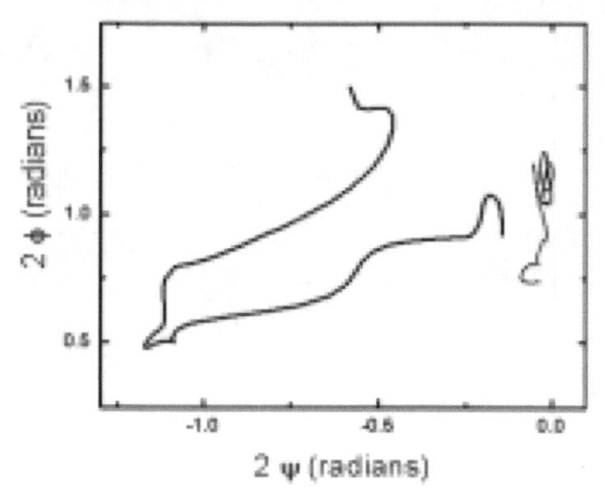

Fig. 8.23. Trajectory of the Stokes vector of an optical pulse propagating in a birefringent fiber grating as a function of the two angles, 2ψ and 2ϕ, that describe the orientation of the Stokes vector. The trajectory starts when the pulse reaches approximately 10% of its peak value and ends when it falls back to 10% of the peak. The trajectories during the pulse are shown as the localized linear trajectory on the right and the looping nonlinear trajectory on the left. The detuning in this experiment is $f = 2.0$ and the intensity in the nonlinear case is $8\,\mathrm{GW/cm^2}$

where E_X is the optical field component along the fast axis and E_Y is the field compnent along the slow axis. The polarization dynamics are visualized by plotting the Stokes parameters on the Poincare sphere, a three dimensional plot using S_1, S_2 and S_3 as axes. Since the total pulse energy is approximately constant, the Stokes vector (S_1, S_2, S_3) dynamics can be plotted on a map of the two vector orientation angles

$$2\psi = \sin^{-1}\left(\frac{S_3}{S_0}\right) \tag{8.39}$$

$$2\phi = \tan^{-1}\left(\frac{S_2}{S_1}\right) \tag{8.40}$$

that give the Stokes vector position on a Poincare sphere with radius

$$S_0 = \sqrt{S_1^2 + S_2^2 + S_3^2}\,. \tag{8.41}$$

An experimental measurement of these angles during a pulse is shown in Fig. 8.23 for both the linear and nonlinear regimes. In the nonlinear case the Stokes vector makes an impressive excursion around the Poincare sphere.

At large detunings (see Fig. 8.7) we are in the regime where the intensity threshold for soliton formation is less than the intensity threshold for polarization instability, $I_{\mathrm{sol}} \ll I_{\mathrm{PI}}$. In this large detuning regime we can excite pulses that retain their shape over long distances and do not exchange energy between the polarization components. This is a regime where solitary wave propagation is similar to that in bare fibers explored experimentally by Silberberg and Barad [50]. At smaller detunings where $I_{\mathrm{sol}} \approx I_{\mathrm{PI}}$ and $I_{\mathrm{sol}} > I_{\mathrm{PI}}$

there is strong exchange of energy between the polarization components. In this regime there may still be solitary wave formation after energy is shed from the the pulse during energy exchange bewteen the polarization components [11,52–54]. This is a very interesting regime that remains to be explored in more detail.

8.6 Conclusions and Future Directions

We have presented detailed experimental and theoretical results on Bragg soliton propagation in apodized fiber gratings. We have confirmed that these phenomena can be well described by the NLCME, and also by a much simpler NLSE that gives many insights into the soliton formation process at wavelengths near the short wavelength edge of fiber grating bandgaps. Some discrepancies between the experiments and theory occurred in the high intensity, two-soliton regime, where the present experiments are difficult to interpret because of time resolution limits and third order dispersion effects begin to play a significant role.

The NLSE description can also be valuable in describing nonlinear pulse propagation through nonuniform Bragg gratings; see for example the recent paper by Slusher et al. [42], which describes nonlinear pulse propagation in chirped fiber Bragg gratings. Lenz and Eggleton [10] recently proposed a pulse compression scheme based on adiabatic evolution of Bragg solitons in a nonuniform Bragg grating. The analysis in this paper [10] was based on the NLSE and good agreement was found between numerical results of the NLCME and the NLSE. We also recently considered interactions between Bragg solitons in the context of the NLCME [43,44]. Numerical simulations indicated that for parameters similar to those in experiments [44], the interaction dynamics resembled those of a NLSE solitons, and thus many of the results of NLSE interactions could be brought to bear [43,45].

We have demonstrated the efficient launching of slow Bragg solitons with a velocity of approximately 50% of the speed of light in bare fiber, leading to delays of greater than 300 ps. One of the limiting factors in launching even slower Bragg solitons is the residual reflections that exist outside of the photonic bandgap, in spite of the apodization. Can very slow soltions be launched with more efficient coupling into the grating? Recent simulations show that tapering the grating apodized regions over several centimeters can lead to much slower soliton velocities near c/10. In these cases there can be a significant portion of the pulse frequencies within the gap. The large pulse delays we have observed in fiber grating experiments are of interest for designing optical delay lines or optical buffers.

The large pulse energies and intensities required to launch Bragg solitons and gap solitons can be dramatically reduced by using materials with much higher nonlinearities than silica. Higher nonlinearities can be obtained by decreasing the material bandgap to the regime where the pulse wavelengths are just below the half-gap wavelengths. For example, in both AlGaAs and the chalcoginide glasses the nonlinearities are as much as a thousand times larger than in silica glass. These large nonlinearities along with reduced waveguide dimensions could lower the thresholds for soliton formation and polarization instabilities by at least four orders of magnitude. However, there are additional losses due to multiphoton absorption in these lower bandgap materials. The ratio of nonlinear index to nonlinear loss is measured by a figure of merit (FOM)

$$\mathrm{FOM} = \frac{n_2}{\beta\lambda}, \tag{8.42}$$

where β is the two photon absorption coefficient. AlGaAs and the chalcogenides have figures of merit in the range from 4 to 40. This means that for the nonlinear phase shifts near π, required for most of the nonlinear effects, there is a nonlinear loss from two-photon absorption of the order of 1 to 10%. This loss will severely limit the soliton propagation distance and is a major problem for future experiments. If the pulse wavelengths are tuned to near the material bandgap in order to achieve even larger nonlinear coefficients, real carriers are generated that limit the response of the system and severely obscure the nonlinear dynamics. Future material engineering advances may well provide the improved parameters required for further nonlinear propagation studies and many new applications.

Acknowledgements. We are grateful to Thomas A. Strasser and Renee Pedrazzani for fabrication of the fiber gratings used in the experiments. We also acknowledge fruitful conversations with Suresh Pereira, John Sipe, C.M. de Sterke, Demetrios Christodoulides, Gadi Lenz, Natalia Litchinitser and Alejandro Aceves.

References

1. B.J. Eggleton, R.E. Slusher, C.M. de Sterke, P.A. Krug, and J.E. Sipe: Phys. Rev. Lett. **76**, 1627 (1996)
2. D. Taverner, N.G.R. Broderick, D.J. Richardson, R.I. Laming, and M. Ibsen: Opt. Lett. **23**, 328 (1998); N.G.R. Broderick, D.J. Richardson, and M. Ibsen: Opt. Lett. **25**, 536 (2000)
3. G.P. Agrawal, *Nonlinear fiber optics*, 2nd Edn. (Academic Press, San Diego 1995)
4. L.F. Mollenauer, R.H. Stolen, and J.P. Gordon: Phys. Rev. Lett. **45**, 1095 (1980)
5. H.G. Winful: Opt. Lett. **11**, 33 (1986)
6. B.J. Eggleton, C.M. de Sterke, and R.E. Slusher: J. Opt. Soc. Am. B **16**, 587 (1999)
7. B.J. Eggleton, C.M. de Sterke, A.B. Aceves, J.E. Sipe, T.A. Strasser, and R.E. Slusher: Opt. Comm. **149**, 267 (1998)

8. R.E. Slusher, S. Spalter, B.J. Eggleton, S. Pereira, and J.E. Sipe: Opt. Lett. **25**, 749 (2000)
9. G. Lenz, B.J. Eggleton, and N. Litchinitser: J. Opt. Soc. Am. B **15**, 715 (1997)
10. G. Lenz and B.J. Eggleton: J. Opt. Soc. Am. B **15**, 2979 (1998)
11. S. Pereira, J.E. Sipe, and R.E. Slusher: submitted to J. Opt. Soc. Am. B
12. T.A. Strasser, P.J. Chandonnet, J. DeMarko, C.E. Soccolich, J.R. Pedrazzani, D.J. DiGiovanni, M.J. Andrejco, and D.S. Shenk, in *Optical Fiber Communication Conference* Vol. 2 of 1996 OSA Technical Digest Series (Optical Society of America, Washington, D.C., 1996): postdeadline paper PD8-1
13. P.S. Cross and H. Kogelnik: Opt. Lett. **1**, 43 (1977)
14. B. Malo, D.C. Johnson, F. Bilodeau, J. Albert, and K.O. Hill: Elect. Lett. **31**, 223 (1995)
15. R. Kashyap: *Fiber Bragg Gratings*, 1st edn. (Academic Press, New York 1999)
16. J. Martin and F. Ouellette: Electron. Lett. **30**, 812 (1994)
17. K.O. Hill, B. Malo, F. Bilodeau, D.C. Johnson, and J. Albert: Appl. Phys. Lett. **62** 1035 (1993)
18. C.M. de Sterke and J.E. Sipe: Phys. Rev. A **42**, 550 (1990)
19. B.J. Eggleton, T. Stephens, P.A. Krug, G. Dhosi, Z. Brodzeli, and F. Ouellette: Elect. Lett. **32**, 1610 (1996)
20. P.S.J. Russell: J. Mod. Opt. **38**, 1599 (1991)
21. N.M. Litchinitser, B.J. Eggleton, and D.B. Patterson: J. Lightwave Technol. **15**, 1303 (1997)
22. N.M. Litchinitser, G.P. Agrawal, B.J. Eggleton, and G. Lenz: Optics Express **3**, 411 (1998)
23. D.N. Christodoulides and R.I. Joseph: Phys. Rev. Lett. **62**, 1746 (1989)
24. A.B. Aceves and S. Wabnitz: Phys. Lett. A **141**, 37 (1989)
25. B.J. Eggleton, C.M. de Sterke, and R.E. Slusher: Opt. Lett. **21**, 1223 (1996)
26. B.J. Eggleton, C.M. de Sterke, and R.E. Slusher: J. Opt. Soc. Am. B **14**, 2980 (1997)
27. C.M. de Sterke, B.J. Eggleton, and P.A. Krug: J. Lightwave Technology **15**, 2908 (1997)
28. D. Taverner, N.G.R. Broderick, D.J. Richardson, M. Ibsen, and R.I. Laming: Opt. Lett. **23**, 259 (1998)
29. H.G. Winful: Appl. Phys. Lett. **46**, 527 (1985)
30. B. Malo, D.C. Johnson, F. Bilodeau, J. Albert, and K.O. Hill: Elect. Lett. **31**, 223 (1995)
31. H. Haus: Opt. Let. **17**, 1134 (1992)
32. H.G. Winful and G.D. Cooperman: Appl. Phys. Lett. **40**, 298 (1982)
33. C.M. de Sterke and J.E. Sipe: "Gap solitons," in *Progress in Optics XXXIII*, E. Wolf, Ed., (Elsevier, Amsterdam 1994), Chap. III-Gap Solitons
34. C.M. de Sterke, K.R. Jackson, and B.D. Robert: J. Opt. Soc. Am B **8**, 403 (1991)
35. M.J. Steel and C.M. de Sterke: Phys. Rev. A **49**, 5048 (1994)
36. C.M. de Sterke: Optics Express **3**, 405 (1998)
37. D. Von der Linde: IEEE J. Quant. Elect. **QE-8**, 328 (1972)
38. B.J. Eggleton, G. Lenz, R.E. Slusher, and N.M. Litchinitser: Appl. Opt. **37** 7055-7061 (1998)
39. B.J. Eggleton, C.M. de Sterke, A.B. Aceves, J.E. Sipe, T.A. Strasser, and R.E. Slusher: Optics Comm. **149**, 267 (1998)

40. J.E. Sipe, B.J. Eggleton, and T.A. Strasser: Opt. Comm. **152**, 269 (1998)
41. K.O. Hill, Y. Fujii, D.C. Johnson, and B.S. Kawasaki: Appl. Phys. Lett. **32**, 647 (1978)
42. R.E. Slusher, B.J. Eggleton, T.A. Strasser, and C.M. de Sterke: Optics Express **3**, 465 (1998)
43. B.J. Eggleton, R.E. Slusher, N.M. Litchinitser, G.P. Agrawal, A.B. Aceves, and C.M. de Sterke, "Experimental observation of interaction between Bragg solitons," in *International Quantum Electronics Conference*, Vol. 7 of 1998 OSA Technical Digest Series (Optical Society of America, Washington, D.C., 1998, paper QTuJ5
44. N.M. Litchinitser, B.J. Eggleton, C.M. de Sterke, A.B. Aceves, and G.P. Agrawal: J. Opt. Soc. Am. B **16**, 18 (1998)
45. J.P. Gordon: Opt. Lett. **8**, 596 (1983)
46. S. Wang, H. Erlig, H. Fetterman, V. Grubsky, and J. Feinberg: Proc. SPIE **3228**, 407 (1997)
47. M. Scalora, R.J. Flynn, S.B. Reinhardt, R.L. Fork, M.J. Bloemer, M.D. Tocci, C.M. Bowden, H.S. Ledbetter, J.M. Bendickson, J.P. Dowling and R.P. Leavitt: Phys. Rev. E. **54**, 1078 (1996)
48. K. Tai, A. Hasegawa, and A. Tomita: Phys. Rev. Lett. **56**, 135 (1986)
49. S. Trillo, S. Wabnitz, R.H. Stolen, G. Assanto. C.T. Seaton and G. Stegeman: Appl. Phys. Lett. **49**, 1224 (1986)
50. Y. Silberberg and Y. Barad: Opt. Lett. **20**, 246 (1995)
51. N.N. Akhmediev and A. Ankiewicz, *Solitons: Nonlinear pulses and beams*, First Edition (Chapman and Hall, London, UK, 1997)
52. N.N. Akhmediev, A.Q. Buryak, and J.M. Soto-Crespo: Opt. Comm. **112**, 278 (1994)
53. N.N. Akhmediev and J.M. Soto-Crespo: Phys. Rev. E **49**, 5742 (1994)
54. J.M. Soto-Crespo, N.N. Akhmediev, and A. Ankiewicz: J. Opt. Soc. Am. B **12**, 1100 (1995)

9 Gap Solitons Experiments within the Bandgap of a Nonlinear Bragg Grating

N.G.R. Broderick

In 1979 Winful et al. showed theoretically that the transmission of a nonlinear Bragg gratings was a multivalued function of the input intensity [1] when the frequency of the incident light lay within the grating's band gap. Since then much work both experimental and theoretical has been done on the nonlinear properties of Bragg gratings and in this chapter we present a brief review of our work concentrating on nonlinear light propagation in Bragg gratings at frequencies within the bandgap.

9.1 Introduction

As is well known a Bragg grating reflects strongly at wavelengths near the Bragg wavelength λ_B given by:

$$\lambda_B = 2\bar{n}d \tag{9.1}$$

where $\bar{n}$ is the average refractive index and d is the period of the refractive index modulation. The grating's bandgap is defined as the frequency region over which there are no propagating solutions to Maxwell's Eqs i.e. the highly reflective region. To lowest order the width of the bandgap is proportional to the depth of the refractive index modulation while the peak reflectivity or strength of the grating depends on the product of the length and the modulation depth. Thus if one wants to fabricate gratings with a narrow bandgap but which are still highly reflective then it is necessary to use long weak gratings. Conversely strong highly reflective gratings with a wide bandgap can be made relatively short. In fibres it is possible to create a refractive index modulation by exposing the fibre to a periodically modulated UV beam [2]. By tailoring the strength and period of the UV beam, gratings can be written with almost any desired impulse response [3,4]. In the linear regime Bragg gratings act as band stop filters in transmission or as band pass filters in reflection. The transmission of light through a Bragg grating depends strongly on the wavelength difference (or detuning) between the input wavelength and the Bragg wavelength and hence change in either the size or position of the grating's bandgap can significantly alter the transmission of the incident light. Bragg gratings can thus function as optical switches.

There are of course many ways to alter the characteristics of a Bragg grating ranging from strain or temperature tuning to using the acousto-optic effect [2]. Alternatively one can use light itself to switch the Bragg grating and here we explore the effects of the Kerr nonlinearity. In many materials the refractive index is given by [5]

$$n(I) = n_0 + n^{(2)} I \tag{9.2}$$

where I is the intensity of light and $n^{(2)}$ describes the strength of the nonlinearity. Since the Bragg wavelength (9.1) is proportional to the refractive index, (9.2) shows that it is possible to manipulate the properties of the grating with light itself, opening the door to effects such as all-optical switching.

This effects of nonlinearity on the reflection of a Bragg grating were first described by Winful et al. [1,6] who showed that nonlinear Bragg gratings were bistable and that for frequencies inside the bandgap the grating could in fact be 100% transmissive at high powers. It was then shown by Chen and Mills [7,8] that this was due to the presence of a localised stationary pulse of light within the structure which they called a gap soliton.[1] Analytic expressions for gap solitons in infinite gratings were found by Christodolous [9] and independently by Aceves and Wabnitz [10]. The theoretical properties of gap solitons are discussed in detail elsewhere (see for example the review by de Sterke and Sipe [11] and Chap. 2 in this current volume). The aim of this chapter is to complement the earlier chapter by Slusher and Eggleton who discussed the nonlinear propagation of light at frequencies outside the bandgap. Here I concentrate on describing experiments performed at the Optoelectronics Research Centre on nonlinear switching in Bragg gratings[2] at frequencies inside the bandgap. This work was done by myself and my colleagues D. Taverner, D. J. Richardson, P. Millar and M. Ibsen. This work falls naturally into three sections, (i) work on 8 cm fibre Bragg gratings, (ii) experiments using 20 cm long fibre Bragg gratings and (iii) experiments using gratings in AlGaAs waveguides.

9.2 Experimental Observation of Nonlinear Switching in an 8 cm Long Fibre Bragg Grating

Our initial experiments [12,13] were performed using an 8 cm long apodised fibre Bragg grating (FBG) whose reflection spectrum is shown in Fig. 9.1a where the two curves correspond to the two polarization axes in the fibre. The wavelength along the horizontal axis is given in terms of the difference from

[1] Note that originally the term gap soliton was used to include all solitary wave solutions to the nonlinear coupled mode equations whether or not their spectrum lay within the bandgap. However more recent terminology has restricted the use of gap soliton to only those solutions which lie within the bandgap and the phrase "grating soliton" is used to describe solutions outside the bandgap.

[2] This chapter is exclusively concerned with self-switching and the effects of cross-phase modulation are described in a later chapter

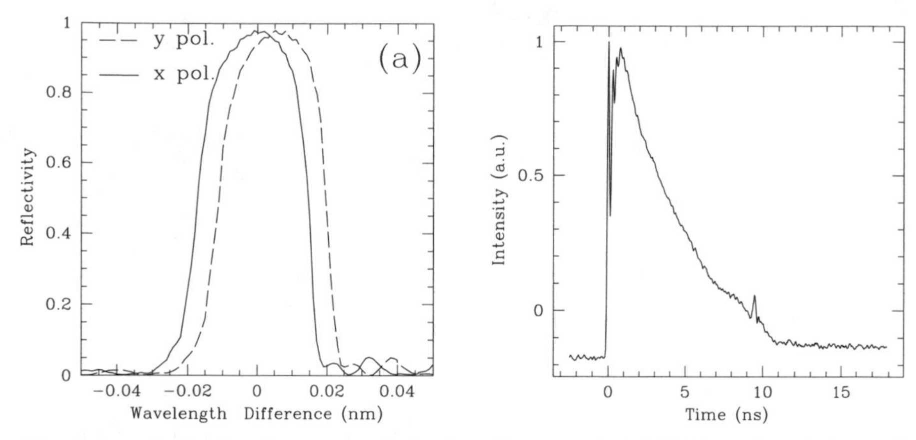

Fig. 9.1. **a** Reflection Spectrum of the 8 cm long apodised FBG used in the initial experiments. The two curves represent the two polarizations. **b** Input pulse profile

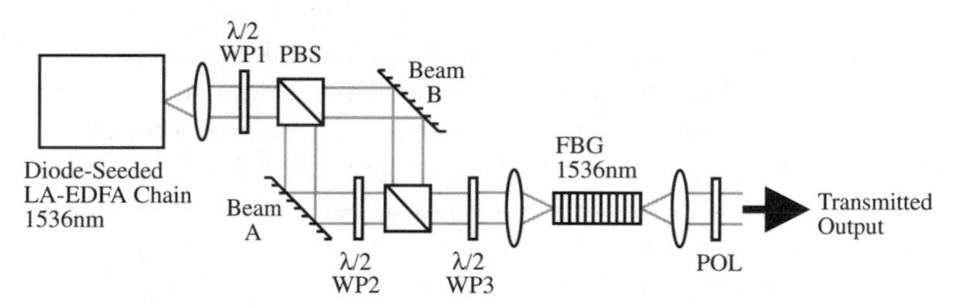

Fig. 9.2. Experimental Setup for nonlinear switching experiments

the Bragg wavelength of 1536 nm. The fibre grating was written into a germanosilicate fibre with a mode-area of 30 μm^2 (N.A. = 0.25, λ_0 = 1250 nm) using a moving fibre/phase-mask scanning beam technique [14]. The grating was 8 cm long, unchirped, with a $0 - \pi$ sinusoidal apodisation profile along its length. The observed wavelength separation between the two reflection peaks was 0.006 nm and corresponds to a fibre birefringence of 4×10^{-6}. The grating had 98% reflectivity at its peak wavelength of 1536 nm and was measured to have a 3 dB bandwidth of 4.1 GHz with strongly suppressed sidelobes (see Fig. 9.1), as expected from the apodisation profile used. The theoretical bandwidth of a perfect grating with such parameters was calculated to be 3.9 GHz, demonstrating the high quality of the fabrication process over such lengths. The grating was mounted in a glass capillary and angle polished front and back to remove unwanted reflections from these surfaces. The lengths of fibre before and after the gratings were less than 2 mm to reduce the effect of nonlinearity prior to the grating.

The experimental setup was as shown in Fig. 9.2 – ignoring the polarization optics for the moment. The transmitted and reflected signals were detected with a single-mode fibre coupled PIN photodiode and sampling os-

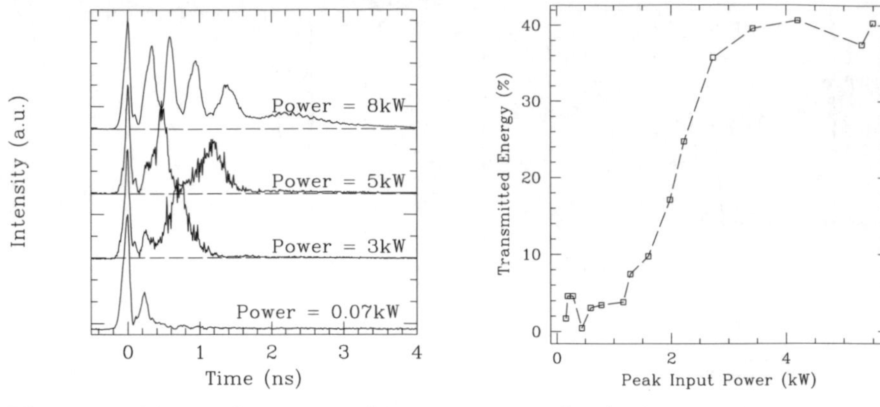

Fig. 9.3. a Measured transmitted intensity traces for the 8 cm grating at varying input power levels. **b** Average transmitted power as a function of the input power

cilloscope. Our temporal resolution was $\sim$ 50 ps. The high power pulses were obtained from a large mode area, erbium doped fibre amplifier chain seeded with 10 ns pulses from a directly-modulated, wavelength-tunable, semiconductor DFB laser. The source was capable of producing nanosecond pulses with energies $>$ 100 µJ and peak powers greater than 100 kW at kHz repetition rates with a linear output polarization state. For the purposes of this experiment the source was operated at 4 kHz repetition rate giving 25 µJ pulses with $\sim$ 2 ns duration and the pulse shape is shown in Fig. 9.1b. By varying the angle of polarisation before the last polarisation beam cube we could vary the power launched into the fibre without changing the pulse shape. The input pulse is highly asymmetric due to amplifier saturation effects and has a sharp 50 ps feature on the leading edge due to chirp on the diode seed pulse. This feature becomes more after transmission through the FBG with the pulse spectrum tuned to lie within the band-gap (see Fig. 9.3 bottom trace). At low powers the chirped spike lies outside the band gap and is transmitted whereas the main body of the pulse is reflected. Although aesthetically undesirable this feature actually proved a valuable calibration aid, giving a direct measure in the transmitted pulse time domain of the input pulse power as it contains $\sim$ 9% of the input pulse energy. The pulse spectrum at the FBG input was measured to have a 3 dB spectral bandwidth of 1.2 GHz, considerably less than the FBG bandwidth. The central wavelength of the source could be continuously and accurately temperature-tuned to wavelengths in and around the FBG band gap.

Tuning the frequency of the pulse to lie in the middle of bandgap the transmitted pulse at varying powers was recorded. These results are shown in Fig. 9.3a. A low power transmitted pulse is shown in Fig. 9.3a [bottom trace] note that it consists of a short 50 ps single peak. This is due to the chirp on the input pulse as discussed earlier. We found that the shape and temporal position of this peak did not change as we tuned the pulse's wavelength and power. Thus it could be used a marker to illustrate the amount of switching in the rest of the pulse. In the linear regime we found that about 4% of the light in the remainder of the pulse was transmitted. As we increased the input power the output pulse shaped changed dramatically as shown in Fig. 9.3a. Here it can be seen that as the power increased a number of pulses were formed and which propagated through the grating. As the input spectrum lay within the grating's bandgap these pulses are likely to have a spectrum lying partially within the grating's bandgap and are thus gap solitons [15].

Figure 9.3b shows the percentage of the pulse energy transmitted as a function of the input pulse power. The values were obtained by integrating over the transmitted pulse shape after neglecting the initial peak which was used to normalise the pulse energy. It can be seen that at high powers about 50% of the pulse energy is transmitted which is a considerable increase on the linear transmission. Also note that there is a definite threshold associated with the formation of a gap soliton which is what one expects from numerical simulations.

9.2.1 Characterisation of an All-optical Grating Based AND Gate

From Fig. 9.3b it can be seen that with such a device simple logic gates could be constructed which utilised the sharp threshold for the creation of a gap soliton. A specific proposal for an all-optical AND gate was put forward by S. Lee and S.T. Ho [16]. The bits in this system are two orthogonally polarised pulses whose frequency lies within the bandgap of the Bragg grating. The AND gate works as the intensity threshold for coupled gap soliton formation is significantly lower than that required for the formation of a gap soliton in an isolated arm. There is thus a range of intensities whereby a coupled gap soliton will form but a single incident pulse will be reflected. Similarly Samir et al. [17] have shown that in the CW regime nonlinear Bragg gratings can perform simple logical functions.

To construct an AND gate we used the polarization optics shown in Fig. 9.2. The linearly polarised incident beam is split with a $\lambda/2$ waveplate (WP1) and polarization beamsplitter (PBS) combination, providing two inputs, A and B, to the FBG. After following equal length paths these beams are recombined at a second PBS and launched into the FBG. The intensity of beam A incident on the grating was controlled by rotation of the $\lambda/2$ waveplate (WP2) preceding this second PBS. Finally, a third $\lambda/2$ waveplate (WP3) allowed rotation of the orthogonally polarised incident beams relative to the FBG birefringence axes.

We next examined operation of the system as an 'AND' gate. The laser's frequency was temperature tuned to the centre of the bandgap and equal powers were launched in each input beam. In Fig. 9.4a we plot the transmitted pulse profiles for the FBG for an estimated launched power of 2.5 kW per

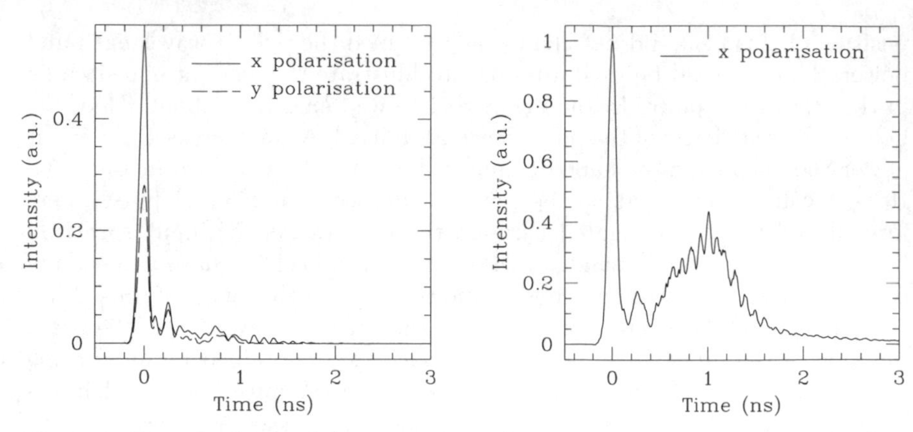

Fig. 9.4. Directly detected (50 ps resolution) transmitted pulse intensity profiles for each individual input beam (**a**) and both beams simultaneously (**b**)

beam. The two curves show the profiles for each input beam, individually incident in isolation. The first two short temporal features seen in Fig. 9.4a were found to be due to the sharp leading edge spike of the incident pulse and could be ignored, leaving only a low level signal accounting for a few percent of the input pulse energy. These cases correspond to $0\cdot1$ and $1\cdot0$ inputs to the AND gate. Fig. 9.4b was obtained with both beams simultaneously launched into the FBG, corresponding to a $1\cdot1$ gate input. In this instance a significant, nonlinearly switched coupled gap soliton pulse of 500 ps duration is formed which exits $\sim$ 1 ns after the spike marking the front edge of the incident pulses. Calculating the energy contained in the pulses for traces (excluding that contained in the leading spikes) shows a switching contrast of 10 dB for the 'AND' gate. The contrast was found to be highly dependent on the wavelength of the incident beams relative to the bandgap of the FBG and the gate operated at this contrast level only over a narrow frequency range ($\sim$ 100 MHz). As the signal was moved further into the bandgap coupled gap soliton formation ceased while moving towards the edge of the bandgap gap soliton formation by the individual channels was observed [12].

Improvements in the switching contrast were obtained by adding a polarizer at the output of the FBG. This allowed us to obtain preferential selection of a single polarization component of the switched pulse. Switching contrasts as high as 17 dB were obtained (once again evaluated by excluding the contribution from the leading edge spike). We also observed soliton polarization rotation similar to that seen by Eggleton et al. in the out of gap regime [18].

The switching behaviour was further studied by varying the power incident in beam A whilst keeping that in beam B constant and recording the transmitted pulse energy. This data is presented in Fig. 9.5. At low (but still nonlinear) powers the output was seen to initially follow a roughly exponential increase with the peak power launched in arm A, however, between

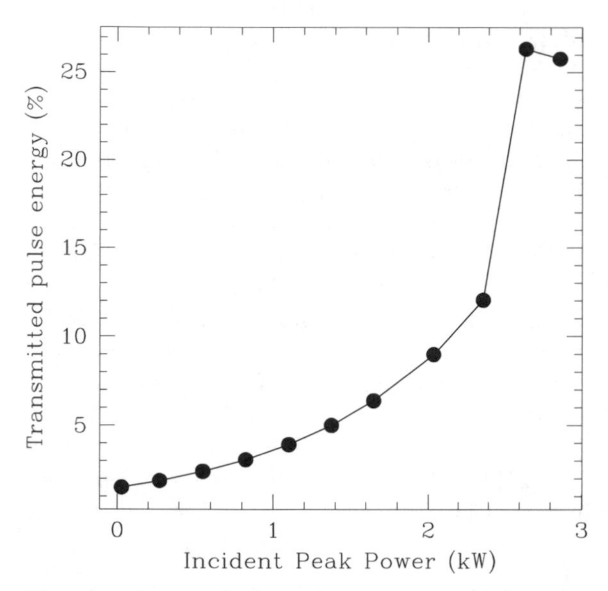

Fig. 9.5. Transmitted pulse energy varying with launched peak power in beam A. The peak power in beam B was held constant at 2.8 kW. The solid line represents an exponential fit to the sub-1.5 kW data points, demonstrating the jump in transmission observed at the point of gap soliton formation (2.4–2.6 kW)

2.4 kW and 2.6 kW a distinct jump in the transmission was observed. The jump was clearly observed to be associated with the formation of gap solitons resulting in the transmitted pulse observed in Fig. 9.4c. For launched powers much higher than this gap soliton formation in each separate arm dramatically reduced the contrast between the off and on states.

We found that in these experiments the main drawback from obtaining a better contrast ratio was the birefringence in the fibre. In Fig. 9.1a it can be seen that there is only an extremely narrow range of frequencies which are inside the bandgap for both polarizations. The original proposal suggested using a non birefringent media as then both arms have the same threshold for soliton formation. In our experiments the threshold for one arm was significantly different from the other due to the fact that the incident light lay in significantly different positions inside the two bandgaps (the power needed to form a soliton increases dramatically with the distance from the band edge [10]). By using stronger gratings the relative effect of the birefringence would be reduced although at the expense of increasing the power needed to form a soliton. We are currently looking at ways to improve the operation of such logic gates in both fibre and semiconductor gratings.

9.2.2 Discussion of Early Results

The results presented in this section were the first results we obtained using a short fibre Bragg grating and do not conclusively demonstrate propagation of light within the bandgap since the apodisation profile of the grating meant that even for light near the centre of reflection peak it was outside the local bandgap for a considerable length of time. Also the chirp on the pulse meant that analysing the results was complicated by the initial spike in transmitted pulse. It was also obvious from the power levels involved that the nonlinearity of silica is far too low for nonlinear fibre Bragg grating devices ever to be practical. This lead us to three main conclusions: firstly that we needed to use longer gratings, secondly that a different pulse source should be used and thirdly different materials should be investigated. The results obtained using longer fibre Bragg gratings [19] are described in Sect. 9.3 while experiments done in AlGaAs waveguides [20,21] which have a nonlinearity ~ 1000 times that of silica are described in Sect. 9.4.

9.3 Switching Experiments in Long Gratings

In these experiments the setup was as discussed in Sect. 9.2 except that instead of using a directly modulated diode we now used an externally modulated diode which significantly reduced the chirp on the leading edge of the pulse making the analysis of the results significantly easier. In addition we fabricated a number of new gratings with lengths of either 20 cm or 40 cm. As before a fraction of the transmitted pulse energy was coupled into an optical fibre and then to a fast sampling scope.

9.3.1 Results for a 20 cm Long Fibre Grating

The reflection spectrum for the 20 cm grating is shown in Fig. 9.6a. The grating was apodised over the initial and final 4 cm and hence there was a uniform grating of length 12 cm through which a pulse could propagate. The grating had a maximum extinction of > 50 dB for low power pulses at frequencies inside the bandgap. This can be seen in Fig. 9.6b which depicts the transmitted pulse spectrum for two different wavelengths, one outside the bandgap and the other inside. The solid vertical line shows where the 2 nd peak should be.

We first examined the high power transmission through the grating as a function of the pulse's frequency. These results are shown in Fig. 9.7 where the scales are identical for all traces. In Fig. 9.7a the pulse is tuned far from the bandgap on the short wavelength side and propagates through without any significant distortion it is thus nearly identical to the input pulse. Moving closer to the short wavelength edge we see in Fig. 9.7b the appearance of modulation instability which is described in more detail in Chap. 8 by Eggleton and Slusher. As expected the period of the modulational instability

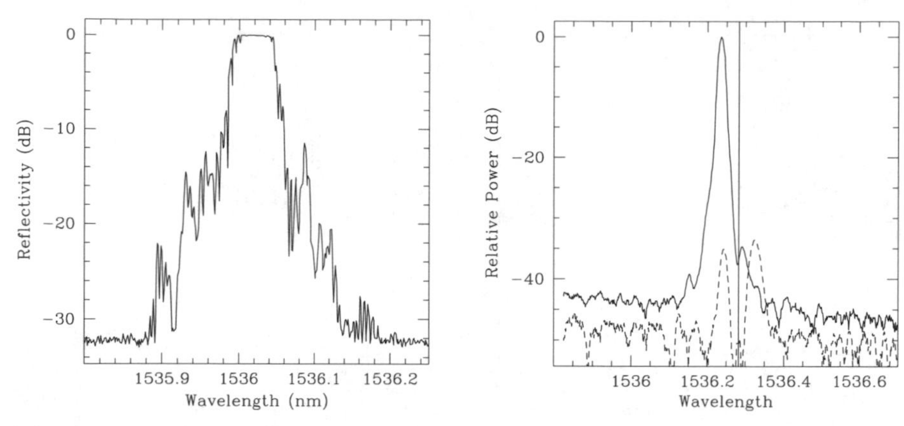

Fig. 9.6. a Reflection Spectrum of the apodisied 20 cm long grating. **b** Transmitted pulse spectrum at low powers for a frequency just outside the bandgap (solid line) and for a frequency just inside the bandgap (dashed line). The vertical line illustrates where the centre of the pulse spectrum should be and it can be seen that the extinction is $> 50\,\mathrm{dB}$

depends on the detuning from the bandgap as can be seen in Fig. 9.7c where the pulse's frequency lies just outside the bandgap and the period of the modulational instability is significantly longer than in Fig. 9.7b. In Fig. 9.7d the pulse's frequency lies just inside the bandgap and we see the formation of gap solitons through the process of modulational instability. Moving further into the bandgap the threshold for the formation of a gap soliton increases until we can only form one or two solitons (Fig. 9.7e) and then finally we are sufficiently far inside the bandgap that it is not possible to form any solitons and the transmitted pulse is essentially a strongly attenuated version of the input pulse (Fig. 9.7f). One noticable feature of the graphs is the differing exit times for the pulses and how strongly this varys with frequency. A central feature of gap solitons is that their velocity can vary from zero to c/n and experimentally Bragg solitons have been observed with a speed as low as 0.5 c/n [22] however no measurements of the speed of a gap soliton has been made. Comparing Fig. 9.7c and d we note that the initial pulse in d arrives 300 ps after the initial pulse in c suggesting that either it has traveled significantly slower or that it took a significant length of time to form. Similarly there is a time delay of 800 ps between the peaks in Fig. 9.7e and f again suggesting that the solitons are propagating significantly slower than the non-soliton components (the transit time for the grating is 1 ns). Our numerical simulations suggest that the correct explanation for the time delay is a combination of the reduced speed and formation time. These graphs provide very strong evidence for reduced propagation of gap solitons even if no definite values can be measured from them.

Next we examined the formation of gap solitons at a fixed wavelength and for varying input powers. We tuned the pulses to lie on the short wavelength

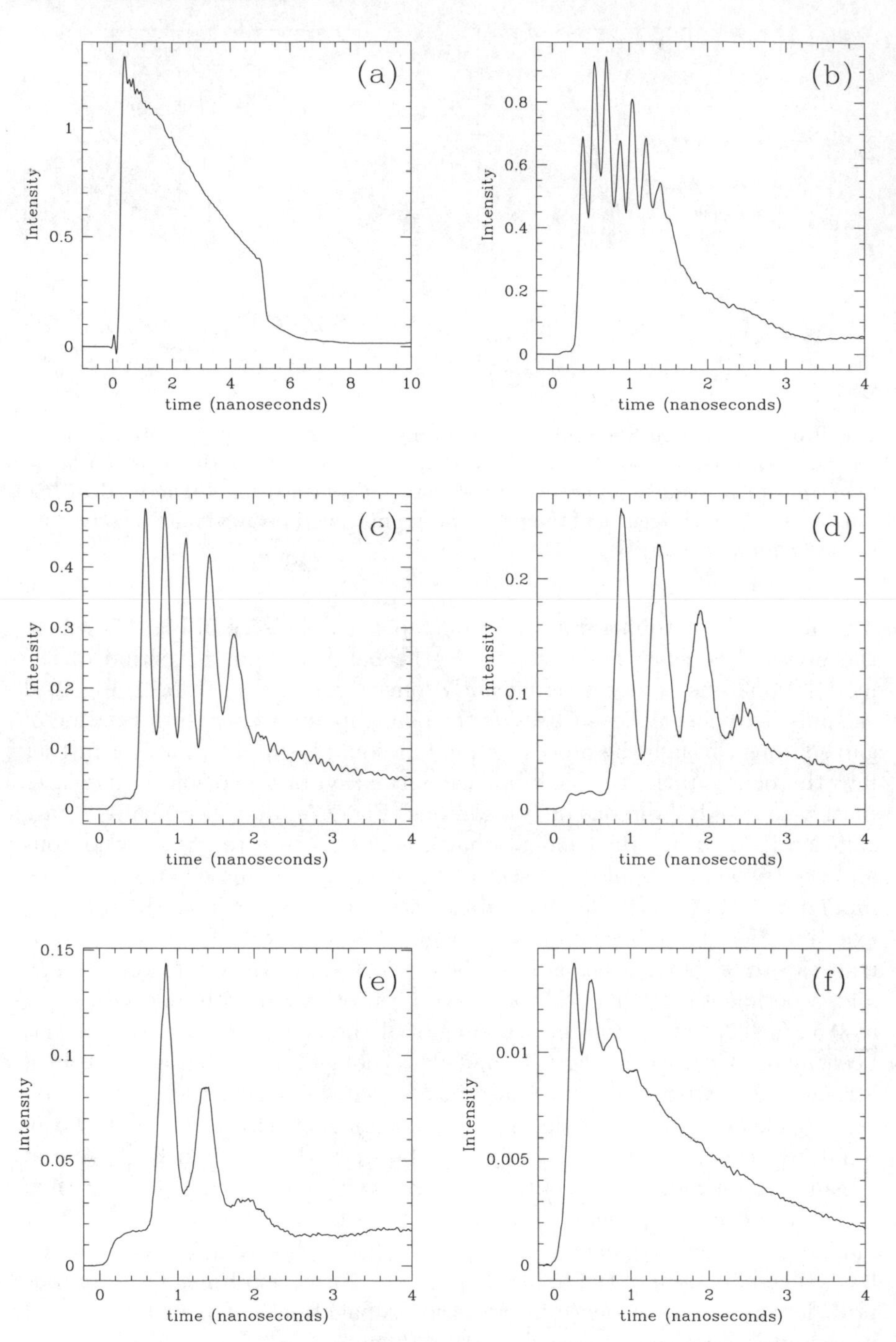

Fig. 9.7. Transmitted pulse shapes for a range of frequency across the bandgap

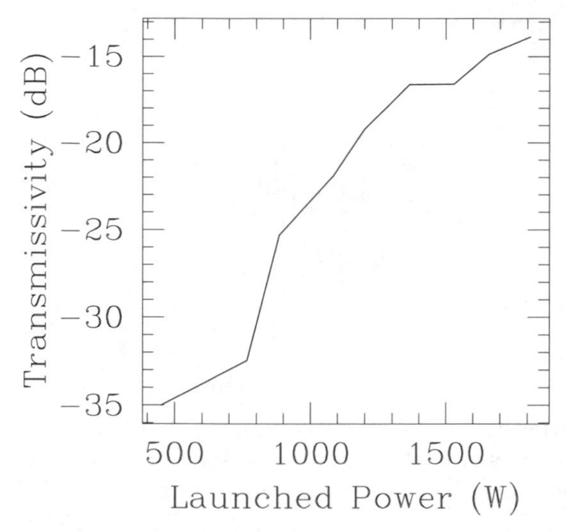

Fig. 9.8. Increase in the transmissivity as a function of the input peak power

edge of the grating where the linear transmission was ~ -35 dB (at a similar frequency to Fig. 9.7d). Measurements of the pulse's energy were done by integrating the sampled pulse shape. By tuning the wavelength of the pulse to lie far from the grating were we could assume 100% transmission relative measurements of the pulse's energy as a function of wavelength and power could be made. These results are shown in Fig. 9.8 and show a clear nonlinear increase in the transmissivity of the grating (note that in a linear system the transmissivity would remain constant).

As is clear from Fig. 9.7 the reason for the increase in the transmissivity at high powers is due to the formation of gap solitons which propagate along the grating. Figure 9.9 shows the output pulse shapes for a range of launched input powers – 463 W and 1081 W for Fig. 9.9a and 1523 W and 1795 W for Fig. 9.9b. In Fig. 9.9b 6 distinct pulses can be seen which correspond to the formation of six gap solitons which then propagate through the grating. The formation of gap solitons through modulational instability is at present poorly understood and is a highly nonlinear problem as can be seen in Fig. 9.9b and Fig. 9.7b where the peak intensities of the 2nd and 3rd peaks are either higher or equal to the intensity of the front peak even though the input pulse is monotonically decreasing. This process agrees qualatatively with our numerical simulations and we expect that better agreement will be obtained as we improve the accuracy of our model especially regarding the input pulse shape.

For the high power trace in Fig. 9.9b the widths of the four main peaks are 166 ps, 130 ps, 190 ps and 210 ps respectively. In addition the time between the peaks is 310 ps, 350 ps and 390 ps. For gap solitons it is known that narrower solitons should have an increased peak power [10] which is the case here.

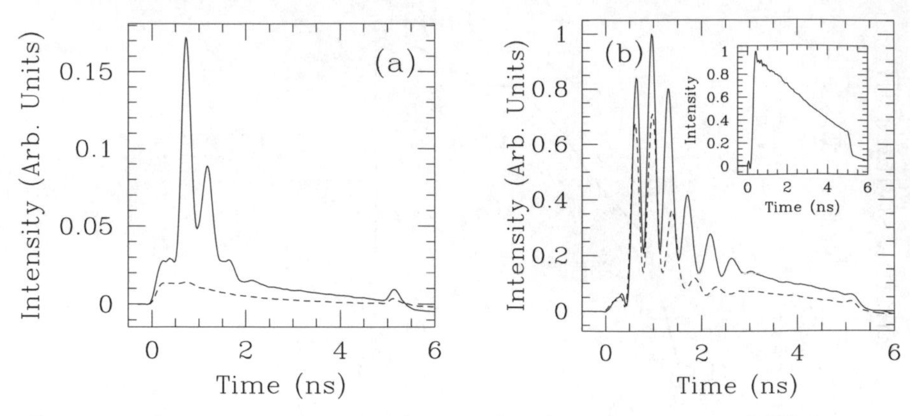

Fig. 9.9. **a** Low power transmitted traces for the 20 cm grating. **b** High power transmitted traces. Note that the units on both graphs are the same. The insert in Fig. 9.9b is the input pulse shape

In addition more intense solitons should move faster than less intense ones which is in agreement with the data on the pulse arrival times shown here. However as we do not know when the solitons were formed it is not possible to determine the actual velocity of the solitons without breaking the grating at some point and looking at the output.

9.3.2 Switching in a 40 cm long Fibre Grating

We also repeated the experiment with a 40cm long grating and our results are shown in Fig. 9.10. Unfortantly the quality of the 40 cm grating was not

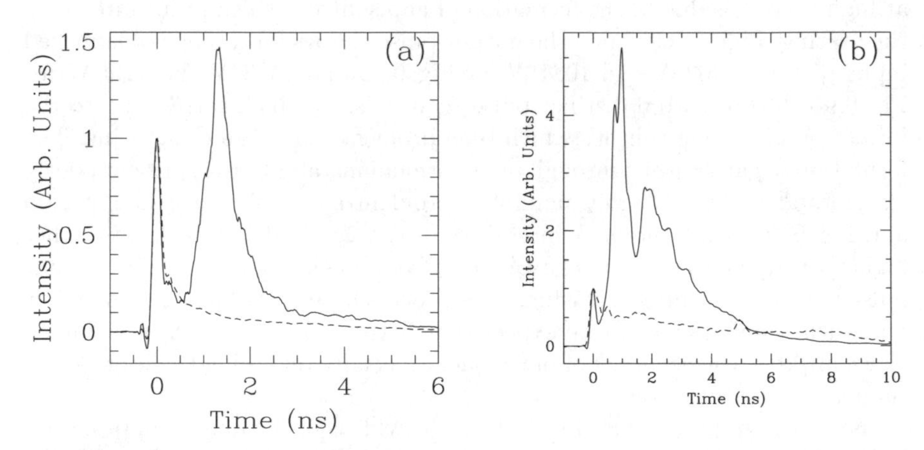

Fig. 9.10. **a** High and low power transmitted pulse shapes for a 40 cm grating. Both curves have been normalised so that the initial peak is of unit height. **b** Similar traces for the case when the pulse is tuned further into the grating's bandgap

as good as that of the 20 cm grating which increased we believe the power needed to see gap soliton formation. As before the grating was apodised over the initial and final 4 cm. In order to observe nonlinear effects the pulse's frequency was tuned to lie very close to the short wavelength edge of the bandgap and hence there was still significant transmission of parts of the input pulse in the linear regime due to a residual chirp of the rising edge. In Fig. 9.10a two traces are shown – a low power trace (dashed line) and a high power trace (solid line). Both curves have been normalised so that the initial spike has the same height in each case (the low power case has be enlarged by approximately a factor of five in this case). This normalisation shows that the front spike is transmitted unchanged at high powers and the only nonlinear effects occur in the rest of the pulse. In this case we were still able to see nonlinear effects although it is not as impressive as in the case of the 20 cm grating. As we moved further into the grating's bandgap less and less of the initial chirp was transmitted. In Fig. 9.10b we show the high and low power transmitted pulses for a wavelength further inside the bandgap. Note that the low power trace has been magnified by a factor of 12 compared to the high power trace. Here we see that in the nonlinear regime two pulses has been formed. However we were unable to observe the formation of more than two pulses in this system.

9.3.3 Discussion of Results

For all the fibre gratings discussed here the maximum refractive index modulation was identical and the only difference was their length. Thus by by comparing the different gratings it should be possible to obtain some information about the propagation of gap solitons. Also the same number of solitons should be formed in each grating which was clearly not the case as at most two solitons where formed in the 40 cm grating while up to six could be seen in the 20 cm grating. We found that Brillouin scattering started to dominate the dynamics in the 40 cm grating and at wavelengths far from the grating stimulated Brillouin scattering restricted the coupled power to be $\sim 10\%$ at high powers while it was greater than 50% at low powers. We believe that this is the main reason why we could not observe the formation of more than two pulses in this grating, while secondary reasons include the slightly decreased quality of the grating compared to the 20 cm grating. Also the interplay between the Kerr nonlinearity and Brillouin scattering appears to be strongly dependent on the pulse's detuning from the bandgap which is why we could form two pulses in case (b) compared to case (a). We note that independent of this work the effects of Brillouin scattering was described theoretically by other researchers [23]. These results suggest that in when describing the generation of gap solitons in fibre Bragg gratings the nonlinear coupled mode equations need to be modified to include the effects of Brillouin scattering.

9.4 Nonlinear Effects in AlGaAs Gratings

Concurrent with the work on long fibre Bragg gratings we have also been examining nonlinear propagation in semiconductor devices. As mentioned earlier the attraction of such devices lies in their enhanced nonlinearity which is approximately 1000 times that of silica. Additionally in AlGaAs one can engineer the electronic bandgap so that any two photon loss is small.

In terms of design space, AlGaAs waveguide gratings offer the opportunity to explore new areas. As the maximum length of a waveguide is limited to 4 cm gratings on AlGaAs must be stronger than fibre Bragg gratings to provide the same peak reflectivity. Using stronger gratings means that the bandgap is wider which allows us to use shorter pulses to interrogate them. The ability to switch and manipulate shorter pulses is important if Bragg grating based devices are ever to be used in telecommunication applications. On a more practical level the use of shorter pulses meant that we could resolve the spectral features of gap solitons using a standard commercial spectrum analyser. This is we believe the first time that the spectral features of gap soliton formation has been observed experimentally [21].

In these experiments a high quality MBE grown AlGaAs wafer was used to fabricate the integrated grating filters. The lower cladding layer was 4 µm thick with 24% Al, the guiding layer was 1.5 µm thick and contained 18% Al and finally the upper cladding layer was 1 µm thick with 24% Al. AlGaAs was chosen as the device material it has an enhanced nonresonant nonlinearity when operating at a wavelength below the half-band gap region, which can be tailored to lie within the 1.55 µm low loss telecommunications window. At this wavelength the detrimental effects of two- and three- photon absorption can be minimized through wafer design. AlGaAs waveguides provide good optical power confinement over long interaction lengths with sufficient power handling capabilities to perform this type of nonlinear experiment. For example in this experiment upto 1.2 kW of power (660 W launched) was coupled into the guides without optical damage occurring. In addition the nonlinear refractive index of AlGaAs is significantly higher than that of silica. Therefore the peak powers required to observe nonlinear effects in AlGaAs grating filters are considerably lower than for comparable effects in optical fibres. The nonlinear refractive index of this material was measured to be $\sim 1.5 \times 10^{-13}\mathrm{cm}^2/\mathrm{W}$. The grating filters were based on a weak grating on a strip-loaded waveguide, to minimize excess scattering losses. A one step electron beam lithography process was used to define the grating and the ridge guides simultaneously. Gratings, 8 mm long were written on 1 cm long single moded waveguides, 5 µm wide. A grating period of 235 nm was selected to position the grating stopband around the maximum gain of the amplifier near 1533 nm. The waveguides were etched using reactive ion etch down to a depth of 0.9 µm. Due to the reactive ion etching lag the grating etch depth was approximately 0.3 µm, resulting in an effective index modulation of 4.4×10^{-4}. All of the gratings used in these experiments were fabricated

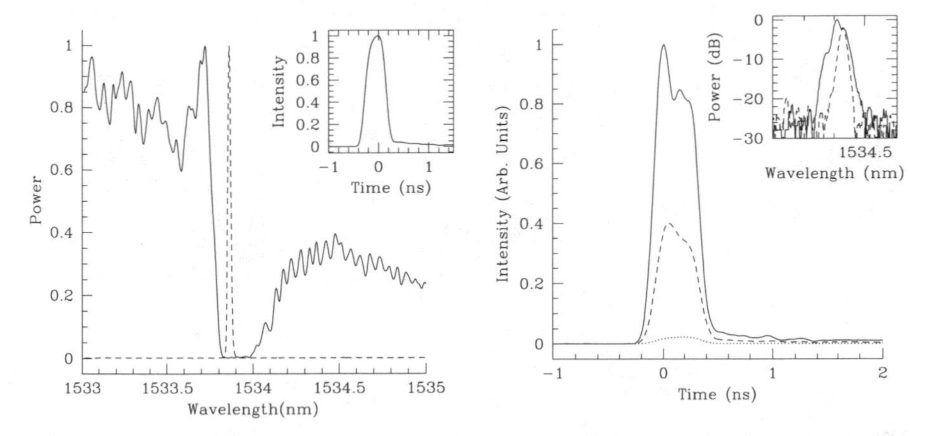

Fig. 9.11. a Reflection Spectrum of the 8 mm long AlGaAs grating (solid line). The dashed line is the linear spectrum of the input pulse measured after the grating. The inset shows the input pulse shape. **b** Transmitted pulse shapes at high, medium and low power at a frequency far detuned from the Bragg grating

at the University of Glasgow while the actual experiments were done at the University of Southampton. To perform the experiments we used the identical setup as used for the fibre Bragg gratings. However as mentioned earlier the increased width of the grating's bandgap meant that shorter pulses could be used and so the pulse width was set to be 450 ps and the pulse repetition rate was 100 kHz.

The grating's reflection spectrum is shown in Fig. 9.11 (solid line) which shows a strong stop band of 0.2 nm width over which the grating is approximately 99% reflecting. This spectrum was taken using the amplified spontaneous emission from the amplifier with the input pulse blocked. This accounts for most of the asymmetry in the spectrum on the long wavelength side. The periodic ripples in the spectrum are Fabry-Perot resonances form the wafer itself – due to the high refractive index of AlGaAs there is a 30% reflection from the end facets.

At low peak powers the input and output pulse spectra were as expected identical [21]. The dashed line in Fig. 9.11a shows the transmitted pulse spectrum which is much less than the width of the bandgap. The insert in Fig. 9.11a shows the input pulse shape which is 450 ps long. We first examined nonlinear propagation through the AlGaAs in both waveguides with and without a grating and at frequencies far detuned from the grating's bandgap. These later results are shown in Fig. 9.11b which shows the transmitted pulse profile at a variety of power levels while the inset shows the spectrum at high and low powers. As expected in a non dispersive Kerr medium there is considerable spectral broadening but very little change in the pulse shape. In this case the pulse's width went from 415 ps at low powers to 420 ps at high powers while the spectral width increased symmetrically from 0.02 nm to 0.04 nm.

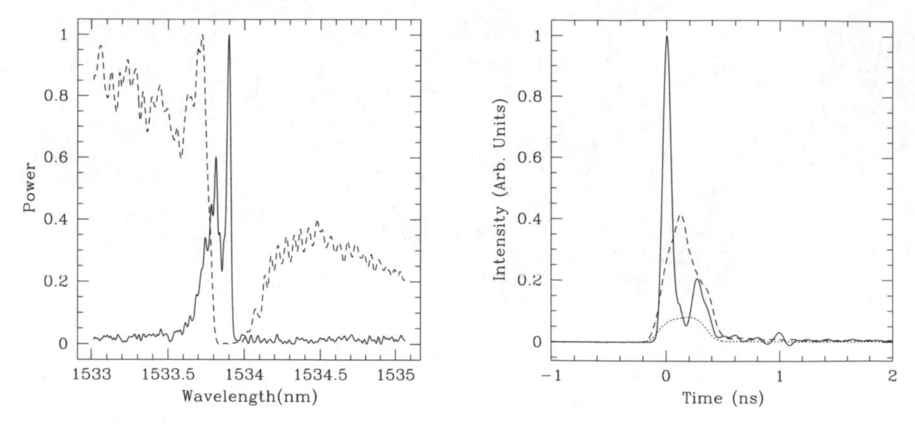

Fig. 9.12. **a** Transmitted pulse spectrum at high powers (solid line) and the grating's transmission spectrum (dashed line). **b** Transmitted pulse shapes at different powers, the final curve corresponds to the spectrum in **a**

We next tuned the pulse to lie near the centre of the grating's bandgap and measured the spectrum and pulse profile at high powers. These results are shown in Fig. 9.12. As the input power was increased temporal compression and associated spectral broaden were observed. Figure 9.12 shows the output spectra (a) and pulse shape data (b) for an incident power of 380 W. In this set of experiments the input pulse was centred at 1533.873 nm which is well inside the grating's bandgap. These graphs clearly show that strong nonlinear effects have occured resulting in the formation of a narrow pulse which propagates through the grating.

The most obvious feature of the high power spectrum in Fig. 9.12a is that it is double peaked with the narrower peak corresponding to the original wavelength while the new peak is shifted towards the short wavelength edge of the grating. Indeed a significant fraction of the pulse's energy now lies outside the grating's stopband. The two peaks are separated by 0.07 nm which is roughly the distance between the input wavelength and the edge of the stop band. Two other features are apparent from Fig. 9.12, firstly the output pulse has been compressed from 415 ps to 80 ps and secondly there is now a delay of 250 ps between the two components of the high power transmitted pulse. Lastly as the power increases the pulse exits the grating earlier: by as much as 155 ps compared to a medium powered pulse.

To ensure the repeatability of this effect we examined a 2nd, nominally identical grating on the same wafer and observed very similar results. For an input wavelength of 1533.900 nm and a launched power of 660 W (i.e. greater than used in Fig. 9.12) the resulting spectrum is as shown in Fig. 9.13. Here again can be seen multiple peaks on the short wavelength side which are widely separated while there is almost no sign of spectral broadening on the long wavelength side. Note also that in this case the power in the new spectral

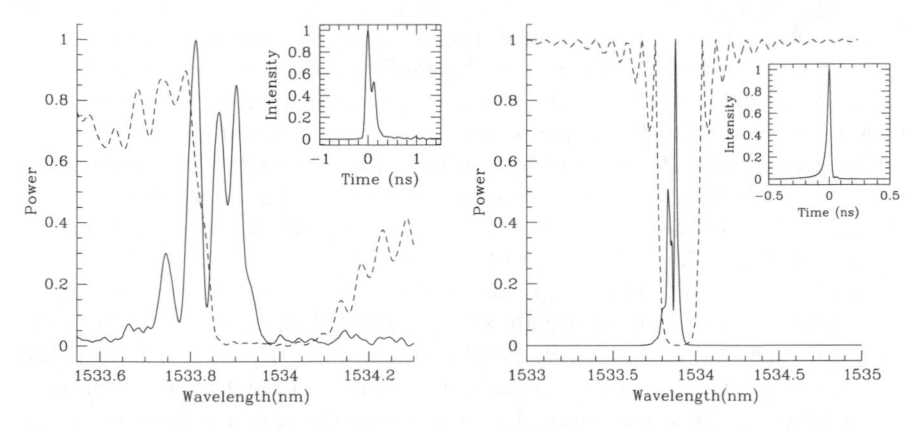

Fig. 9.13. a Transmitted pulse spectrum at high powers (solid line) and the grating's transmission spectrum (dashed line). **b** Theoretical model of the initial formation of a gap soliton in AlGaAs at a power corresponding to that used in Fig. 9.12

peaks is higher than the power left at the original wavelength. The insert in Fig. 9.13 shows the transmitted pulse shape.

A central feature of the measured spectra in the fact that the gap soliton forms at a different central wavelength to the initial pulse. This was unexpected by us and in the current literature although there has been a considerable discussion on the formation of gap solitons from an incident pulse no one has looked at the output spectrum or at least we did not find any mention of anyone doing so. Hence we did some simple modelling of the nonlinear coupled mode equations [1,11] which describe this system. These results are shown in Fig. 9.13b which shows the same trends as we observed experimentally.

9.5 Discussion and Conclusions

In this chapter I have reviewed our work on nonlinear propagation in Bragg gratings at frequencies inside the bandgap. To do this I have showed results obtained with both fibre and semiconductor gratings and with both short and long gratings. In each case the different gratings enable one to sample a different part of the phase space giving a more complete picture. These results along with those from outside the bandgap demonstrate the wide range of effects that can be seen in a nonlinear Bragg grating.

The initial results presented here used an 8 cm apodised fibre Bragg grating and were the first experimental observation of nonlinear propagation at frequencies inside the photonic bandgap. Using such a grating we showed that simple logic gates could be constructed however in order to improve the contrast longer gratings were needed. We have subsequently obtained switching characteristics for both 20 cm and 40 cm gratings and saw clear evidence of gap soliton formation in both cases. However by moving to longer gratings we were limited by the effects of stimulated Brillouin scattering which reduced the amount of light we could couple into the grating. Noting that others have shown that it is possible to use stimulated Raman scattering to create gap solitons [24] it could be possible in the future to use the Brillouin effect to couple light from frequencies outside the grating to frequencies inside the grating. More work is needed to fully understand the effect of Brillouin scattering and on how best to avoid or ulitise it.

Lastly I discussed work on gratings written in semiconductor waveguides. This work was particularly fruitful since it allowed us to simultaneous measure the spectrum and pulse shape of the transmitted pulses. This was the first time to our knowledge that any information about the spectrum of a gap soliton has been presented. Understanding the spectral features of the pulses is important for the operation of devices such as logic gates especially if we try to cascade them. AlGaAs gratings also offer additional degrees of freedom since the size and speed of the nonlinearity can be tuned by altering the width of the electronic bandgap during fabrication. In the near future we hope to better characterise nonlinear effects in these gratings and obtain better agreement between the theoretical and experimental results.

One common problem to all these experiments has been how to efficiently launch light into the grating at frequencies where the grating is naturally highly reflective. In the case of the fibre gratings we have apodised the initial and final sections. However earlier theoretical work suggests that a combination of a step grating and a sech shaped pulse is highly efficient [25]. This also allows control over the launched soliton's parameters whereas currently there is little control over the process. It is comparatively easy to create such a grating however creating the required pulse shape is more difficult but can be managed. Being able to controllable create gap solitons would enable one to study their properties such as stability and speed in more detail.

References

1. H.G. Winful, J.H. Marburger, and E. Garmire: Applied Physics Letters **35**, 379 (1979)
2. R. Kashyap, *Fiber Bragg Gratings* (Academic Press, San Diego 1999)
3. M.I.P.C. Teh, P. Petropoulos, and D.J. Richardson: Electron. Lett. **37**, 190 (2001)
4. L. Poladian: Opt. Fib. Tech. **5**, 215 (1999)
5. G.P. Agrawal, *Nonlinear Fibre Optics*, 3rd edn. (Academic Press, San Diego 2001)
6. H.G. Winful and G.D. Cooperman: Appl. Phys. Lett. **40**, 298 (1982)
7. W. Chen and D.L. Mills: Phys. Rev. B **35**, 524 (1987)
8. W. Chen and D.L. Mills: Phys. Rev. Lett. **58**, 160 (1987)
9. D.N. Christodoulides and R.I. Joseph: Phys. Rev. Lett. **62**, 1746 (1989)
10. A.B. Aceves and S. Wabnitz: Phys. Lett. A **141**, 37 (1989)

11. C.M. de Sterke and J.E. Sipe, in *Progress in Optics*, E. Wolf, Ed. (North Holland, Amsterdam 1994) Vol. XXXIII, Chap. III Gap Solitons, pp. 203–260
12. D. Taverner, N.G.R. Broderick, D.J. Richardson, R.I. Laming, and M. Isben: Opt. Lett. **23**, 259 (1998)
13. D. Taverner, N.G.R. Broderick, D.J. Richardson, R.I. Laming, and M. Isben: Opt. Lett. **23**, 328 (1998)
14. M.J. Cole, W.H. Loh, R.I. Laming, M.N. Zervas, and S. Barcelos: Elec. Lett. **31**, 1488 (1995)
15. D. Taverner, N.G.R. Broderick, D.J. Richardson, M. Isben, and R.I. Laming: Opt. Lett. **23**, 328 (1998)
16. S. Lee and S.-T. Ho: Optics Letters **18**, 962 (1993)
17. W. Samir, S.J. Garth, and C. Pask: J. Opt. Soc. Am. B **11**, 64 (1994)
18. R.E. Slusher, S. Spalter, B.J. Eggleton, S. Pereira, and J. Sipe: Opt. Lett. **25**, 749 (2000)
19. N.G.R. Broderick, D.J. Richardson, and M. Isben: Opt. Lett. **25**, 536 (2000)
20. P. Millar, R.M.D.L. Rue, T.F. Krauss, J.S. Aitchson, N.G.R. Broderick, and D.J. Richardson: Opt. Lett. **24**, 685 (1999)
21. N.G.R. Broderick, P. Millar, R.M.D.L. Rue, T.F. Krauss, J.S. Aitchson, and D.J. Richardson: Opt. Lett. **25**, 740 (2000)
22. B.J. Eggleton, C.M. de Sterke, and R.E. Slusher: J. Opt. Soc. B **16**, 587 (1999)
23. K. Ogusu: J. Opt. Soc. Am. B **17**, 769 (2000)
24. W. Winful and V. Perlin: Phys. Rev. Lett. **84**, 3586 (2000)
25. N. Broderick, C.M. de Sterke, and J.E. Sipe: Op. Comm. **113**, 118 (1994)

10 Pulsed Interactions in Nonlinear Fiber Bragg Gratings

M.J. Steel and N.G.R. Broderick

Multiple frequency interactions in fiber Bragg gratings provide fertile ground for new nonlinear effects. Exploiting the unusual dispersive properties of fiber gratings permits new techniques for both pulse compression and frequency conversion. Combined with cross-phase modulation, the grating allows rapid compression and acceleration of a weak pulse in a grating while parametric amplification of a weak pulse is automatically phase-matched, regardless of the underlying material dispersion. We present theoretical descriptions of both these effects and analyze the first series of experiments which have successfully demonstrated compression due to cross-phase modulation in a grating – the Optical Pushbroom.

10.1 Introduction

The nonlinear behavior of light in a periodic waveguide is now established as a fundamental branch of nonlinear physics that includes numerous interesting effects. These range from bistability [1] and optical switching to the paradigmatic gap solitons [2–6], all of which are discussed in other chapters of this volume. With this catalog of effects for single-frequency propagation, one might ask how periodicity affects the behavior of *multiple* frequency nonlinear interactions. In fact, the problem of second-harmonic generation in gratings has generated a small but steady stream of papers over the years, largely concerned with the novel phase-matching possibilities introduced by gratings [7–14] and more recently much theoretical work has looked at cascading nonlinearities in gratings [15–17]. However, with the exception of work by Yariv and Yeh [18], who treated phase-matching of continuous-wave processes, the influence of a grating on other multiple-frequency interactions, and particularly $\chi^{(3)}$ effects has been largely neglected. In fact, as we show below, nonlinear gratings are at least as rich in new effects for multiple-frequency problems as for quasi-monochromatic ones.

We concentrate on two specific problems: pulse compression by cross-phase modulation in a grating, a concept known as the "Optical Pushbroom", and $\chi^{(3)}$ parametric amplification in gratings. In the first case, we show that a weak slow-moving or stationary probe pulse may be compressed and ejected from the grating by a high-intensity control pump pulse. In the second, we explain how the grating allows us to avoid the perennial problem of phase-matching coherent nonlinear processes. Instead, the system automatically tunes itself into a phase-matched regime and produces high gain.

The key to these effects is of course the unusual dispersion introduced by gratings, just as it is responsible for the remarkable behavior of gap solitons and the other single-frequency effects. In the present case, however, the dispersion has a significance beyond merely its large magnitude. For the optical pushbroom, it is the large variation in *group* velocity over a small bandwidth rather than the *phase* velocity that is critical. For parametric amplification, both group velocity and phase matching effects play important roles.

To lay the background for understanding both effects, in Sect. 10.2 we first summarize the coupled mode theory of nonlinear gratings and then review the elements of standard schemes for pulse compression in Sect. 10.3. We introduce the optical pushbroom through theoretical simulations and discuss the power requirements for demonstration in Sect. 10.4. In Sect. 10.5 we present the results of a number of experiments that realized the pushbroom and explore the additional freedom that is provided by nonuniform gratings. Finally in Sect. 10.6, we return to the theoretical presentation and include coherent effects, culminating in the prediction of spontaneous phase-matching of parametric amplification in fiber gratings.

10.2 Mathematical Description

To explain the multiple-frequency effects of interest, we require a number of results from the coupled mode theory of one-dimensional nonlinear waveguides, which we now review. The effective linear refractive index of the fiber grating is well-approximated as

$$n(z) = n_0 + \Delta n(z)\cos(2k_B z)\,, \tag{10.1}$$

where n_0 is the background index and $\Delta n(z)$ the position dependent amplitude of the grating. We assume a constant grating period $d = \pi/k_B$ corresponding to a Bragg wavelength $\omega_B = \pi c/d$. We further assume that the electric field is composed of two parts – a weak probe field of frequency ω that is close to the grating resonance, and a strong pump field. The pump frequency ω_p is sufficiently far detuned that it is unaffected by the grating and that coherent effects between the pump and probe are negligible (this approximation is remedied in Sect. 10.6). Neglecting transverse effects, the electric field may then be written

$$\begin{aligned}\boldsymbol{E}(z,t) = {} & [f_+(z,t)\exp(\mathrm{i}k_B z) + f_-(z,t)\exp(-\mathrm{i}k_B z)]\exp(-\mathrm{i}\omega_B t)\boldsymbol{e_1} \\ & + P(z,t)\exp[\mathrm{i}(k_p z - \omega_p t)]\boldsymbol{e_2} + \mathrm{c.c.}\,,\end{aligned} \tag{10.2}$$

where $f_\pm(z,t)$ are slowly varying amplitudes for the forward and backward-traveling probe fields and $P(z,t)$ is the pump field with associated propagation constant k_p and frequency ω_p. The polarization of the fields are indicated by the vectors $\boldsymbol{e_1}$ and $\boldsymbol{e_2}$.

Using standard slowly-varying envelope approximations we obtain the governing equations [19,20]

$$i\frac{\partial f_+}{\partial z} + \frac{i}{v_g}\frac{\partial f_+}{\partial t} + \kappa(z) f_- + \delta f_+ + \sigma\Gamma|P(z,t)|^2 f_+ = 0\,,$$
$$-i\frac{\partial f_-}{\partial z} + \frac{i}{v_g}\frac{\partial f_-}{\partial t} + \kappa(z) f_+ + \delta f_- + \sigma\Gamma|P(z,t)|^2 f_- = 0\,, \tag{10.3}$$

where

$$\delta = \frac{\omega - \omega_B}{v_g}\,, \quad \kappa(z) = \frac{k_B n(z)}{2}\,, \quad \Gamma = \frac{2k_B n_0}{Z} n^{(2)}\,, \tag{10.4}$$

denote the detuning of the probe from the grating, the coupling strength of the grating and the effective nonlinear constant respectively. Further, v_g is the group velocity of the bare fiber at ω_B, $Z = \sqrt{\mu_0/\varepsilon_0}$ is the vacuum impedance and $n^{(2)}$ is the nonlinear refractive index coefficient. The coefficient σ accounts for the relative polarization of the pump and probe. It takes the value $\sigma = 2$ for parallel polarization and $\sigma = 2/3$ for orthogonal polarizations. Note that no explicit dispersive terms are included in (10.3) due to large grating dispersion which is discussed more later. To close the system (10.3), we assume that the pump moves with a constant profile such that $P(z,t) = P(z - v_g t)$. This simplification relies on the fact that since the grating dispersion is many orders stronger than the material dispersion, the effects of interest should occur in distances much less than the dispersion length of the bare fiber. Since the pump is also detuned from the grating, it should not change shape to any significant degree. Taking typical values $n^{(2)} = 2.3 \times 10^{-20}\,\mathrm{m^2W^{-1}}$, and $n_0 = 1.45$, we find that for $\lambda = 1536\,\mathrm{nm}$, the nonlinear coefficient $\Gamma = 7.2 \times 10^{-16}\,\mathrm{mW^{-1}\Omega^{-1}}$. For pulse compression by XPM, the system (10.3) provides a complete description. In Sect. 10.6 we show how the equations can be extended to describe coherent nonlinear effects.

10.2.1 Linear Properties

Before studying the nonlinear response, we first review the linear properties of the coupled mode equations. Consider the simplest case of a probe with harmonic time dependence in a uniform grating of length L in the absence of the pump. The coupled mode equations are now linear and easily solved to give

$$\begin{bmatrix} f_+(z) \\ f_-(z) \end{bmatrix} = M \begin{bmatrix} f_+(0) \\ f_-(0) \end{bmatrix}, \tag{10.5}$$

where M is the transfer matrix

$$M = \frac{1}{\beta}\begin{bmatrix} i\delta\sin(\beta z) + \beta\cos(\beta z) & i\kappa\sin(\beta z) \\ -i\kappa\sin(\beta z) & \beta\cos(\beta z) - i\delta\sin(\beta z) \end{bmatrix}, \tag{10.6}$$

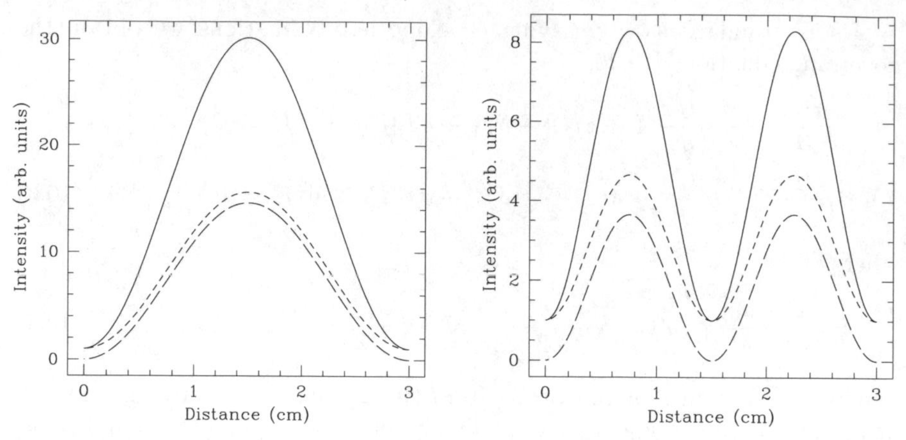

Fig. 10.1. First and second Fabry-Perot resonances of a finite grating of length $L = 3\,\mathrm{cm}$ and $\kappa = 4\,\mathrm{cm}^{-1}$. Line styles denote $|f_+|^2$ (short dash), $|f_-|^2$ (long dash) and $|f_+|^2 + |f_-|^2$ (solid)

with $\beta = \sqrt{\delta^2 - \kappa^2}$. For $|\delta| < \kappa$, this solution of course consists of exponentially decaying waves corresponding to the photonic band gap inside which the incident light suffers strong Bragg reflection. Of more interest here is the response *outside* the band gap: for $\delta^2 = \kappa^2 + N^2\pi^2/L^2$, the reflection coefficient $r = -M_{21}/M_{22}$ vanishes and energy is stored within the grating in Fabry-Perot-like resonances of the finite structure. Figure 10.1 illustrates the field structures of the first two for an incoming field $f_+(0) = 1$. At the grating boundaries, the forward and backward fields have the values unity and zero respectively, indicating a state of perfect transmission. The peak intensity inside the grating far exceeds the intensity of the input field, showing that a large quantity of energy is stored inside the grating at resonance.

Turning to the properties of the infinite system, we next obtain the dispersion relation for the grating. Writing

$$\begin{bmatrix} f_+ \\ f_- \end{bmatrix} = \begin{bmatrix} u \\ v \end{bmatrix} \exp[\mathrm{i}(qz - \delta t)] \,, \tag{10.7}$$

and substituting in (10.3) we find the relation

$$\delta = \pm\sqrt{q^2 + \kappa^2} \,, \tag{10.8}$$

where q and δ are related to the actual wavenumber and frequency by $k = k_B + q$ and $\omega = \omega_B + c/n\delta$. The associated eigenstates of the grating are given by the Bloch vectors

$$\begin{bmatrix} f_+ \\ f_- \end{bmatrix} = \begin{bmatrix} \kappa \\ q \mp \sqrt{q^2 + \kappa^2} \end{bmatrix} . \tag{10.9}$$

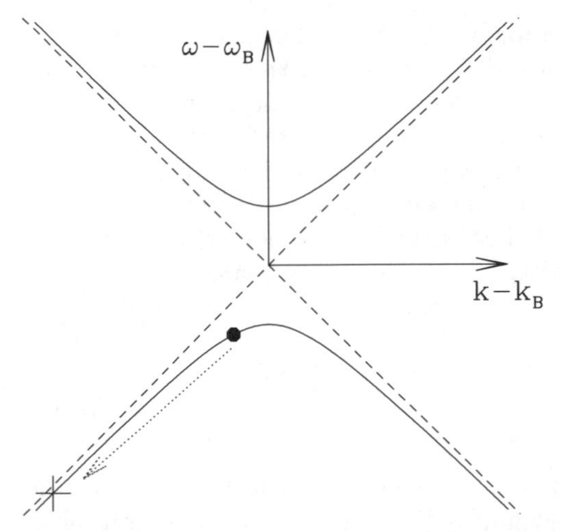

Fig. 10.2. Dispersion relation of optical fiber grating. Solid lines show the grating dispersion, dotted lines show the dispersion relation in the absence of the grating

The dispersion relation is plotted in solid lines in Fig. 10.2 superimposed on the background dispersion relation in dashed lines, which over the small frequency range shown is assumed to be dispersionless. The photonic band gap is visible as the frequency gap around the Bragg resonance ω_B, and has a width proportional to the coupling strength. Traveling waves within this frequency range cannot exist inside the grating.

The normalized group velocity of light in the grating $V = 1/v_g \, \mathrm{d}\omega/\mathrm{d}q$ satisfies

$$V = \pm \frac{q}{\sqrt{\kappa^2 + q^2}} \,. \tag{10.10}$$

The group velocity vanishes at the band edges and asymptotically approaches the speed of light far from the grating resonance. The curvature of the dispersion relation yields the group velocity dispersion. From Fig. 10.2 we can infer that the dispersion is anomalous above the bandgap and normal below it. Close to the bandgap, the dispersion is enormous. Indeed for a typical fiber grating with $\kappa \approx 1\,\mathrm{cm}^{-1}$, the grating dispersion is of order 10^6 times larger than the background material dispersion. The omission of any terms describing material dispersion in (10.3) is thus clearly justified, though we return to this issue in Sect. 10.6.

10.2.2 Nonlinear Effects

We now consider the effects of the nonlinear terms in (10.3), treating first the simplest case of switching by a CW pump, and then discussing time-dependent nonlinear phase shifts in preparation for our treatment of pulse compression.

Continuous Wave Switching. Suppose that to the linear system discussed above, we add a pump pulse with a width much longer than the grating. The pump may then be regarded as a CW source. Then the solution (10.5) is unchanged except for a nonlinear shift in the detuning: $\delta \to \delta + \delta_{\rm NL}$ where $\delta_{\rm NL} = \sigma\Gamma|P|^2$. The band gap is essentially pushed downwards by the quantity $\delta_{\rm NL}$ so that a frequency which lies in the band gap in the linear regime, may be transmitted for sufficiently high intensities. In particular, a pump field with intensity $|P|^2 = \kappa/(\sigma\Gamma)$ switches a frequency at band center to the first transmission resonance outside the band gap. A critical quantity is the required switching intensity

$$I = \frac{2n_0}{Z}|P|^2 = \frac{\kappa\lambda}{2\pi\sigma n^{(2)}} . \tag{10.11}$$

The corresponding peak power $P_0 = IA_{\rm eff}$ where the effective index of the optical fiber is typically $A_{\rm eff} = 80\,\mu{\rm m}^2$.[1] For a relatively weak grating of $\kappa = 0.45\,{\rm cm}^{-1}$ we obtain $P_0 \approx 60\,{\rm kW}$, and discover that nonlinear effects in gratings require significant optical power.

Experiments on CW switching by XPM by La Rochelle et al. [21] marked the very first examination of two-wave interactions in nonlinear fiber gratings. Using a 3.5 cm grating at 512 nm and 100 ps pulses from a mode-locked Q-switched Nd:YAG laser at 1060 nm they observed an increase in the average transmitted power from 4% to 6%. Later analysis of the experiments suggest that complete switching was most likely occurring, but was obscured by the available temporal resolution. These experiments highlighted some of the advantages of using XPM for switching in fiber Bragg gratings: as there are almost no wavelength restrictions on the pump pulse, the wavelength can be chosen so as to maximize the available power. Moreover, the pump pulse need not be transform limited since XPM is not a coherent effect.

Recently Melloni et al. [22] have revisited La Rochelle's work, but instead of a Hill grating, they used an externally written grating with a phase-shift to give a narrow transmission band in the center of the bandgap. As the switching power depends on the width of the reflection feature this is a simple way to reduce the switching power. Similarly in Sect. 10.5 we present more recent results by one of us on CW switching.

SPM and XPM for Optical Pulses. For the remainder of the theoretical discussion, we are concerned with pulsed interactions. The two most familiar pulsed nonlinear effects are the frequency chirps imposed by self-phase modulation (SPM) and cross-phase modulation (XPM) – the nonlinear phase changes induced by a pulse on itself and another pulse respectively. In fact,

[1] The expression (10.11) can also be found more straightforwardly by equating the wavenumber detuning associated with the nonlinear index change $\Delta k = n^{(2)}I2\pi/\lambda$ with the grating strength κ. This provides a useful rule of thumb for ther power requirements of all the effects described below.

since the probe is assumed to be too weak to cause nonlinear effects, the coupled mode equations (10.3 contain only XPM terms due to the pump acting on the probe. The effects of pulsed XPM are best understood however, by comparison to those of SPM, so for the moment we consider the purely forward-going equation

$$i\frac{\partial f_+}{\partial z} + \frac{i}{v_g}\frac{\partial f_+}{\partial t} + \Gamma |f_+(z,t)|^2 f_+ = 0\,, \tag{10.12}$$

where we have restored the SPM term for the forward wave and dropped the grating coupling to the backward wave f_-. The general solution is

$$f_+(z,t) = a(\zeta)\exp\left[\mathrm{i}\Gamma a^2(\zeta)\,t/v_g\right]\,, \tag{10.13}$$

where $a(\zeta)$ is an arbitrary real envelope function of $\zeta = z - v_g t$. For a pulse $a(\zeta)$, SPM thus produces a wavevector shift that varies across the pulse. Defining the instantaneous wavevector as the spatial derivative of the accumulated nonlinear phase ϕ_{NL}:

$$\begin{aligned} \Delta k &= \frac{\partial}{\partial z}\,\phi_{\mathrm{NL}} \\ &= \frac{\Gamma t}{v_g}\,\frac{\partial f^2(z - v_g t)}{\partial z}\,, \end{aligned} \tag{10.14}$$

we see that the shift Δk is proportional to the local intensity gradient of the pulse. For a positive nonlinearity, the shift is negative on the leading edge of the pulse and positive on the trailing edge. Equation (10.13 also has the important implication that although the pulse *spectrum* is broadened on propagation, its temporal shape is unchanged by the nonlinearity (in the absence of dispersion).[2]

Cross phase modulation, which is our chief concern here, behaves differently to SPM in a number of ways. Trivial differences are the polarization dependent strength of the interaction described in (10.3) by the quantity σ and the fact that XPM can occur between fields of different frequencies and polarization states. A more critical difference is that since the two pulses may be non-coincident and/or possess different group velocities the chirp imposed by XPM may be much more complicated than with SPM [23,24]. For example, (10.14) shows that under SPM, a symmetric pulse develops an

---

[2] The interpretation of SPM as a wavevector shift is convenient in grating problems. In general fiber problems, a more common and equivalent viewpoint is in terms of the frequency shift

$$\Delta\omega = -\frac{\partial}{\partial t}\,\phi_{\mathrm{NL}} = -\Gamma z\,\frac{\partial f^2(t - z/v_g)}{\partial t}\,. \tag{10.15}$$

For gratings, the vanishing of the group velocity at the band edges makes the wavevector choice essential [20].

antisymmetric wavevector shift. With XPM, the chirp is only antisymmetric for coincident pulses of the same speed. This turns out to be critical when we come to the pushbroom.

10.3 Elements of Optical Pulse Compression

With the understanding of pulsed nonlinear phase shifts in hand, we are now in a position to present the basic principles of optical pulse compression, leading to the new method and advantages of the optical pushbroom.

The minimum temporal width of any optical pulse is controlled by the time-bandwidth uncertainty product of Fourier analysis $\Delta\omega\Delta t \geq c_0$, where $\Delta\omega$ and Δt are characteristic widths in frequency and real space respectively, and c_0 is a constant that depends on the shape of the pulse.

Thus pulse compression in time requires an increase in the bandwidth of the frequency content. It was realized long ago that this can be achieved through nonlinear optical effects. As we saw in the previous section, however, a nonlinear element alone is insufficient to produce compression, since the pulse becomes chirped but its intensity profile is fixed.

Gires and Tournois [25], and Giordmain et al. [26] realized that a dispersive element is needed to rearrange the frequency components such that the final pulse is as close as possible to unchirped. The general approach for pulse compression is thus as indicated in Fig. 10.3 – a nonlinear element broadens the spectrum but introduces a chirp; a linear dispersive element removes the chirp as to narrow the pulse temporally. In practice, the chirps induced by the nonlinearity and dispersion are not identical and perfect cancellation does not occur.

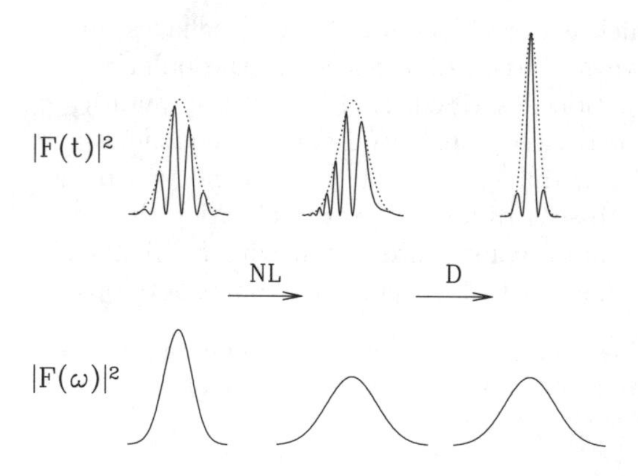

Fig. 10.3. Schematic of pulse compression. The top row shows the temporal profile of a pulse in its initial state, after the action of the nonlinearity (NL), and after the action of the dispersion (D). The bottom row shows the corresponding spectra

10.3.1 Conventional Pulse Compression

Many schemes have been developed to exploit these ideas. The majority use SPM in fibers as the nonlinear element but numerous sources of dispersion have been suggested. The fiber itself can of course provide dispersion. To compensate the positive nonlinearity, anomalous dispersion at $\lambda > 1.27\,\mu\text{m}$ is required. Such "soliton-effect" compression was demonstrated by Mollenauer et al. [27,28]. External diffractions gratings [29,30] or prisms allow a much more flexible compensation of dispersion – the careful cancellation of third and fourth order dispersion [31–35] has led to pulses as short as 5 fs [36].

Jaskorzynska and Schadt [37] pointed out that XPM between a pump and probe would allow compression of pulses too weak to produce useful SPM on their own. For coincident pulses, XPM imposes an antisymmetric chirp, and compression should occur in the anomalous dispersion regime. XPM-induced compression was first demonstrated by Rothenberg [38] with the two pulses in orthogonal polarization states. In this case, the probe trailed the pump and was influenced only by its trailing edge. The chirp thus induced by XPM is of opposite sign to that experienced at the center of the pump and hence in-fiber compression was produced in the *normal* dispersion regime for the first time.

10.3.2 Compression in Fiber Bragg Gratings

Observing that fiber Bragg gratings provide a convenient source of dispersion of both signs, Winful [39] proposed the use of fiber gratings for compression. Two advantages accrue: the huge magnitude of the grating dispersion should permit strong compression over greatly reduced propagation lengths; and a grating can provide dispersion of the opposite sign to the intrinsic material dispersion at a given wavelength. This was confirmed in several experiments by Eggleton et al. [5,40]. Using a 6.5 cm grating with $\kappa L \approx 21.5$, they compressed pulses from a mode-locked Q-switched Nd:YLF laser by a factor of over five, from 80 ps to 15 ps. Here, the nonlinear element was the YLF rod itself rather than the grating or its short pigtails. The pulse wavelength for this experiment was $1.052\,\mu\text{m}$, and hence gratings indeed allow compression in the normal dispersion regime.

10.4 XPM Compression in Fiber Gratings: The Optical Pushbroom

The optical pushbroom approach to compression, suggested by de Sterke [41], is an amalgam of the previous two schemes: XPM is the nonlinearity and a grating is the dispersive element. Consider Fig. 10.4, which shows snapshots in time of the fields in a Bragg grating with $\kappa = 1\,\text{cm}^{-1}$ and length $L = 10\,\text{cm}$. The plots were obtained by simulation of (10.3) [42]. Initially the grating is

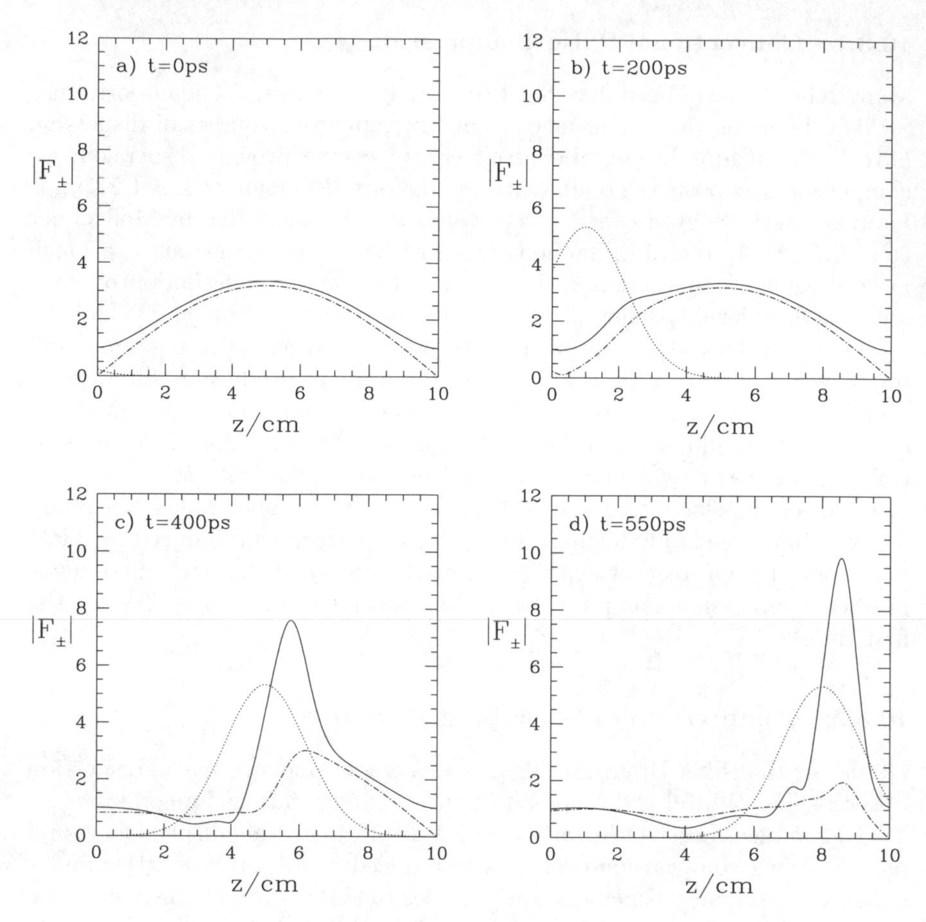

Fig. 10.4. Schematic of optical pushbroom: Moduli of field amplitudes during push-broom process for an optical fiber grating with $\kappa = 1\,\mathrm{cm}^{-1}$, $L = 10\,\mathrm{cm}$. The pump has width 60 ps and peak intensity $35\,\mathrm{GWcm}^{-2}$. Fields are shown at **a**: $t = 0\,\mathrm{ps}$; **b**: $t = 200\,\mathrm{ps}$; **c**: $t = 400\,\mathrm{ps}$; **d**: $t = 550\,\mathrm{ps}$. The line styles denote the forward-moving probe $|F_+|$ (solid), backward-moving probe $|F_-|$ (dot-dashed), pump $|F_p|$ (dotted)

occupied by a probe field tuned to the first Fabry-Perot resonance below the band gap (see Sect. 10.2.1). Since the probe is at resonance it experiences perfect transmission. At the left of Fig. 10.4a, the pump pulse (dashed) enters the grating. Due to its detuning, it propagates with a constant shape at a velocity c/n and so (10.3) apply. From the discussion of Sect. 10.2.2, as the pump approaches the probe, by XPM the leading edge of the pump induces a *negative* frequency shift on the trailing edge of the probe. As the probe lies on the lower branch of the dispersion relation (see dot in Fig. 10.2) where the grating dispersion is normal, the rear of the probe is shifted away from the band gap and increases in velocity. As the XPM and velocity increase continue, the rear of the probe begins to pile up on the front of the pump (Fig. 10.4b). With sufficient pump power, the frequency shift and consequent velocity increase is so rapid that this probe energy remains ahead of the peak of the pump and catches up with the probe energy that was initially towards the rear of the grating, producing pulse compression (Fig. 10.4c). As time progresses, the pump influences an ever greater proportion of the probe, so that ultimately almost all of the initial probe energy is compressed and swept out of the grating on the front of the pump (Fig. 10.4d), in a fashion that inspires the name "Optical Pushbroom".

The pushbroom approach to compression is unique in that the process does not rely on a simple chirp of the probe, but also a substantial downward shift in the mean frequency. Without this shift, the mean velocity of the probe would be unchanged, and it would quickly be left behind by the faster moving pump pulse. So the increase in velocity allows the pump to act on the probe throughout the grating, maximizing the compression. The flushing of energy out of the grating can also be seen to have broader application than just pulse compression. In essence, the pump performs a (destructive) measurement of the quantity of probe energy in the grating. The presence or absence of energy in the grating could be used to store a single bit of information in an optical logic unit. The contents of this bit is then determined by flushing out the contents with a pump.

10.4.1 Transmitted Field

In Fig. 10.5 we show simulations of the pump and probe intensity transmitted from the grating as a function of time. These are the measureable quantities in an experiment. In this case the parameters correspond to results presented in Sect. 10.5. In particular, the grating is apodised, that is, the grating strength vanishes smoothly at the edges, producing a smooth transmission spectrum. The apodization function is $\kappa(x) = 0.519 \sin(\pi x/L)\,\mathrm{cm}^{-1}$. Due to the apodization, there are no sharp transmission peaks, but a significant quantity of energy may still be trapped in the grating, associated with the reduced group velocity near the band edges. In addition, the pump (dashed) has an unusual temporal profile, with a short 30 ps rise time and a long 3-ns decay time. The peak power is ≈ 25 kW. Initially, the probe intensity (solid) in Fig. 10.5 is close to unity, corresponding to almost perfect transmission. At $t = 0$ ns, the pump exits the grating and the probe intensity exhibits a sharp spike of compressed energy. Moreover, the transmitted probe intensity immediately after drops to zero and builds up slowly over 10 ns, indicating that essentially all the energy initially in the grating was carried out in the compressed pulse. Note from the inset, that the probe width is approximately 0.03 ns. This corresponds to a pulse width of 0.62 cm from a grating of length 8 cm, or a compression factor of greater than 10.

We have described the pushbroom such that a CW finite grating resonance plays the role of the probe, since that corresponds to the experimental geom-

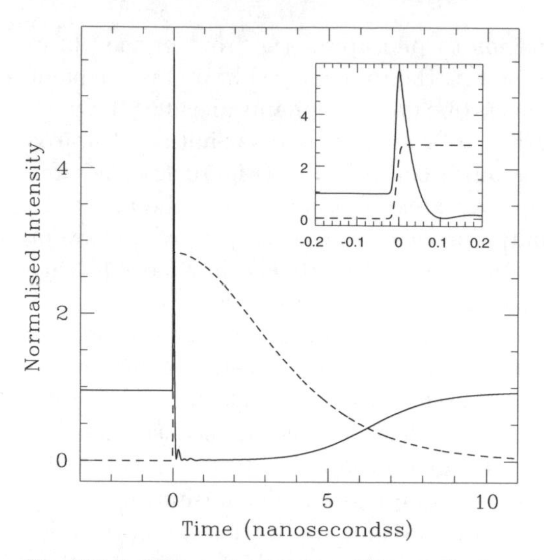

Fig. 10.5. Theoretical trace of the optical pushbroom. The solid line is the transmitted probe intensity, while the dashed line shows the pump profile. The insert is a blowup of the front spike in the transmission. The parameters chosen match those used in the actual experiment

etry we discuss presently. However, the effect can also be modeled for infinite or very long gratings, in which the pump is incident upon a slowly-moving probe pulse rather than the stationary field structure of a finite grating CW resonance [20,43]. For experimental purposes, such a configuration is much more challenging since it requires precise relative timings of the pump and probe pulses. We return to this kind of geometry in the final section when we discuss parametric amplification.

10.4.2 Minimum Power Requirements

For experimental purposes, it is important consider the power requirements for the pushbroom. Using semi-analytic techniques, it may be shown that to accelerate the probe from rest to the speed of light requires a nonlinear index change $\Delta s = \kappa_{\max}$, and a corresponding peak pump power $P_0 = \lambda A_{\text{eff}} \Delta s / (2\pi n^{(2)})$. With a weaker pump, its leading edge begins to accelerate the probe, but at such a small rate that the pump overtakes the probe pulse, and its rear edge acts in reverse, shifting the probe back to its original frequency and velocity. The *power required to sweep up the probe can not be decreased by reducing the width of the pump pulse*, which might have been expected from (10.14).

This surprising result has a natural explanation which again hinges on the fact that the probe suffers a shift in its mean frequency as well as a nonlinear chirp. The velocity of the probe is dependent only on its position on the

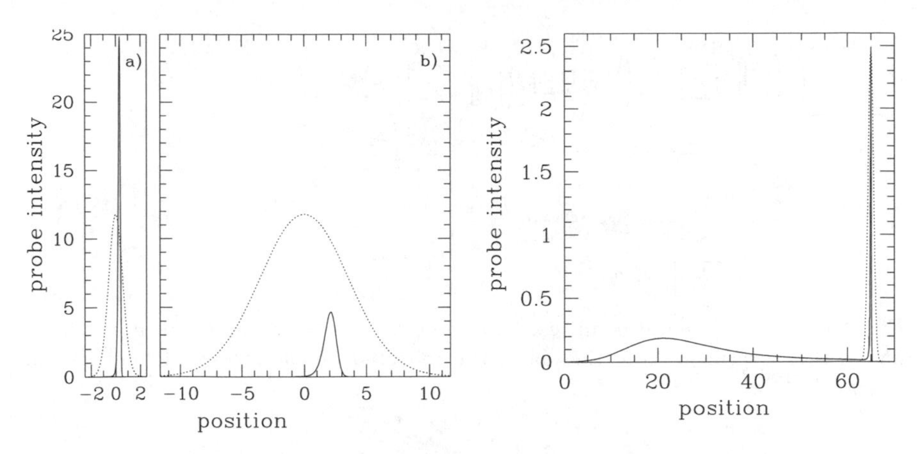

Fig. 10.6. **a** Compressed probe (solid) and pump (dashed) for a pump width $\sigma = 0.8$, $A = 3.3$, $\kappa = 3$. **b** Same but with pump width $\sigma = 5$. **c** Pump parameters $\sigma = 0.8$, $A = 0.85\kappa$ with $\kappa = 2.55$

dispersion relation, and thus the total change in velocity is given by the total change in probe frequency. As the instantaneous frequency shift due to XPM is proportional to the gradient of the pump intensity (see Sect. 10.2.2), the total frequency shift is given by the integral of the intensity gradient, that is, the difference in the final and initial intensities, regardless of the pump shape. This point is illustrated by the results of numerical simulations in Fig. 10.6. In Figs. 10.6a and b the peak nonlinear detuning $A > \kappa$ and the probe (solid) is swept up (moving to the right) and compressed. The narrower pump (dotted) in Fig. 10.6a produces stronger compression. In part c, however, the peak detuning $A < \kappa$, and only a small fraction of the probe energy is captured by the pump. Thus for optimum results, we require large pump powers to capture the probe, and narrow pump widths for strong compression.

10.5 Experimental Results

We now turn from the theoretical discussion to present experimental results obtained by one of us using a high power fiber amplifier as the pump source and an 8 cm apodised fiber Bragg grating [44–46]. Our experimental setup is shown in Fig. 10.7. High power pump pulses at 1550 nm are used to switch a low-power (1 mW), narrow-linewidth (< 10 MHz) probe that could be temperature tuned right across the bandgap which is centered near 1536 nm. The pump pulses, derived from a directly modulated DFB laser, were amplified to a high power (> 10 kW) in an erbium-doped fiber amplifier cascade based on large mode area erbium-doped fiber and had a repetition frequency of 4 kHz. Figure 10.8 shows the intensity profile of the pump pulse. Its shape is asymmetric due to gain saturation effects within the amplifier chain with

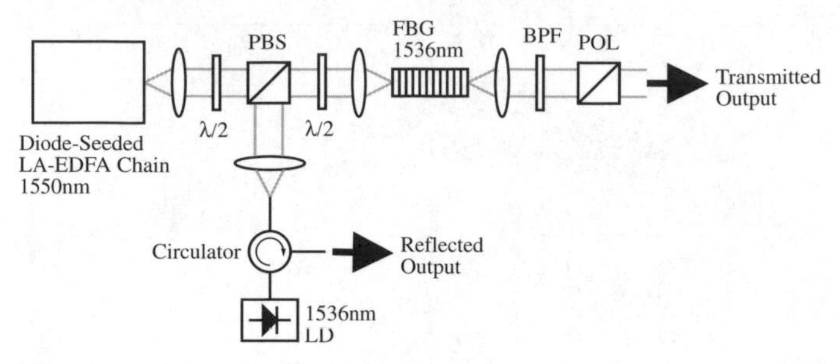

Fig. 10.7. Experimental setup used to observe cross phase modulation effects in fiber Bragg gratings. Note that both the reflected and transmitted light can be observed

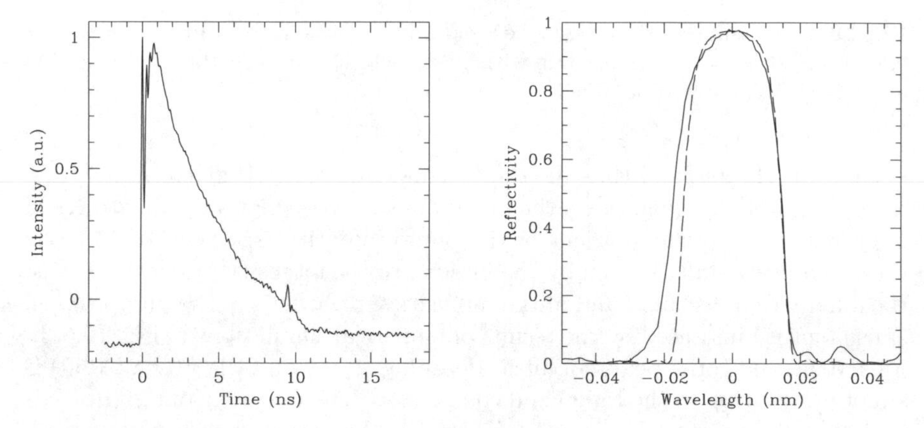

Fig. 10.8. **a** Measured intensity profile of the pump pulse. **b** Measured reflection spectrum of the Bragg grating (solid line) and theoretical profile (dashed line)

a 30 ps rise time and a 3 ns half-width. The spectral half-width of the pulses at the grating input was measured to be 1.2 GHz, as defined by the chirp on the input seed pulses.

The pump and probe were polarization coupled into the FBG and were thus orthogonally polarized within the FBG. A half-wave plate was included within the system allowing us to orient the beam along the grating birefringence axes. The time response of both the reflected and transmitted probe signals could be measured with a temporal resolution of ≈ 50 ps after filtering out the pump.

The FBG was 8 cm long with an apodised profile resulting in almost complete suppression of the side-lobes. The grating had a peak reflectivity of 98% and a measured width of less than 4 GHz. The measured reflection spectrum is shown in Fig. 10.8b (solid line), along with a theoretical reflection spectrum for an idealised grating with identical parameters. In Fig. 10.8b the wavelengths along the X-axis are given in terms of the difference from the center wavelength of 1535.930 nm. The grating was mounted in a section of capillary tube, angle polished at both ends so as to eliminate reflections from the grating end faces and was appropriately coated to strip cladding modes.

It should be noted that the pump pulse shape requirements for CW switching and the optical pushbroom are somewhat incompatible. The pushbroom requires pulses with a large intensity gradient while to see CW effects the intensity gradient should be zero. However with our pulse we are able to have our cake and eat it too. The exceedingly rapid rise time of the pulse allows us to see the optical pushbroom yet the fact that the pulse is longer than the grating allows CW effects to be seen. As we discuss in detail below the frequency of the *probe* pulse determines whether we are in a pushbroom or CW regime. In some cases, we see a combination of both effects.

We performed a series of measurements looking at both the transmitted and reflected light as a function of both the probe wavelength and the pump power [46]. These experiments split neatly up into three parts: (i) CW switching results; (ii) the optical pushbroom; and (iii) effects in reflection.

10.5.1 CW Switching

As discussed in Sect. 10.2 the transmitted probe shape is expected to be a strong function of its detuning from the bandgap. This can be seen in Fig. 10.9 which shows the probe's averaged transmitted waveform as we tune across the bandgap from the short wavelength side to the center of the bandgap, keeping the pump power constant. In these graphs the transmitted intensity has been scaled so that the normalized linear intensity at the center of the bandgap is 0.04 while the normalized intensity at wavelengths well outside the bandgap is unity. This in fact slightly underestimates the true transmission for long wavelengths due to a slight wavelength dependence of the detection process. In all traces, the origin of time is set to coincide with the start of the transmitted pump pulse.

The common feature in Fig. 10.9 is that the transmission increases in the presence of the pump. This is due to the nonlinear index change induced by the pump which shifts the Bragg wavelength to lower frequencies. Thus the probe is effectively further from the Bragg resonance and its transmission increases. In this regime, our results are similar to those of La Rochelle et al. [21], the most important difference being that we are able to resolve the temporal shape of the transmitted light. In addition we have a cleaner source and a better grating, giving significantly clearer results. Note that in cases (a) and (b) the transmission follows very closely the pump profile in Fig. 10.8 due to the fact that the grating's slope is almost linear near the short wavelength edge. However as we move closer towards the center of the Bragg grating only the peak of the pulse is sufficiently intense to switch the probe. Hence instead of seeing the broad switched pulses there is only a narrow pulse corresponding in time to the peak of the pump pulse.

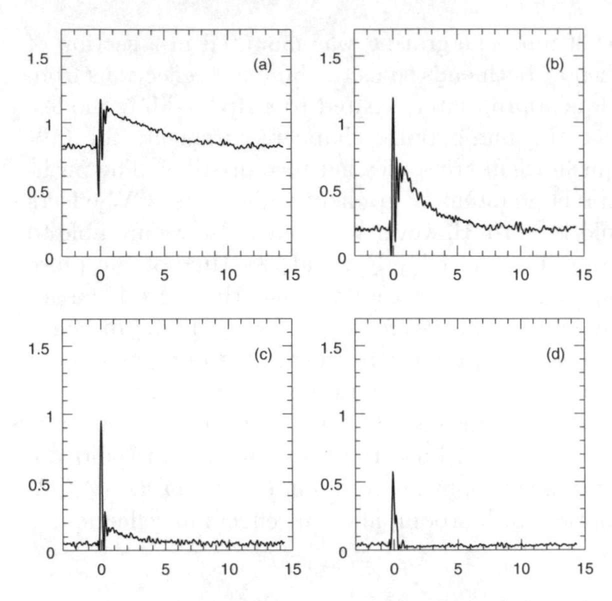

Fig. 10.9. Measured transmitted intensity profiles of the probe pulses as a function of time in nanoseconds. The detunings are $0.662\,\mathrm{cm}^{-1}$, $0.466\,\mathrm{cm}^{-1}$, $0.049\,\mathrm{cm}^{-1}$ and $-0.197\,\mathrm{cm}^{-1}$ for figures **a**–**d** respectively. The probe intensity is normalized to the peak of the transmission at wavelengths far from the grating

10.5.2 Pulsed Experiments: Realizing the Optical Pushbroom

As the probe wavelength moves below the center of the bandgap, the transmitted output changes significantly. This can be seen in Fig. 10.10 which shows analogous traces to Fig. 10.9 but for frequencies below the bandgap. Again we see a strong dependence on the probe wavelength. As we are now below the Bragg resonance, the pump actually moves the probe wavelength *closer* to the center of the bandgap, decreasing the transmission. This is the cause of the long dips in the transmission which can be seen in all the traces. In addition to these dips, it can also be seen that the transmission initially increases due to the presence of the pump. This is precisely the optical pushbroom effect explored in such detail above! As predicted by the earlier analysis, it can only be seen for frequencies close to the long wavelength edge of the grating. However, as discussed in Sect. 10.5.2, for our apodised grating, this wavelength range is much broader than that of a comparable uniform grating.

Optimising the wavelength for the optical pushbroom results in the trace shown in Fig. 10.11. In this case, the probe was detuned by 0.02 nm from the Bragg resonance. This is right on the very edge of the grating resonance, where the transmission is near unity in the absence of the probe. Note that the parameters used in the theoretical trace (Fig. 10.5) correspond as closely as possible to the experimental parameters of Fig. 10.11. There is clearly excellent agreement between the theoretical and experimental traces, which

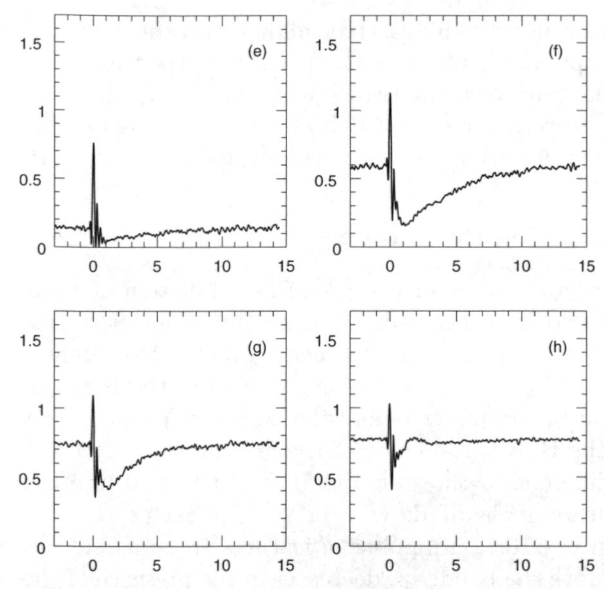

Fig. 10.10. Transmitted intensity profiles for frequencies below the center of the bandgap. The derived detunings are $-0.444\,\mathrm{cm}^{-1}$, $-0.592\,\mathrm{cm}^{-1}$, $-0.643\,\mathrm{cm}^{-1}$ and $-0.891\,\mathrm{cm}^{-1}$ for figures **e**–**h** respectively

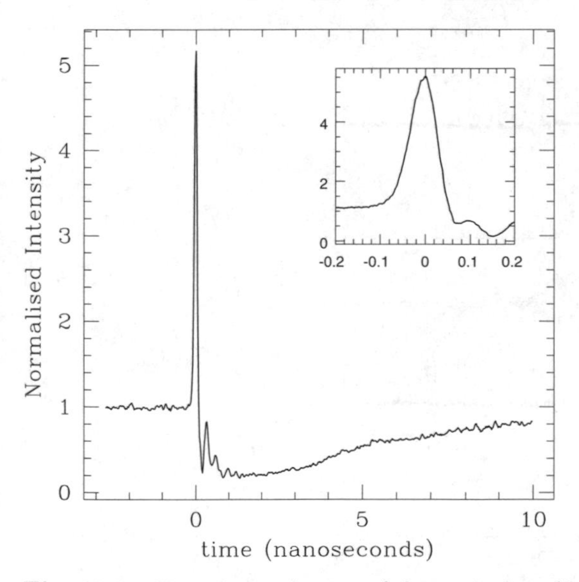

Fig. 10.11. Experimental trace of the optical pushbroom when optimised for maximimum energy storage

is matched by agreement at other detunings. This allows us to be confident that we are observing the optical pushbroom and not some other nonlinear effect. The main disagreement between the two comes from the difference in pulse shapes between the theoretical model which assumed a triangular-like pulse and the actual pump profile which is more complicated (see Fig. 10.8).

10.5.3 Backward-Propagating Pushbrooms

Figure 10.12 shows the reflected traces of the probe as a function of time. These traces are normalized so that the peak reflection in the linear regime corresponds to a value of 0.96. This normalisation again underestimates slightly the actual reflectivity for frequencies near the edge of the bandgap due to the uneven gain from the amplifier before the detector. We note that due to a thermal shift in the Bragg resonance of the grating the actual detunings in Fig. 10.12 are different to those in Fig. 10.9 and Fig. 10.10, even though the actual temperature of the diode was the same in each case.

Looking at the traces in Fig. 10.12 a number of features are apparent. As expected, the reflectivity above the bandgap, decreases in the presence of the pump. This is due to the simple effect of XPM as discussed in Sect. 10.5.1, whereby the effective frequency of the center of bandgap decreases in the

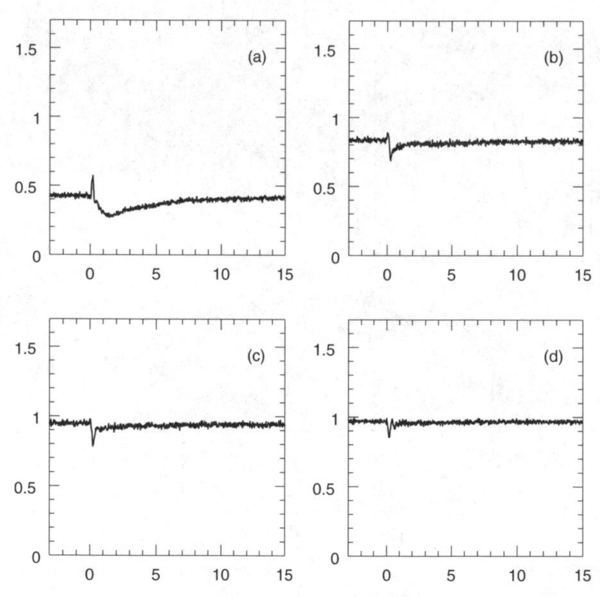

Fig. 10.12. Measured reflected intensity profiles of the probe pulses as a function of the time in nanoseconds (horizontal axis). The detuning of the probe is $0.588\,\mathrm{cm}^{-1}$, $0.490\,\mathrm{cm}^{-1}$, $0.246\,\mathrm{cm}^{-1}$ and $0.000\,\mathrm{cm}^{-1}$ for traces **a**–**d** respectively. The probe intensity is normalized so that the peak reflection in the linear regime corresponds to a value of 0.96

presence of the pump. Compared to the transmitted cases, however, the effects of cross-phase modulation are not as apparent (e.g. compare Fig. 10.9b with Fig. 10.12b). We are not fully confident of the reason for this but it is most likely simply due to a drop in the pump power during the course of the measurements.

The other main feature of interest in the reflected traces can be seen in Fig. 10.12a. Note that initially the reflected intensity increases in a manner similar to the traces of the optical pushbroom. This is a new effect caused by the apodised profile of the grating and which could not have been seen if one used a uniform grating [45]. The reason behind this peak is very similar to the explanation for the optical pushbroom, in that a nonlinear shift in frequency is responsible for the appearance of the peak. In our situation, due to the apodisation, light propagates through almost half the grating before being reflected. This means that compared to a uniform grating of equal maximum strength, significantly more energy is stored in the grating at frequencies within its bandgap. The pump pulse acts on this stored energy by lowering its frequency through cross phase modulation. The apodisation profile then ensures that lower frequencies are reflected earlier in the grating and thus creates the right amount of dispersion to compress the reflected pulse.

Unlike the forward pushbroom this effect can only be seen in apodised gratings and it takes place over a narrower frequency range as can be seen from the experimental traces. We have performed numerical simulations which show that it is a robust phenomenon which occurs in a wide variety of nonuniform gratings including linearly chirped gratings. It is also possible to increase the size of the effect by appropriately designing the grating however, it will always remain a relatively small effect compared to the optical pushbroom as less energy is stored in the grating at frequencies within the bandgap compared to frequencies outside the bandgap.

10.5.4 Discussion on Experimental Results

The results discussed in detail in this paper along with the earlier work by La Rochelle et al. clearly demonstrate that XPM can be used to modulate a signal beam in a fiber Bragg grating resulting in significant all-optical switching in the telecommunication windows. Furthermore, these results are in excellent agreement with the simple theory which assumes that there is no coherent processes which exchange energy between the pump and the probe. However this assumption which is obviously true for the experiments of La Rochelle in which the pump and probe were seperated by over 500 nm starts to look slightly suspect in our more recent experiments in which the pump and probe were separated by 20 nm. The advantage of our pump source was that it was based around fiber amplifiers and a tuneable diode, and hence the separation between pump and probe could be continuously varied. If the separation is reduced, at some point coherent effects such as four wave mixing would start to dominate. These effects are the subject of the next section.

10.6 Parametric Amplification in Fiber Gratings

We have now presented theoretical and experimental results on the pulse compression aspects of the two-frequency nonlinear grating problem. To this point, our discussion of both theory and experiment has made two major approximations: the pump has been treated not as a full dynamical field, but as a "hard-wired" inert region of moving refractive index to which the probe is exposed; and we have ignored all effects of material dispersion. In this final section, we improve upon these restrictions. We suppose that the frequency separation between the pump and probe is reduced somewhat so that we must include the coherent nonlinear interaction of $\chi^{(3)}$ parametric amplification — the process by which two pump photons at ω_p are nonlinearly converted into one photon each at the probe or "signal" ω_s and idler ω_i frequencies, such that

$$\omega_p + \omega_p = \omega_s + \omega_i \,. \tag{10.16}$$

We assume the undepleted pump assumption so that the reverse process may be neglected. The frequencies are separated by the quantity

$$\Delta\omega = \omega_s - \omega_p = \omega_p - \omega_i \,, \tag{10.17}$$

and the general layout of the different components is indicated in Fig. 10.13a. Note that although the fields are now more closely spaced in frequency, we still assume that only the signal frequency is affected by the grating. We demonstrate below that is possible for optical fibers.

10.6.1 Material Dispersion

With a coherent nonlinear process, we must now account for material dispersion between the different fields by introducing the wave vector mismatch parameter

$$\Delta = k_s + k_i - 2k_p \,, \tag{10.18}$$

where k_s, k_i, k_p are the wave vectors corresponding to the frequencies ω_s, ω_i, ω_p. In the absence of the grating, it is well-known that one finds amplification of the signal for a limited range of the mismatch Δ, depending on the pump power [23]. Specifically, gain occurs for

$$-4\mu \leq \Delta \leq 0 \,, \tag{10.19}$$

where the nonlinear detuning $\mu = \Gamma |P|^2$. The gain coefficient is

$$g = \sqrt{-\Delta\mu - \Delta^2/4} \,,$$

which is maximized for

$$\Delta = -2\mu \,. \tag{10.20}$$

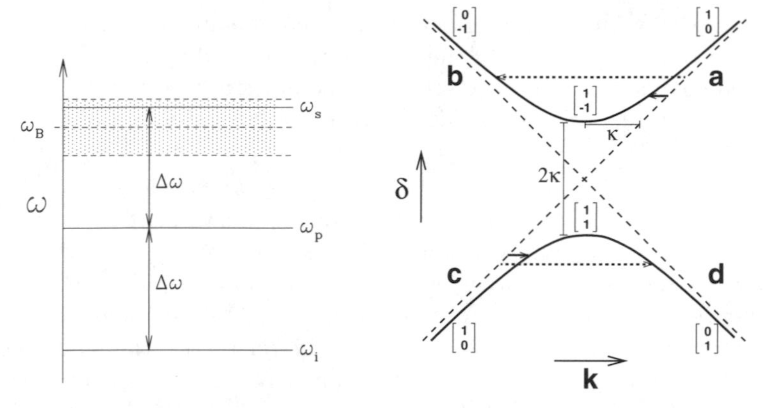

Fig. 10.13. a Schematic of the frequencies involved in the parametric amplification system. **b** Dispersion relation for a Bragg grating (solid) superposed on uniform medium dispersion relation (dotted). Column vectors indicate the Bloch vectors $\boldsymbol{f} = (f_+, f_-)$ describing the relative strength of forward (f_+) and backward (f_-) plane waves in the Bloch functions at different points on the dispersion relation. Solid horizontal lines indicate wave vector shifts to the near branches (labeled **a** and **c**), of the dispersion relation. Dotted horizontal lines indicate shifts to the far branches (labeled **b** and **d**)

This equation represents an exact balance between the wave vector mismatch that results from the material dispersion, and the nonlinear wave vector mismatch induced by cross-phase modulation [23]. With the addition of a grating, the phase-matching characteristics are completely altered and the system displays very rich behavior. In what follows we illuminate the role played by "grating-assisted" phase-matching and show that for essentially any input conditions (permitted by the underlying approximations,) the system exhibits a self-locking or "spontaneous phase-matching" with large gain over a much wider parameter space. Providing Δ is not too large, the purely compressive effect seen in the optical pushbroom [20,43] is swamped by amplification of the signal. Moreover, in contrast to the optical pushbroom where the signal sits on the leading edge of the pump pulse, when parametric amplification is able to operate, the signal becomes located on the trailing edge of the pump.

We note that while grating-enhanced phase-matching of second-harmonic generation has attracted considerable attention [8–12,18,47,48], for the more complex $\chi^{(3)}$ problem there has been only a little work [47] apart from our own [49–51].

10.6.2 Grating-Assisted Continuous Wave Frequency Conversion

For interpretation of the pulsed results, it is useful to first summarize the main results of cw parametric amplification in gratings. A full analysis may be found in [49]. We begin by repeating in Fig. 10.13b, the grating dispersion relation of Fig. 10.2 with extra information. For phase-matching phenomena, the most important consequence of the grating is indicated by the horizontal arrows. When compared to its wave vector in the uniform medium, light of a particular frequency experiences a shift in wave vector of size [8,9,18,47,49]

$$\Delta k_s = -\delta \pm \sqrt{\delta^2 - \kappa^2}\,. \tag{10.21}$$

The choice of sign reflects the fact that the grating allows light to exist on either of two branches for any detuning, with the upper sign taken for the branches to the right of the Bragg wave number $k_\mathrm{B} = \omega_\mathrm{B}\bar{n}/c$. The shift Δk_s is negative for frequencies above the band gap ($\delta > \kappa$), and positive for frequencies below the gap ($\delta < -\kappa$). Further, in theory the shift may be of any size: shifts of $|\Delta k_s| < \kappa$ are obtained on the branch near the uniform medium line (indicated by the solid arrows and the labels **a** and **c** in Fig. 10.13b); shifts of $|\Delta k_s| > \kappa$ are obtained on the far branch (indicated by the dotted arrows and the labels **b** and **d** in Fig. 10.13b). By taking $|\delta|$ large enough, a wave vector shift of arbitrary size and sign is obtained regardless of the grating strength! Thus although the phase-matching condition is not satisfied in the uniform medium, the grating shifts the signal wavevector so that phase-matching becomes possible [9]. From (10.21) we find that for any given Δk_s, the signal is phase-matched at the unique detuning

$$\delta = -\frac{(\Delta k_s)^2 + \kappa^2}{2\Delta k_s}\,. \tag{10.22}$$

In principle then, at the correct detuning the grating produces phase-matching for *any* initial mismatch Δ!

This is at first sight a ridiculous claim – a weak grating should not enhance the gain at a frequency far detuned from the Bragg resonance. The resolution lies in noting that phase-matching is not of itself sufficient to produce gain – there must be a reasonable longitudinal mode overlap between the pump and signal modes. Recall that whereas the pump is a forward-traveling plane wave, the signal occupies a Bloch function given by (10.9). The column vectors in Fig. 10.13b show the (unnormalized) Bloch functions at representative points. Representing the pump by the column $(1, 0)$, the overlap may be expressed as the angle α between the two vectors. If we assume perfect phase-matching according to (10.21), we find

$$\tan\alpha = \left|\frac{\Delta k_s}{\kappa}\right|\,. \tag{10.23}$$

So for $|\Delta k_s| \ll \kappa$, α is small and the two vectors are almost parallel giving large coupling – this corresponds to a small wavevector shift onto one of the "near" branches **a** or **c**. The result is strong gain. For $|\Delta k_s| \gg \kappa$, which requires a shift to one of the far branches **b** or **d**, we have $\alpha \lesssim \pi/2$ so the vectors are almost orthogonal and there is negligible coupling and negligible gain. All these ideas play central roles in the pulsed results for parametric amplification as shown in the next section.

10.6.3 Pulsed Parametric Amplification

To study the pulsed regime we must extend the original coupled mode equations (10.3). The electric field is written

$$\begin{aligned} E = & [f_+ \exp(\mathrm{i}k_s z) + f_- \exp(-\mathrm{i}k_s z)] \exp(-\mathrm{i}\omega_s t) \\ & + P \exp[\mathrm{i}(k_p z - \omega_p t)] + I \exp[\mathrm{i}(k_i z - \omega_i t)] + \mathrm{c.c.}\,, \end{aligned} \tag{10.24}$$

in terms of the signal ($f_\pm$), pump and idler fields. Using the assumption that the idler and pump are far detuned from the grating (see Fig. 10.13), (10.24) do not contain backward-moving fields for the pump and idler. On substitution of (10.24) into the wave equation, we find after some effort [49,51], the complete coupled mode equations

$$+i\frac{\partial f_+}{\partial z} + i\frac{\partial f_+}{\partial T} + \kappa f_- + \Gamma\left(2|P|^2 f_+ + \exp(-\mathrm{i}\Delta z) P^2 I^*\right) = 0\,, \tag{10.25a}$$

$$-i\frac{\partial f_-}{\partial z} + i\frac{\partial f_-}{\partial T} + \kappa f_+ + 2\Gamma|P|^2 f_- = 0\,, \tag{10.25b}$$

$$+i\frac{\partial I}{\partial z} + i\frac{\partial I}{\partial T} + \Gamma\left(2|P|^2 I + \exp(-\mathrm{i}\Delta z) P^2 f_+^*\right) = 0\,, \tag{10.25c}$$

$$+i\frac{\partial P}{\partial z} + i\frac{\partial P}{\partial T} + \Gamma|P|^2 P = 0\,. \tag{10.25d}$$

In keeping with the level of approximation up till now, we have neglected any difference in the group velocities and nonlinear coefficients of the three fields, and have neglected the second derivatives associated with material dispersion. We also work with the normalized time $T = v_g t$. Critically, however, while we have neglected material dispersion over the bandwidth of the individual fields, dispersion over the much larger bandwidth *between* the fields is included through the mismatch parameter Δ. Due to the undepleted pump assumption, (10.25d) is uncoupled from the other three. Also, f_- experiences gain only indirectly through the grating coupling to f_+.

If once more we neglect the grating for a moment, f_- is decoupled, and we find spatially-dependent analogs to the cw results (10.19)–(10.20). For a fixed pump profile $|P(z,T)| = f(\zeta)$, we introduce the position-dependent detuning

$$\mu(\zeta) = \Gamma f^2(\zeta)\,, \tag{10.26}$$

which has the maximum value $\mu_{\max} \equiv \mu(0) = \Gamma f^2(0)$. The signal gain is now

$$g(\zeta) = v\sqrt{-\Delta\Gamma f^2(\zeta) - \Delta^2/4}\,, \tag{10.27}$$

and is real for

$$-4\mu(\zeta) < \Delta < 0\,. \tag{10.28}$$

It is helpful to define the parameter

$$\varepsilon = -2\mu_{\max} - \Delta , \tag{10.29}$$

which is a measure of how well the system would be phase-matched in the absence of the grating at the peak of the pump – perfect phase-matching corresponds to $\varepsilon = 0$. Note that for a particular value of Δ, the gain increases monotonically with pump power and thus the gain is strongest at the peak of the pump. So provided $|\varepsilon| < 2\mu_{\max}$, the signal width decreases with time and its peak becomes *coincident* with that of the pump, irrespective of the initial relative positions of the two pulses. This contrasts with the pushbroom for which the signal lies on the *leading edge* of the pump. For $|\varepsilon| > 2\mu_{\max}$, g is imaginary and the pulses develop rapid oscillations but do not grow in amplitude.

10.6.4 Results of the Full System

Although in presenting results we cannot span the entire parameter space which includes pulse widths, grating strengths, pump powers and wave vector mismatch, our extensive simulations have shown that the qualitative behavior of the system is robust to significant variations in parameters, and we show cases which are representative of the general behavior. In dimensionless units, for the pump, we take a Gaussian pulse with width $w_p = 2$ and peak detuning $\mu_{\max} = 2$. The initial signal field is also a Gaussian of width $w_s = 3$ coincident with the pump, while the idler field is intially empty. Thus without the grating we would expect gain for $|\varepsilon| < 4$ [see (10.28) and (10.29)]. If we take the unit of length as 1 cm, these parameters describe an experimental regime similar to that discussed in Sect. 10.5. The pump and initial signal widths are 100 ps and 150 ps respectively and we perform propagation over grating lengths of $L = 15$–20 cm. The peak pump power is equivalent to an actual power of 30 kW. The grating strength is of order $10\,\text{cm}^{-1}$. The frequency separations corresponding to the wavevector mismatches considered depend on the details of the material dispersion of the fiber. Typical separations are $\Delta\omega = 0.05 \times 10^{15}\,\text{s}^{-1}$ to $0.25 \times 10^{15}\,\text{s}^{-1}$ [51].

Figure 10.14a–f shows the fields as a function of position at six different times for the parameters $\kappa = 8$ and $\varepsilon = 12$ (so that the signal is well outside the gain band $|\varepsilon| < 4$ – in a uniform fiber there would be no amplification). The initial configuration appears in Fig. 10.14a with the pump and forward signal f_+ coincident, and the backward signal f_- and idler set to zero. Figure 10.14b–d shows the fields at $T = 2.5$, $T = 5$ and $T = 10$ respectively. The idler begins to be generated by interaction of f_+ and P while the grating couples energy between the forward and backward signal. The bulk of the energy in the signal and idler gradually becomes concentrated on the rear of the pump. In this period the field structures are complicated and change rapidly with time, and only modest growth occurs. By $T = 12.5$ (Fig. 10.14e),

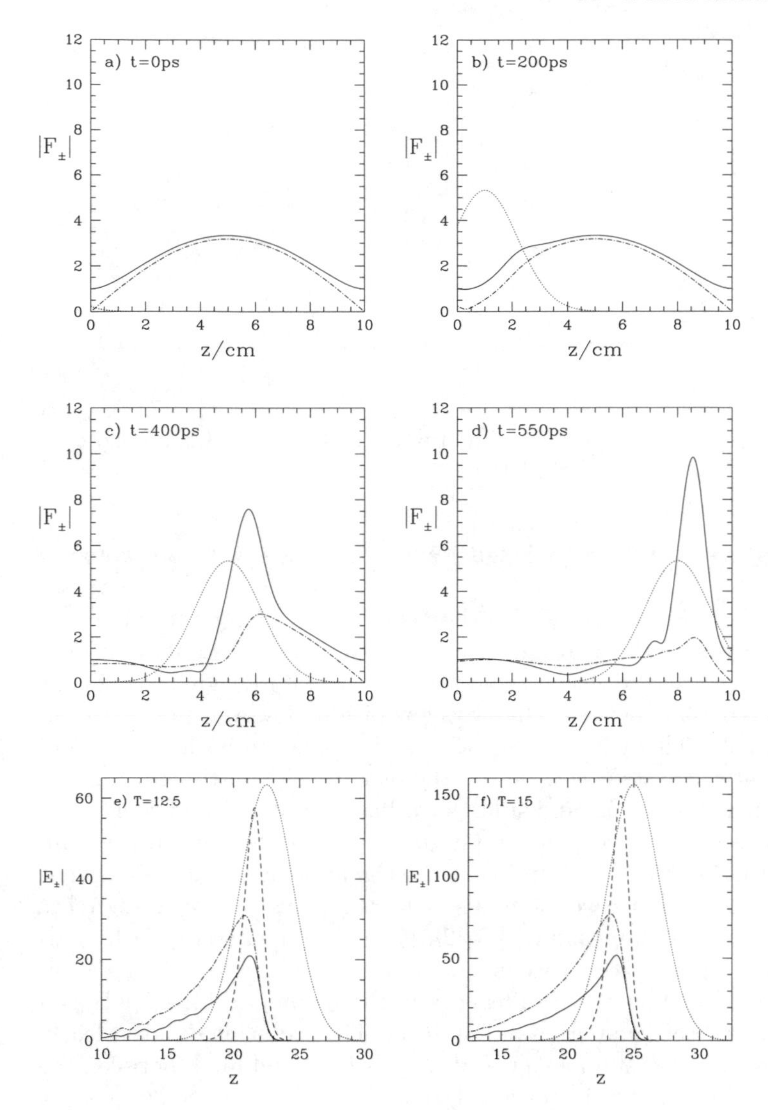

Fig. 10.14. a–f Time sequence of pulse evolution for $\Delta = -16$, $\kappa = 8$. The fields are E_+ (solid), E_- (dash-dot), I (dash), P (dotted). The y axis corresponds to the signal and idler fields. The fixed pump intensity is many times larger

substantial growth has occurred and all three weak fields are localized on the rear of the pump and have a regular single-peaked shape. As propagation continues to $T = 15$ (Fig. 10.14f), the signal and idler intensities grow further, but ther shapes remain virtually unchanged. Thus in a non-growth dispersion regime, which would normally prohibit gain, we have seen very strong growth

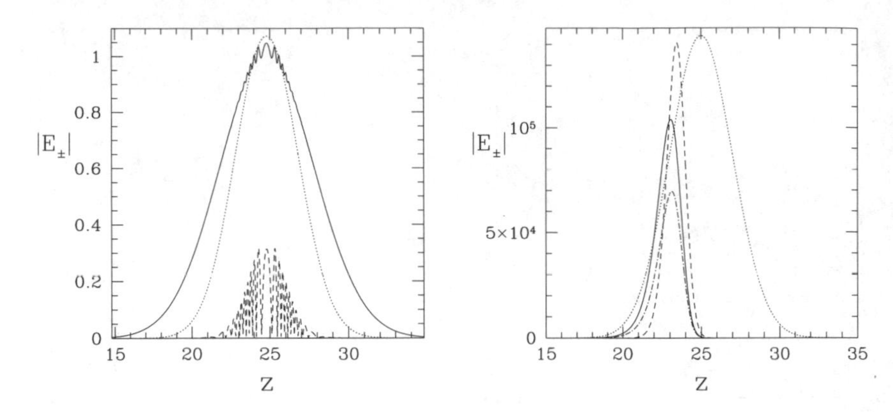

Fig. 10.15. Fields at $T = 15$ for $\Delta = 8$ and **a** $\kappa = 0$, **b** $\kappa = 14$. Lines denote f_+ (solid), f_- (dash-dot), I (dash), P (dotted)

and the emergence of well-shaped signal and idler pulses with no temporal sidelobes.

For comparison, Fig. 10.15a shows the fields at $T = 15$ for the same configuration but without the grating. Now, both the forward signal and idler develop oscillatory features with no gain while the backward signal vanishes as there is now no coupling between the two signal fields. However, in Fig. 10.15b we show the final fields at $T = 15$ for a second simulation with $\kappa = 14$ and $\varepsilon = -12$, so that the system now lies on the *opposite* side of the grating-free gain band. Again the signal and idler have been amplified and located in well-formed pulses on the rear of the pump. Note that the gain in this case is larger still than in Fig. 10.14 and that the signal fields $E_\pm$ are somewhat narrower than in the first case. Gain also occurs for wavevector mismatches *inside* the grating-free gain band $|\epsilon| < 2\mu$, though of course for such cases the appearance of gain in this case is less striking.

In Fig. 10.16a we show the total energy in the signal and idler fields as a function of time for the simulation in Fig. 10.14. The evolution divides clearly into two stages. In the first stage, energy couples back and forth between the two signal fields but the total energy is essentially unchanged. Subsequently, the oscillations decay and exponential gain sets in. Comparison with the field snapshopts in Fig. 10.14 indicates that the onset of growth occurs just as the field structures begin to become simpler and concentrated at the rear of the pump. The examples in Figs. 10.14 and 10.15b are typical of a very broad range of parameters. The initial pulse evolution is highly involved but with sufficient time, the weak fields always experience growth and become located on the rear of the pump regardless of the relative sizes of Δ and κ. For all the simulations, however, the *rate* of growth and detail of the pulse *shapes* do depend on the parameters κ and Δ, and for $\Delta \gg \kappa$ the initial period before amplification begins may be quite long. In this regime, the parametric amplification can be so poorly phase-matched that it essentially plays no

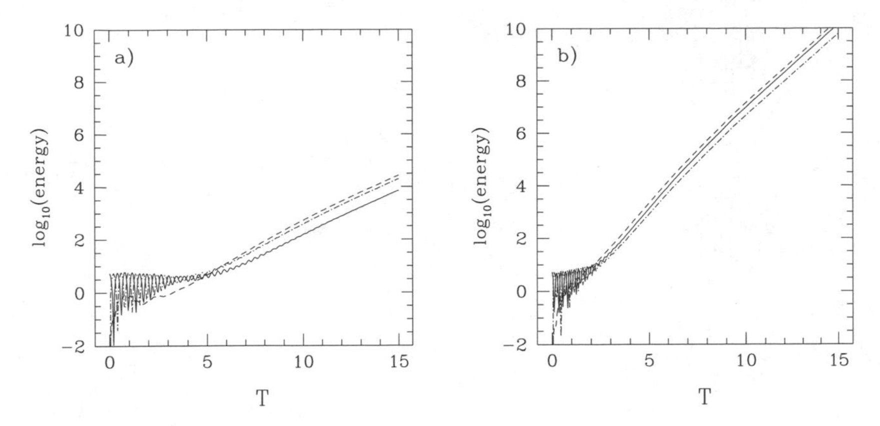

Fig. 10.16. Energy in fields as a function of time for **a** the simulation in Fig. 10.14 ($\Delta = -16$, $\kappa = 8$) and **b** the simulation in Fig. 10.15b ($\Delta = 8$, $\kappa = 14$). Line styles are E_+ (solid), E_- (dash-dot) and I (dash)

role and pulse-shaping effects such as the optical pushbroom described earlier which rely only on XPM dominate. A large number of trends such as the final signal width, dependence of gain on different parameters and the time before strong gain begins can all be explained in terms of simple grating physics [51]. Here we will have space only to explain the basic gain mechanism itself.

10.6.5 Mechanism for Gain

The pulsed gain mechanism is similar to the CW gain mechanism from Sect. 10.6.2. In the CW case, the gain is enhanced for frequencies at which the grating introduces a wave vector shift compensating for the original mismatch Δ. In the pulsed simulations, however, we did not initially tune the signal fields to the correct frequency for phase-matching, and yet the gain still occurred automatically. Recall however from Sect. 10.2.2, that XPM and SPM impose frequency shifts and chirps upon the pulses. Thus we may postulate that even if the initial frequencies are not phase-matched by the grating, ultimately they may become so as new frequencies are generated. We now demonstrate that this is in fact the case.

It is natural to suppose that the gain is largest if the signal spectrum moves to a point on the grating dispersion relation at which the signal wave vector is shifted to the center of the gain band in the grating-free case. In other words, the signal should experience optimum gain if the grating induced wave vector shift produces an effective mismatch

$$\Delta_{\text{eff}} \equiv \Delta + \Delta k_s = -2\mu_{\max} \,, \tag{10.30}$$

or equivalently, if

$$\Delta k_s = \varepsilon \,. \tag{10.31}$$

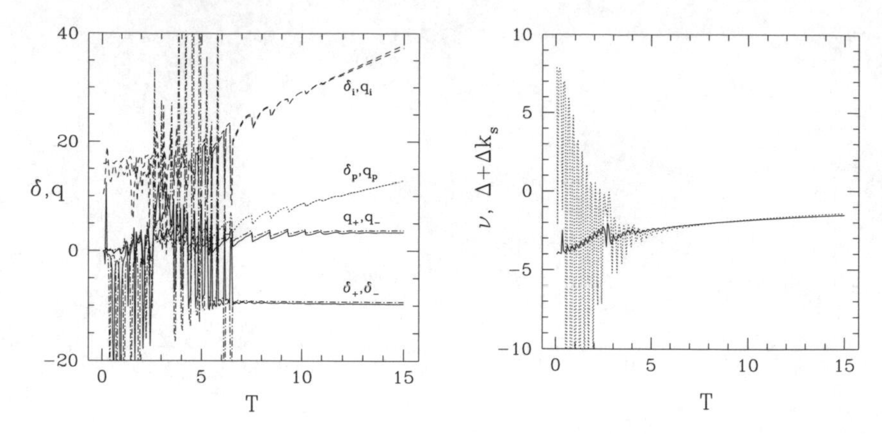

Fig. 10.17. a Detuning and wave number of fields as a function of time for the time sequence in Fig. 10.14. Line styles indicate the same fields as in Fig. 10.14. **b** Effective wave number mismatch $\Delta_{\text{eff}} = \Delta + \Delta k_s$ and $\nu = -2\Gamma|P(z_{\max})|^2$ as a function of time for simulation with $\Delta = 8$, $\kappa = 14$ (Fig. 10.15b)

However, since the weak fields become located on the rear of the pump (see Fig. 10.14f) rather than at its peak, we should consider the strength of the pump at the peak of the signal. Thus we introduce the modified detuning

$$\mu_s(T) = \Gamma f^2\big(z_s(T) - T\big) , \tag{10.32}$$

where $z_s(T)$ is the position of the signal peak at time T, and predict that the gain should be maximized if $\Delta_{\text{eff}} \approx -2\mu_s$.

This claim is confirmed by examining the evolution of the frequencies of the different pulses. For each field, we define the instantaneous frequency by the expressions

$$\delta_j = -\frac{\partial}{\partial T}\phi_j(z,T) + \sigma_j \mu_s(T) , \tag{10.33}$$

where the subscript $j = +$, $-$, i and p for the signal, pump and idler fields respectively, $\sigma_p = 1$, and $\sigma_\pm = \sigma_i = 2$. The phases $\phi_\pm(z,T)$ satisfy $f_\pm(z,T) = |f_\pm(z,T)| \exp[\mathrm{i}\phi_\pm(z,T)]$, and similar expressions hold for the pump and idler. The final term in (10.33) corrects for the effects of SPM and XPM. In addition we monitor the local wave numbers of the fields

$$q_j = \frac{\partial}{\partial z}\phi_j(z,T) . \tag{10.34}$$

Figure 10.17a shows the detunings and wave vectors for all the fields as a function of time for the simulation of Fig. 10.14. Consider first the pump traces. Both the pump detuning and wave number increase monotonically with time, since on the rear of the pump (where the signal peak is found and the quantities are "measured"), SPM induces a positive frequency shift.

The detuning and wave number are exactly coincident ($\delta_p = q_p$) because the pump obeys the uniform medium dispersion relation $\omega_p = k_p v_g$. The idler shows similar behavior: for early times where the fields are disordered (see Fig. 10.14a–b), δ_i and q_i have no simple relationship, but once strong amplification begins, they become coincident since the idler also satisfies $\omega_i = k_i v_g$. The signal fields behave in a totally different fashion. Initially, as the energy oscillates between f_+ and f_-, the signal detunings also fluctuate wildly, since for such complicated field structures, the simple definition of detuning (10.33) is not well-defined. However, for $T \gtrsim 6.5$, when the field oscillations die away and the gain begins, both the detuning and wave vectors parameters become constant and remain so for the rest of the simulation. Due to the interchange of energy between the two signal fields through the grating $\delta_+ \approx \delta_-$ and $q_+ \approx q_-$. Noting that the signal detunings settle at a negative value and the wave vectors converge to a positive value, we see that the signal lies on the bottom right branch (**d**) of the dispersion relation in Fig. 10.13b and hence that the induced wave vector shift satisfies $0 < \kappa < \Delta k_s$. Recall from the first paragraph of Sect. 10.6.4 that for this case, $\kappa = 8$ and the "desired" shift is $\varepsilon = 12 > \kappa$. Thus the location of the signal spectrum on branch **d** is in the correct quadrant of the dispersion relation to satisfy (10.31) and achieve phase-matching.

This argument is quantitatively confirmed in Fig. 10.17b for the simulation corresponding to Fig. 10.15b where we take into account the location of the signal on the rear of the probe [see discussion following (10.31]. The solid line shows the nonlinear detuning at the signal peak $\nu = -2\mu_s(t)$. We expect gain when the effective wave vector $\Delta_{\text{eff}} \approx -2\mu_s(t)$. Now in the uniform medium with a constant phase velocity, we would have $\delta_+ = q_+$, so with the grating included, $q_+ - \delta_+$ represents the grating-induced wave vector shift Δk_s. Therefore we have $\Delta_{\text{eff}} = \Delta + (q_+ - \delta_+)$. This quantity is shown with the dotted line. The convergence of the two lines ν and Δ_{eff} at about the same time as gain is observed in Fig. 10.16b is a dramatic confirmation that the onset of gain is indeed associated with phase-matching mediated by the grating.

In summary the gain occurs as follows. Initially the center frequency of the signal does not permit phase-matching and the field evolution is complicated. Over time, XPM broadens the signal spectrum and shifts its center frequency until part of the signal spectrum lies at the phase-matched detuning. These frequencies are then amplified and quickly grow in a well-behaved fashion to dominate the field. While appealing in its own right, this effect may also simplify experiments since it removes the requirement for precise tuning of the signal laser – a signal anywhere near the band gap should ultimately experience gain.

Finally, we should clarify when we expect to see parametric amplification, and when the optical pushbroom effect dominates. Essentially, this is controlled by the relative size of wavevector mismatch Δ and the grating strength. As discussed earlier, for large mismatches, while the grating can formally provide phase-mismatching, the poor overlap function between the pump and probe leads to minimal gain. As a rough guide, typical fiber parameters should preclude the parametric gain effect for wavelength separations of greater than $\approx 20\,\mathrm{nm}$ [51]. The parametric gain can also be suppressed if the input signal frequency is far from the phase-matched point on the dispersion relation. In that case, the time taken for the XPM chirp to introduce energy at the phase-matched frequencies may exceed the transit time of the grating, and the pushbroom will dominate.

10.7 Discussion

In this chapter we have presented an overview of multiple wave interactions in a $\chi^{(3)}$ medium with a grating. This work draws on the experiments of La Rochelle et al. [21] and those of Broderick et al. as well as the theory of de Sterke and Steel. Such effects range from the intitutively obvious XPM mediated CW switching to the more surprising pulse effects such as the optical pushbroom and parametric amplification.

Key to all these effects is the dispersion induced by the grating, as indeed it is the key to other nonlinear effects such as enhanced second harmonic generation in Bragg gratings [14]. The large grating dispersion completely dwarfs the material dispersion allowing these effects to be seen in any material system which includes a grating. Although all of the experiments presented here were done in fiber, in the future we would expect similar results in semiconductor systems such as AlGaAs. AlGaAs has the advantage of having a nonlinearity some 1000 times greater than silcia and already researchers have demonstrated the possibility of nonlinear pulse compression in an AlGaAs Bragg grating [52], and the promise of integration on heterostructure chips is especially exciting.

For the fiber system discussed earlier in the context of the optical pushbroom we would expect to see parametric amplication when the wavelength seperated between the pump and the probe was reduced to less than 20 nm, which is readily achievable and hopefully might be done in the near future.

References

1. H.G. Winful, J.H. Marburger, and E. Garmire: Appl. Phys. Lett. **35**, 379 (1979)
2. W. Chen and D.L. Mills: Phys. Rev. B **35**, 524 (1987)
3. A.B. Aceves and S. Wabnitz: Phys. Lett. A **141**, 37 (1989)
4. C.M. de Sterke and J.E. Sipe, in *Progress in Optics XXXIII*, E. Wolf, Ed. (Elsevier, Amsterdam 1994), Chap. III Gap Solitons, pp. 203–260
5. B.J. Eggleton, R.E. Slusher, C.M. de Sterke, P.A. Krug, and J.E. Sipe: Phys. Rev. Lett. **76**, 1627 (1996)
6. D. Taverner, N.G.R. Broderick, D.J. Richardson, R.I. Laming, and M. Isben: Opt. Lett. **23**, 328 (1998)

7. A. Ashkin and A. Yariv, "Bell Labs. Tech. Memo No. MM-61-124-46," 13 November 1961
8. N. Bloembergen and A.J. Sievers: Appl. Phys. Lett. **17**, 483 (1970)
9. C.L. Tang and P.P. Bey: IEEE J. Quantum Electron. **QE-9**, 9 (1973)
10. J.P. van der Ziel and M. Ilegems: Appl. Phys. Lett. **28**, 437 (1976)
11. V.A. Belyakov and N.V. Shipov: Phys. Lett. **86A**, 94 (1981)
12. V.A. Belyakov, *Diffraction Optics of Complex-Structured Periodic Media* (Springer, Berlin Heidelberg New York 1992) Chap. 6, pp. 188–205
13. J. Martorell and R. Corbalán: Opt. Commun. **108**, 319 (1994)
14. J. Trull, R. Vilaseca, J. Martorell, and R. Corbalán: Opt. Lett. **20**, 1746 (1995)
15. Y.S. Kivshar: Phys. Rev. E **51**, 1613 (1995)
16. C. Conti, G. Assanto, and S. Trillo: Optics Express **3**, 389 (1998)
17. C. Conti, S. Trillo, and G. Assanto: Phys. Rev. E **57**, R1251 (1998)
18. A. Yariv and P. Yeh: J. Opt. Soc. Am. **67**, 438 (1977)
19. C.M. de Sterke: Phys. Rev. A **45**, 8252 (1992)
20. M.J. Steel and C.M. de Sterke: Phys. Rev. A **49**, 5048 (1994)
21. S. La Rochelle, Y. Hibino, V. Mizrahi, and G.I. Stegeman: Electron. Lett. **26**, 1459 (1990)
22. A. Melloni, M. Chinello, and M. Martinelli: IEEE Photonics Technology Letters **12**, 42 (2000)
23. G.P. Agrawal, *Nonlinear Fiber Optics* (Academic Press, San Diego 1989)
24. P.L. Baldeck, P.P. Ho, and R.R. Alfano, "Cross-phase modulation: A new technique for controlling the spectral, temporal, and spatial properties of ultrashort pulses," in *The Supercontinuum Laser Source*, R.R. Alfano, Ed. (Springer, Berlin Heidelberg New York 1989) Chap. 4, pp. 117–183
25. F. Gires and P. Tournois: Comptes Rendus Acad. Sci. (Paris) **t.258**, 6112 (1964)
26. J.A. Giordmaine, M.A. Duguay, and J.W. Hansen: IEEE J. Quantum Electron. **QE-4**, 252 (1968)
27. L.F. Mollenauer, R.H. Stolen, and J.P. Gordon: Phys. Rev. Lett. **45**, 1095 (1980)
28. L.F. Mollenauer, R.H. Stolen, J.P. Gordon, and W.J. Tomlinson: Opt. Lett. **8**, 289 (1983)
29. E.B. Treacy: Phys. Lett. A **28A**, 34 (1968)
30. E.B. Treacy: IEEE J. Quantum Electron. **QE-5**, 454 (454)
31. C.V. Shank, R.L. Fork, R. Yen, R.H. Stolen, and W.J. Tomlinson: Appl. Phys. Lett. **40**, 761 (1982)
32. B. Nikolaus and D. Grischkowsky: Appl. Phys. Lett. **42**, 1 (1983)
33. J.G. Fujimoto, A.M. Weiner, and E.P. Ippen: Appl. Phys. Lett. **44**, 832 (1984)
34. J.-M. Halbout and D. Grischkowsky: Appl. Phys. Lett. **45**, 1281 (1984)
35. W.H. Knox, R.L. Fork, M.C. Downer, R.H. Stolen, C.V. Shank, and J.A. Valdmanis: Appl. Phys. Lett. **46**, 1120 (1985)
36. J.-C. Diels and W. Rudolph, *Ultrashort laser pulse phenomena* (Academic Press, San Diego, 1996)
37. B. Jaskorzynska and D. Schadt: IEEE J. Quantum Electron. **24**, 2117 (1988)
38. J.E. Rothenberg: Opt. Lett. **15**, 495 (1990)
39. H.G. Winful: Appl. Phys. Lett. **46**, 527 (1985)
40. B.J. Eggleton, G. Lenz, R.E. Slusher, and N.M. Litchinitser: Appl. Opt. **37**, 7055 (1998)

41. C.M. de Sterke: Opt. Lett. **17**, 914 (1992)
42. C.M. de Sterke, K.R. Jackson, and B.D. Robert: J. Opt. Soc. Am. B **8**, 403 (1991)
43. M.J. Steel, D.G.A. Jackson, and C.M. de Sterke: Phys. Rev. A. **50**, 3447 (1994)
44. N.G.R. Broderick, D. Taverner, D.J. Richardson, M. Isben, and R.I. Laming: Phys. Rev. Lett. **79**, 4566 (1997)
45. N.G.R. Broderick, D. Taverner, D.J. Richardson, M. Isben, and R.I. Laming: Opt. Lett. **22**, 1837 (1997)
46. N.G.R. Broderick, D.J. Richardson, D. Taverner, and M. Isben: J. Opt. Soc. Am. B **16**, 345–353 (2000)
47. P.S.J. Russell and J.-L. Archambault: J. Phys. III France **4**, 2471 (1994)
48. M.J. Steel and C.M. de Sterke: Applied Optics **35**, 3211 (1996)
49. M.J. Steel and C.M. de Sterke: J. Opt. Soc. Am. B **12**, 2445 (1995)
50. M.J. Steel and C.M. de Sterke: Opt. Lett. **21**, 420 (1996)
51. M.J. Steel and C.M. de Sterke: Phys. Rev. E. **54**, 4271 (1996)
52. P. Millar, R.M.D.L. Rue, T.F. Krauss, J.S. Aitchson, N.G.R. Broderick, and D.J. Richardson: Opt. Lett. **24**, 685 (1999)

Part III

Novel Nonlinear Periodic Systems

11 Chalcogenide Glasses

G. Lenz and S. Spälter

11.1 Introduction

Observing nonlinear phenomena in periodic structures has been mostly confined to fiber Bragg gratings (FBG's) in silica fiber because they are straightforward to produce and use. However, the Kerr nonlinearity of silica is extremely small ($n_2 = 2.6 \times 10^{-16}\mathrm{cm}^2/\mathrm{W}$) and many experimental regions are not accessible due to the impractically long gratings that are required. Combining the ease of use of fiber or waveguide gratings with a large Kerr nonlinearity is therefore very attractive. Attempts to increase the Kerr nonlinearity by adding Germanium have been successful in increasing the Kerr coefficient by factors of three to five, but achieving factors of hundreds or even thousands requires non-oxide glasses, as will be discussed below. In this chapter we will review recent work, theoretical as well as experimental, on the Kerr nonlinearity in chalcogenide glasses. These are typically sulfide, selenide or telluride based glasses (in contrast to the oxide based silica glass), that can be engineered to have a smaller "bandgap" and hence a larger nonlinearity than silica. Both chalcogenide fiber gratings and chalcogenide waveguide gratings have been successfully fabricated and measured. In fact, the measured Kerr nonlinearity at a wavelength of $\lambda = 1.55\,\mu\mathrm{m}$ in bulk As_2Se_3 is about 500 times larger than in silica [1]. This means that in a 10 cm long waveguide with an effective mode area of $1\,\mu\mathrm{m}^2$, an optical pulse with only 1 W of peak power will experience a nonlinear phase shift of about $\pi/2$. This sort of performance enables the observation of a host of nonlinear phenomena and even some practical devices, such as nonlinear pulse compressors, that will be discussed later in the chapter.

The chapter is organized as follows: Sect. 11.2 will introduce the Kerr nonlinearity in glasses and its analytical derivation. We will also discuss some of the tradeoffs involved in engineering a large Kerr nonlinearity (larger nonlinear loss and larger chromatic dispersion). In Sect. 11.3 we will review some of the recent experimental results measured mostly in sulfide and selenide glasses. Section 11.4 will be devoted to a nonlinear optical pulse compression scheme and a pulse train generation scheme, which are feasible using current FBG technology in conjunction with chalcogenide glass waveguides or fibers. Section 11.5 will conclude this chapter.

11.2 General Considerations

Achieving a large nonlinearity (of any order) requires careful material engineering. Typically, resonant enhancement is employed, i.e., the material is one that has an energy transition, that closely matches the wavelength of interest. Structural or geometrical resonant enhancement, such as placing the material in a resonant cavity can also be used. In both cases this enhancement comes at a price: the larger the resonance the smaller the bandwidth. This may be of interest to narrowband applications, but the applications we will consider are broadband, i.e., requiring sub-picosecond response times. Another problem with operation close to resonance is the creation of electronic carriers in the upper level resulting in absorption of light and long response times related to the relatively slow recombination times (or spontaneous emission times). In some cases, the absorbed light is converted to heat, which becomes problematic under high repetition rate operation. It is clear from this discussion that the nonlinear material should have a bandgap larger than the one-photon energy, leading to negligible linear loss. What about nonlinear loss? Two-photon absorption (TPA) is a third order nonlinearity, the same as the order of the Kerr nonlinearity. Indeed, when we solve the propagation equation in a Kerr medium in the presence of TPA, the nonlinear phase is given by [1]:

$$\Phi_{\mathrm{NL}}(L) = 2\pi \frac{n_2}{\beta\lambda} ln(1 + \beta I_0 L_{\mathrm{eff}}) . \tag{11.1}$$

Here n_2 is the Kerr coefficient, β is the TPA coefficient, λ is the wavelength, I_0 is the input intensity and L_{eff} is the effective length given by

$$L_{\mathrm{eff}} = \frac{1}{\alpha}(1 - e^{-\alpha L}) \tag{11.2}$$

with α the linear loss (like waveguide scattering, since one photon absorption is assumed to be negligible) and L the length of the device. From (11.1) we see that the nonlinear phase depends on the ratio of the Kerr effect to TPA, which we define as a figure of merit

$$F = \frac{n_2}{\beta\lambda} . \tag{11.3}$$

Intuitively, we are trying to accumulate nonlinear phase while propagating in the waveguide, but at the same time are losing intensity to TPA making it ever harder to accumulate additional nonlinear phase. Note also that the term in brackets in (11.1) can be rewritten as follows

$$1 + \beta I_0 L_{\mathrm{eff}} = \frac{1}{I(L)/I_0} = \frac{1}{T_{\mathrm{NL}}(L)} , \tag{11.4}$$

where T_{NL} is the nonlinear transmission. This means that to achieve some nonlinear phase in length L, we have to tradeoff the figure of merit to total intensity transmission as shown in Fig. 11.1.

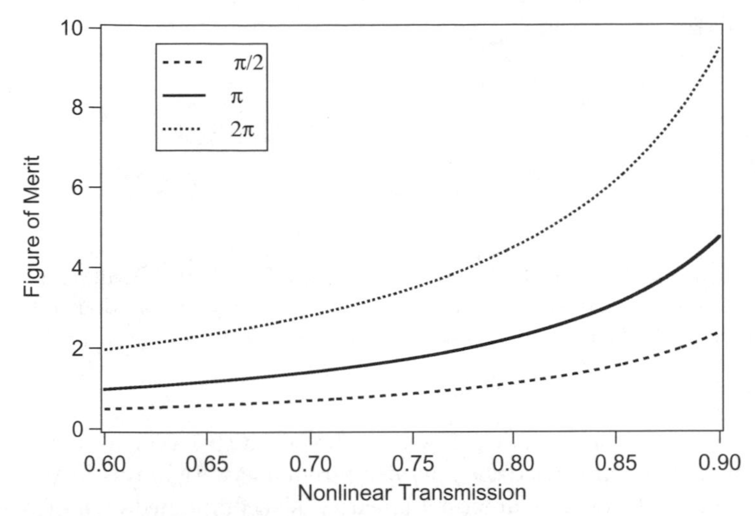

Fig. 11.1. The required figure of merit as a function of the nonlinear transmission for different total nonlinear phase shifts **a** $\pi/2$ (dashed line). **b** π (solid line). **c** 2π (dotted line). To get more light through the device (less nonlinear loss) a higher figure of merit is required to achieve the needed nonlinear phase shift

From this discussion we see the importance of TPA and the need to minimize it in the nonlinear material. Since TPA is caused by the absorption of two photons, we will aim to engineer the nonlinear material such that the bandgap is larger than twice the photon energy

$$E_g > 2\hbar\omega \tag{11.5}$$

$$\hbar\omega < \frac{E_g}{2}, \tag{11.6}$$

that is the frequency of operation should be below the half-gap. We have now established the main design rule, namely look for a material that has the largest possible Kerr effect below the half-gap of that material. Note however that as the one-third-gap is approached three-photon absorption is resonantly enhanced and may degrade performance. In fact, it has been shown that the best results are achieved when the frequency of operation is between the one-third-gap and the half-gap of the material [2] (see Fig. 11.2).

It should also be remembered that the propagating pulse has finite bandwidth, which for pulses of very short duration may become quite large. In this case, n_2 cannot be considered frequency independent and its dispersion has to be taken into account. In comparison, at a wavelength of 1.55 μm in silica we operate at about $0.1E_g$ and the dispersion of n_2 can be safely ignored even for very short pulses. Additionally, in silica the material dispersion of the linear refractive index n_0 is very small, since we operate so far from the

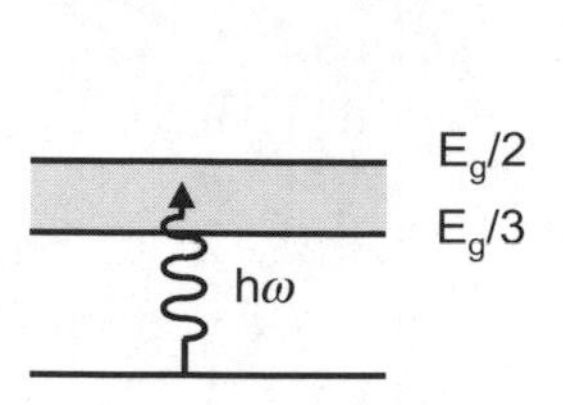

Fig. 11.2. The ideal band structure maximizing the Kerr nonlinearity while minimizing the nonlinear loss. By keeping the operating frequency ω less than the half-gap we minimize TPA. By staying above the third-gap, three-photon loss is minimized

bandgap. We expect the linear index dispersion to be much larger in materials with a gap that is only twice the operating frequency. Again, as in the case of TPA, there is a tradeoff – larger nonlinearity is accompanied by larger index dispersion [3]. It therefore makes sense to define another figure of merit to quantify this tradeoff. We start by defining some relevant length scales: the nonlinear length – the length over which we accumulate a 1-radian nonlinear phase shift and the dispersion length is the length over which a pulse is dispersively broadened by a factor of $\sqrt{2}$ [4]. These lengths are given by

$$L_{\mathrm{NL}} = \frac{\lambda A_{\mathrm{eff}}}{2\pi n_2 P} \tag{11.7}$$

$$L_D = \frac{\tau^2}{\beta''}\,, \tag{11.8}$$

where P is the peak power, τ is the input pulse width (assumed chirp free) and β'' is the quadratic dispersion. Typically, interesting nonlinear phenomena occur at a nonlinear phase shift of π, so that the device length L should be

$$L = \pi L_{\mathrm{NL}}\,, \tag{11.9}$$

if we want small dispersion over this length, the dispersion length should be much larger than this length

$$L_D > \pi L_{\mathrm{NL}} \tag{11.10}$$

$$\tau > \sqrt{\frac{\lambda\beta''}{2n_2 I_0}} = \frac{1}{\sqrt{G I_0}}\,, \tag{11.11}$$

where we have defined a figure of merit

$$G = \frac{2n_2}{\lambda\beta''}\,. \tag{11.12}$$

This shows that the shortest pulses that we can use are not necessarily determined by the material nonlinearity response time. The shortest input pulse depends on the figure of merit but also on the input intensity. This can be understood as follows: higher input intensities require shorter distance to achieve a π nonlinear phase shift, and the pulse undergoes less dispersive broadening, so shorter input pulses may be used. It should be pointed out that in some applications, such as soliton propagation, smaller dispersion length might be desirable (for fundamental solitons $L_D = L_{\mathrm{NL}}$).

Finally, we consider the nonlinearity time response. Since, by design, we are far away from any material resonance, there are no real transitions involved in the nonlinear response, only virtual transitions. These virtual transitions are very fast since they scale as the inverse of the energy detuning from the band edge. The Kerr nonlinearity has two major contributions, namely the ultrafast electronic contribution and the slower nuclear contribution (related to the Raman response of the material). It has been shown theoretically and experimentally that in glasses, 80 to 85 percent is due to the electronic contribution. Time-resolved measurements of n_2 have shown an instantaneous contribution followed by damped "ringing" due to the nuclear response [5]. The total response time is typically on the order of 100 femtoseconds or less. In the following we will apply some of these results to the chalcogenide glasses of interest.

11.3 Chalcogenide Glass

We will now apply the above results to glasses and more specifically to chalcogenide glasses. Chalcogenide glasses are amorphous compounds, which are based on column VI materials (sulfides, selenides, tellurides), in contrast to oxide-based glasses such as silica. These glasses are typically used for mid-infrared optics, because of their low loss in this spectral region. These glasses are of interest to us because a "bandgap" of 1.6 eV (twice the 1.55 µm photon energy) can be achieved, which is not possible with any oxide composition. It was shown recently that the Kerr coefficient for a large class of glasses (including chalcogenides) is given by [1,6]

$$n_2 = 1.7 \times 10^{-16}(n_0^2 + 2)^2(n_0^2 - 1)\left(\frac{d}{n_0 E_s}\right)^2 F\left(\frac{\hbar\omega}{E_g}\right) , \qquad (11.13)$$

where n_2 is in units of cm^2/W, n_0 is the linear refractive index, d (in Å) is the average bond length of the oscillators contributing to the response, E_s is the Sellmeier gap (in eV) which is an energy gap related to the optical bandgap E_g and is typically 2 to 2.5 times the size of the optical gap. F is the dispersion function of the Kerr nonlinearity and at the half-gap takes the value of F(1/2) = 3.3 [7]. If we use some typical values for chalcogenide glasses: $n_0 = 2.5$, $d = 2.4\,$Å, $E_s = 4\,$eV and we operate at the half gap, we get for $n_2 \approx 10^{-13}\,\mathrm{cm^2/W}$ about 400 times the value of silica, explaining why chalcogenides are attractive as Kerr media.

The linear refractive index is given by [8]

$$n_0^2 - 1 = \frac{E_d E_s}{E_s^2 - (\hbar\omega)^2} - \frac{E_l^2}{(\hbar\omega)^2} \,. \tag{11.14}$$

Here E_d is a characteristic energy related to the electronic oscillator strength and E_l is related to the lattice oscillator strength. The first term is therefore the electronic contribution to the index and the second term is the vibrational contribution. When operating near the half-gap in chalcogenides the vibrational contribution is negligible and we will therefore drop the second term. From (11.14) we can compute the quadratic dispersion and find that

$$\beta'' = \frac{\lambda^3}{2\pi c^2}\frac{d^2 n_0}{d\lambda^2} = \frac{3h^2}{2\pi}\frac{E_d}{E_s^3 \lambda n_0} = 8166\frac{E_d}{E_s^3 \lambda n_0} \tag{11.15}$$

with c the speed of light in vacuum, h is Planck's constant, E_d and E_s in eV and λ in µm and the quadratic dispersion is in units of $\mathrm{ps^2/km}$. As an example, using the parameters for $GeSe_3$ at 1.55 µm, (11.15) yields $545\,\mathrm{ps^2/km}$ and the measured value was found to be $563\,\mathrm{ps^2/km}$. For the figure of merit G we find using (11.13) and (11.15)

$$G = \frac{2n_2}{\lambda\beta''} \approx \frac{1}{120}\frac{E_s}{E_d} d^2 n_0^7 F\left(\frac{\hbar\omega}{E_g}\right) , \tag{11.16}$$

where we have approximated the polynomial in n_0 by $2n_0^7$, which results in an error of at most 18 percent over the range of interest $n_0 = 2.4 - 3$. The units of G in the above expression are $\mathrm{\mu m^2/ps/pJ}$ such that when it is multiplied by intensity in units of $\mathrm{W/\mu m^2}$ the result is in $\mathrm{ps^{-2}}$. This expression shows the strong dependence of G on the linear index n_0, so that going to larger index materials not only improves the Kerr nonlinearity, but also the figure of merit. In other words, for a device achieving π nonlinear phase shift with a fixed input intensity, higher index materials will allow shorter input pulses. As we will see, some of the selenide glasses have bandgaps very close to the ideal 1.6 eV with linear refractive index that can approach 3. The ultimate limit is predicted to be about 1000 times the Kerr nonlinearity of silica and an input pulse of 300 fs for an intensity of $100\,\mathrm{MW/cm^2}$, which should be achievable with some of the As-Se based compounds.

Finally, we address TPA in chalcogenides: there is presently no good model for TPA in glasses and it is therefore difficult to predict the value of TPA analytically. In contrast to crystalline semiconductor materials with very sharp absorption edges, glasses have a "smeared" edge due to the Urbach tail component of the absorption (in fact, it is difficult to define the bandgap and typically it is defined as the energy where the absorption is about $1000\,\mathrm{cm^{-1}}$)

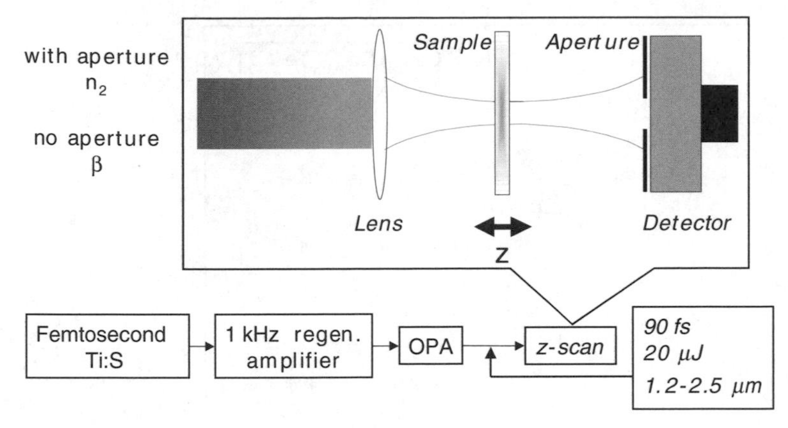

Fig. 11.3. The experimental setup for measuring n_2 and TPA in bulk chalcogenide glass samples, using the z-scan method

[9,10]. This edge is then "reflected" in the half-gap – since photons of energy less than the half-gap can reach states in the conduction band via TPA. The Urbach edge in glasses is virtually temperature independent (in contrast to crystalline semiconductors), and depends mainly on the composition of the glass. Further, impurity tails even deeper in the band can also contribute available states for TPA. In engineering a "good" glass it will be important to eliminate impurities and "sharpen" the Urbach tail.

Recently there has been great interest in chalcogenide glass for ultrafast all-optical switching, using As_2S_3 [17]. This glass, however, has a bandgap of about 2 eV and a linear index $n_0 = 2.4$ and the Kerr nonlinearity in this glass is about 100 times the value in silica. In the following we will describe recent measurements of bulk selenide-based glasses, where the Kerr nonlinearity is predicted to be up to 1000 larger than silica. Both n_2 and TPA were measured and therefore the figure of merit F could also be determined.

In order to measure n_2 and TPA in bulk chalcogenide samples, we employed the Z-scan technique [11]. This technique is simple and allows the measurement of both n_2 and TPA with a trivial adjustment in the experimental setup. Since the direct result of the measurement is the nonlinear phase and the total nonlinear absorption, we need to know the peak intensity of the beam to get at n_2 and TPA. This was accomplished using a reference glass with a well-known Kerr coefficient. Also, in the case of samples that were not flat or had faces that were not parallel, a low-intensity background scan was required to correct for these linear effects.

The source for the measurement was a regeneratively amplified femtosecond Ti:Sapphire laser pumping an optical parametric amplifier (OPA). This source routinely produced 100 fs pulses with an energy of 10–20 μJ (per pulse), which were wavelength tuned from 1.3 μm to about 1.7 μm (Fig. 11.3). Mea-

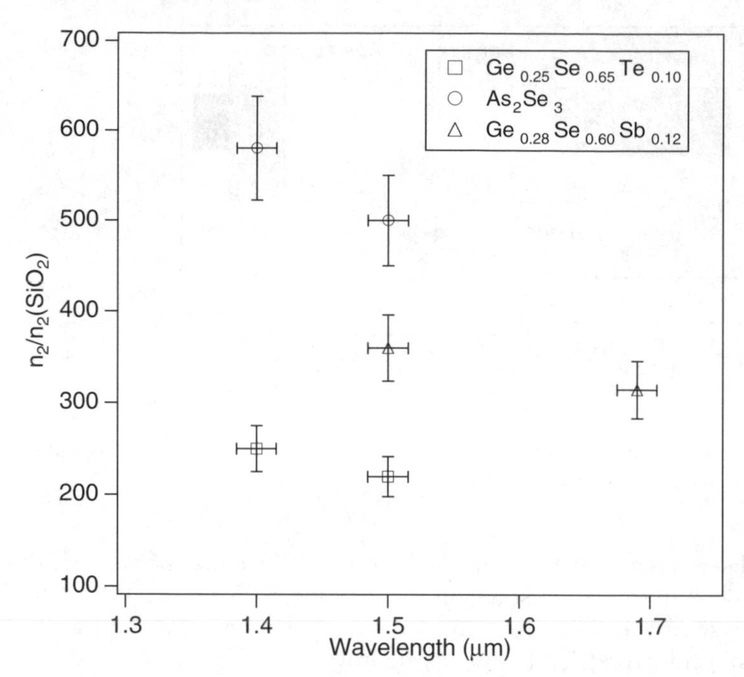

Fig. 11.4. Summary of n_2 results for three different chalocgenide material compositions at different wavelengths. As_2Se_3 shows the largest n_2 of about 500 times the value of silica in good agreement with the theoretical prediction of 470

surements of a number of Se-based glasses were performed and the results are summarized in Fig. 11.4.

We found two compounds, which showed great promise, As_2Se_3, and a Ge-Se-Sb compound (also known as AMTIR-3). As_2Se_3 has a bandgap of about 1.75 eV and a linear index of about 2.7 and it is therefore not surprising that it produced the largest Kerr nonlinearity. The figure of merit for most of the measured glasses was between 1 and 3 and could be impurity limited. We believe that purifying the glasses would yield much larger figure of merits. We also note that further enhancements of the Kerr nonlinearity are achievable: for example, $As_{.5}Se_{.5-x}Cu_x$ with $x = 0.1$–0.15 has a bandgap close to the required 1.6 eV with an index of 3 [12]. Because n_2 scales as a large power of n_0 going from an index of 2.7 to 3 increases n_2 by almost a factor of 2, getting us to the range of 1000 times the nonlinearity of silica.

An optical pulse of 1 pJ energy and of 1 ps duration, at a wavelength of 1.55 μm, propagating through a 10-cm long As_2Se_3 waveguide with effective mode area of 1 μm^2, will undergo a peak nonlinear phase shift of about 1.7π. This demonstrates that even at very low peak powers (1 W) large nonlinear phase shifts are possible even in waveguides of only a few centimeters. Note that a peak nonlinear phase shift of 1.6 pi has been demonstrated in GeSe-based chalcogenide glass waveguides [19]. In the next section we will look at some specific device examples that would greatly benefit from such a nonlinear glass.

11.4 Application to Pulse Compressors and Pulse Train Generators

In this section we will give an example of an FBG-based device that can be used as a pulse compressor (or reshaper) or as a pulse train generator. This device when implemented in silica fiber requires input peak powers of approximately 1 kW – much too high to be practical. Using chalcogenide fiber would bring that power level down to the 1 W range. The device is called an adiabatic soliton compressor and was originally suggested and demonstrated using kilometers of dispersion decreasing fiber [13]. This approach requires long sections of specialty fiber. An alternative was proposed using FBG's that are inherently very dispersive and that can easily be tailored to yield very complex grating profiles [14]. We also note that this is a true pulse compressor – it compresses transform-limited pulses to shorter transform-limited pulses, rather than just compensating chirp.

The basic operation principle of this device is as follows: fundamental solitons satisfy the condition [4]

$$N = \frac{L_D}{L_{\mathrm{NL}}} = 1 \qquad (11.17)$$

$$\frac{\tau}{\beta''}\frac{2\pi n_2 E}{\lambda A_{\mathrm{eff}}} = 1\,, \qquad (11.18)$$

where E is the pulse energy at the input and we have used the fact that $P = E/\tau$. We also neglect losses, which are insignificant over the distances we will consider. If one of the "geometrical" parameters (i.e., effective area, dispersion or nonlinearity) is varied along the propagation path slowly enough (adiabatically), the pulse duration will self-adjust to continue satisfying (11.18). In particular, if only the quadratic dispersion is adiabatically decreased the soliton will compress, while staying a soliton. The compression factor is given by

$$C = \frac{\beta''(0)}{\beta''(L)}\,, \qquad (11.19)$$

that is, the compression factor depends only on the local dispersion at the endpoints of the device. The exact dispersion profile along the device is not important as long as the adiabatic condition is satisfied, namely

$$\frac{L}{L_D(0)}\frac{2}{lnC} \gg 1\,. \qquad (11.20)$$

The inherent tradeoff is apparent here: larger compression factors make it more difficult to satisfy the adiabatic condition. Larger device length is desirable, but, in our case, will be limited by the length of gratings that can be reliably produced. We now apply these general results to the case where the propagation medium is a FBG. We will consider propagation in the passband of the grating, where the transmission is close to unity and rely on the large dispersion close to the grating band edge. To avoid group delay ripple (which may lead to varying dispersion across the pulse bandwidth), we will require apodized grating. These are routinely manufactured and result in very smooth monotonously varying group delay.

In this case, it can be shown that the quadratic dispersion is given by [15]

$$\beta'' L = -\frac{q}{\pi^2(\Delta\nu)^2}\frac{1}{(x^2-1)^{3/2}}, \tag{11.21}$$

where

$$q = \kappa L = \frac{\pi\eta(\Delta n)}{\lambda_B}L \tag{11.22}$$

$$\Delta\nu = 2\kappa\frac{c}{n} \tag{11.23}$$

$$x = \frac{\nu-\nu_B}{\Delta\nu}. \tag{11.24}$$

q is the grating strength, κ is the coupling constant, η is the overlap of the mode with the grating, Δn is the grating index modulation depth, λ_B is the Bragg wavelength, $\Delta\nu$ is the bandwidth of the grating stopband, c is the speed of light, n is the average index, ν is the frequency of operation and $\nu_B = c/\lambda_B$ is the Bragg frequency. The cubic dispersion is given by [15]

$$\beta''' L = 3\frac{q}{\pi^3(\Delta\nu)^3}\frac{x}{(x^2-1)^{5/2}}. \tag{11.25}$$

In order for the quadratic dispersion to be constant over the pulse bandwidth we need to have small cubic dispersion relative to the quadratic dispersion. This can be characterized by a figure of merit

$$M = \frac{|\beta''|}{\beta'''(2\pi\Delta f)} = \frac{3}{2}\frac{\Delta\nu}{\Delta f}\left(x-\frac{1}{x}\right), \tag{11.26}$$

where Δf is the spectral width of the pulse. For M to be large (i.e., for the cubic dispersion to be negligible over the bandwidth of interest) we need the detuning, x, to be large. Again, we are faced with a tradeoff, since increased detuning will reduce the dispersion.

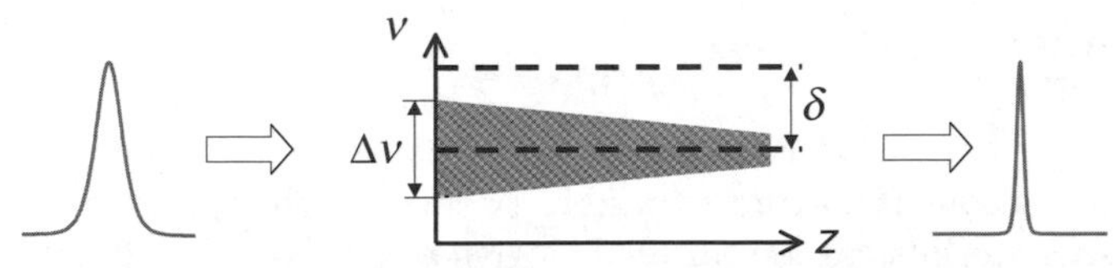

Fig. 11.5. Adiabatic soliton compression using an apodized FBG with linearly varying index modulation depth. The stopband (shaded region) shrinks because the coupling constant is decreasing because of the decreasing modulation depth. The detuning relative to the Bragg frequency is kept constant

Changing the dispersion along the grating can be accomplished by varying one or more of the following parameters: $\lambda_B = 2n\Lambda$ (where Λ is the pitch of the grating) through n and/or Λ, or by varying Δn. In the following example we chose to vary Δn linearly along the grating, such that our pulse propagating at fixed frequency ν, is getting further away (spectrally) from the band edge, since the stopband is shrinking along the propagation path (see Fig. 11.5).

In this configuration compression factors of about 5 are easily achievable with input pulses of 10–20 ps in FBG's that are a few tens of centimeters long (easily produced with current technology). Using silica FBG's however requires input peak powers on the order of a kilowatt, dictated by using the n_2 in (11.18). Increasing n_2 by a factor of 1000 will allow us to scale down the pulse energy required at the input to the compressor by the same amount. This sort of device would be practical for producing short pulses in optical time division multiplexed (OTDM) systems, or even just reshaping them back to short solitons.

Another application of this device is a pulse train generator. It was shown that if the input to this device is the beat signal between two CW sources, it evolves into a soliton train. If one of the lasers is tunable relative to the other we have a soliton train source with tunable repetition rate [16]. Here too, the device would benefit from a chalcogenide FBG implementation. Recently, both FBG's and waveguide Bragg gratings (WBG's) in As_2S_3 [18] and WBG's in $GeSe_3$ [20] have been fabricated and measured. Both interferometrically written gratings and relief gratings were produced, demonstrating the feasibility of making these devices. The linear spectral response showed the expected stop band and bandwidth, but still lacks the high quality of UV written silica gratings.

In summary, using chalcogenide FBGs (or WBGs) results in short, versatile devices requiring low input powers. Note also that FBG's are devices that can be tuned by applying heat or strain to the grating.

11.5 Conclusions

Se-based chalcogenide glasses have been shown to be excellent materials for nonlinear optical applications requiring a large Kerr effect in the spectral region of 1.5 μm. These materials can be tailored to give a variety of bandgaps, and therefore large n_2 for a variety of wavelengths in the mid-IR. This will become important as the spectral window for optical communications starts to include wavelengths longer and shorter than $1.5\mu m$. Additionally, these glasses have acceptable nonlinear loss and have an inherently ultrafast response. One should keep in mind that in many cases the material dispersion will determine the limits on the temporal response, although dispersion managing the device is sometimes possible (dispersion managed ultrafast all-optical switches operating with 400 fs pulses have been demonstrated).

Producing gratings in chalocgenide glasses is very interesting, not only for making some devices practical, but also for observing nonlinear phenomena that are not possible to observe in practical silica FBG's. Even though at present producing high quality, high-complexity chalcogenide FBG's or WBG's (e.g., a 10-cm apodized chirped grating) is not yet possible, the widespread interest in this area is sure to lead to these devices in the near future.

References

1. G. Lenz, J. Zimmermann, T. Katsufuji, M.E. Lines, H.Y. Hwang, S. Spälter, R.E. Slusher, S.-W. Cheong, J.S. Sanghera, and I.D. Aggarwal: Opt. Lett. **25**4, 254 (2000)
2. M.N. Islam, C.E. Soccolich, R.E. Slusher, A.F.J. Levi, W.S. Hobson, and M.G. Young: J. Appl. Phys. **71**4, 1927 (1992)
3. M. Asobe, K. Naganuma, T. Kaino, T. Kanamori, S. Tomaru, and T. Kurihara: Appl. Phys. Lett. **64**22, 2922 (1994)
4. G.P. Agrawal, *Nonlinear Fiber Optics*, 2nd edn. (Academic Press, San Diego 1995)
5. S. Smolorz, I. Kang, F. Wise, B.G. Aitken, and N.F. Borrelli: J. of Non-Crystalline Solids **256–257**, 310 (1999)
6. M.E. Lines: J. Appl. Phys. **69**10, 6876 (1991)
7. M. Sheik-Bahae, D.C. Hutchings, D.J. Hagan, and E.W. Van Stryland: IEEE J. Qunat. Electron. **27**6, 1296 (1991)
8. S.H. Wemple: Appl. Opt. **18**1, 31 (1979)
9. N.F. Nott and E.A. Davis, *Electronic Processes in Non-Crystalline Materials*, 2nd edn. (Oxford University Press, Oxford 1979)
10. J. Tauc, in *Amorphous and Liquid Semiconductors*, J. Tauc, Ed. (Plenum, London 1974) pp. 171–206
11. M. Shiek-Bahae, A.A. Said, T.-H. Wei, D J. Hagan, and E.W. Van Stryland: IEEE J. Quant. Electron. **26**4, 760 (1990)
12. E. Marquez, J.M. Gonzales-Leal, R. Jimenez-Garay, S.R. Lukic; and D.M. Petrovic: J. Phys. D: Appl. Phys. **30**, 690 (1997)

13. S.V. Chernikov and P.V. Mamyshev: J. Opt. Soc. Am. B **8**, 1633 (1991)
14. G. Lenz and B.J. Eggleton: J. Opt. Soc. Am. B **15**, 2979 (1998)
15. G. Lenz, B.J. Eggleton, R.E. Slusher, and N.M. Litchinitser, "Nonlinear Pulse Compression in Fiber Bragg Gratings", in *Bragg Gratings, Photosensitivity, and Poling in Glass Waveguides*, T. Erdogan, A.J. Friebele, and R. Kashyap, Eds. TOPS **33**, 421 (2000)
16. N.M. Litchinitser, G.P. Agrawal, B.J. Eggleton, and G. Lenz: Opt. Express **3**11, 411 (1998)
17. M. Asobe: Opt. Fiber Tech. **3**, 142 (1997)
18. C. Meneghini and A. Villeneuve, "Two-photon induced photodarkening in As_2S_3 for gratings and self-written channel waveguides", in *Bragg Gratings, Photosensitivity, and Poling in Glass Waveguides*, T. Erdogan, A.J. Friebele and R. Kashyap, Eds. TOPS **33**, 433 (2000).
19. S. Spälter, H.Y. Hwang, J. Zimmermann, G. Lenz, T. Katsufuji, S.-W. Cheong, and R.E. Slusher: Optics Lett. **27**, 363–365, (2002)
20. S. Spälter, H.Y. Hwang, J. Rogers, T. Katsufuji, G. Lenz, S.-W. Cheong, R.E. Slusher, A. Kelsey, B. MacLeod, Digest of Conference on Lasers and Electro-Optics, San Francisco 2000 (Optical Society of America, Washington, D.C., 2000), paper CTuH2

12 Optical Properties of Microstructure Optical Fibers

J.K. Ranka and A.L. Gaeta

12.1 Introduction

The development of optical-fiber cables and communications technology has undergone a dramatic revolution over the past decade [1]. The most basic design of an optical fiber consists of silica cladding that surrounds a silica core doped with germanium (GeO_2), which increases the index of refraction by up to 2% above that of pure silica, allowing light to be guided by total internal reflection at the core-cladding interface. Over the years, however, this basic geometry has evolved to substantially alter the waveguide properties, especially the dispersion in the 1.3 to 1.6 micron wavelength region. Design variations have included shaping the core index profile and adding down-doped and up-doped rings as well as elliptical cores for polarization-preserving fibers. New fiber structures that incorporate numerous air holes within the cladding region have emerged (Fig. 12.1) allowing for additional degrees of freedom that do not exist in conventional waveguide designs. The use of microstructure air holes around the fiber core have enabled new capabilities that have and will have an impact on many fields, from metrology to medicine. Microstructure or holey fibers and photonic crystal fibers can guide light either through total internal reflection between a solid silica core and air cladding or through a photonic bandgap (i.e., via diffraction) created by a periodicity in a microstructure cladding.

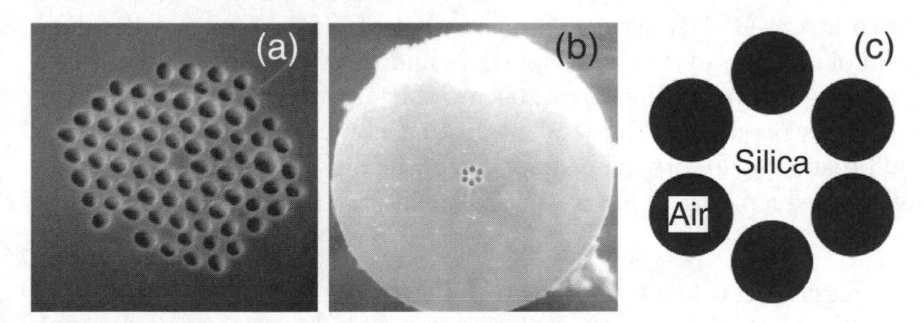

Fig. 12.1. Electron micrograph image of the inner cladding and core of the air-silica microstructure fiber with **a** a honeycomb array and **b** a single ring of air holes. Both fibers have similar optical properties. **b** Simulated microstructure fiber consisting of a 1.7-micron diameter core surrounded by a ring of 1.4-micron diameter air holes

Many of the novel properties of microstructure fibers involve consideration of the group-velocity dispersion (GVD) of the material, which causes the different spectral components of a pulse to travel through the medium with slightly different velocities, resulting in temporal broadening of the pulse. The net GVD of a single-mode optical fiber consists of two contributions, that from the material itself and that due to the waveguide. The waveguide contribution is a result of the wavelength dependence of the fraction of the light confined within the fiber core. At longer wavelengths, light is less confined by the higher-index core and expands further into the lower index cladding region, lowering the effective index of the guided mode. By properly modifying the core-cladding index difference and the core diameter, the waveguide dispersion contribution can be tailored to shift the zero-dispersion wavelength of a single-mode silica fiber to wavelengths greater than 1300 nm, the material zero-dispersion wavelength of bulk silica. Fibers can also be designed to have large normal dispersion ($>$ 100 ps/(nm-km) at 1.55 microns) or a net GVD of less than a few ps/(nm-km) across several hundred nanometers. As a general rule of thumb for a step-index fiber, the larger the core-cladding index contrast and the smaller the core diameter, the larger the magnitude of the waveguide contribution; i.e. a small wavelength-dependent variation in the mode-field diameter will result in a large effective index change.

The development of these novel microstructured fibers has had perhaps its most important impact in the area of nonlinear-optical effects in optical fibers. Although fused silica possesses a relatively low nonlinear susceptibility, single-mode optical fibers are excellent nonlinear media because of their low loss, small effective areas, and long interaction lengths. A myriad of nonlinear interactions, including spectral broadening and continuum generation, stimulated Raman and Brillouin scattering, and parametric amplification, have been efficiently demonstrated in optical fibers [1]. Since the strength of these nonlinear interactions is often limited by pulse spreading (and hence intensity reduction) due to the chromatic dispersion of the fiber, dispersion management through fiber design can be used to enhance these effects. Unfortunately in the single-mode regime, no conventional step-index silica fiber has been shown with a zero-GVD wavelength below that of bulk silica. As ultrashort pulse technology is most fully developed in and near the visible wavelength region, such as Ti:sapphire lasers at 800 nm, the large fiber dispersion in conventional fibers precludes a host of interesting nonlinear effects. Dispersive temporal broadening lowers the peak power of the pulses over short distances, substantially reducing the effective nonlinear interaction lengths.

12.2 Microstructure Optical Fibers

The basic air-silica microstructure fiber, originally known as single-material fibers, was first demonstrated in the early 1970's. Recent developments with photonic crystal fibers and "honeycomb" microstructured fibers add significant complexity, requiring air voids arranged in a periodic hexagonal lattice [2–7]. With just a solid silica core surrounded by an air-silica cladding, such fibers have been shown to exhibit single-mode behavior with negligible bend loss over broad spectral ranges [8], and theoretical calculations have indicated that these fibers can exhibit unusual dispersion characteristics such as anomalous group-velocity dispersion at wavelengths as short as 1-micron. Using an effective-index model to simulate the air-silica cladding, single mode propagation has been calculated to be due to the wavelength dependence of the cladding index. Light at shorter wavelengths becomes more confined in the silica core, avoiding the air holes and increasing the effective index of the cladding. For air-filling fractions (the percentage of the cladding that is air) below a few percent, the fiber can maintain single-mode operation over a broad wavelength range with minimal bend losses. As the air-filling fraction is increased, the fiber becomes highly multimode.

Microstructure optical fibers with large difference between the core and cladding refractive indices ("high delta"), where the air-fill fraction of the cladding is greater than $\sim 60\%$ can be considered simply as step-index waveguides, guiding light through total internal reflection. With these fibers, which are similar to the earlier single-material optical fibers [10,11], effective core-cladding index differences of nearly $\Delta n \sim 0.45$ can be realized, which is substantially higher than what can be achieved in conventional step-index optical fibers. These fibers will support numerous transverse spatial modes for any reasonable core diameter due to the large index contrast. Such simple step-index multimode fibers however have unique properties not possible in conventional optical fibers. The large index contrast and small core diameters result in a large waveguide contribution to the fiber's dispersion and hence dramatically altered dispersive characteristics, such as the realization of anomalous dispersion at visible wavelengths [12].

Since light is guided in these high-delta structures by the large index difference between the silica core and the surrounding air-hole layer, the mode-propagation and dispersive properties can be obtained directly through comparison to a simple high-delta step index fiber. The calculated waveguide contribution to the net GVD ($D = \frac{d}{d\lambda}(1/v_g)$, where v_g is the group velocity] for the lowest-order mode of a step-index fiber with a core diameter of 1-micron and core-cladding index difference of 0.1 and single-mode cutoff at $\sim$ 800-nm is shown in Fig. 12.2a. In the regime in which the fiber supports only a single mode, the waveguide contribution to the GVD is negative (normal dispersion), resulting in a shift of the fiber zero-GVD wavelength from the corresponding 1.28-micron zero-GVD wavelength of bulk silica to longer wavelengths. In the multimode region below $\sim 580\,\mathrm{nm}$, the waveguide GVD contribution to the structure is positive (anomalous); however the magnitude is substantially smaller than the corresponding dispersion of bulk silica. The maximum anomalous waveguide GVD for the fundamental mode increases in magnitude and shifts to longer wavelengths as the core-cladding index

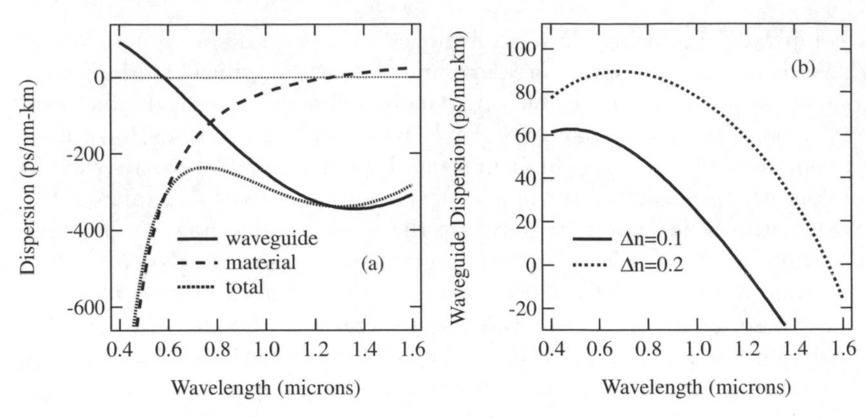

Fig. 12.2. **a** Calculated dispersion properties for a silica fiber with core-cladding index difference of 0.1 and core diameter of 1 micron. **b** Calculated waveguide GVD for a fiber with a core diameter of 2 micron and core-cladding index difference of 0.1 (solid line) and 0.2 (dotted line)

difference and the core diameter are increased. The waveguide dispersion contribution for a step-index fiber with a core diameter a 2-micron diameter and core-cladding index difference of 0.1 and 0.2 are shown in Fig. 12.2b. The magnitude of the anomalous contribution increases in value and shifts to longer wavelengths as the index difference is increased. As the index difference is increased to 0.3, the waveguide GVD contribution increases to 130 ps/(nm-km) at 760 nm, balancing the GVD of bulk silica of ~ -130 ps/(nm-km) at this wavelength.

12.3 Optical Properties

The high-delta microstructure fiber in Fig. 12.1a consists of a $\sim$ 1.7-micron diameter silica core surrounded by an array of 1.3-micron diameter air holes in a hexagonal close-packed arrangement. A small ellipticity in the fiber core results in a strong polarization-maintaining birefringence. Light is completely confined within the first ring of air holes of the structure and the outer layers do not appear to influence the waveguide properties. Similar versions of the fiber (Fig. 12.1b) incorporating only a single layer of large air holes with similar dimensions show identical waveguide properties. The background loss of these fibers is typically under 50 dB/km at 1 micron, while substantially higher at the water absorption peaks (Fig. 12.3a). Through proper manufacturing techniques, fibers with losses less than 10 dB/km are possible. As this is still approximately 50 times larger than for ordinary silica fibers, present day microstructure fibers are primarily suited for applications where short lengths ($\sim$ few hundred meters or less) are needed. Several kilometers of fiber are readily drawn from a single preform.

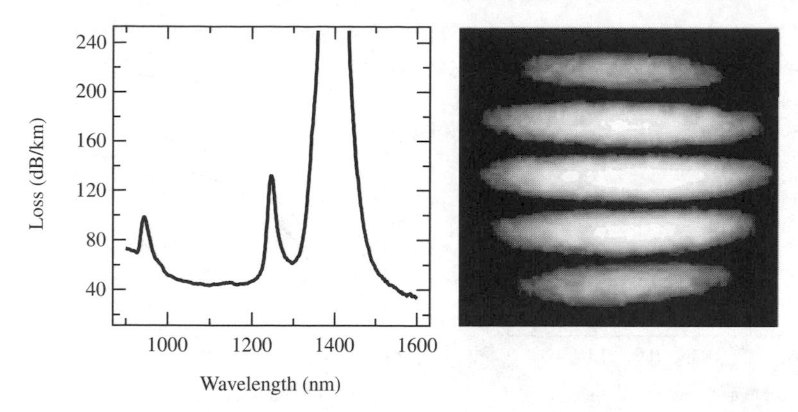

Fig. 12.3. a Loss spectrum of the microstructure fiber in Fig. 12.1a, **b** Spatial interference pattern formed by interfering the collimated output of the microstructure fiber with the output of a single-mode fiber. The clear fringe pattern indicates that only a single transverse mode is excited in the microstructure fiber

Experimentally it is seen that only a single mode can be excited for wavelengths from 500 nm to 1600 nm, and it is not possible to excite higher-order modes by bending or twisting the fiber as is easily done with conventional multimode fibers. Furthermore, the fiber is insensitive to bend loss at wavelengths less that 1600 nm for bends as tight as 0.5 cm in diameter. The single-mode nature of the fiber is determined by performing a spatial interference measurement between the output mode of the microstructure fiber with the output of a standard single-mode fiber. In Fig. 12.3b an example of a spatial interference fringe pattern formed with 633-nm light showing a clear fringe pattern indicates that only a single transverse mode is present. Additional measurements of the nonlinear refractive-index coefficient for linearly polarized light indicate that n_2 is $\sim 3 \times 10^{-16}\,\mathrm{cm}^2/\mathrm{W}$, similar to that of bulk silica.

The dispersion properties for the fundamental mode of high-delta microstructure fibers closely follow the predictions of analogous step-index waveguides. The experimentally measured and numerically calculated dispersion values for this fiber is shown in Fig. 12.4a. The GVD is experimentally determined by measuring the group delay of a 50-cm section of fiber in a Michelson interferometer with an air path reference. A standard single mode fiber measured using this technique exhibits a large normal dispersion, similar to bulk silica $[D \sim 110\,\mathrm{ps/(nm\,km)}]$ indicating a negligible waveguide contribution. The microstructure fiber is measured to have net anomalous dispersion across the entire spectral range, with a calculated zero-dispersion wavelength of 767 nm [12].

In the numerical calculation, a microstructure fiber with a 1.7-micron diameter silica core surrounded by a single ring of six 1.4-micron diameter air holes (Fig. 12.1c), similar to the inner layer of the measured fiber is con-

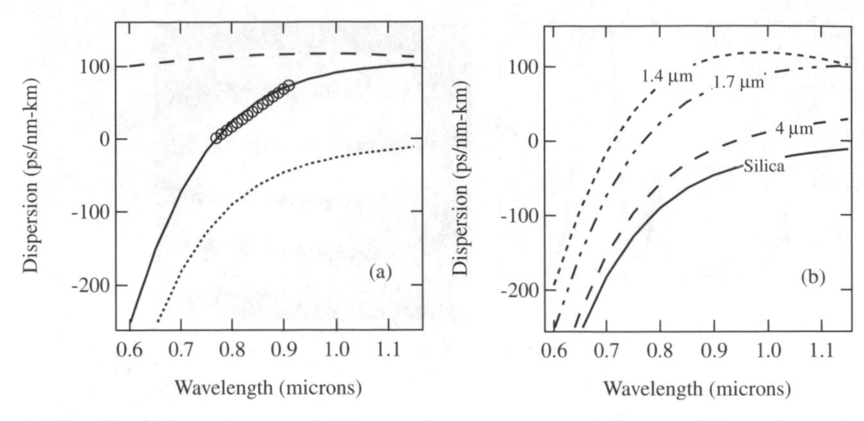

Fig. 12.4. **a** Calculated waveguide GVD contribution (dashed line), the material dispersion of bulk silica (dotted line), and the resulting net GVD (solid line) for the microstructure fiber. The experimentally measured values from reference 4 are shown by the circles. **b** Calculated net GVD of the microstructure fiber as the fiber dimensions are scaled for a 1.4, 1.7, and 4-micron core diameter

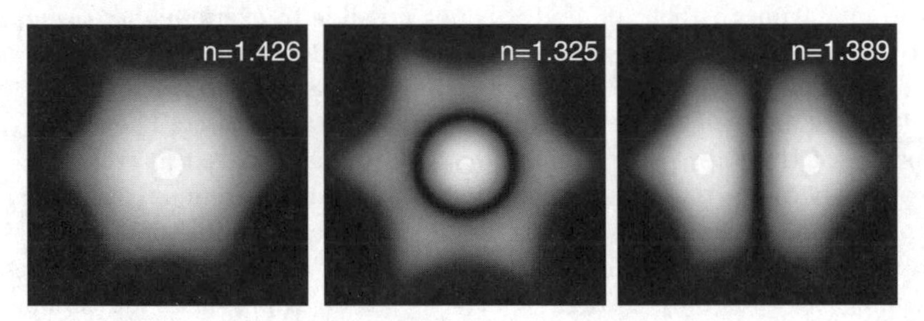

Fig. 12.5. Calculated field amplitude and effective index for the lowest-order modes of the microstructure fiber

sidered. An imaginary-distance beam propagation method that includes the polarization effect is used to calculate the field shapes and propagation constants of the transverse spatial modes of the fiber. The numerical simulation is in excellent agreement with the experimentally measured values. The simulation also shows that by simply by scaling the fiber dimensions for larger or smaller core diameters, fibers with zero-GVD anywhere between 650 nm and 1300 nm are possible (Fig. 12.4b).

In order to understand the experimental observation of single-mode behavior of this fiber, the calculated properties of the lowest order spatial modes are examined. The field amplitudes and the corresponding effective indicies of the three lowest-order modes are shown in Fig. 12.5. The simulations reveal that the individual guided modes can have substantially different propagation constants. The difference in the calculated effective index between the lowest two modes of the microstructure fiber, $\Delta n_{\text{eff}} \sim 2.6\%$, is significantly larger than even the core-cladding index difference in standard silica optical fibers. As a result, coupling between the fundamental mode and higher-order modes will be precluded even in the presence of substantial perturbations. Similar behavior is found in high-delta, small-area step-index fibers. Experimentally, it is also found that, owing to the small core size, only the fundamental mode is excited when the input coupling efficiency is maximized; excitation of higher-order modes is difficult. Hence, although the fiber can support numerous transverse spatial modes, single-mode excitation and propagation can easily be achieved. Increasing the fiber-core diameter reduces the difference in propagation constant between the modes and increases the potential of intramodal coupling. However, even when the fiber dimensions are scaled to a 4-micron core diameter at a wavelength of 1-micron the difference in effective index between the lowest two modes is still relatively large, $\sim 0.78\%$. Hence, although the fiber can support numerous transverse spatial modes, single-mode excitation (not necessarily the fundamental mode) and propagation is easily be achieved [13].

12.4 Nonlinear Interactions and Visible Continuum Generation

The unique dispersion characteristics, small core size, and novel mode characteristics of high-delta microstructure fibers open up a world of unique nonlinear and ultrashort pulse interactions at visible wavelengths [12,13]. Optical soliton propagation in these structures was first demonstrated by propagating 100-fs duration (τ_0) pulses at 790 nm through a 20-meter section of fiber. The fiber length corresponds to $\sim$ 15 dispersive lengths ($L_{\text{DS}} = \frac{2\pi c}{\lambda^2}\frac{\tau_0^2}{D}$). In the linear propagation regime the GVD of the fiber results in the pulse's temporal broadening to 400 fs-in duration. At 55-W input peak power a transmitted pulse of 175-fs duration is measured (Fig. 12.6). The transmitted spectrum is measured to shift towards the red, as is characteristic of intrapulse Raman scattering in the anomalous dispersion regime. Many other classic effects such as pulse breakup at high input powers are also observed [12] (Fig. 12.7).

As the input pulse wavelength is tuned closer to the zero-GVD wavelength of the fiber, the regime in which $L_{\text{fiber}} \ll L_{\text{DS}}$, substantial modification to the pulse spectrum are observed. The spectrum of a pulse after it propagates through a 10-cm section of the fiber as the peak power of the input pulse is varied is shown in Fig. 12.8. The input pulses are 110-fs in duration and centered at 770 nm, near the zero-GVD wavelength of the fiber. The spectra are measured with a calibrated optical spectrum analyzer with 1-nm resolution. As the peak power is increased to 220 W, the spectrum is seen to broaden considerably, with an asymmetric blueshifted spectral feature indicating a steepening of the trailing edge of the pulse. The spectrum broadens to more than 500 nm in width as the input peak power is further increased

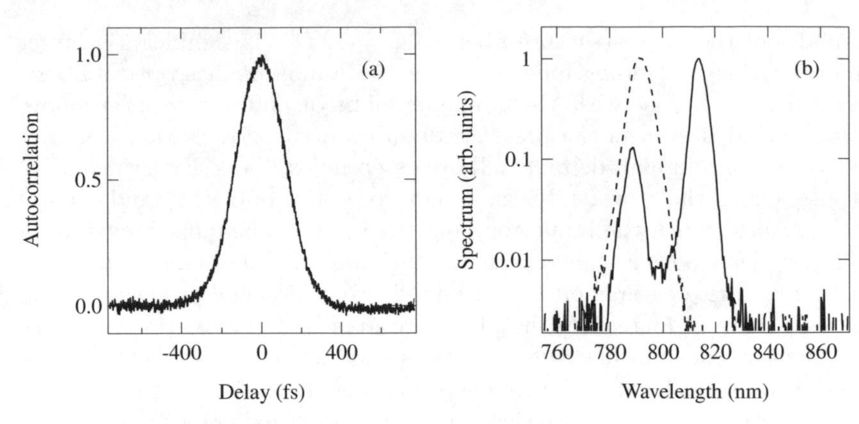

Fig. 12.6. Autocorrelation **a** and spectrum **b** of an initial 100-fs pulse after propagating through a 20-meter section of microstructure fiber. The input pulse spectrum is shown by the dashed line

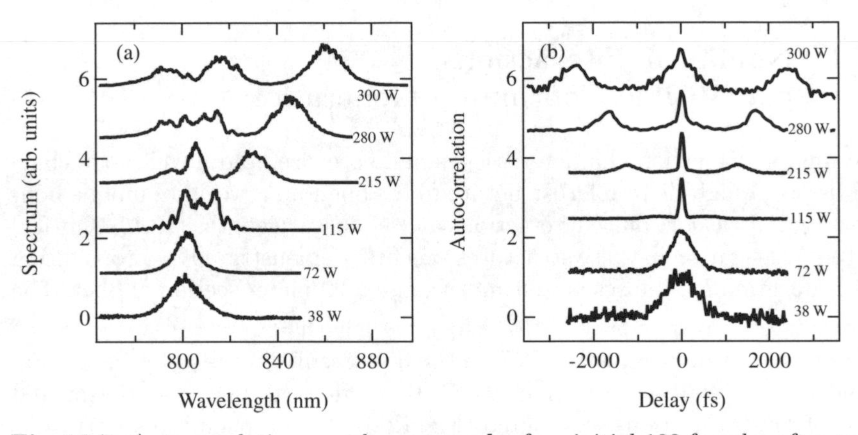

Fig. 12.7. Autocorrelation **a** and spectrum **b** of an initial 100-fs pulse after propagating through a 1-meter section of microstructure fiber as the input power is increased

to 1.6 kW. From the calibrated spectral measurements, the calculated mean photon energy of the transmitted pulse remains the same as the input in all cases, indicating a negligible contribution of Raman scattering to the process.

The most remarkable result seen to date however has been the generation of an ultrabroadband single-mode optical continuum. By injecting pulses of 100-fs duration, 800-pJ energy, and a center wavelength of 790 nm into a 75-cm section of fiber, we were able to create a continuum extending from 390 to 1600 nm (Fig. 12.9). Here the combined effects of self-phase modulation, four-wave mixing, second-harmonic generation, and Raman scattering (a virtual textbook of nonlinear phenomena) in the long length of fiber produce a broad,

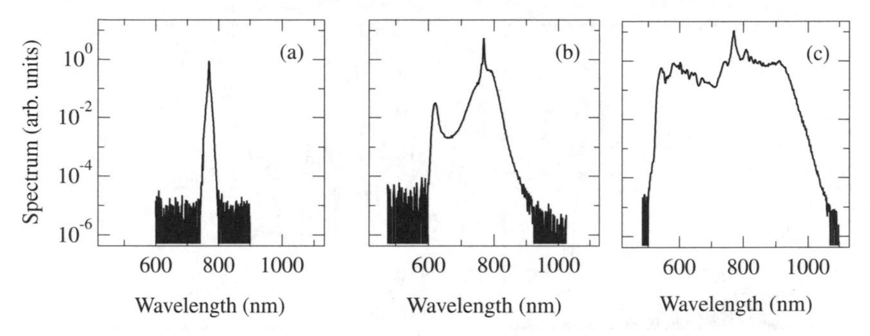

Fig. 12.8. Spectrum generated in a 10-cm section of microstructure fiber using 100-fs, 770-nm input pulses with peak powers of **a** 20 W, **b** 220 W, and **c** 1.6 kW

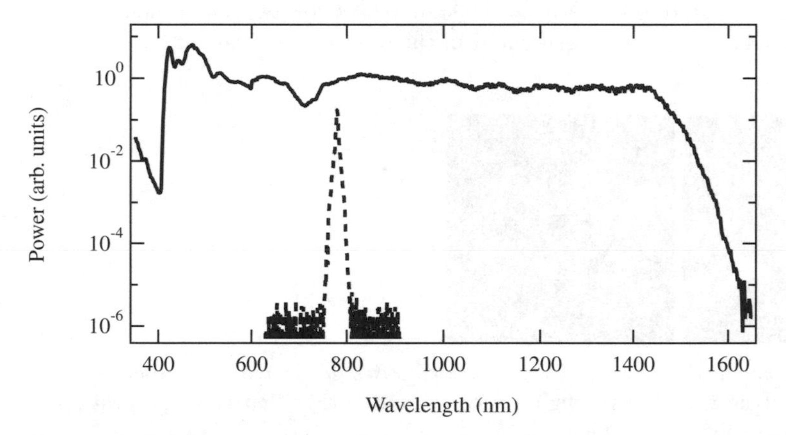

Fig. 12.9. Optical spectrum of the single-mode continuum generated in a 75-cm section of microstructure fiber. The blue curve shows the spectrum of the initial 100-fs duration, 800-pJ pulse

flat spectrum. By using input pulses of < 30-fs in duration and shorter (∼ few cm) sections of fiber, the Raman contribution can be suppressed and a stable continuum, generated primarily through self-phase modulation and the effects of dispersion is possible. Previous work in continuum generation has required pulses with megawatt peak power or microjoules of energy for generation of similar spectra. The small core diameter and small dispersion at this wavelength allow such a continuum to be achieved with 3 orders of magnitude less energy, specifically, pulses from a fairly simple unamplified Ti:Sapphire laser oscillator.

A number of nonlinear effects that utilize birefringent and multimode phasematching in the fiber have also been demonstrated using the fiber of Fig. 12.1a. By launching 100-fs-duration pulses at 895-nm with only 250 W of peak power along a birefringent axis of the fiber an orthogonally polarized

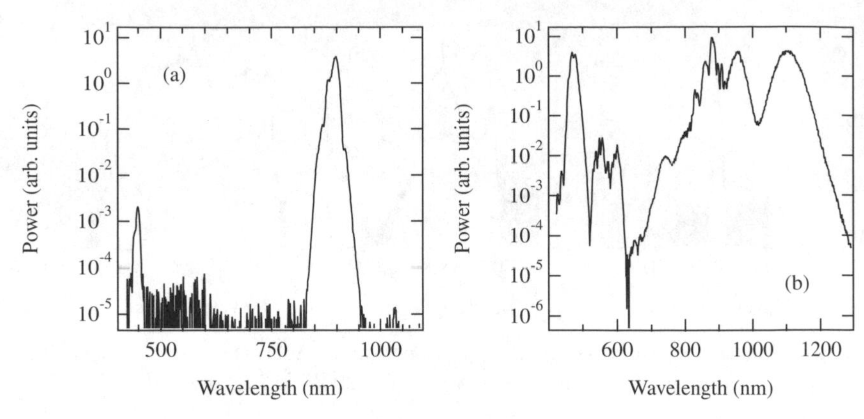

Fig. 12.10. Spectrum generated in a 10-cm section of microstructure fiber by use of linearly polarized, 100-fs duration, 895 nm pulses for two input powers. The second-harmonic component is generated in the orthogonal polarization

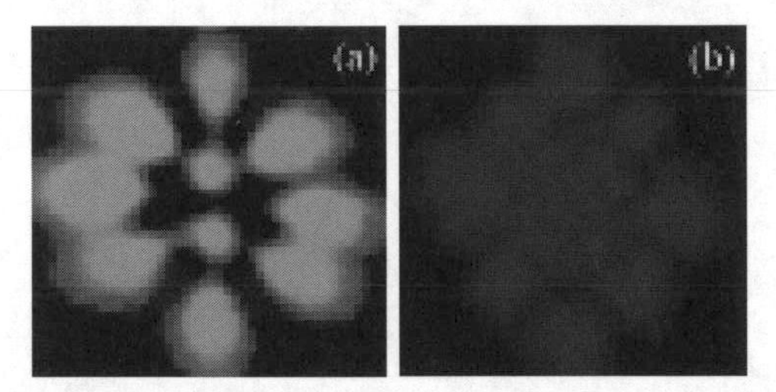

Fig. 12.11. Far-field patterns of the spatial modes generated in a 50-cm section of the microstructure fiber through **a** second- and **b** third-harmonic generation of 1064 nm input light

second-harmonic component is generated within the first few centimeters (Fig. 12.10). Multimode phase matching is seen when 1-kW pulses from a Q-switched Nd:YAG laser operating at 1064 nm are launched into the fundamental mode of a 50-cm section fiber consisting. The far-field mode patterns of the generated second- and third-harmonic components are shown in Fig. 12.12. Conversion efficiencies well above 5% are achieved. Experimentally, the generated higher-order modes appear insensitive to bend loss and do not couple to other transverse spatial modes even in the presence of strong perturbations to the single-ring microstructure fiber. The second-order nonlinear effects in the fiber are not clearly understood, and may be a result of stress-induced birefringence or surface effects.

12.5 Numerical Analysis

As discussed above, propagation of ultrashort pulses in optical fibers near the zero-dispersion wavelength allows for relatively long nonlinear interaction lengths and can result in the generation of an extremely broad continuum of radiation. Here we present theoretical results on the propagation of femtosecond pulses tuned near the zero-dispersion point in a microstructured fiber. It is expected that near the zero-GVD point higher-order diseprsion, such as third-order dispersion, plays an important role. There have been several experimental and theoretical studies of nonlinear propagation of ultrashort pulses near the zero-GVD point [14–17]. For a fiber in which positive TOD dominates pulse propagation, it has been observed that strong oscillations occur at the rear edge of the pulse [21] and that in combination with self-phase modulation (SPM) complex waveforms are produced which consist of a very rapid modulation at the rear half of the pulse [14,17]. However, in none of these studies did the interaction correspond to the the highly nonlinear operating conditions of these microstructure fiber experiments in which the combination of relatively high energy femtosecond pulses and the tight mode confinement results in supercontinuum generation.

We model pulse propagation inside the fiber by using the nonlinear envelope equation (NEE) [19,20] with the inclusion of the effects of stimulated Raman scattering. We assume that the pulse propagates along the z-axis with a propagation constant $\beta_0 = n_0\omega_0/c$, where n_0 is the linear refractive index of the material at the central frequency ω_0 of the pulse. We take the input pulse at $z = 0$ to be Gaussian in space and time such that $A(z = 0, t) = A_0 \exp[-t^2/2\tau_p^2]$. In this case, the equation for the normalized amplitude $u(z, t) = A(z, t)/A_0$ can be expressed as

$$\frac{\partial u}{\partial \xi} = -i \sum \frac{L_{\text{ds}}}{n! L_{\text{ds}}^{(n)}} \frac{\partial^2 u}{\partial \tau^2} + i \left(1 + \frac{i}{\omega \tau_p} \frac{\partial}{\partial \tau}\right) p^{\text{nl}} , \tag{12.1}$$

where $L_{\text{ds}}^{(n)} = \tau_p^n/\beta_{\text{n}}$ is the nth-order dispersion length, β_n ($n \geq 2$) is the nth-order dispersion constant [22] [e.g., $\beta_2 = (2\pi c/\omega_0^2)D$ represents the group-velocity dispersion (GVD) constant], $L_{\text{ds}} = L_{\text{ds}}^{(2)}$ is the dispersion length, $\zeta = z/L_{\text{ds}}$ is the normalized distance, $\tau = (t - z/v_g)/\tau_p$ is the normalized retarded time for the pulse traveling at the group velocity v_g, and p_{nl} is the suitably normalized nonlinear polarization. The presence of the operator $(1 + i\partial/\omega\tau_p\partial\tau)$ gives rise to self-steepening effects [18] and allows for the modeling of pulses with spectral widths comparable to the optical frequency ω_0. In the expression for the nonlinear polarization we include the effects of both the instantaneous (i.e., electronic) and non-instantaneous (i.e., nuclear)

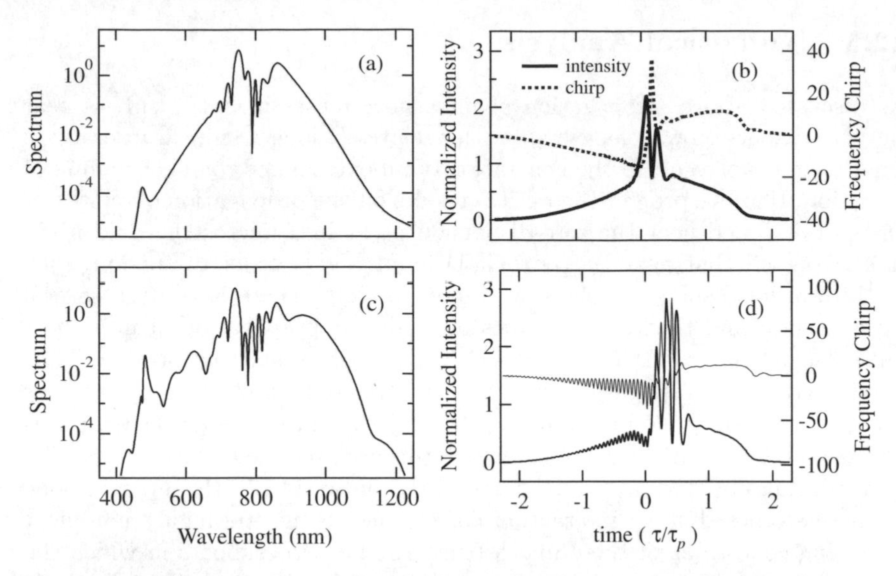

Fig. 12.12. Theoretically predicted spectra for the case described in the text in which the peak power of the input pulse is 10 kW and the propagation distance is **a** $\zeta = 0.0016$ and **c** $\zeta = 0.002$. The corresponding temporal profile and frequency chirp are shown in **b** and **d**

nonlinear refractive index change such that

$$p_{\mathrm{nl}}(\zeta,\tau) = \frac{L_{\mathrm{ds}}}{L_{\mathrm{nl}}} \times \left[(1-f)|u(\zeta,\tau)|^2 + f\gamma_R \int_{-\infty}^{\tau} d\tau' R(\tau-\tau')|u(\zeta,\tau')|^2\right] u(\zeta,\tau)\,, \tag{12.2}$$

where $L_{\mathrm{nl}} = (c/\omega n_2 I_0)$ is the nonlinear length, $I_0 = n_0 c|A_0|^2/2\pi$ is the peak input intensity, f is the fraction of the Raman contribution to the nonlinear refractive index, and $R(\tau) = \{[1 + (\Omega_R\tau_R)^2]/\Omega_R\tau_R\}\exp(-\tau/\tau_R)\sin(\Omega_R\tau)$ is the Raman response function.

In these simulations we use parameters that correspond to the case of a propagation of an initially 100 fs pulse at 770 nm inside a fused-silica microstructure fiber in which the core size is 1.7 μm in diameter, that is, $L_{\mathrm{ds}} =$ 500 cm, $\omega_0\tau_p = 140$, $L_{\mathrm{ds}}/L_{\mathrm{ds}}^{(3)} = 5$, $L_{\mathrm{ds}}/L_{\mathrm{ds}}^{(4)} = -0.2$, $f = 0.15$, $\tau_p/\tau_R = 1$, and $\Omega_R\tau_R = 4$. For the case in which the peak power of the pulse is 1 kW, $L_{\mathrm{ds}}/L_{\mathrm{nl}} = 500$. The qualitiative nature of the results described here occur over a range of parameters and does not depend sensitively precise values. We first consider the case in which the central wavelength of the pulse is tuned slightly to the red side of the zero-GVD point (i.e., $\beta_2 < 0$). At low powers and short propagation distances, the pulse exhibits asymmetric spectral broadening as a result of TOD, SRS, and self-steepening. At sufficiently high powers ($P = 10$ kW) and long propagation distances ($L_{\mathrm{ds}}/L = 0.016$), a

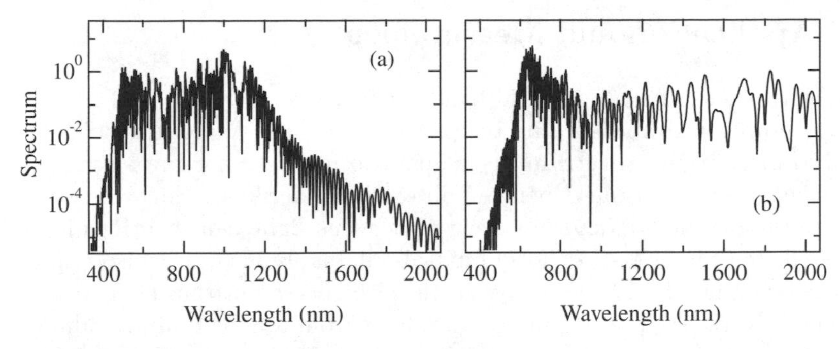

Fig. 12.13. Theoretically predicted spectra for the case described in the text in which the peak power of the input pulse is 80 kW, the propagation distance is $\zeta = 0.0032$, and the third-order dispersion is **a** positive and **b** negative

peak appears in the spectrum towards the blue side (see Fig. 12.12a). There is no corresponding peak on the red of the pulse, and thus the appearance of this feature cannot be attributed simply to four-wave mixing. The corresponding time-domain behavior is shown in Fig. 12.12b, and it is apparent that the abrupt phase drop that occurs between the two peaks give rise to the observed blue spectral features. At slightly longer propagation distances ($L_{\mathrm{ds}}/L = 0.02$) (see Fig. 12.12c) the pedestal on the short-wavelength side becomes significantly broader than that at long-wavelength side. As shown in Fig. 12.12d, the corresponding time-domain profile of the light exhibits increasingly complicated structure.

At higher powers the wings of the spectrum become comparable in amplitude to the central portion (see Fig. 12.13a), and the envelope is similar to the shape of the supercontinuum observed in experiments; the blue edge is sharp whereas the red edge falls off more gradually. At these powers, changes in the pulse energy do not significantly alter the spectral enevelope, however the substructure in the simulations is sensitive to the initial pulse energy. Similar substructure has been observed in recent experiments [23]. Although SRS and self-steepening are included in this model, we find that in the limit where the GVD is small, the basic shape depends primarily on the amount of TOD and to a lesser extent the amount of fourth-order dispersion. In the time domain, the initial pulse has broken up into numerous pulses, some of which are much shorter than the initial pulse.

Since TOD primarily determines the dynamics that leads to the supercontinuum, manipulation of the magnitude will play a critical role in optimizing the shape and width of the spectrum for particular applications. For example, by changing the sign of the TOD (i.e., of β_3), but keeping all other parameters the same, the red edge of the spectrum can be substantially extended whereas the extent of the blue edge is reduced. Suitable designs of microstructured fibers offer the possibility of further increases in the total spectral bandwidths via reduction of the TOD and the flattening of the dispersion profile.

12.6 Application and Measurement

A pulse with such a broad spectral bandwidth poses a serious challenge for anyone attempting to measure its temporal properties. While it is straightforward to measure its spectrum, techniques do not exist for measuring the spectral phase or the time-dependent intensity and phase. Such measurements are important for determining whether this light can in principle be compressed to its Fourier-transform limit-about 1 fs! Indeed, it is also important to determine whether these quantities are the same from shot-to-shot or are highly variable. For example, white light from a light bulb has almost as broad a spectrum, but it has a random spectral phase and hence is uncompressible. Fortunately, improvements are being developed to existing sophisticated pulse-measurement techniques such as frequency-resolved-optical-gating (FROG), to allow the measurement of the ultrabroadband continuum. Researchers have shown the spectral phase to be quite stable by performing a spectrally-resolved cross-correlation measurement (often referred to as XFROG) between a $\sim$ 30-fs reference pulse (characterized separately by the FROG technique) and the continuum, [24]. The delay vs. wavelength is measured to be approximately quadratic, indicating that compression should be possible.

The broadband continuum generated by femtosecond pulses propagating through this fiber has enabled a revolution in optical frequency metrology. While it may be surprising that femtosecond pulses are useful in optical frequency metrology, the very stable train of pulses generated by a modelocked laser corresponds to a ł‘comb” of regularly spaced frequencies. If this comb is broad enough, it can be used to subdivide an optical frequency into intervals that are small enough to be directly compared to a cesium clock, which is the primary standard for the definition of the second. Using microstructure fiber to generate the necessary bandwidth, this technique has recently been demonstrated [25]. This frequency domain technique can also be used to stabilize the phase shift between pulses, which is of interest for time-domain measurements.

Another promising application of this broadband continuum will be in optical coherence tomography (OCT) [26]. OCT is a new imaging modality that is the optical analog of ultrasound and enables the in situ, real-time cross-sectional imaging of materials and biological tissues. In OCT, imaging is performed using low coherence interferometry and the longitudinal resolution is inversely proportional to the bandwidth of the light. In the past, ultrahigh resolution OCT measurements have been performed using state-of-the art Ti:sapphire lasers that utilize double-chirped mirrors to generate 5 femtosecond pulse durations with 300–400 nm bandwidths. The microstructure fibers promise to be powerful sources for OCT imaging since they can generate extremely broadband light that spans essentially the entire visible as well as the near infrared wavelength range. These bandwidths could enable submicron resolution imaging. In addition, the broad spectral bandwidths will be powerful for spectroscopically resolveed OCT imaging as well as imaging in wavelength regimes that were previously inaccessible.

12.7 Conclusion

Recently demonstrated high-delta microstructure optical fibers have unique properties that are not possible with conventional optical fiber designs. The high-delta and small core diameter of the structure provide for a large waveguide dispersion contribution that can result in net anomalous GVD of the fiber at wavelengths down to the visible regime. Similar results can also be achieved in many other structures, such as ribbed planar waveguides, where unique highly nonlinear materials can be used. Recent work has examined properties similar to those discussed here using tapered optical fibers [27,28]. Low loss microstructure fibers, where the air holes provide a small perturbation to the waveguide have been demonstrated as a potential dispersion-compensating device, with large anomalous dispersion, for optical communication systems [29]. While it is uncertain that microstructure fibers will be used as a long-haul transmission fiber, the nonlinear properties of such fiibers are well suited for device applications. The ability to easily generate a broad, stable continuum has already revolutionized fields such as frequency metrology and biomedical imaging in a short period of time. We have only begun to see the potential and benefits of novel microstructure optical fibers [30].

References

1. G.P. Agrawal, *Optical Fiber Communications Systems* (Academic Press, Boston, MA 1995); G.P. Agrawal, *Nonlinear Fibers Optics* (Academic Press, Boston, MA 1995)
2. J.D. Joannopoulos, R.D. Meade, and J.N. Winn, *Photonic Crystals: Molding the Flow of Light* (Princeton University Press, Princeton, NJ 1995); S. John: Physics Today 1990
3. J.C. Knight, J. Broeng, T.A. Birks, and P.S.J. Russell: Science **282**, 1476 (1998)
4. S.E. Barkou, J. Broeng, and A. Bjarklev: Opt. Lett. **24**, 46 (1999)
5. A. Ferrando, E. Silvestre, J.J. Miret, and P. Andres: Opt. Lett. **24**, 276 (1999)
6. T.M. Monro, D.J. Richardson, N. Broderick, and P.J. Bennett: J. Lightwave Technol. **17**, 1093 (1999)
7. D. Mogilevtsev, T.A. Birks, and P.S.J. Russell: Opt. Lett. **23**, 1662 (1998)
8. T.A. Birks, J.C. Knight, and P.S.J. Russell: Opt. Lett. **22**, 961 (1997)
9. T.M. Monro, P.J. Bennett, N.G.R. Broderick, and D.J. Richardson: Opt. Lett. **25**, 206 (2000)
10. E.A.J. Marcatili: Bell Syst. Tech. J. **54**, 645 (1974)
11. P. Kaiser, E.A.J. Marcatili, and S.E. Miller: Bell Syst. Tech. J. **52**, 265 (1973)
12. J.K. Ranka, R.S. Windeler, and A.J. Stentz: Opt. Lett. **25**, 25 (2000); in *Conference on Lasers and ElectroOptics, 1999* OSA Technical Digest Series (Optical Society of America, Washington, DC 1999) postdeadline paper CD-8

13. J.K. Ranka, R.S. Windeler, and A.J. Stentz: Opt. Lett. **25**, 798 (2000)
14. G.P. Agrawal and M.J. Potasek: Physical Review A **33**, 1765 (1986)
15. P. Beaud, W. Hodel, B. Zysset, and H.P. Weber: Journal of Quantum Electronics, **QE-23**, 1938 (1987
16. V. Yanofsky and F.W. Wise: Optics Letters, **19**, 1547 (1994)
17. G. Boyer: Optics Letters **24**, 1547 (1999)
18. G. Yang and Y.R. Shen: Optics Letters, **9**, 510 (1984)
19. P.V. Mamyshev and S.V. Chernikov: Optics Letters **15**, 1076 (1990)
20. T. Brabec and F. Krausz: Physical Review Letters **78**, 3283 (1997)
21. See, for example, G.P. Agrawal, *Nonlinear Fiber Optics*, 3rd edn., Chap. 3 (Academic Press, San Diego 2001)
22. See, for example, G.P. Agrawal, *Nonlinear Fiber Optics*, 3rd edn., Chap. 1 (Academic Press, San Diego 2001)
23. L. Xu, M.W. Kimmel, P. O'Shea, R. Trebino, J.K. Ranka, R.S. Windeler, and A.J. Stentz, *Ultrafast Phenomena XII*, ed. by T. Elsaesser, S. Mukamel, M.M. Murnane, and N.F. Scherer (Springer, Berlin Heidelberg New York 2001) p. 129–131
24. M. Kimmel and R. Trebino, *Conference on Lasers and ElectroOptics, 2000* OSA Technical Digest Series (Optical Society of America, Washington, DC 2000) paper CFL7
25. D.J. Jones, S.A. Diddams, J.K. Ranka, A. Stentz, R.S. Windeler, J.L. Hall, and S.T. Cundiff: Science **288**, 635 (2000)
26. I. Hartl, X.D. Li, C. Chudoba, J.G. Fujimoto, J.K. Ranka, and R.S. Windeler: Opt. Lett. **26**, 9 (2001)
27. T.A. Birks, W.J. Wadsworth, and P.S.J. Russell: Optics Letters, **25**, 1415 2000
28. J.K. Chandalia, B.J. Eggleton, R.S. Windeler, and S.G. Kosinski, *Optical Fiber Communications, 2001* OSA Technical Digest Series (Optical Society of America, Washington, DC 2001) paper TuC2-1
29. T. Hasegawa, E. Sasaoka, M. Onishi, M. Nishimura, Y. Tsuji, and M. Koshiba, *Optical Fiber Communications, 2001* OSA Technical Digest Series (Optical Society of America, Washington, DC 2001) postdeadline paper PD5-1
30. J.K. Ranka and R.S. Windeler: Optics and Photonics News **11**, 20 (2000)

13 Semiconductor Optical Amplifiers with Bragg Gratings

G.P. Agrawal and D.N. Maywar

13.1 Introduction

Nonlinear periodic structures can be made using a variety of materials including fibers and semiconductors. Fiber Bragg gratings are the subject of much research and, indeed, are the focus of several chapters in this book. In this chapter, we consider a Bragg grating fabricated within a semiconductor optical amplifier (SOA). The strong carrier-induced nonlinearity and the resulting nonlinear shift of the Bragg resonances are discussed in Sect. 13.2. The origin and characteristics of bistable switching, a common nonlinear response that occurs at powers as low as $1\,\mu W$, are discussed in Sect. 13.3. Section 13.4 is devoted to a simple theoretical model for investigating the nonlinear response of Bragg-grating SOAs. In Sect. 13.5, we consider Bragg-grating SOAs in the context of a broader class of devices referred to as resonant-type SOAs. In Sect. 13.6, potential applications in the domain of fiber-optic communication systems are discussed, emphasizing the features of all-optical switching and memory.

13.2 Active Bragg Resonances

13.2.1 Semiconductor Optical Amplifiers

SOAs exhibit many features that make them advantageous as nonlinear functional devices [1,2]. They are compact ($< 500\,\mu m^3$ active volume), integrateable with other devices, and operable at any wavelength used in optical communication systems. SOAs also provide optical gain and therefore allow features such as fan-out and cascadability, which are general requirements for multi-component lightwave systems and large photonic circuits [3]. Most importantly, SOAs exhibit a strong carrier-induced nonlinearity with an effective value of $n_2 \sim 10^{-9}\,cm^2/W$ [4], seven orders of magnitude larger than that of silica fiber.

Both the nonlinear refractive index and gain are functions of the carrier density N. For time scales longer than the intraband relaxation time ($<$ 0.1 ps), the dynamics of the carrier density can be modeled by a rate equation of the form [5]

$$\frac{\partial N}{\partial t} - D\nabla^2 N = \frac{J}{ed} - \frac{N}{\tau} - a(N - N_0)\frac{I}{\hbar\omega}\,, \tag{13.1}$$

where D is the diffusion coefficient, J is the current density injected into the active layer of thickness d, e is the electron charge, and τ is the carrier lifetime. The last term accounts for the stimulated recombination of electron-hole pairs by an optical signal of frequency ω and intensity I.

Optical signals passing through an SOA experience gain; the gain coefficient g is well approximated as a linear function of the carrier density N [5]:

$$g(x, y, z, t) = a\Gamma[N(x, y, z, t) - N_0]\,, \tag{13.2}$$

where the differential gain $a = dg/dN$ is evaluated at the value of carrier density required to achieve transparency N_0. The optical confinement factor Γ, with a typical value ~ 0.2, represents the fraction of the transverse-intensity profile that falls within the SOA active region. The wavelength dependence of the gain is suppressed as it does not significantly alter the nonlinear response considered in this chapter. Due to fast-acting intraband relaxation, the gain medium is homogeneously broadened for times scales longer than 0.5 ps.

SOAs exhibit a refractive index that is significantly dependent on the carrier density N. This dependence is embodied by the linewidth enhancement factor α, which couples changes in the real part of the refractive index n to changes in the imaginary part, or equivalently, to gain [5]:

$$\alpha = -\frac{4\pi}{\lambda_0}\frac{(dn/dN)}{(dg/dN)}\,. \tag{13.3}$$

The changes in the refractive index n and gain g with respect to the carrier density N are opposite in sign, yielding a positive-valued α; typical values of α range from 2–3 for quantum-well devices to 5–8 for bulk active regions. The linewidth enhancement factor α is the basis of the nonlinear response considered in this chapter; although the coupling between the refractive index and gain is dependent on many physical quantities, including the carrier density N and signal-wavelength λ_0, the constant-valued α is often sufficient to capture the qualitative behavior of nonlinear phenomena in SOAs.

13.2.2 Distributed Feedback

The propagation of the optical signal through the SOA is significantly altered by introducing a built-in grating parallel to the SOA gain region, as shown in Fig. 13.1. This device is commonly refered to as a distributed feedback (DFB) SOA since the Bragg grating provides feedback along the device. The optical field inside a DFB SOA can be expressed as

$$\begin{aligned}\boldsymbol{E}(x, y, z, t) = \mathrm{Re}\{\hat{\epsilon}F(x, y)[&A(z, t)\exp(i\beta_B z)\\ &+B(z, t)\exp(-i\beta_B z)]\exp(-i\omega t)\}\,,\end{aligned} \tag{13.4}$$

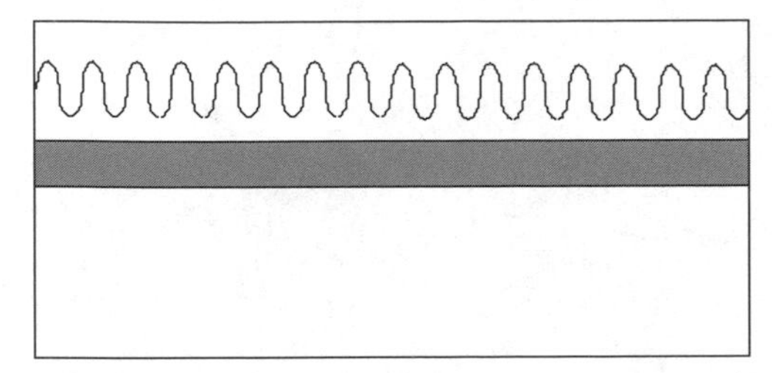

Fig. 13.1. Schematic of a Bragg-grating SOA. The grating is typically fabricated outside of the gain region (shaded grey)

where Re represents the real part, $\hat{\epsilon}$ is the unit vector along the transverse-electric (TE) orientation of polarization, and $F(x, y)$ is the transverse mode distribution. The field scatters off of the Bragg grating in the longitudinal direction (z coordinate) and is therefore conveniently expressed as two counterpropagating terms; A and B are the slowly varying field envelopes for the forward and backward propagating fields, respectively, and the Bragg wavenumber $\beta_B = \pi/\Lambda$ is related inversely to the grating period Λ.

The field envelopes A and B are governed by the following coupled-mode equations [5]:

$$\frac{\partial A}{\partial z} + \frac{1}{v}\frac{\partial A}{\partial t} = i\Delta A + i\kappa B\,, \tag{13.5}$$

$$-\frac{\partial B}{\partial z} + \frac{1}{v}\frac{\partial B}{\partial t} = i\Delta B + i\kappa A\,, \tag{13.6}$$

where v is the group velocity of the optical signal, and κ is the coupling coefficient that characterizes scattering by the Bragg grating. Spontaneous emission into the modes is not included since the amplified optical signal is expected to be much stronger [6].

The Δ term in the coupled-mode equations (13.5) and (13.6) represents the "self" coupling exhibited by each mode, and contains terms that are particular to SOAs [7]:

$$\Delta = \delta - i\frac{g}{2}(1 - i\alpha) + i\frac{\alpha_{\text{int}}}{2}\,. \tag{13.7}$$

The quantity α_{int} accounts for internal losses within the SOA waveguide, and the detuning parameter $\delta = \beta_0 - \beta_B$ accounts for the mismatch between the carrier-density-independent portion of the signal's wavenumber β_0 and the Bragg wavenumber β_B. The carrier-density-*dependent* portion of the real part of the optical wavenumber is given by $\beta_N = -\alpha g/2$.

13.2.3 Nonlinear Bragg Resonances

Feedback from the Bragg grating gives rise to a spectral region of low transmission called the photonic bandgap, found in the center of Fig. 13.2. Located on either side of the photonic bandgap are strong Bragg resonances

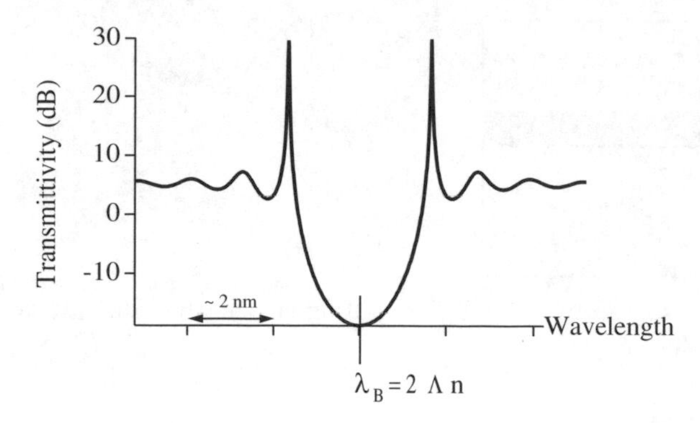

Fig. 13.2. Bragg-resonance structure of a DFB SOA, centered about the Bragg wavelength (for a uniform grating without facet reflections)

that each provide high gain within a localized spectral region. In the case shown in Fig. 13.2, a pump level of $g_0L = 1.2$ provides only 5 dB of gain for signal wavelengths unaffected by the Bragg grating (i.e., tuned away from the photonic bandgap), whereas a grating of strength $\kappa L = 3$ yields Bragg resonances having 30-dB peak gain.

The Bragg resonances and photonic bandgap in Fig. 13.2 are centered about the Bragg wavelength λ_B. For a first-order grating, this quantity is given by [5]:

$$\lambda_B = 2\Lambda n\,. \tag{13.8}$$

As noted earlier, the refractive index is highly nonlinear. Thus, optical signals cause the Bragg-resonance spectrum to shift. The nonlinear response considered in this chapter occur as a Bragg resonance, whose spectral location depends on the refractive index, shifts onto and off of an optical signal as a function of the optical power within the DFB SOA.

13.3 Optical Bistability

13.3.1 Physical Origin

In the mid 1980s, researchers began to explore the nonlinear response of Bragg-grating SOAs. In 1985, a bistable hysteresis was demonstrated in the transmitted power as a function of the input-signal wavelength [8]. Within a year, another report showed a nonlinear input–output transfer function exhibiting hysteresis [9]. These demonstrations confirmed an earlier prediction that nonlinear Bragg structures can support dispersive optical bistability [10].

Optical bistability is typically characterized by a steady-state input-output transfer function that doubles back on itself, as shown in Fig. 13.3 for the average power within a DFB SOA. Bistable switching within a DFB SOA

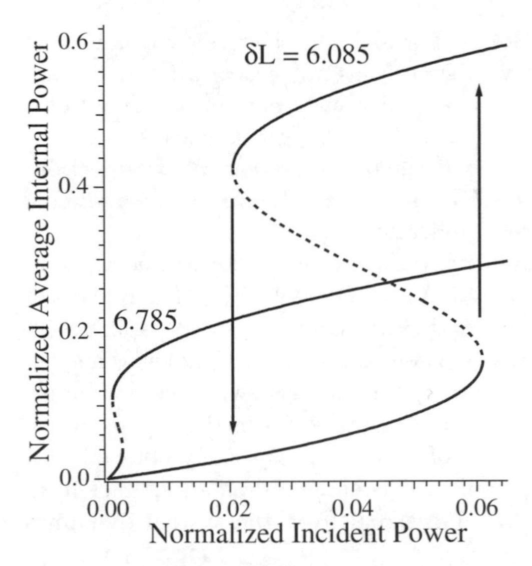

Fig. 13.3. Average-internal-power hysteresis curve for two signal wavelengths. The longer signal wavelength ($\delta L = 6.085$) exhibits higher switching thresholds. Dashed portions of curves are unstable

occurs via a positive-feedback loop involving the gain-dependent refractive index, a Bragg resonance, and the internal optical power. An optical signal enters the amplifier with a wavelength longer than that of a Bragg resonance. The signal saturates the gain and shifts the photonic bandgap and the associated Bragg resonances to longer wavelengths. If a Bragg resonance shifts onto the signal wavelength, the internal optical power increases even more. As a result, the refractive index continues to increase, and the resonance shifts even farther. This positive feedback loop for the internal optical power moves the Bragg resonance fully through the signal wavelength. The resulting jump experienced by the internal power is indicated by the up-arrow in Fig. 13.3.

The reverse process occurs near the down-arrow. If the incident power is lowered so that the signal wavelength returns to the peak of the Bragg resonance, a subsequent decrease in incident power allows the gain to partially recover, thereby decreasing the refractive index. The Bragg resonance, in turn, shifts to shorter wavelengths and away from the signal wavelength, decreasing the internal power further. This positive feedback loop shifts the resonance off of the signal wavelength and the internal power switches downward. These upward and downward switching processes give rise to a S-shaped hysteresis curve, a common shape exhibited by many bistable systems.

13.3.2 Switching Characteristics

Switching powers in SOAs have been measured as low as 1 µW [6]. Bistable switching occurs at such low powers that it is likely to manifest itself in any experiment using Bragg-grating SOAs. For example, bistability occurred in studies that sought to use the device strictly as an optical filter with gain [11,12]; bistable switching occurred at submicrowatt powers in simulations, and the associated self-tuning effect shifted the Bragg resonance far enough away from the signal wavelength to limit the output power. A low injection-current level and a long device length were suggested as means for reducing this nonlinear effect for active filter applications [11].

Switching from the lower to higher level has been demonstrated with an on–off contrast ratio in Bragg-grating SOAs as high as 10:1 [9]. The process of switching is often accompanied by a spiking behavior. Physically, spiking occurs as the peak of the Bragg resonance passes through the signal wavelength. Investigations demonstrated how a spike during upward switching dominates the output-pulse shape as the signal wavelength is detuned away from the Bragg resonance [13]. Spiking behavior of this sort is especially noticeable in simulations of dispersive bistability [6], and is enhanced as an artifact from the adiabatic elimination of the bistable signal from the system dynamics [14].

A variety of shapes of the hysteresis curve has been reported. For example, the contrast ratio differs for bistability occurring at the Bragg resonances at either side of the photonic bandgap. Although the resonance spectrum is symmetric about the Bragg wavelength, the optical signal is initially tuned to the long-wavelength side and thus experiences a different resonance shape at either side of the photonic bandgap (see Fig. 13.2). The hysteresis curve varies in shape even more drastically on reflection [15,16]. As a function of the injection current, or signal wavelength, the hysteresis curve warps from a standard S-shape to a loop-shape and eventually to an inverted S shape. The origin of this variety lies in the behavior of the Bragg resonance on reflection; as the gain is lowered, the reflectivity resonance flips from a high peak to a low dip [16].

13.4 Theoretical Model

The nonlinear response of Bragg-grating SOAs has been modeled using several methods. Most theoretical investigations use a carrier-density rate equation and describe the optical field using couple-dmode equations; the optical intensity distribution along the amplifier length is approximated either as a single average value [6,13] or as a an array of average values representing its longitudinal distribution [16,17]. The bistable optical signal has also been modeled using a single-mode rate equation [12].

In this section, we discuss a simple model based on the carrier-density rate equation (13.1) and a single average value of the optical power derived from the coupled-mode equations (13.5) and (13.6) discussed in Sect. 13.2. The model incorporates a number of approximations common to the study of optical bistability and were first applied to Bragg-grating SOAs in 1987

[6]. Most of these approximations are based on two distinct features of SOAs, namely a relatively small size of the gain region and a relatively slow speed of the carrier-induced nonlinearity.

13.4.1 Average-Gain Equation

The carrier rate equation (13.1) can be simplified considerably by noting that the diffusion length is larger than the gain-region thickness ($d \sim 0.15\,\mu\mathrm{m}$) and is of the same order as the width $W \sim 2\,\mu\mathrm{m}$ (for an index-guided waveguide). An average value of carrier density is therefore used in the transverse dimensions; averaging (13.1) over the width W and thickness d of the gain region yields

$$\frac{dN}{dt} = \frac{J}{ed} - \frac{N}{\tau} - \frac{a}{\hbar\omega}(N - N_0)\frac{\Gamma\sigma}{Wd}(|A|^2 + |B|^2)\,, \tag{13.9}$$

where N is now understood as the average value. The confinement factor Γ and the mode cross section σ are given by

$$\Gamma = \int_0^W \int_0^d dx dy \frac{|F(x,y)|^2}{\sigma}\,, \qquad \sigma = \int_{-\infty}^{\infty} \int_{-\infty}^{\infty} dx dy\, |F(x,y)|^2\,.$$

Carrier diffusion also smoothes out the spatial holes burned by the counter-propagating fields (typical period $\approx 0.2\,\mu\mathrm{m}$), allowing the interference term in the intensity to be neglected.

Since the carrier density enters the coupled-mode equations through the modal gain g, it is convenient to rewrite (13.9) as a gain rate equation:

$$\tau\frac{dg}{dt} = g_0 - \left[1 + \bar{P}\right] g\,. \tag{13.10}$$

The quantity $g_0 = \Gamma a N_0(\bar{J} - 1)$ is the small-signal value of g with $\bar{J} = J\tau/edN_0$. The normalized optical power $\bar{P}$ is given by

$$\bar{P} = \frac{[|A(z)|^2 + |B(z)|^2]\sigma}{P_{\mathrm{sat}}} = \frac{P_A + P_B}{P_{\mathrm{sat}}}\,, \tag{13.11}$$

where $P_A = |A|^2\sigma$ and $P_B = |B|^2\sigma$ are the optical powers of the individual envelopes, and $P_{\mathrm{sat}} = \hbar\omega W d/(\tau a \Gamma)$ is the saturation power. This equation also provides information on phase changes experienced by the signal because the carrier-density dependent portion of the wavenumber is given by $\beta_N = -\alpha g/2$.

Equation (13.10) can be simplified by assuming that the *average* value of the optical power is sufficient to calculate the saturated gain [6]. This is a sensible approximation for a uniform-grating DFB SOA because the summed intensity of the coupled modes can result in a nearly uniform saturated-gain distribution [6]. Using the mean power and averaging the rate equation (13.10) over the length of the amplifier L, we obtain

$$\tau \frac{d\langle g\rangle}{dt} = g_0 - \left[1 + \frac{\langle P_A\rangle + \langle P_B\rangle}{P_{\text{sat}}}\right] \langle g\rangle , \tag{13.12}$$

where the angled brackets indicate longitudinal averaging. We have also assumed $\langle gP_j\rangle \approx \langle g\rangle\langle P_j\rangle$ for $j = A$ or B. Such a factoring scheme is referred to as the mean-field approximation [18].

13.4.2 Analytic Solution for the Optical Power

The optical powers P_A and P_B found in (13.12) are obtained from the coupled-mode equations (13.5) and (13.6). Analytic solutions for the optical powers can be obtained under certain approximations. Assuming that the average optical power is sufficient to calculate the saturated gain profile (as discussed above), the gain becomes uniform along the amplifier, and thus we may use its average value $\langle g\rangle$ in the coupled-mode equations [6]. Furthermore, the small SOA length L ($L \sim 300\,\mu\text{m}$) results in a short, unobstructed signal-transit time ($L/v \approx 3\,\text{ps}$). As this transit time is much shorter than the carrier lifetime ($\tau \approx 0.2$–$1\,\text{ns}$), and the rise and fall times of the input-field envelope $h(t)$, the field envelopes A and B can adjust to changes in the SOA gain almost instantaneously [6,19]. This adiabatic approximation allows the time derivatives to be dropped in (13.5) and (13.6). The system dynamics are then determined solely by the rate equation (13.12) for the gain.

Under the adiabatic and uniform-gain approximations, the coupled-mode equations (13.5) and (13.6) become ordinary differential equations with constant coefficients. Applying the boundary conditions $A(z = -L/2, t) = h(t)$ and $B(z = L/2, t) = 0$, and assuming negligible facet reflections (through antireflection coatings), the counterpropagating fields inside the amplifier are given by

$$A(\xi) = h \frac{\gamma \cos(\gamma\xi) + i\Delta \sin(\gamma\xi)}{\gamma \cos(\gamma L) - i\Delta \sin(\gamma L)} , \tag{13.13}$$

$$B(\xi) = h \frac{i\kappa \sin(\gamma\xi)}{\gamma \cos(\gamma L) - i\Delta \sin(\gamma L)} , \tag{13.14}$$

where $\xi = z - L/2$, $\gamma = \sqrt{\Delta^2 - \kappa^2}$, and Δ, given by (13.7), is understood to incorporate the average gain $\langle g\rangle$.

Using these solutions and integrating over the device length L, we obtain the following expressions for the *average* powers within the SOA:

$$\langle P_A \rangle = P_0 \zeta \left\{ \frac{\sinh(2\gamma_i L)\theta_1 + [\cosh(2\gamma_i L) - 1]\theta_2}{2\gamma_i L} \right\}$$

$$+ P_0 \zeta \left\{ \frac{\sin(2\gamma_r L)\theta_3 + [1 - \cos(2\gamma_r L)]\theta_4}{2\gamma_r L} \right\}, \quad (13.15)$$

$$\langle P_B \rangle = P_0 \zeta |\kappa|^2 \left[\frac{\sinh(2\gamma_i L)}{2\gamma_i L} - \frac{\sin(2\gamma_r L)}{2\gamma_r L} \right], \quad (13.16)$$

where $P_0(t) = |h(t)|^2 \sigma$ is the input power, $\gamma = \gamma_r + i\gamma_i$, and

$$\theta_1 = \gamma\gamma^* + \Delta\Delta^*, \quad \theta_2 = \gamma\Delta^* + \gamma^*\Delta, \quad (13.17)$$

$$\theta_3 = \gamma\gamma^* - \Delta\Delta^*, \quad \theta_4 = i(\gamma\Delta^* - \gamma^*\Delta), \quad (13.18)$$

$$\zeta = [\cosh(2\gamma_i L)\theta_1 + \sinh(2\gamma_i L)\theta_2 + \cos(2\gamma_r L)\theta_3 + \sin(2\gamma_r L)\theta_4]^{-1}. \quad (13.19)$$

The analytic expressions for the transmitted power $T(t) = P_A(z = L/2, t)$ and reflected power $R(t) = P_B(z = -L/2, t)$ are found to be

$$T = P_0 \zeta 2(\gamma_r^2 + \gamma_i^2), \quad (13.20)$$

$$R = P_0 \zeta |\kappa|^2 [\cosh(2\gamma_i L) - \cos(2\gamma_r L)]. \quad (13.21)$$

Equations (13.20) and (13.21) provide the output powers for a given input power P_0. The dynamics are completely determined by the average gain $\langle g \rangle$, found using (13.12). Since the expressions for the internal optical powers (13.15) and 13.16) are substituted into (13.12), thereby adiabatically eliminating the optical power, the resulting ordinary differential equation has only a single dependent variable, $\langle g \rangle$, and is relatively simple to solve numerically.

13.5 Resonant-Type SOAs

The nonlinear behavior discussed in the previous sections is based on the interaction of an optical signal with a Bragg resonance. Similar nonlinear behavior is found in other SOA structures that support resonances. The most common example is optical bistability in a Fabry–Perot SOA [20,21], where a Fabry–Perot cavity resonance plays the same role as the Bragg resonance discussed in Sect. 13.3 and thus gives rise to bistable switching. Fabry–Perot SOAs and Bragg-grating SOAs are examples of *resonant-type* SOAs [22–24], a general class of devices with common applications in fiber-optic communications. Before discussing these applications, we first discuss the structural differences between the two kinds of resonators.

Fabry–Perot devices are easy to fabricate; cavities can be formed simply by cleaving the semiconductor, and the semiconductor material can be grown without stopping to create the feedback structure. DFB SOAs require a more complicated fabrication procedure; Bragg gratings are created using techniques such as interferometric exposure or electron-beam lithography

[25], and the growth of the semiconductor material is typically arrested to create the grating.

The more difficult-to-fabricate DFB SOA, however, has advantages. Bragg gratings can be incorporated directly into a larger waveguide structure, allowing integration onto a single substrate with other photonic gates [6]. In addition, gratings also have a wider range of parameters that can be tailored in search of improved performance. For example, the uniformity of the grating period and depth can be altered [7]; the former can be used to increase the spectral range of bistality [17]. Grating phase shifts add an additional degree of freedom by controlling the location of the Bragg resonances; bistability has been demonstrated using a $\lambda/4$-shifted grating, where the resonance occurs in the center of the photonic bandgap [12]. Grating uniformities come, however, at the expense of increasing the difficulty of fabrication.

Bragg-grating and Fabry-Perot SOAs also differ in the number of strong resonances occurring within the SOA gain curve. Bragg gratings provide only a few strong modes, whereas Fabry-Perot cavities generally support many. This inherent spectral filtering property of DFB SOAs reduces the background noise which appears as a DC offset and lowers the on-off switching ratio [3]. Also, since Fabry-Perot cavities have many modes, the strongest modes are determined by the gain curve and lie at the gain peak. The wavelengths of the dominant resonances of DFB SOAs, however, can be fabricated to occur anywhere along the SOA gain spectrum. This is advantageous since the strength of the carrier-induced nonlinearity varies along the gain curve; DFB SOAs thus allow for a tunable nonlinearity strength.

The technologies of Bragg-grating and Fabry–Perot SOAs are very well developed; these feedback structures are commonly used to make commercial DFB and Fabry–Perot lasers, respectively. Indeed, a resonant-type SOA can be realized simply by driving a semiconductor laser *below* the lasing threshold. In practice, however, semiconductor lasers are engineered to have a small nonlinearity (small α) to prevent linewidth enhancement – they use a multi-quantum well active region and often fabricate the Bragg wavelength on the short-wavelength side of the gain peak. Nonlinear signal processing using resonant-type SOAs, however, utilizes the linewidth enhancement factor α to advantage; so these device should be engineered to have a large value of α by fabricating the dominant Bragg resonance on the *long*-wavelength side of the gain peak and by using a *bulk* semiconductor active region.

Distributed feedback lasers are often sold with an anti-reflection coating on one facet and a high-reflection coating on the other facet. This configuration is beneficial to lasers because the finite facet reflection alters the phase condition of feedback and breaks the degeneracy of equal-strength modes on either side of the photonic bandgap (this degeneracy is evident in Fig. 13.2). The dominant Bragg resonance can occur anywhere within the photonic bandgap. Since many experiments on the nonlinear response of Bragg-grating SOAs use commercial diode lasers driven below lasing threshold [26,27], the

Bragg resonances are altered by the facet reflections and are not like the simple case shown in Fig. 13.2. In addition, a non-zero reflectivity at the SOA facets can be used to enhance bistable switching at one edge of the photonic bandgap [28].

13.6 All-Optical Signal Processing

This section discusses applications of resonant-type SOAs to fiber-optic communication systems. Using an analogy to electronic processing, we have divided the section into two parts covering applications based on combinational and sequential logic [35]. The first part covers optical switching, demultiplexing, logic, and limiting; these application are based on *combinational* logic in that the output is determined by the existing state of the input signals. The output signal from devices exhibiting *sequential* logic is determined by the existing state of input signals *and* the state of past input signals—it exhibits memory. Demonstrations of sequential logic are discussed in the second part. Advantages and disadvantages of resonant-type SOAs for these applications are discussed throughout.

13.6.1 Combinational Logic

The nonlinear response of resonant-type SOAs can be used for optical switching. Here, signals with enough power to exceed the upward switching threshold pass through the device with high gain. Weaker input signals do not initiate switching and pass with a much lower output power. The low threshold powers of resonant-type SOAs discussed earlier allow these devices to be easily operated at power levels available in fiber-optic communication systems.

Optical switching, however, only occurs for signals tuned near the SOA resonance, and thus the wavelength range of operation is very limited [22]. This limited wavelength range has been quantified for devices biased near 98% of lasing threshold; upward switching below 0.1 mW spanned a spectral range of less than 0.02 nm [29]. Sensitivity to operating wavelength generally limits the application of resonant-type SOAs [15].

The narrow wavelength range of switching, however, can be used to demultiplex a single wavelength from several others [24]. Wavelength demultiplexing was achieved up to a wavelength separation of 12 GHz at a bit rate of 140 Mb/s [30]). This application is useful in wavelength-division multiplexed (WDM) system, where multiple signal wavelengths are used as a means of increasing total network capacity.

In WDM systems, the capability to transfer data from one signal wavelength to another is expected to be a key technology. Data transfer (commonly called "wavelength conversion") was demonstrated in resonant-type SOAs in 1987 [31]; as the data-carrying signal underwent bistable switching, the resonance pulled away from a CW probe signal. Using this technique, polarity-inverted wavelength conversion from the data signal at 1546 nm to the probe signal at 1531.5 nm was achieved at 800 Mb/s.

Wavelength conversion has also been demonstrated using a data signal that does *not* encounter a Bragg resonance, but instead shifts the resonance onto the probe signal via cross-phase modulation (XPM). Conversion by this means was performed over 30 nm in the 1550-nm communication spectral window. Cross-phase modulation can also be used for data-wavelength conversion from the 1310-nm band to the 1550-nm band [27]. For these kinds of data-transfer schemes, the wavelength range of operation is very wide because only the probe signal, which can be integrated with the SOA, needs to be tuned near the SOA resonance.

Multiple signals passing through a resonant-type SOA can be used for optical logic. If the power of each signal is less than the upward switching threshold, but more than half of its value, then the AND gate operation can be achieved [3]. If the power of each signal surpasses the upward switching threshold, then the device functions as an OR gate. In addition, the inverted-S shape of the hysteresis curve that occurs on reflection can be used for NAND and NOR gate operation [15].

The high-input power region beyond the switching thresholds can also be used for signal processing. Here, the output power tends to be independent of the input power. Optical limiting was demonstrated to transform a train of optical pulses of different heights (by a factor of 3) to a pulse train of nearly constant height [14]. Physically, the output power remains nearly constant through two effects: 1) gain saturation reduces the amount by which the signal is amplified; 2) the Bragg resonance shifts away from the signal wavelength, thus suppressing the output power. Optical limiting has also been suggested as a potential application for the loop-shape hysteresis curves that occur on reflection [15].

Optical switching, demultiplexing, logic, and limiting are all limited by the speed of the nonlinear SOA. The shortest rise and fall times of the switched output power have been measured to be 0.5 ns [32], which are on the order of the carrier lifetime ($\tau \sim 1$ ns). Although switching times faster than the carrier lifetime were predicted for high-finesse cavities driven near 98% of lasing threshold, the repeatability of the system is limited by the carrier lifetime τ [19]. For example, even if the initial fall time is fast, is will be followed by a relatively slow recovery to the initial carrier density [32]. The repeatability of the bistable system can also be slowed by an *upward*-switching delay [6] (from the moment of pulse impact). Such "critical slowing down" is on the order of the carrier lifetime and can be decreased below τ by choosing a high input optical power, or a small spectral detuning between the optical signal and the amplifier resonance [14].

Opportunities for optical processing exist at data rates greater than 10 Gb/s, for which electronic processing is prohibitively expensive. However, signal processing in resonant-type SOAs is limited to 1–10 Gb/s by the carrier lifetime. Since this rate does not surpass that achievable by electronic processing, resonant-type SOAs are unlikely to be implemented for these applications in their current state. One technique for effectively reducing the carrier lifetime uses a strong gain-saturating signal to sweep away charge carriers [33]; this technique has produced an effective τ of 10 ps. It may be possible to apply this technique to achieve high-speed processing.

13.6.2 Sequential Logic

Bistability is distinguished from other types of nonlinear switching in that it exhibits memory. For an input power $P_0 = P_H$ that falls within the bistable hysteresis curve, as shown in Fig. 13.4, two possible output states P_{on} and P_{off} are realizable; the actual state of the output power depends on the previous state of the system. Memory is a potent feature that can be exploited for new applications.

A 3R optical regenerator (re-amplification, regeneration, and retiming) was demonstrated in 1987 at 140 Mb/s using the memory capability of resonant-type SOAs [34]. For this application, the bistable hysteresis is exhibited by a clock signal. The peak power of the clock signal, however, is slightly lower than the upward switching threshold. Data signals provide the extra optical power needed to surpass the switching threshold, thereby regenerating and re-amplifing the data. Once switched, the high output state is maintained by the properly-timed clock signal, thereby retiming the output data.

Applications such as optical buffering and data-format conversion can in principle be enabled by optical memory. The basic building block of mem-

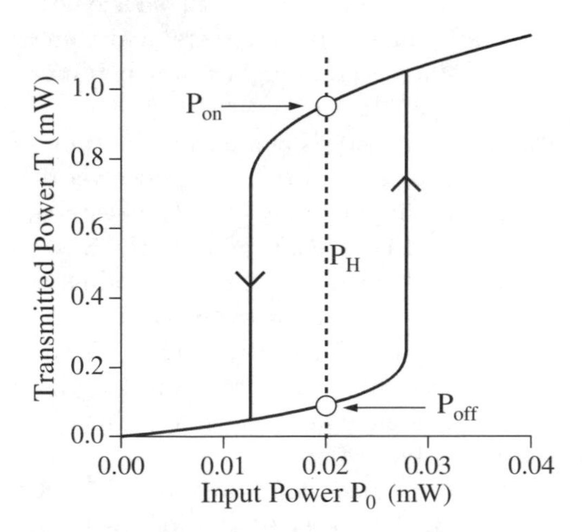

Fig. 13.4. Flip–flop output states. Two stable transmission states, P_{on} and P_{off}, occur for a single input power P_{H}

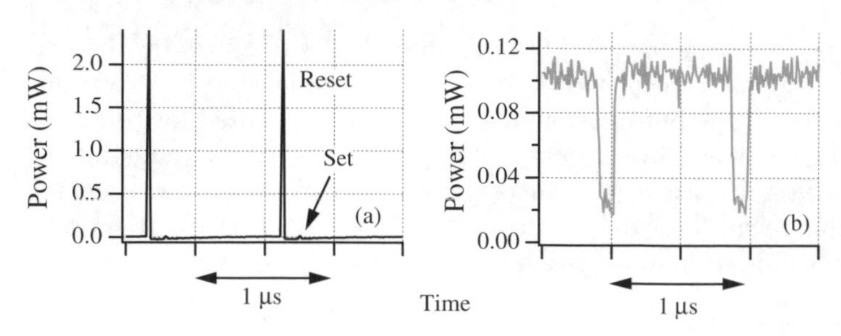

Fig. 13.5. All-optical flip–flop operation. (**a**) 1567-nm set and 1306-nm reset signals controlling (**b**) the output power of the 1547-nm holding beam

ory in electrical systems is called a flip–flop [35], in which a bistable output signal is controlled by auxiliary signals. Resonant-type SOAs have been used to create *all-optical* flip–flops, where the output and control signals are optical. For an input power P_0 that is initially located between the switching thresholds, like P_H in Fig. 13.4, optical set can be performed by increasing the input power beyond the upward switching threshold, and has been demonstrated using Fabry–Perot SOAs [14,36]. Likewise, reset can be performed by reducing the signal power below the downward switching threshold [14].

Optical set can also be achieved by signal wavelengths other than that of the bistable signal [26]. Signals that fall within the SOA gain spectrum deplete the carrier density and increase the refractive index even *at the holding-beam wavelength.* This cross-phase modulation (XPM) pushes the photonic bandgap and Bragg resonances to longer wavelengths. Set occurs when the Bragg resonance has been shifted at least enough to seed the positive feedback loop for upward bistable switching. In terms of the hysteresis curve, using XPM to shift the Bragg resonance toward the holding-beam wavelength corresponds to pushing the switching thresholds to lower powers. Switching occurs once the upward-switching threshold has been brought to the holding-beam input power P_H.

The flip–flop can be reset by pushing the hysteresis curve to higher powers (in the opposite direction than for optical set), allowing the downward-switching threshold to reach the holding-beam input power [26]. The hysteresis curve can be shifted in this way by signals that are *absorbed* by the SOA, giving their energy to electrons that are then excited into the semiconductor conduction band. This gain pumping is accompanied by a decrease in the refractive index and optical phase at the bistable-signal wavelength. As the refractive index decreases, the Bragg resonance shifts to *shorter* wavelengths; reset occurs when the Bragg resonance shifts enough to cause the positive feedback loop that results in downward bistable switching.

Using the complementary XPM control techniques in tandem, the bistable holding beam is toggled as shown in Fig. 13.5. Another option for resetting the flop–flop while leaving the input-holding beam power fixed is to use a closely tuned signal to interfer with the holding beam, thereby lowering the internal optical power [36]. The interfering-signal technique shifts the bistable switching thresholds using relatively low reset-beam powers, but requires a close matching between the holding and reset wavelengths of about 0.008 nm. The XPM-based reset technique, on the other hand, can be achieved over a wide wavelength range ($>$ 160 nm), but requires higher powers ($>$ 0.7 mW).

Switching and flip–flop operation both occur at small optical power levels. Since typical switching times and powers are $\sim$ 1 ns and $\sim$ 1 µW, respectively, femtojoule switching energies are expected. The lowest energy reported thus far has been 500 attojoules (0.5 ns $\times$ 1 µW) [32], which corresponds to about 3000 photons for a signal wavelength of 1310 nm [24]. Fabry–Perot SOAs also exhibit a high optical switching energy per gain ($\sim$ 100 fJ), and large switching energy per unit surface area ($\sim$ 1 fJ/µm^2 [24]).

Although the optical energy for switching is low, a substantial amount of energy is still needed to operate the SOA. Namely, an electrical bias is required to maintain the device near lasing threshold [22]. For an injection current of 10 mA, a semiconductor bandgap energy of 1 eV, and a carrier lifetime of 1 ns, the required electrical energy is 10 pJ. This energy must be dissipated, and this dissipation limits the number density of devices that can be placed within a given area of a single substrate. For a power consumption of 10 mW per SOA, and assuming a practical heat-sink power density of 1 W/cm^2 [22], the maximum number density is 100 cm^{-2}. Although this density does not rival that of electronic gates, it is sufficient for many applications for lightwave systems where as few as a single resonant-type SOA is required.

13.7 Concluding Remarks

Semiconductor optical amplifiers with built-in Bragg gratings exhibit many characteristics that make them intriguing candidates for optical-processing applications in lightwave systems. Their inherent optical gain and strong nonlinearity result in low switching powers ($\sim$1 microwatt) and energies ($\sim$1 fJ). Moreover, Bragg grating SOA technology is already mature, permits compatibility with all wavelengths found in fiber-optic communications, and allows monolithic integration with other optical gates.

The main disadvantages of Bragg-grating SOAs are a slow switching repeatability $\sim$ 1–10 Gb/s, and a limited wavelength range of operation in the vicinity of a Bragg resonance. The latter drawback can be overcome in some applications by using cross-phase modulation (XPM); all-optical flip–flop operation, for example, has recenlty been achieved using two, complementary XPM techniques that control the Bragg resonance with respect to the bistable signal. The serious limitation in speed remains to be overcome with future research.

References

1. H. Kawaguchi: IEE Proc., Pt. J **140**, 3 (1993)
2. R.J. Manning, A.D. Ellis, A.J. Poustie, and K.J. Blow: J. Opt. Soc. Am. B **14**, 3204 (1997)
3. W.F. Sharfin and M. Dagenais: Appl. Phys. Lett. **48**, 1510 (1986)
4. G.P. Agrawal and N.A. Olsson: IEEE J. Quantum Electron. **25**, 2297 (1989)
5. G.P. Agrawal and N.K. Dutta, *Semiconductor Lasers*, 2nd edn. (New York, Van Nostrand Reinhold 1993)
6. M.J. Adams and R.J. Wyatt: IEE Proc., Pt. J **134**, 35 (1987)
7. D.N. Maywar and G.P. Agrawal: IEEE J. Quantum Electron. **33**, 2029 (1997)
8. H. Kawaguchi, K. Inoue, T. Matsuoka, and K Otsuka: IEEE J. Quantum Electron. **21**, 1314 (1985)
9. R. Wyatt and M.J. Adams, *CLEO '86 MB5* (San Fransisco, CA 1986)
10. H.G. Winful, J.H. Marburger, and E. Garmire: Appl. Phys. Lett. **35**, 379 (1979)
11. K. Magari, H. Kawaguchi, K. Oe, and M. Fukuda: IEEE J. Quantum Electron. **24**, 2178 (1988)
12. G.H.M. van Tartwijk, H. de Waardt, B.H. Verbeek, and D. Lenstra: IEEE J. Quantum Electron. **30**, 1763 (1994)
13. R. Calvani, M. Calzavara, and R. Caponi: Proc. SPIE **1787**, 134 (1992)
14. N. Ogasawara and R. Ito: Jpn. J. Appl. Phys. **25**, L739 (1986)
15. M.J. Adams: Opt. Quantum Electron. **19**, S37 (1987)
16. D.N. Maywar and G.P. Agrawal: IEEE J. Quantum Electron. **34**, 2364 (1998)
17. D.N. Maywar and G.P. Agrawal: Optics Express **3**, 440 (1998)
18. P. Meystre: Optics Comm. **26**, 277 (1978)
19. W.F. Sharfin and M. Dagenais: IEEE J. Quantum Electron. **23**, 303 (1987)
20. T. Nakai, N. Ogasawara, and R. Ito: Jpn. J. Appl. Phys. **22**, 1184 (1983)
21. K. Otsuka and S. Kobayashi: Electronic Lett. **19**, 262 (1983)
22. M.J. Adams: Int. J. Electron. **60**, 123 (1986)
23. M.J. Adams: Solid-State Electron. **30**, 43 (1987)
24. M. Dagenais and W.F. Sharfin: Proc. SPIE **881**, 80 (1988)
25. N. Chinone and M. Okai, "Distributed Feedback Semiconductor Lasers," in *Semiconductor Lasers: Past, Present, and Future*, G.P. Agrawal, Ed. (New York, AIP Press, 1995) Chap. 2
26. D.N. Maywar, G.P. Agrawal, and Y. Nakano: Optics Express **6**, 75 (2000)
27. D.N. Maywar, Y. Nakano, and G.P. Agrawal: IEEE Photonics Tech. Lett. **12**, 858 (2000)
28. R. Hui and A. Sapia: Opt. Lett. **15**, 956 (1990)
29. Z. Pan, H. Lin, and M. Dagenais: Appl. Phys. Lett. **58**, 687 (1991)
30. Z. Pan and M. Dagenais: IEEE Photon. Tech. Lett. **4**, 1054 (1992)
31. K. Inoue: Electron. Lett. **23**, 921 (1987)
32. W.F. Sharfin and M. Dagenais: Appl. Phys. Lett. **48**, 321 (1986)
33. R.J. Manning, and D.A.O. Davies, and J.K. Lucek: Electron. Lett. **30**, 1233 (1994)
34. R.P. Webb: Opt. Quantum Electron. **19**, S57 (1987)
35. P. Horowitz and W. Hill, *The Art of Electronics*, 2nd edn. (New York, Cambridge University Press 1989) Chap. 8
36. K. Inoue: Opt. Lett. **12**, 918 (1987)

14 Atomic Solitons in Optical Lattices

S. Pötting, P. Meystre, and E.M. Wright

14.1 Introduction

The experimental demonstration of Bose–Einstein condensation in atomic vapors [1–3] has rapidly lead to spectacular new advances in atom optics. In particular, it has enabled its extension from the linear to the nonlinear regime, very much like the laser lead to the development of nonlinear optics in the 1960s. It is now well established that two-body collisions play for matter waves a role analogous to that of a Kerr nonlinear crystal in optics. In particular, it is known that the nonlinear Schrödinger equation which describes the condensate in the Hartree approximation supports soliton solutions. For the case of repulsive interactions normally encountered in BEC experiments, the simplest solutions are dark solitons, that is, 'dips' in the density profile of the condensate. These dark solitons have been recently demonstrated in two experiments [4,5] which appear to be in good agreement with the predictions of the Gross–Pitaevskii equation.

While very interesting from a fundamental physics point-of-view, dark solitons would appear to be of limited potential for applications such as atom interferometry, where it is desirable to achieve the dispersionless transport of a spatially localized ensemble of atoms, rather than a 'hole'. In that case, bright solitons are much more interesting. However, the problem is that large condensates are necessarily associated with repulsive interactions, for which bright solitons might seem impossible since the nonlinearity cannot compensate for the kinetic energy part (diffraction) in the atomic dynamics. While this is true for atoms in free space, this is however not the case for atoms in suitable potentials, eg. optical lattices. This is because in that case, it is possible to tailor the dispersion relation of the atoms in such a way that their effective mass becomes negative. For such negative masses, a repulsive interaction is precisely what is required to achieve soliton solutions. This result is known from nonlinear optics, where such soliton solutions, called gap solitons, have been predicted and demonstrated [6–9].

The article is organized as follows: In Sect. 14.2 we briefly review the basic formalism for nonlinear atom optics with a focus on Bose condensed systems. This leads to the Gross–Pitaevskii equation, the nonlinear mean-field equation of motion for the atomic condensate. In Sect. 14.3 we introduce the bright and dark soliton solutions of these equations and motivate the concept of gap solitons. In Sect. 14.4 we then describe a specific model for gap solitons in a spinor Bose–Einstein condensate. We demonstrate how to launch and control these solitons in Sect. 14.5. Finally, Sect. 14.6 is a summary and conclusion.

14.2 Nonlinear Atom Optics

Early experiments and theories concerning atom optics considered the low-density regime where atom-atom interactions are negligible. However, as the prospect of atomic BEC became more tangible prior to 1995, Lenz et al. [10,11] and Zhang et al. [12] theoretically explored the higher density regime where atom-atom interactions become relevant. In particular, they view the cold atomic collisions as nonlinear mixing processes analogous to those in nonlinear optics, and the new area of nonlinear atom optics was born. These initial theoretical works considered the dipole-dipole interaction between atoms, and derived a mean-field equation for the macroscopic wave function or order parameter for the system of cold atoms analogous to the well-known nonlinear Schrödinger equation from nonlinear optics [13], or the Gross–Pitaevskii equation (GPE) from BEC theory [14,15]. It was shown that these equations admit soliton solutions under suitable conditions [10,12]. The experimental demonstrations of atomic BEC in 1995 in both Rubidium [1] and Sodium [2] ushered in the era of experimental nonlinear atom optics, where the cold atomic collisions are now mediated via the Van der Walls interaction, with experimental demonstrations of atomic four-wave mixing [16], dark atomic solitons [4,5], atomic vortices [17,18], and mode-locking of an atom laser [19].

To set the stage for our discussion, we first briefly review the physics and derivation of the mean-field GPE for an atomic BEC, which underlies nonlinear atom optics. We assume that an ensemble of ultracold atoms of mass m experience a single-particle Hamiltonian of the form

$$H_0 = -\frac{\hbar^2}{2m}\nabla^2 + V(\boldsymbol{R}), \tag{14.1}$$

where the external potential $V(\boldsymbol{R})$ could be for example an optical dipole potential [20]. In the language of second-quantization this translates to the Hamiltonian operator

$$\hat{\mathcal{H}}_0 = \int d\boldsymbol{R}\hat{\Psi}^\dagger(\boldsymbol{R})H_0\hat{\Psi}(\boldsymbol{R}), \tag{14.2}$$

where $\hat{\Psi}^\dagger(\boldsymbol{R})$ and $\hat{\Psi}(\boldsymbol{R})$ denote the creation and annihilation operators (in the Schrödinger picture) for an atom with center-of-mass position vector $\boldsymbol{R}$. Assuming bosonic atoms, the field operators obey the commutation relation $[\hat{\Psi}(\boldsymbol{R}), \hat{\Psi}^\dagger(\boldsymbol{R}')] = \delta(\boldsymbol{R} - \boldsymbol{R}')$. Next we add atom-atom interactions by introducing a two-body potential $V(\boldsymbol{R}, \boldsymbol{R}')$, which in second-quantization leads to the total Hamiltonian operator $\hat{\mathcal{H}} = \hat{\mathcal{H}}_0 + \hat{\mathcal{V}}$ with

$$\hat{\mathcal{V}} = \frac{1}{2} \int d\boldsymbol{R} d\boldsymbol{R} \hat{\Psi}^\dagger(\boldsymbol{R}) \hat{\Psi}^\dagger(\boldsymbol{R}') V(\boldsymbol{R}, \boldsymbol{R}') \hat{\Psi}(\boldsymbol{R}') \hat{\Psi}(\boldsymbol{R}) . \tag{14.3}$$

For a system of N cold atoms, the many-body state vector may be related to the N-particle Schrödinger wave function using the general relation

$$\begin{aligned} |\phi_N(t)\rangle = & \frac{1}{\sqrt{N!}} \int d\boldsymbol{R}_1 \dots d\boldsymbol{R}_N \phi_N(\boldsymbol{R}_1, \dots, \boldsymbol{R}_N, t) \\ & \times \hat{\Psi}^\dagger(\boldsymbol{R}_N) \dots \hat{\Psi}^\dagger(\boldsymbol{R}_1) |0\rangle . \end{aligned} \tag{14.4}$$

Since for a BEC at zero temperature the atoms are predominantly in the same Bose-condensed state, we employ the Hartree approximation in which the N-particle wave function $\phi_N(\boldsymbol{R}_1, \dots, \boldsymbol{R}_N, t)$ is written in the form of a product

$$\phi_N(\boldsymbol{R}_1, \dots, \boldsymbol{R}_N, t) = \prod_{i=1}^{N} \vartheta(\boldsymbol{R}_i, t) , \tag{14.5}$$

where $\sqrt{N}\vartheta(\boldsymbol{R}, t) = \langle \phi_{N-1} | \hat{\Psi}(\boldsymbol{R}) | \phi_N \rangle$ is the effective single-particle, or Hartree, wave function, and is assumed normalized [21]. We note that the Hartree approximation automatically satisfies the symmetry condition for bosons that interchanging any two particle indices produces the same wave function. To obtain an equation of motion for the effective single-particle wave function, we employ Dirac's variational principle [22]

$$\frac{\delta}{\delta\vartheta^*(\boldsymbol{R}, t)} \left[\langle \phi_N | i\hbar \frac{\partial}{\partial t} - \left(\hat{\mathcal{H}}_0 + \hat{\mathcal{V}} \right) | \phi_N \rangle \right] = 0 , \tag{14.6}$$

which yields the GPE

$$i\hbar \frac{\partial \vartheta}{\partial t} = H_0 \vartheta + (N-1) \int d\boldsymbol{R}' V(\boldsymbol{R}, \boldsymbol{R}') |\vartheta(\boldsymbol{R}', t)|^2 \vartheta(\boldsymbol{R}, t) . \tag{14.7}$$

Here the nonlinear term describes the mean-field effect of the $(N-1)$ other atoms on the effective single-particle evolution.

For the particular case of atomic BECs the cold atomic collisions may be treated in the s-wave scattering limit. The effective two-body potential can then be approximated by a contact potential of the form [23,24]

$$V(\boldsymbol{R}, \boldsymbol{R}') = U_0 \delta(\boldsymbol{R} - \boldsymbol{R}') , \tag{14.8}$$

where the coefficient $U_0 = 4\pi\hbar^2 a_{\mathrm{sc}}/m$, with a_{sc} the s-wave scattering length. The GPE equation (14.7) for an atomic BEC then becomes

$$i\hbar \frac{\partial \vartheta}{\partial t} = -\frac{\hbar^2}{2m} \nabla^2 \vartheta + V(\boldsymbol{R}) \vartheta + N U_0 |\vartheta|^2 \vartheta , \tag{14.9}$$

which applies for large N. This equation forms the basis of the following discussion of atomic solitons. Here we have concentrated on the case of a single atomic species, but the formalism is straightforward to generalize to multi-component or vector condensates [25,26] involving, for example, several different Zeeman sublevels of a given atom.

14.3 One-dimensional Atomic Solitons

14.3.1 One-dimensional GPE

In this section we briefly discuss the one-dimensional (1D) soliton solutions of the GPE (14.9). In particular, we assume strong transverse confinement of the BEC in the (X,Y) plane such that the transverse mode $u_g(X,Y)$ is the ground state of the potential $V(X,Y)$,

$$E_g u_g = -\frac{\hbar^2}{2m}\left(\frac{\partial^2}{\partial X^2}+\frac{\partial^2}{\partial Y^2}\right)u_g + V(X,Y)u_g\,. \tag{14.10}$$

This may be realized, for example, using a red-detuned optical dipole trap around the focus of a Gaussian laser beam centered on the origin [20], thereby giving approximately harmonic transverse confinement. Thus, the BEC becomes essentially one-dimensional with cross-sectional area

$$A_T = \int dXdY|u_g(X,Y)/u_g(0,0)|^2\,, \tag{14.11}$$

transverse to the unbound Z-axis. We remark that the quasi-1D nature of the system does not preclude BEC along with the associated GPE description, as discussed by Petrov et al. [27]. Setting

$$\sqrt{N}\vartheta(\boldsymbol{R},t) = e^{-iE_g t/\hbar}\left(\frac{u_g(X,Y)}{u_g(0,0)}\right)\varphi(Z,t)\,, \tag{14.12}$$

so that φ is normalized to the number of atoms N, and projecting onto the ground transverse mode, the reduced 1D GPE becomes (we ignore the variation of the optical potential along the Z-axis for simplicity)

$$i\hbar\frac{\partial\varphi}{\partial t} = -\frac{\hbar^2}{2m}\frac{\partial^2\varphi}{\partial Z^2} + U|\varphi|^2\varphi\,, \tag{14.13}$$

with

$$U = U_0 A_T^{-1}\int dXdY|u_g(X,Y)/u_g(0,0)|^4\,. \tag{14.14}$$

14.3.2 Dark Solitons

Soliton solutions are non-spreading solutions of (14.13) which preserve their shape under propagation [28]. The kinetic energy term in the GPE tends to spread wavepackets, and the nature of the solitons depends on the sign of the nonlinearity. For the case of repulsive interactions ($a_{sc} > 0$), both the kinetic energy and nonlinearity terms in (14.13) tend to broaden localized wavepackets, so we do not expect localized, or bright soliton, solutions for that case. However, dark solitons describing localized density dips in an otherwise constant background can arise and are given by [4,5,29] (with analogous solutions being well known in nonlinear fiber optics e.g. [30])

$$\varphi(Z,t) = n^{1/2}\sqrt{1 - \left(1 - \frac{v^2}{v_0^2}\right)\operatorname{sech}^2\left(\frac{(Z-vt)}{l_0}\left(1 - \frac{v^2}{v_0^2}\right)^{1/2}\right)} \times e^{i(\phi(Z,t,v) - \mu t/\hbar)}, \tag{14.15}$$

where n is the background density away from the dark soliton core, $\mu = n|U|$, $l_0 = \sqrt{\hbar^2/m\mu}$ is the correlation length which determines the width of the soliton core, $v_0 = \sqrt{\mu/m}$ is the Bogoliubov speed of sound, v is the dark soliton velocity, whose magnitude is bounded by v_0, and the soliton phase ϕ is given by

$$\phi(Z,t,v) = -\arctan\left[\left(\frac{v_0^2}{v^2} - 1\right)^{1/2}\tanh\left(\frac{(Z-vt)}{l_0}\left(1 - \frac{v^2}{v_0^2}\right)^{1/2}\right)\right]. \tag{14.16}$$

These solutions show that the dark solitons are characterized by the presence of a phase step δ across the localized density dip. It can be related to the velocity v and the density n_{bot} at the bottom of the atomic density dip [4,5]. In particular, one finds the relation

$$\cos\left(\frac{\delta}{2}\right) = \frac{v}{v_0} = \frac{n_{\text{bot}}}{n}, \tag{14.17}$$

so that for a stationary soliton $v = 0$, $n_{\text{bot}} = 0$, and there is a $\delta = \pi$ phase step. Only stationary solitons have a vanishing density at their center, so they are also referred to as 'black' solitons, whereas for the $n_{\text{bot}} > 0$ case the expression 'grey' soliton is used.

Dark matter-wave solitons have recently been observed experimentally in a Na BEC by Denschlag et al. [4], and in a cigar-shaped ^{87}Rb condensate by Burger et al. [4,5]. In both experiments dark solitons of variable velocity were launched via the phase imprinting of a BEC by a light-shift potential. By applying a pulsed laser to only half of the BEC and choosing the laser intensity and duration to select a desired phase step δ, the soliton velocity could be selected according to (14.17). Figure 14.1 shows a schematic of the

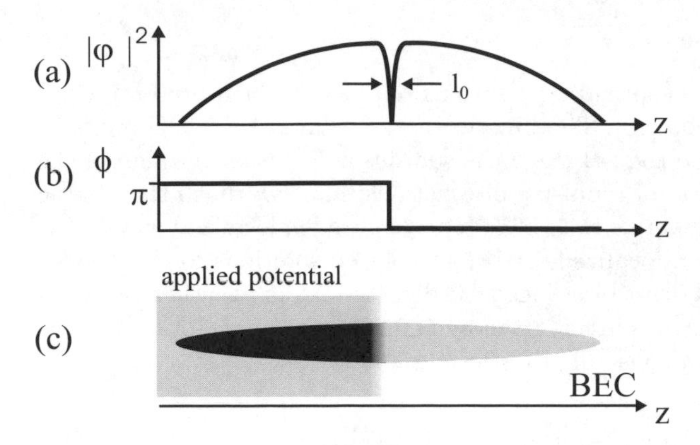

Fig. 14.1. Density distribution (**a**) and spatial phase (**b**) of a stationary dark soliton with $\delta = \pi$. The dip in the density has a width $\sim l_0$. The scheme for the generation of dark solitons by phase imprinting is shown in **c**

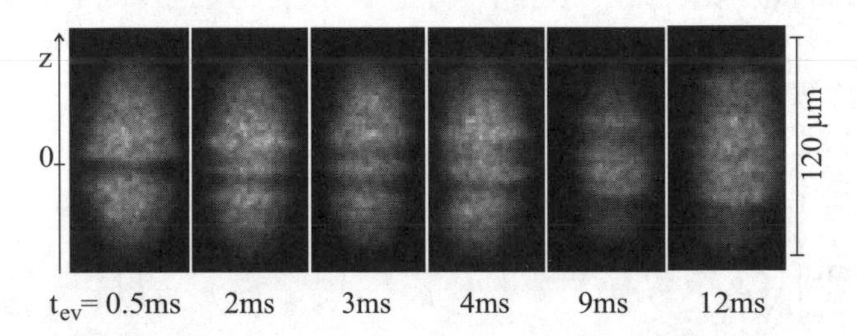

Fig. 14.2. Absorption images of BECs with dark soliton structures propagating along the long condensate axis for different evolution times t_{ev}

basic phase imprinting idea for a cigar-shaped BEC, and Fig. 14.2 shows the experimental results of Burger et al.

The motion of the density dip is seen for a phase-step of π and a variety of evolution times. Ideally this value of the phase step should lead to a stationary dark soliton, but the experiments are carried out in a harmonic trap [31] and dissipative effects were shown to occur that accelerated the solitons [32,33].

14.3.3 Bright Solitons

For the case of attractive interactions (i.e. $a_{\mathrm{sc}} < 0$), the kinetic energy of the BEC can be balanced by the nonlinearity yielding non-spreading wave packets. These solutions correspond to spatially localized bright solitons [29,30]. The one-parameter solution to (14.13) then reads

$$\varphi(Z,t) = n^{1/2}\sqrt{2-\frac{v^2}{v_0^2}}\,\mathrm{sech}\left[\frac{(Z-vt)}{l_0}\left(2-\frac{v^2}{v_0^2}\right)^{1/2}\right] e^{i(\phi(Z,t,v)+\mu t/\hbar)}\,, \tag{14.18}$$

with v the velocity parameter and μ, l_0 and v_0 as in the dark soliton case. The phase ϕ is given by

$$\phi(Z,t,v) = \frac{(Z-vt)}{l_0}\left(\frac{v}{v_0}\right)\,. \tag{14.19}$$

Such 1D bright solitons could in principle be realized in cigar shaped BECs with negative scattering lengths, for example, in ^{7}Li [3] or ^{85}Rb by using Feshbach resonances [34–36] to tune the scattering length. To the best of our knowledge such experiments have not been attempted[1]. In two or more dimensions negative scattering lengths can lead to catastrophic collapse in homogeneous systems for large enough particle numbers. However, in quasi-1D systems with strong transverse confinement the solitons are rendered stable [37].

14.3.4 Gap Solitons in Optical Lattices

So far in our discussion dark solitons arise for positive scattering lengths and bright solitons for negative scattering lengths. However, it is possible to extend the range of options, e.g. bright solitons with a positive scattering length, by considering a 1D BEC in a periodic optical lattice [38,39]. This situation was first discussed by Zhang et al. [40] for a scalar or single component condensate, and yields atomic gap solitons. Historically, gap solitons were first considered in nonlinear optics as arising from the combination of optical nonlinearity and a periodic spatial refractive-index distribution [6–9]. As an introduction to our treatment of gap solitons in spinor condensates, we briefly review the physics underlying gap solitons in scalar condensates, referring the reader to the literature for more details [40].

Consider the 1D GPE (14.13) with an additional periodic potential, or optical lattice, produced by periodic optical fields and applied along the Z-axis, so that

$$i\hbar\frac{\partial\varphi}{\partial t} = -\frac{\hbar^2}{2m}\frac{\partial^2\varphi}{\partial Z^2} + V_{\mathrm{opt}}\cos^2(k_{\mathrm{opt}}Z)\varphi + U|\varphi|^2\varphi\,. \tag{14.20}$$

Here V_{opt} determines the strength of the optical lattice, and k_{opt} is the wave vector characteristic of the optical lattice.

[1] Note added: Since submission of this paper, two experiments have demonstrated bright soliton generation in condensates of ^{7}Li (Lithium-7): K.E. Strecker, G.B. Partridge, A.G. Truscott, R.G. Hulet, Nature **417**, 150 (2002) and L. Khaykovich, F. Schreck, G. Ferrari, T. Bourdel, J. Cubizolles, L.D. Carr, Y. Castin, C. Salomon, Science **296**, 1290 (2002).

Let us first consider the linear eigensolutions of (14.20) neglecting the nonlinear term. Introducing the band-index n and the wave vector k within the band, we can write the atomic wave function in terms of the periodic Bloch functions $u_{n,k}(Z)$ according to Bloch's theorem [41]

$$\varphi(Z,t) = \varphi_{n,k}(Z)e^{-iE_{n,k}t/\hbar} = u_{n,k}(Z)e^{i(kZ-E_{n,k}t/\hbar)} .$$

Using this expression in (14.20) with $U = 0$, the eigenvalue equation associated with the cosine potential then reads

$$E_{n,k}\varphi_{n,k} = -\frac{\hbar^2}{2m}\frac{d^2\varphi_{n,k}}{dZ^2} + V_{\rm opt}\cos^2(k_{\rm opt}Z)\varphi_{n,k} , \tag{14.21}$$

the eigenvectors and eigenvalues of which are well known [42]. For our discussion the important fact is the occurrence of an atomic band structure given by the dispersion relation $E_{n,k} = \hbar\omega_{n,k}$. Consider an atomic wavepacket for a given band index n and a narrow spread of wave vectors Δk around a central value k_0 that is slowly varying on the scale of the periodicity. Then we can write the atomic wave function as a product of the slowly varying envelope $\chi(Z,t)$ with the fast oscillating Bloch part

$$\varphi(Z,t) = \chi(Z,t)\varphi_{n,k_0}e^{-iE_{n,k_0}t/\hbar} . \tag{14.22}$$

In this approximation the group velocity v_g and effective mass $m_{\rm eff}$ for the wavepacket will be

$$v_g = \frac{1}{\hbar}\frac{\partial E_{n,k}}{\partial k}|_{k_0} , \quad \frac{1}{m_{\rm eff}} = \frac{1}{\hbar^2}\frac{\partial^2 E_{n,k}}{\partial k^2}|_{k_0} , \tag{14.23}$$

in terms of which the approximate dynamics of the slowly varying envelope $\chi(Z,t)$ may be described by

$$i\hbar\left(\frac{\partial}{\partial t} + v_g\frac{\partial}{\partial Z}\right)\chi = -\frac{\hbar^2}{2m_{\rm eff}}\frac{\partial^2\chi}{\partial Z^2} + U|\chi|^2\chi , \tag{14.24}$$

where we have re-introduced the nonlinearity.

We take for illustration the case $k_0 = 0$, that is zone center, so that $v_g = 0$. Then (14.24) becomes identical with (14.13), so that they have the same bright and dark soliton solutions but with $m \to m_{\rm eff}$. The key idea is that the effective mass $m_{\rm eff}$ can assume both positive and negative signs by a suitable choice of the band index n and k_0, whereas for the spatially homogeneous case the atomic mass m is always positive. So now, for example, bright solitons can arise for a positive scattering length if the effective mass is negative, the general condition for bright solitons being $m_{\rm eff}U < 0$, and conversely for dark solitons $m_{\rm eff}U > 0$. We remark that the applied optical lattice can modify the effective mass along the Z-axis, but the effective mass along the transverse dimensions is still m. This leaves open the question of higher-dimensional gap solitons, a point that we do not further address here [43].

The solitons formed in optical lattices are referred to as gap solitons [7–9]. This name derives from the fact that the soliton energies actually lie in the energy gaps of the atomic band structure [6]. The significance of gap solitons in the context of atomic BEC is that they provide a means to realize bright solitons, that is, spatially localized atomic packets, *even for positive scattering lengths.* This may have important applications, e.g. for the coherent transport of atoms. In the following section we further develop the theory of atomic gap solitons in spinor condensates to quantify some of their most important properties.

14.4 Atomic Gap Solitons

We now consider a system of two condensates in different Zeeman sublevels coupled by a spatially periodic two-photon interaction. This spatially modulated coupling again leads to an atomic band structure for the spinor condensate, and hence by our general arguments of the previous section, to gap solitons. The advantage of multicomponent condensates is that magnetic fields can now be used to phase-imprint the two Zeeman components differently, and hence to manipulate and control the gap solitons to a high degree. We demonstrate this flexibility in Sect. 14.5 by simulating an atomic interferometer based on gap solitons.

14.4.1 The Physical Model for a Spinor Condensate

The two-component Bose–Einstein condensate interacts with two counter-propagating, focused Gaussian laser beams of equal frequencies ω_l but opposite circular polarizations. The optical dipole potential associated with the applied laser beams is assumed to provide tight transverse confinement for the BEC in the (X, Y) plane, thereby forming a cigar-shaped condensate of transverse cross-sectional area A_T. As in Sect. 14.3, we confine our discussion to the one-dimensional dynamics of the BEC along the Z-axis for simplicity.

In addition to supplying a transverse optical potential, the laser beams can drive two-photon transitions between different Zeeman sublevels of the atomic ground state. For illustrative purposes we consider the case of Sodium and the two-photon coupling of the Zeeman sublevels $|-1\rangle = |F_g = 1,\, M_g = -1\rangle$ and $|1\rangle = |F_g = 1,\, M_g = 1\rangle$. For example, starting in the $|-1\rangle$ state this process involves the absorption of a σ_+ photon from the right propagating laser beam followed by emission of a σ_- photon into the left propagating laser beam. We must of course assume that the excited states involved in the atom-field interaction are far-detuned from the applied laser frequency, a necessary requirement to avoid the detrimental effects of spontaneous emission.

By restricting our attention to the coupled states $|\pm1\rangle$ the effective single-particle Hamiltonian for our model system can be written as [44,45]

$$H_{\rm eff} = \frac{P_Z^2}{2m} + g\hbar\delta' \left[|1\rangle\langle -1| e^{2iK_l Z} + |-1\rangle\langle 1| e^{-2iK_l Z} \right] , \tag{14.25}$$

where we have omitted constant light-shift terms. Here P_Z is the atomic center-of-mass momentum operator along Z, $K_l = \omega_l/c$ is the magnitude of the field wave vector along Z, g is a coupling constant between the ground and excited states characteristic of the atom and transition involved, $\delta' = \mathcal{D}^2\mathcal{E}^2/\hbar^2\delta$, with the detuning $\delta = \omega_l - \omega_a$, $\mathcal{E}$ is the laser field amplitude, and $\mathcal{D}$ is the reduced electric dipole moment for the $3S_{1/2}$-$3P_{3/2}$ transition.

The second term of the effective Hamiltonian (14.25) describes the effective coupling of the two Zeeman sublevels via the applied laser fields. The exponential terms $\exp(\pm 2iK_l Z)$ arise from the fact that the two-photon transitions involve the absorption of a photon from one light field and reemission into the other. Introducing a spinor macroscopic condensate wave function $\varphi(Z,t) = [\varphi_1(Z,t), \varphi_{-1}(Z,t)]^T$ normalized to the number of atoms N, and including the many-body effects via a mean-field nonlinearity, we obtain the coupled GPE equations

$$i\hbar\frac{\partial\varphi}{\partial t} = H_{\rm eff}\varphi + U|\varphi|^2\varphi , \tag{14.26}$$

where $U = 4\pi\hbar^2 a_{\rm sc}/m$ as in Sect. 14.2 and 14.3, $|\varphi|^2 = |\varphi_1|^2 + |\varphi_{-1}|^2$, and we have assumed that the magnitude of the self- and cross-nonlinearities are equal for simplicity.

It is convenient to re-express (14.26) in dimensionless form by introducing the scaled variables $t = \tau t_c$, $Z = z l_c$ and $\varphi_j = \psi_j\sqrt{\rho_c}$ where

$$t_c = \frac{1}{g\delta'} , \quad l_c = \frac{t_c\hbar K_l}{m} , \quad \rho_c = \left|\frac{g\hbar\delta'}{U}\right| . \tag{14.27}$$

Equations (14.26) then become

$$\begin{aligned} i\frac{\partial}{\partial\tau}\begin{pmatrix}\psi_1\\ \psi_{-1}\end{pmatrix} &= \begin{pmatrix}-M\nabla^2 & e^{2ik_l z}\\ e^{-2ik_l z} & -M\nabla^2\end{pmatrix}\begin{pmatrix}\psi_1\\ \psi_{-1}\end{pmatrix} \\ &\quad + \mathrm{sgn}\left(g\delta'/U\right)|\psi|^2\begin{pmatrix}\psi_1\\ \psi_{-1}\end{pmatrix} , \end{aligned} \tag{14.28}$$

where $M = g\delta' m/2\hbar K_l^2$ is a mass-related parameter such that $k_l = K_l l_c = 1/2M$. For a discussion of characteristic values for the case of a Sodium condensate we refer the reader to [46].

14.4.2 Soliton Solutions

The spatially modulated coupling between the optical fields and the condensate induces a single-particle band structure with regions of negative effective

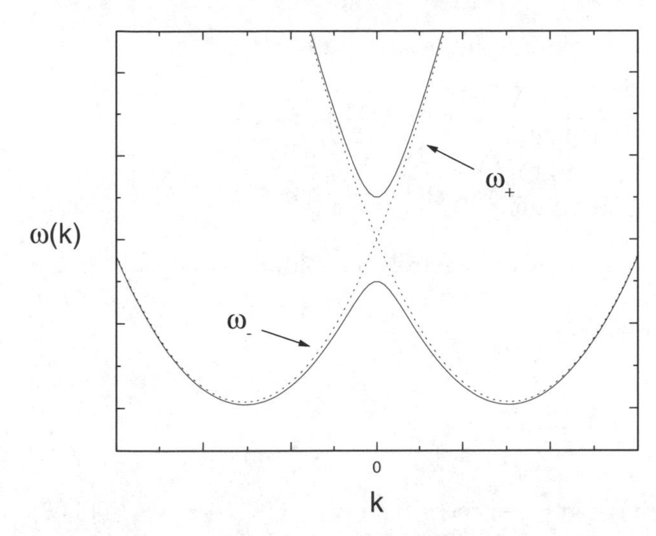

Fig. 14.3. Linear dispersion curve (arbitrary units): Shown are the two branches of the dispersion relation exhibiting an avoided crossing at $k = 0$ (solid line). The negative curvature in the lower branch around $k = 0$ defines a negative effective mass that enables gap soliton solutions. The free dispersion when no linear coupling is present corresponds to two displaced parabolas (dashed line)

mass. As mentioned before, this leads to the possibility of bright atomic solitons even for repulsive interactions [44]. Approximate analytic expressions for these solitons can be obtained by expressing the spinor condensate components as

$$\psi_{\pm 1}(z,\tau) = e^{\pm i k_l z} e^{-i\tau/4M} \phi_{\pm 1}(z,\tau)\,, \tag{14.29}$$

where the field envelopes $\phi_{\pm 1}$ are assumed to be slowly varying in space compared to $1/k_l$. Neglecting the second-order spatial derivatives yields the coupled partial differential equations

$$\begin{aligned} i\left(\frac{\partial}{\partial \tau} \pm \frac{\partial}{\partial z}\right) \begin{pmatrix} \phi_1 \\ \phi_{-1} \end{pmatrix} &= \begin{pmatrix} 0\,1 \\ 1\,0 \end{pmatrix} \begin{pmatrix} \phi_1 \\ \phi_{-1} \end{pmatrix} \\ &\quad \pm (|\phi_1|^2 + |\phi_{-1}|^2) \begin{pmatrix} \phi_1 \\ \phi_{-1} \end{pmatrix}, \end{aligned} \tag{14.30}$$

where the choice $\pm 1 = \mathrm{sgn}(g\delta'/U)$. For a red-detuned laser and our choice of g this becomes $\pm 1 = \mathrm{sgn}(U)$. Neglecting the nonlinearity one finds a linear dispersion curve as sketched in Fig. 14.3.

In this particular case it is the linear coupling induced by the laser field that creates the avoided crossing at $k = 0$ and the negative effective mass. Aceves and Wabnitz [47] have shown that the full equations (14.30) have the explicit two-parameter gap soliton solutions (see also [8])

$$\phi_1 = \pm \frac{\sin(\eta)}{\beta\gamma\sqrt{2}} \left(-\frac{e^{2\theta} + e^{\mp i\eta}}{e^{2\theta} + e^{\pm i\eta}} \right)^v \operatorname{sech}\left(\theta \mp \frac{i\eta}{2} \right) e^{\pm i\sigma} ,$$
$$\phi_{-1} = -\frac{\beta \sin(\eta)}{\gamma\sqrt{2}} \left(-\frac{e^{2\theta} + e^{\mp i\eta}}{e^{2\theta} + e^{\pm i\eta}} \right)^v \operatorname{sech}\left(\theta \pm \frac{i\eta}{2} \right) e^{\pm i\sigma} . \tag{14.31}$$

Here $-1 < v < 1$ is a parameter which controls the soliton velocity, $0 < \eta < \pi$ is a shape parameter, and

$$\beta = \left(\frac{1-v}{1+v} \right)^{\frac{1}{4}} , \quad \gamma = \frac{1}{\sqrt{1-v^2}} , \tag{14.32}$$

$$\theta = -\gamma \sin(\eta)(z - v\tau) , \quad \sigma = -\gamma \cos(\eta)(vz - \tau) . \tag{14.33}$$

Since we are interested in creating bright solitons in the presence of repulsive interactions we restrict ourselves to $\operatorname{sgn}(U) = +1$, corresponding to the choice of the upper sign in the analytic solutions. The characteristic length scale associated with the solitons is l_c, so that the approximate solitons (14.31) are valid for $K_l l_c = 1/2M \gg 1$.

14.4.3 Soliton Properties

From the dependence of the hyperbolic-secant on $\theta = -\gamma \sin(\eta)(z - v\tau)$ in (14.31), we identify the gap soliton parameter $v = V_g/V_R$ as the group velocity V_g of the soliton in units of the recoil velocity $V_R = l_c/t_c = \hbar K_l/m$. Since $-1 < v < 1$, the magnitude of the group velocity is bounded by the recoil velocity. From (14.31), one can extract further important soliton properties, such as the number N_s of atoms in the soliton and the soliton width $W_s = w_s l_c$. Specifically, the number of atoms in a particular gap soliton is given by

$$N_s = A_T \int dZ [|\varphi_1(Z,t)|^2 + |\varphi_{-1}(Z,t)|^2] , \tag{14.34}$$

where A_T is the effective transverse area.

We can gain further insight into the analytical structure of the gap solitons by taking the extreme limit $\eta \ll 1$ of (14.31). Using the definitions (14.29) and returning to dimensional units, we have then

$$\varphi_1(Z,0) = \frac{\eta}{\beta\gamma} \sqrt{\frac{\rho_c}{2}} \operatorname{sech}(Z/W_0)(-1)^v e^{i(K+K_l)Z} ,$$
$$\varphi_{-1}(Z,0) = \frac{\beta\eta}{\gamma} \sqrt{\frac{\rho_c}{2}} \operatorname{sech}(Z/W_0)(-1)^v e^{i[(K-K_l)Z+\pi]} , \tag{14.35}$$

where

$$W_s = 3.44 W_0 = 3.44 \left(\frac{l_c \sqrt{1-v^2}}{\eta} \right), \quad K = -\frac{\gamma v}{l_c}. \tag{14.36}$$

Here $W_s = 3.44 W_0$ is the soliton width, the factor 3.44 being the numerical conversion from the width of the hyperbolic secant to the $1/e^2$ width of the distribution, and $K = k/l_c$ a velocity dependent wave vector shift. Obviously the width decreases with increasing η and v. The soliton atom number obtained by combining (14.34) and (14.35),

$$N_s = 2(A_T l_c \rho_c)\eta \cdot (1 - v^2), \tag{14.37}$$

predicts that increasing the shape parameter η increases the atom number, and faster solitons have lower atom numbers.

An essential point to keep in mind is that the gap solitons are coherent superpositions of the two Zeeman sublevels. The approximate solutions (14.35) contain important information on the phase and amplitude relations that need to be created between them to successfully excite and manipulate gap solitons. In particular, they show that there is always a spatially homogeneous π phase difference between the two states. In addition, the two components have the spatial wave vectors

$$K_{\pm 1} = K \pm K_l, \tag{14.38}$$

with $K = -\gamma v / l_c$.

Finally, it follows from dividing the amplitudes of the two components that

$$\left| \frac{\varphi_1(Z,t)}{\varphi_{-1}(Z,t)} \right|^2 = \frac{1}{\beta^4} = \left(\frac{1+v}{1-v} \right), \tag{14.39}$$

which shows that their relative occupation depends on the soliton velocity parameter v. For $v = 0$ the sublevels are equally populated, but as $v \to 1$ the $|1\rangle$ sublevel has a larger population, and vice versa for $v \to -1$.

The characteristic time scale for the evolution of the gap solitons can be determined from the plane-wave exponential factors in (14.31). Converting back to dimensional form the soliton period t_s is defined as the time to accumulate a 2π phase, or in the limit $\eta \to 0$

$$t_s = 2\pi t_c \sqrt{1 - v^2}. \tag{14.40}$$

Physically, t_s corresponds to the internal time scale for the gap soliton. In order to observe a soliton-like behavior, it is therefore necessary to investigate the atomic propagation over several periods.

We conclude this section by noting that Aceves and Wabnitz [47] have shown that the gap soliton solutions (14.31) are stable solutions of (14.30) in that they remain intact during propagation, even when perturbed away from the exact solutions. However, one should remember that (14.30) is only an approximation to the exact system of equations (14.28), so that in general the gap solitons are solitary wave solutions only. As such, they are not guaranteed to be absolutely stable.

14.5 Magneto-optical Control

Summarizing the main results of the preceding section, we have seen that gap solitons require the right population and density distribution in each Zeeman sublevel, a certain phase difference between these sublevels, and appropriate plane-wave factors $e^{iK_{\pm 1}Z}$. Based on these properties, we now show that a magneto-optical scheme involving a combination of pulsed coherent optical coupling and of phase-imprinting using spatially inhomogeneous magnetic fields turn out to be an adequate and convenient tool to manipulate these solitons.

14.5.1 Manipulation Tools

The shapes of the Hartree wave functions corresponding to the two Zeeman sublevels are hyperbolic-secant, which we approximate by a Gaussian in the following. They could for example be initialized in an optical dipole trap [20]. Manipulating the gap solitons reduces therefore to the problem of controlling the populations and phases throughout the spinor condensate. The coherent optical coupling can be achieved e.g. by a laser pulse of frequency ω_l propagating perpendicularly to the Z-axis and with linear polarization perpendicular to that axis. For sufficiently short pulses, one can neglect changes in the center-of-mass motion of the atoms during its duration, leading to a very simple description. We assume for simplicity a plane-wave rectangular pulse of duration t_p and of spatial extent large compared to the soliton. The Hamiltonian describing the coupling between this pulse and the condensate is then the same as in (14.25), but without the linear momentum exchange terms $\exp(\pm 2iK_l Z)$ and the kinetic energy term, and with $\delta' \to \delta'_p$. The state of the system after the pulse is then easily found to be

$$\begin{aligned} \begin{pmatrix} \varphi_1(t_p) \\ \varphi_{-1}(t_p) \end{pmatrix} &= \begin{pmatrix} \cos\chi & i\sin\chi \\ -i\sin\chi & \cos\chi \end{pmatrix} \begin{pmatrix} \varphi_1(0) \\ \varphi_{-1}(0) \end{pmatrix} \\ &\equiv M_L(\chi) \begin{pmatrix} \varphi_1(0) \\ \varphi_{-1}(0) \end{pmatrix} . \end{aligned} \tag{14.41}$$

where $\chi = g\delta'_p t_p$ is the excitation pulse area and the operator M_L can be used to control the population transfer by an appropriate choice of χ.

The required phase relationship between the two states can be achieved via Zeeman splitting. Considering for concreteness a spatially inhomogeneous rectangular magnetic field pulse of duration t_B we have, neglecting again all other effects,

$$i\hbar\frac{\partial}{\partial t}\varphi_{\pm1}(Z,t) = \pm\mu_B g_F\,(B_0 + B'Z)\,\varphi_{\pm1}(Z,t)\,, \tag{14.42}$$

where g_F is the Landé g-factor of the hyperfine ground state, μ_B is the Bohr magneton, B_0 the spatially homogeneous component of the magnetic field, and B' its gradient, the direction of the magnetic field being along the Z-axis. The application of this field results in the state

$$\begin{aligned}\begin{pmatrix}\varphi_1(t_B)\\ \varphi_{-1}(t_B)\end{pmatrix} &= \begin{pmatrix}e^{i(\vartheta+K_B Z)} & 0\\ 0 & e^{-i(\vartheta+K_B Z)}\end{pmatrix}\begin{pmatrix}\varphi_1(0)\\ \varphi_{-1}(0)\end{pmatrix}\\ &\equiv M_B(\vartheta,K_B)\begin{pmatrix}\varphi_1(0)\\ \varphi_{-1}(0)\end{pmatrix}\,,\end{aligned} \tag{14.43}$$

where $\vartheta = -(\mu_B g_F/\hbar)B_0 t_B$ and $K_B = -(\mu_B g_F/\hbar)B' t_B$ are the imprinted phase shift and phase gradient (or wave vector), respectively. That is, the application of the magnetic pulse results in a phase difference of 2ϑ between the two Zeeman sublevels, and in addition it imparts them wave vectors $\pm K_B$.

14.5.2 Excitation and Application

To illustrate how stationary and moving solitons can be excited using this scheme, we start from a scalar condensate in the $|-1\rangle$ state, $\varphi(Z,0) = [0,\varphi_0(z)]^T$, with spatial mode

$$\varphi_0(Z) = \frac{N_s}{\sqrt{A_T}}\left(\frac{2}{\pi W_s^2}\right)^{1/4} e^{-Z^2/W_s^2}\,, \tag{14.44}$$

with W_s and N_s the width and atom number of the gap soliton desired. This Gaussian is chosen to approximate the hyperbolic-secant structure of the analytic gap soliton solution in (14.35). For a stationary solution we need to prepare the Zeeman sublevels with equal populations and with a π phase difference. We further need to impose wave vectors which are equal in magnitude but opposite in sign, $K_{\pm1} = \pm K_l$. This can be achieved by applying a laser pulse of area $\chi = \pi/4$, followed by a magnetic pulse with $\vartheta = \pi/4$ and $K_B = K_l$. The state then transforms as $(t = t_p + t_B)$

$$\begin{aligned}\varphi(Z,t) &= M_B(\pi/4,K_l)M_L(\pi/4)\varphi(Z,0)\\ &= \frac{e^{\frac{3i\pi}{4}}}{\sqrt{2}}\begin{pmatrix}e^{iK_l Z}\\ e^{-i\pi}e^{-iK_l Z}\end{pmatrix}\varphi_0(Z)\,.\end{aligned} \tag{14.45}$$

Figure 14.4 shows the resulting stable evolution of the total density $|\varphi(Z,t)|^2$. As a result of the Gaussian approximation to the exact solution there are some slow oscillations imposed on the motion, but the solution remains centered at

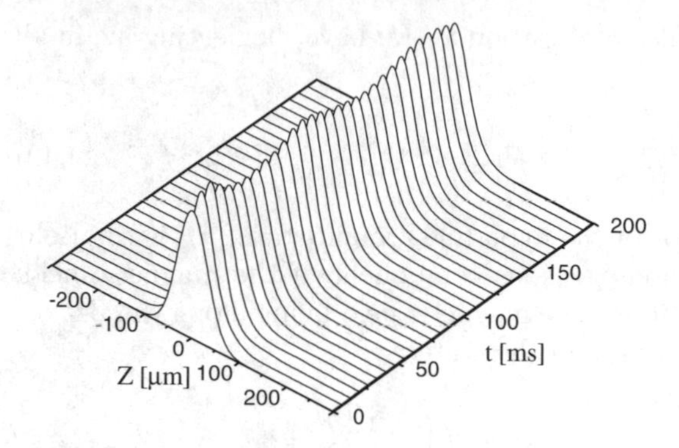

Fig. 14.4. The propagation of the total density of a stationary soliton ($v = 0$) over around 50 soliton periods. Peak density oscillations are due to imperfect initial conditions

$Z = 0$ and stationary over a time $t = 200$ ms, much longer than the soliton period for the chosen values. We note that although Sect. 14.5.1 treated the pulsed excitations in an impulsive manner to intuitively understand their action, our simulations do not make this approximation. The numerics confirm the accuracy of the impulsive approximation for the parameters at hand; a consequence of the fact that we consider pulse durations significantly shorter than the soliton period (14.40).

The excitation of a moving soliton is slightly more complicated since the velocity dependent wave vector $K = -\gamma v/l_c$ in (14.35) is no longer zero and the two components have different populations, see (14.39). However, our simulations show that the proposed scheme performs well on this task, as well as on splitting solitons and reversing their directions [46].

Solitons present some advantages for atom interferometry in that they are many-atom wavepackets which are immune to the effects of spreading, hence allowing longer path lengths, and also increased signal-to-noise for large atom numbers. Typically many-body effects limit the utility of high-density wavepackets due to spatially varying mean-field phase shifts, but solitons have the cardinal virtue that they have fixed spatial phase variations. Thus they may provide a key to making maximal use of high density sources for atom interferometry. Indeed, due to these very properties they have long been advocated for all-optical switching applications.

14.6 Conclusion

In this paper, we have reviewed important aspects of the matter-wave solitons that can be launched in weakly interacting Bose–Einstein condensates. So far, only dark solitons have been demonstrated experimentally, see, however, the note added in proof, but the rapid progress in the optical manipulation and

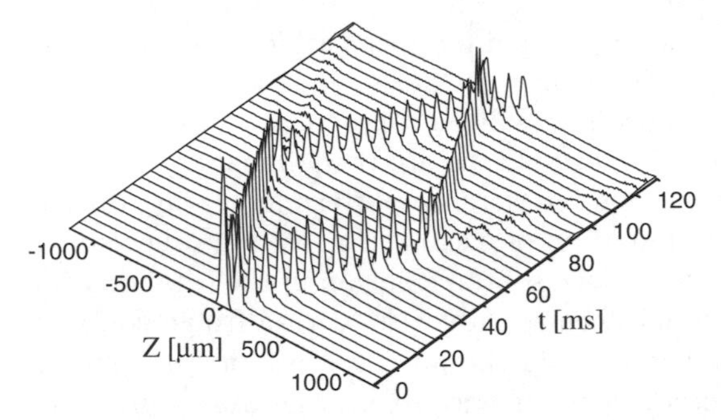

Fig. 14.5. Demonstration of an atomic Mach–Zehnder interferometer: The initial condensate is split into two counterpropagating solitons, then their direction is reversed and they collide (shown is the total density)

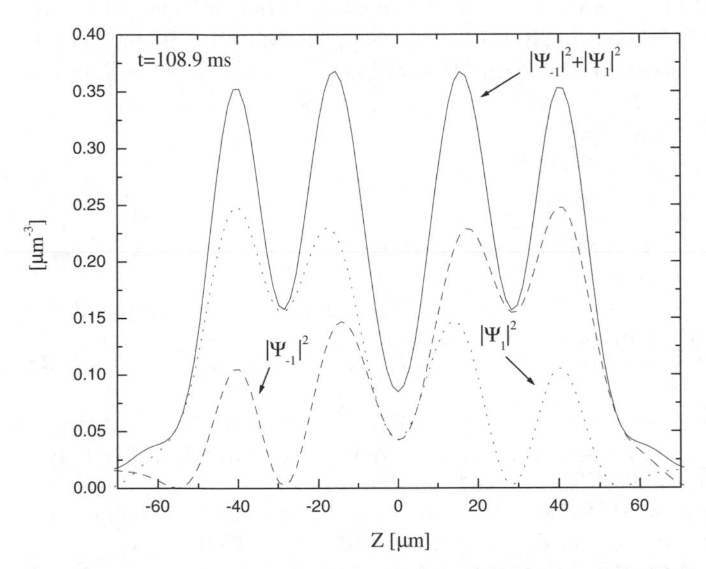

Fig. 14.6. Interference pattern when two solitons collide in the Mach–Zehnder configuration: The total density (solid line) is symmetric, whereas the interference pattern of each of the two orthogonal Zeeman states (dotted and dashed line) is asymmetric due to the shape of the colliding wavepackets. The contrast in the total density pattern is around 30%

control of condensates is expected to lead to the demonstration of bright gap solitons.

As an illustration of the potential use of these objects, we consider a soliton-based nonlinear atomic Mach–Zehnder interferometer. Our specific demonstration involves an initial scalar condensate that is split into two oppositely moving solitons along Z, see Fig. 14.5 for $t < 60$ ms.

At $t = 60\,\mathrm{ms}$ laser and magnetic pulses are applied which act to reverse the direction of the two solitons. This process causes some loss of atoms in both solitons, as before. The two reversed soliton components come together again at $t = 120\,\mathrm{ms}$. Since the colliding solitons are predominantly in opposite orthogonal Zeeman sublevels, the interference pattern appearing during the collision is due to the contamination of each soliton by the other state. Figure 14.6 shows the interference at the soliton collision in the total density (solid line) and also the individual Zeeman sublevels, with a fringe contrast of around 30%. These results demonstrate the potential use of gap solitons for realizing nonlinear atom interferometers with high brightness sources.

Acknowledgments. The authors would like to thank the Hannover group for providing Figs. 14.1 and 14.2 as well as K.M. Hilligsøe for poiting out a mistake in (14.23). This work is supported in part by Office of Naval Research Contract Nos. 14-91-J1205 and N00014-99-1-0806, the National Science Foundation Grant PHY98-01099, the Army Research Office and the Joint Services Optics Program.

References

1. M.H. Anderson, J.R. Ensher, M.R. Matthews, C.E. Wieman, E.A. Cornell: Science **269**, 198 (1995)
2. K.B. Davis, M.-O. Mewes, M.R. Andrews, N.J. van Druten, D.S. Durfee, D.M. Kurn, W. Ketterle: Phys. Rev. Lett. **75**, 3969 (1995)
3. C.C. Bradley, C.A. Sackett, J.J. Tollett, R.G. Hulet: Phys. Rev. Lett. **75**, 1687 (1995)
4. J. Denschlag, J.E. Simsarian, D.L. Feder, C.W. Clark, L.A. Collins, J. Cubizolles, L. Deng, E.W. Hagley, K. Helmerson, W.P. Reinhardt, S.L. Rolston, B.I. Schneider, W.D. Phillips: Science **287**, 97 (2000)
5. S. Burger, K. Bongs, S. Dettmer, W. Ertmer, K. Sengstock, A. Sanpera, G.V. Shlyapnikov, M. Lewenstein: Phys. Rev. Lett. **83**, 5198 (1999)
6. W. Chen, D.L. Mills: Phys. Rev. Lett. **58**, 160 (1987)
7. D.N. Christodoulides, R.I. Joseph: Phys. Rev. Lett. **62**, 1746 (1989)
8. C.M. de Sterke, J.E. Sipe: 'Gap Solitons'. In: *Progress in Optics XXXIII.* Ed. by E. Wolf (Elsevier, Amsterdam 1994) pp. 203–260
9. B.J. Eggleton, R.E. Slusher, C.M. de Sterke, P.A. Krug, J.E. Sipe: Phys. Rev. Lett. **76**, 1627 (1996)
10. G. Lenz, P. Meystre, E.M. Wright: Phys. Rev. Lett. **71**, 3271 (1993)
11. G. Lenz, P. Meystre, E.M. Wright: Phys. Rev. A **50**, 1681 (1994)
12. W. Zhang, D.F. Walls, B.C. Sanders: Phys. Rev. Lett. **72**, 60 (1994)
13. A.C. Newell, J.V. Moloney: *Nonlinear Optics* (Addison-Wesley, Redwood City 1992)
14. L.P. Pitaevskii: Sov. Phys. JETP **13**, 451 (1961)
15. E.P. Gross: J. Math. Phys. **4**, 195 (1963)
16. L. Deng, E.W. Hagley, J. Wen, M. Trippenbach, Y. Band, P.S. Julienne, J.E. Simsarian, K. Helmerson, S.L. Rolston, W.D. Phillips: Nature **398**, 218 (1999)

17. M.R. Matthews, B.P. Anderson, P.C. Haljan, D.S. Hall, C.E. Wieman, E.A. Cornell: Phys. Rev. Lett. **83**, 2498 (1999)
18. K.W. Madison, F. Chevy, W. Wohlleben, J. Dalibard: Phys. Rev. Lett. **84**, 806 (2000)
19. B.P. Anderson, M.A. Kasevich: Nature **282**, 1686 (1998)
20. D.M. Stamper-Kurn, M.R. Andrews, A.P. Chikkatur, S. Inouye, H.-J. Miesner, J. Stenger, W. Ketterle: Phys. Rev. Lett. **80**, 2027 (1998)
21. E.M. Lifshitz, L.P. Pitaevskii: *Statistical Physics: Part 2* (Pergamon Press, Oxford 1980)
22. P.A.M. Dirac: Proc. Camb. Phil. **26**, 376 (1930)
23. E.M. Lifshitz, L.P. Pitaevskii: *Quantum Mechanics: Part 3* (Pergamon Press, Oxford 1991)
24. K.Huang: *Statistical Mechanics* (Wiley, New York 1987)
25. E.V. Goldstein, P. Meystre: Phys. Rev. A **55**, 2935 (1997)
26. H.Pu, N.P. Bigelow: Phys. Rev. Lett. **80**, 1130 (1998)
27. D.S. Petrov, G.V. Shlyapnikov, and J.T.M. Walraven, Phys. Rev. Lett. **85**, 3745 (2000)
28. A.C. Scott, F.Y.F. Chu, D.W. McLaughlin: Proc. IEEE **61**, 1443 (1973)
29. W.P. Reinhardt, C.W. Clark: J. Phys. B **30**, L785 (1997)
30. G.P. Agrawal: *Nonlinear Fiber Optics* (Academic Press, San Diego 1995)
31. R. Dum, J.I. Cirac, M. Lewenstein, P. Zoller: Phys. Rev. Lett. **80**, 2972 (1998)
32. P.O. Fedichev, A.E. Muryshev, G.V. Shlyapnikov: Phys. Rev. A **60**, 3220 (1999)
33. T. Busch, J.R. Anglin: Phys. Rev. Lett. **84**, 2298 (2000)
34. S.L. Cornish, N.R. Claussen, J.L. Roberts, E.A. Cornell, C.E. Wieman: Phys. Rev. Lett. **85**, 1795 (2000)
35. J.L. Roberts, N.R. Claussen, J.P. Burke, Jr., C.H. Greene, E.A. Cornell, C.E. Wieman: Phys. Rev. Lett. **81**, 5109 (1998)
36. S. Inouye, M. R. Andrews, J. Stenger, H.-J. Miesner, D.M. Stamper-Kurn, W. Ketterle: Nature **392**, 151 (1998)
37. Y.S. Kivshar, T.J. Alexander: *Trapped Bose–Einstein Condensates: Role of Dimensionality* (cond-mat/9905048)
38. K. Berg-Sørensen, K. Mølmer: Phys. Rev. A **58**, 1480 (1998)
39. D.-I. Choi, Q. Niu: Phys. Rev. Lett. **82**, 2022 (1999)
40. M.J. Steel, W. Zhang: *Bloch function description of a Bose–Einstein condensate in a finite optical lattice* (cond-mat/9810284)
41. C. Kittel: *Introduction to Solid State Physics* (Wiley & Sons, New York Chichester 1986)
42. M. Abramowitz, I.A. Stegun: *Handbook of Mathematical Functions* (Dover, New York 1970)
43. N. Aközbek, S. John: Phys. Rev. E **57**, 2287 (1998)
44. O. Zobay, S. Pötting, P. Meystre, E.M. Wright: Phys. Rev. A **59**, 643 (1999)
45. C. Cohen-Tannoudji: 'Atomic Motion in Laser Light'. In: *Fundamental Systems in Quantum Optics*. Ed. by J. Dalibard, J.M. Raimond, J. Zinn-Justin (North-Holland, Amsterdam 1992) pp. 1–164
46. S. Pötting, O. Zobay, P. Meystre, E.M. Wright: J. Mod. Opt. (*to be published*)
47. A.B. Aceves, S. Wabnitz: Phys. Lett. A **141**, 37 (1989)

Part IV

Spatial Solitons in Photonic Crystals

15 Discrete Solitons

H. Eisenberg and Y. Silberberg

15.1 Introduction

Optics deals with continuous objects. The optical electro-magnetic fields are continuous functions. Optical elements like mirrors and lenses can be presented as continuous operators. Discrete optical fields, which have values only at specific locations, are somewhat unnatural. Nevertheless, there are situations where optical systems can be represented as discrete fields. In particular, the interaction of several coupled modes is usually approximated by a discrete set of coupled mode equations. One such case is that of coupled one-dimensional waveguide array [1,2]. In a waveguide array, large number (infinite in principal) of one-dimensional single-mode waveguides (e.g. optical fibers) are laid one near the other such that their individual modes overlap. The transversal propagating field is now described by an infinite set of complex amplitudes of the individual modes.

The problem of light propagation in an infinite coupled array of waveguides was treated first theoretically by Jones [1], and later demonstrated by Yariv and co-workers [2], who fabricated such an array in GaAs. The nonlinear problem was discussed only years later by Christodoulides and Joseph [3]. They showed that the equations describing propagation in a nonlinear Kerr array are formally a discrete version of the nonlinear Schrödinger equation (NLSE), and that they may have solitary solutions equivalent to the NLSE solitons. The study of the discrete nonlinear Schrödinger equation (DNLSE) became quite popular in the early 90's and many of the properties of discrete solitons were discovered then. Among the topics investigated were instabilities and solitary solutions of discrete structures [4–6], phenomena of steering and switching [7–9], nonlinear dynamics in non-uniform arrays [10–13], temporal effects in discrete solitons [14–17], dark solitary solutions [18,19], vector discrete solitons [20] and discrete soliton bound states ("twisted modes") [21,22].

It should be noted that similar discrete equations appear in other area of physics. In particular, in the context of energy transport in molecular systems [23], localized modes in molecular systems such as long proteins [24], polarons in one dimensional ionic crystals [25], and localized modes of nonlinear mechanical [26] and electrical lattices [27]. The broader area of discrete nonlinear equations and discrete breathers has been reviewed recently [28].

The study of discrete optical solitons has remained primarily a theoretical area until recently, when several experimental studies were reported. Experiments in GaAs waveguides demonstrated the formation of discrete solitons [29] and later studied many of their properties [11,30,31]. These studies led to a renewed interest in the linear properties of discrete optical structures. For example, recent reports on optical realization of Bloch oscillations in waveguide arrays [32–35], and engineering of diffraction using discrete optics [36].

15.2 Theory

We begin by analyzing the basic phenomena that are related to propagation of light in an infinite set of coupled optical waveguides. When light is injected into such an array, it couples to more and more waveguides as it propagates. This broadening of the spatial distribution of the filed is analogous to diffraction in continuous structures, hence we refer to it as discrete diffraction. We discuss this linear phenomenon first, and then proceed to discuss nonlinear structures, when discrete diffraction is accompanied by the optical Kerr effect. This combination leads to the formation of discrete solitons and to their dynamic properties. This theoretical section is not exhaustive, and the reader is referred to the bibliographic list for more detailed publications on this topic.

15.2.1 Discrete Diffraction

Consider an infinite uniform array of identical waveguides. We denote the set of amplitudes of the modes of the individual waveguides as $\{E_n\}$, where n is the waveguide index, $n = 0$ being the central waveguide. The overlap between nearest-neighbors modes has a dominant effect on propagation. It results in a phase sensitive linear coupling between the waveguides, leading to power exchange among them.

The basic linear set of equations, which describes the light propagation in a waveguide arrays using coupled-mode theory formulation is [37]:

$$\mathrm{i}\frac{\mathrm{d}E_n}{\mathrm{d}z} = \beta E_n + C\left(E_{n-1} + E_{n+1}\right) . \tag{15.1}$$

z is the spatial coordinate in the direction of propagation, β is the propagation constant of the individual waveguides and C is the coupling coefficient. If only two waveguides are coupled, an arrangement known as an optical directional coupler [37], light is completely transferred from one waveguide to the other after propagating a distance of $z_c = \pi/2C$, the *coupling length*. There is a known mathematical solution for this infinite set of equations [1]. In the case of a single unity input into the central waveguide and no power in the rest, the electrical field evolution is given by:

$$E_n = (\mathrm{i})^n \mathcal{J}_n\left(2Cz\right) , \tag{15.2}$$

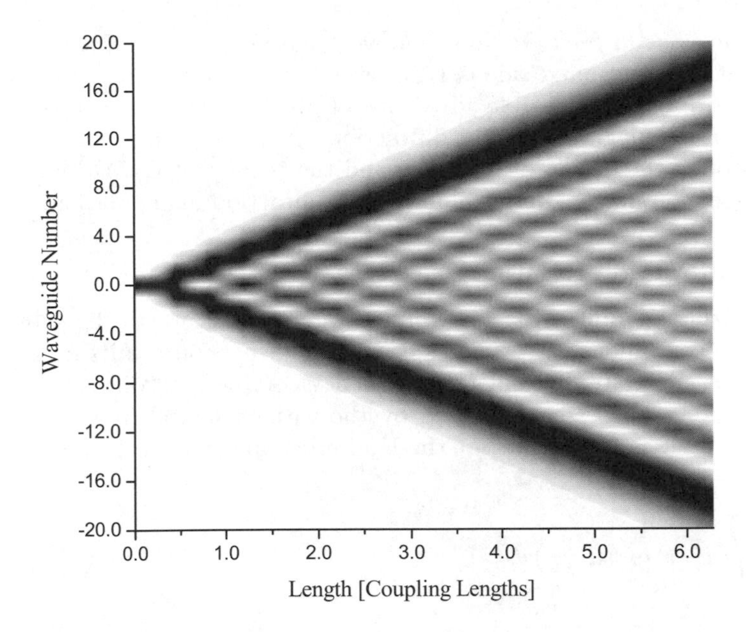

Fig. 15.1. Solution for a linearly coupled array of 41 waveguides, when light is injected into the central waveguide, with $E_0 = 1$. The intensity is shown in gray scale. The scale is chosen such that the peak intensity for every propagation distance along the waveguide is represented by black. The energy is spread mainly into two lobes. The number of central peaks is an indication to how many coupling lengths the light has propagated

where $\mathcal{J}_n(x)$ is the nth order Bessel function. Because the evolution equations are linear, any other solution can be constructed from this solution by a linear superposition. The single input linear evolution is depicted in Fig. 15.1. Notice how most of the light is concentrated in two distinct lobes.

An interesting limit of (15.1) is when the excitation is varying slowly between neighboring waveguides ($E_{n+1} - E_n \ll E_n$) [3,38]. Applying the transformation $E_n = A_n \exp\left[\mathrm{i}\left(2C + \beta\right) z\right]$, we get propagation equations of the form:

$$\mathrm{i}\frac{\mathrm{d}A_n}{\mathrm{d}z} = C\left(A_{n-1} - 2A_n + A_{n+1}\right) . \tag{15.3}$$

Changing the discrete coordinate n to a continuous one, $x = nd$ where d is the distance between the centers of two adjacent waveguides, we notice that the right-hand side of (15.3) is a discrete version of a second derivative, hence:

$$\mathrm{i}\frac{\partial A}{\partial z} = Cd^2\frac{\partial^2 A}{\partial x^2} . \tag{15.4}$$

Equation (15.4) describes, under the assumption of paraxiallity, propagation of light in free two-dimensional space. We identify the coupling terms in (15.1) as the cause for *discrete diffraction*. Note, however, that in paraxial diffraction the coefficient of the right-hand side is $1/2k$, with k the wavenumber in the medium. Hence, we see one possible advantage of discrete optics: it enables the engineering of the magnitude of the diffraction, normally fixed at a given wavelength, through the design of the array and the coefficient C. We shall see in the next section even more possibilities for diffraction manipulation.

15.2.2 The Diffraction Relation

In order to understand diffraction in discrete systems, it is worthwhile to examine first the continuous case. Just as chromatic dispersion results from different phases accumulated by different frequency components, diffraction results from different phases accumulated by the various spatial frequency components [36]. These components are the Fourier components of the beam profile:

$$\widetilde{E}\left(k_x\right) = \int_{-\infty}^{\infty} E\left(x\right) \exp\left(\mathrm{i}k_x x\right) dx \,, \tag{15.5}$$

where we define x as the transverse dimension and z as the propagation dimension. The third direction, y, is omitted for simplicity, but can be added in a straightforward manner. Each of the spatial components denoted by k_x is accumulating optical phase differently while propagating. The amount of phase gained by each k_x component after propagating distance z is $\phi\left(k_x\right) = k_z\left(k_x\right) \cdot z$. A group of transverse components centered at k_x is transversely shifted by an amount of

$$\Delta x = \frac{\partial \phi}{\partial k_x} = \frac{\partial k_z}{\partial k_x} z \,.$$

The beam broadens because of the divergence between the different displacements $\Delta x(k_x)$. The propagation direction in real space of the power contained in each component, the Pointing vector, points perpendicular to the $k_z(k_x)$ curve at k_x [39]. We define the function $k_z(k_x)$ as the *diffraction relation* and the divergence

$$D = \frac{1}{z}\frac{\partial^2 \phi}{\partial k_x^2} = \frac{\partial^2 k_z}{\partial k_x^2}$$

as *diffraction*, in analogy to the definition of dispersion.

The diffraction relation for a specific system can be derived from the evolution equation by assuming a plane wave solution of the form $E\left(\boldsymbol{r}\right) = E^0 \exp\left(\mathrm{i}\boldsymbol{k} \cdot \boldsymbol{r}\right)$ where $\boldsymbol{k}$ is a vector in spatial frequency space whose components are k_x and k_z. For a scalar propagation in free two-dimensional space, according to the standard time-independent wave equation, the diffraction relation is:

$$k_z\left(k_x\right) = \sqrt{k^2 - k_x^2} \,, \tag{15.6}$$

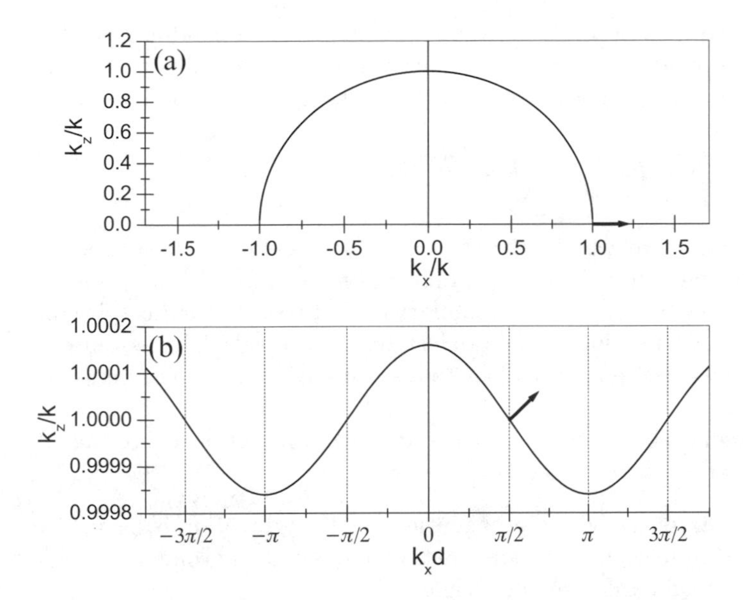

Fig. 15.2. Spatial diffraction curves showing phase vs. spatial frequency for **a** continuous and **b** discrete models. The arrows mark the largest possible angle of energy propagation for each model. Only in the discrete case, an inversion of curvature (beyond $k_x d = \pi/2$) leads to anomalous diffraction

where $k = \frac{2\pi n_0}{\lambda_0}$ is the wavenumber in the medium and λ_0 is the wavelength in vacuum. The half-circular diffraction relation for forward propagating free space beams is depicted in Fig. 15.2a. There are a few useful observations that can be made from this graph:

1. A beam can not have a component $k_x > k$. This results in a maximal resolution for a specific wavelength. For example, the minimal spot size of a focused beam.
2. Light can propagate in all π forward angles. For any direction, there is always a point on the diffraction curve, where the Pointing vector is directed along it.
3. The diffraction value around any component of k_x is always negative ($D_{\text{cont}} < 0$). This fact will have important implications for nonlinear evolution.

Let us now consider the diffraction relation for the discrete propagation equation [39,40]. We apply the same plane wave solution and the result is:

$$k_z(k_x) = \beta + 2C\cos(k_x d) . \tag{15.7}$$

The discrete diffraction relation is presented in Fig. 15.2b. This relation is periodic, hence there are infinite number of components k_x propagating for each k_z, spaced equally by $2\pi/d$. Therefore, we should use a modified function base to span the beam profiles, where each component contains all of these equidistant plane waves. These are the Floquet-Bloch wave functions [41]

$$E(\boldsymbol{r}) = E^0 \sum_{m=-\infty}^{\infty} \Gamma_m \exp\left[\mathrm{i}\left(\boldsymbol{k} + \frac{2\pi m}{d}\widehat{k_x}\right)\cdot \boldsymbol{r}\right],$$

where Γ_m is a factor related to the individual waveguide mode shape. In a true discrete system, there is only meaning for values of $\boldsymbol{r} = (nd, z)$ and Γ_m becomes always unity. We note the similarities between the optical discrete model and the tight binding model of electrons in crystals. The assumption of single-mode waveguide is equivalent to a system of atoms with one S-level electron.

Let us examine the properties of the discrete diffraction curve and its differences from the continuous case:

1. There is a maximal angle of propagation to the z-direction $\alpha_{\max} = 2Cd$. Even if the input light contains spatial frequencies beyond this angle, they will propagate at a smaller angle.
2. The meaningful range of k_x is $2\pi/d$. Beyond this range, i.e. the first Brillouin zone, any k_x has an equivalent value that is inside it, sharing the same Floquet–Bloch wave function. It is convenient to choose $-\pi/d < k_x < \pi/d$.
3. The curvature of the discrete diffraction curve, or the diffraction value, is negative for $|k_x| < \pi/2d$, zero for $k_x = \pi/2d$ and positive for $\pi/2d < k_x < \pi/d$. This is in contrast with the continuous case, where zero and positive values of diffraction are impossible.
4. The diffraction curve has a maximum value as well as a minimum, while the continuous relation has no minimum.

15.2.3 Nonlinear Excitations – Discrete Solitons

Consider now propagation of light in a Kerr-like nonlinear waveguide array. The Kerr effect is the linear dependency of the refractive index on the local light intensity in a dielectric material [37]. The refractive index varies as $n = n_0 + n_2 I$ where n_0 is the refractive index at low intensities, I is the light intensity and n_2 is the material nonlinear parameter, proportional to the nonlinear electrical susceptibility term $\chi^{(3)}$.

Equation (15.1) is now modified to be [3]:

$$\mathrm{i}\frac{\mathrm{d}E_n}{\mathrm{d}z} = \beta E_n + C\left(E_{n-1} + E_{n+1}\right) + \gamma\left|E_n\right|^2 E_n\,, \tag{15.8}$$

where $\gamma = \frac{k_0 n_2}{A_{\mathrm{eff}}}$ and A_{eff} is the common effective area of the waveguide modes. This equation is known as the Discrete Nonlinear Schrödinger Equation (DNLSE). The Kerr nonlinearity in the presence of continuous diffraction (or dispersion) may lead to the formation of spatial (or temporal) solitons [42–45]. These are stable light beams (or pulses), propagating without diffraction (or dispersion). Unlike its continuous counterpart, the DNLSE is not integrable, and therefore it does not posses exact soliton solutions. However, numerical and analytic studies show that it does have solitary-wave solutions that propagate without diffraction. In particular, for a wide excitation, one can use the analogy with the continuous case to derive an approximate shape distribution [3]:

$$E_n(z) = \frac{A_0 \exp\left[\mathrm{i}\left(2C+\beta\right)z\right]}{\cosh\left(nd/w_0\right)}, \tag{15.9}$$

where w_0 is the width constant of the light distribution, assumed to be much larger than d. The amplitude is given by $A_0 = \sqrt{\frac{2Cd^2}{\gamma w_0^2}}$. Narrow solitary wave solutions are also possible, but they deviate from this shape function. It is quite obvious that at very high powers, all the light is concentrated in a single waveguide, increasing its index of refraction so that it is completely decoupled from its neighbors. Although, formally speaking, all of these solutions are solitary waves, we shall use the term *discrete solitons* to describe all non-diffracting waves.

Before we move to discuss other properties of discrete solitons, it is worthwhile to discuss how one may infer their existence from the diffraction relations. The effect of the Kerr nonlinearity on the diffraction relations is similar for continuous and discrete cases. These relations are, respectively:

$$k_z\left(k_x\right) = \sqrt{k^2\left[1+\left(\frac{n_2 I_0}{n_0}\right)^2\right] - k_x^2}\,, \tag{15.10a}$$

$$k_z\left(k_x\right) = \beta + 2C\cos\left(k_x d\right) + \gamma I_0\,. \tag{15.10b}$$

Here I_0 is a measure of the optical power contained in the beam and it is proportional to $\left|E^0\right|^2$. In both cases the light intensity shifts the excited modes upward along the k_z axis.

There is a useful a phenomenological way of understanding solitary waves [28]. As discussed above, diffraction results from the curvature of the diffraction relation. In the presence of significant nonlinear Kerr effect the wave increases its effective propagation constant, and, eventually, shifts it out of the diffraction zone, eliminating the option for beam broadening. For diffraction to happen, the beam spatial components have to be able to radiate along the x-axis. When they are outside the linear spectrum, this radiation is eliminated. It is now clear that in continuous homogeneous media, where the diffraction curve has a maximum, bright solitons are possible only when $n_2 > 0$. Negative values of n_2 will only lower the value of the propagating k_z frequencies, and actually enhance the rate of diffraction.

In contrast, the discrete diffraction curve has both negative and positive curvatures, which enable the formation of bright discrete solitons for both positive and negative n_2 [3,5,28–30,46,47]. Furthermore, this curve is limited both from top and bottom to the range $\{\beta - 2C < k_z < \beta + 2C\}$ resulting in more options for solitary waves. For example, in a positive n_2 material, an excitation experiencing positive curvature at the optical band bottom, and therefore nonlinear defocusing, can still be solitary if enough power is supplied such that it is pushed upward enough so it emerges out of the optical band. In fact, any excitation in any type of material will be eventually solitary in the discrete optics case because of the bounded band.

15.2.4 Dynamics of Light in a Waveguide Array

When we consider the dynamics of light in a waveguide array, it is important to understand the direction of energy flow. It is easy to show that in the two waveguides coupler the phase relation between the two individual modes is always $\pm\pi/2$ and that power is always flowing from the mode, whose phase is retarded relative to the other. Similarly, from the linear solution for the propagation of a single waveguide input, we find that the same phase relation between waveguides of $\pm\pi/2$ occurs in an infinite waveguide array. At the outer regimes of the diffracting light, power transfer is outwards, while in the center, a more complex phase structure evolves. For different input conditions this phase structure disappears, but power still flows from waveguides, which their phase is retarded relative to their neighbors.

This is not the case for discrete solitons. By definition, the phase across a soliton is flat. A flat phase front results in no power coupling, therefore no diffraction, exactly what we expect from a soliton. The Kerr effect is the cause of this flatness, by increasing the refractive index of some waveguides, hence accelerating their phase evolution.

Continuous solitons possess two fundamental geometrical symmetries of space. Under rotations of the axes and translation of their origin, exactly the same mathematical solution is reproduced. On the other hand, for discrete optics both symmetries are broken due to the direction and position of the array waveguides [6]. Rotational symmetry is gone all together while translational symmetry is reduced. There is still option to translate solutions to other z values, while in the x direction only a discrete translational symmetry is left because of the periodicity d in that direction. Nevertheless, discrete solitons can be centered either on a waveguide or anywhere in between two adjacent waveguides [5]. Even more, solutions of discrete solitons that propagate with an angle to the z-axis do exist [7–9]. The discrete nature though, is still reflected in some new discreteness-induced effects [31].

In order to learn about the effect of launching discrete solitons centered at different lateral positions, we look at the generalized Hamiltonian of (15.8):

$$H_{\text{disc}} = \sum_n \left[C\left|E_n - E_{n-1}\right|^2 - \frac{\gamma}{2}\left|E_n\right|^4 \right] . \tag{15.11}$$

The first term is a generalized kinetic energy term and the second is an interaction energy term. Plotting H_{disc} as a function of the lateral center of the soliton beam shows an oscillating function. The Hamiltonian minima are when the soliton center is on a waveguide site and its maxima are when the soliton is centered between two adjacent waveguides. This discreteness induced periodic potential is known from solid state physics as the Peierls–Nabarro (PN) potential [6]. It is a result of the combination of nonlinearity and discreteness. The depth between the minima and maxima points is power dependent. Solitons experiencing this potential will be stable for small lateral shift when centered on a site and unstable if centered in between two sites.

If a discrete soliton which is wider than one waveguide is excited, it can be forced to hop sideways while propagating in the z direction [7–9]. A linear phase gradient across its profile such as $\phi_n = n\Delta\phi$ will allow the light distribution to keep its solitary properties, but the soliton will also move away from the retarded phase side. This is understood when examining the direction of energy flow. However, increasing the power of the excitation leads to a new, discreteness induced effect. At high power the waveguides that contain light are decoupled from their neighborhood. Therefore, a discrete soliton will be locked at high powers to its input waveguides and will not travel sideways anymore [8,9,31]. The power dependence of the propagation angle is called discrete soliton *power steering*. It can be also thought of as interplay between the PN potential and the generalized kinetic energy of the soliton.

15.3 Experiments

While many have contributed to the theoretical understanding of discrete solitons, the experimental studies have been carried out primarily at our group at the Weizmann Institute, in close collaboration with the Aitchison group at the University of Glasgow. In the following we shall review some of our main results.

15.3.1 Experimental Considerations

Two ways were proposed in the past for realizing an optical discrete system. One is to form the array by patterning a slab planar waveguide [1,2]. The other is a multi-core optical fiber where the cores are arranged in a circular shape [7,48]. There are a few advantages for each of the configurations. Slab waveguides can be made from a variety of materials while optical fibers are typically made of glass variants. There are many standard techniques for patterning planar configurations, while multi-core fibers are not so common and easy to make. Integrating arrays with other elements is easier on a planar layout. On the other hand, very long fibers can be formed compared to only a few inches of planar waveguides. The periodic boundary conditions that are achieved with the circular layout of multi-core fibers are very attractive as it can better simulate an infinite array.

In all our experiment, we chose to work with waveguide arrays that were etched of a planar slab waveguide made ofAlGaAs [49]. There are a few reasons for this choice. AlGaAs is an effective nonlinear material, about 500 times more nonlinear then fused silica glass [50]. This enables us to use very short waveguides of a few millimeters and modest optical powers. Working at the communication standard wavelength of 1.5 μm, we are below the half of the band gap. Thus, not only linear absorption is minimized but nonlinear two-photon absorption as well. The technology for fabricating AlGaAs waveguides is quite advanced and waveguides of relatively low loss are possible.

There are some details we should be aware of, though. Coupling light into AlGaAs is relatively ineffective due to its high linear refractive index. With index of 3.3, almost 30 percents are reflected back at the input facet. Normal dispersion is also a factor, broadening an initial pulse of about 100 fs after propagating a typical sample length of 6 mm by about a factor of two. At the intensities of our interest, three-photon absorption starts to play a role as well.

15.3.2 Experimental Setup

We shall review now experiments performed in our group which demonstrate some of the properties of discrete solitons. Our light source was a commercially available optical parametric oscillator (*OPAL*), pumped by a 810 nm Ti:Sapphire (*Tsunami*). 4 nJ pulses were produced at a repetition rate of 80 MHz. Their average power is 300 mW while each has a peak power of about 40 kW. The pulse length is about 100fs and the wavelength is tunable between 1450 nm to 1570 nm. We usually worked at a wavelength of 1530 nm in order to minimize both two- and three-photon absorption.

The AlGaAs waveguides are patterned by either photolithography or electron–beam lithography. The core–clad index difference is achieved by different Aluminum concentration. The waveguides are typically 4 μm wide and 6 mm long. The separation between waveguides in the array is varied from 8 μm to 11 μm in order to have samples of different coupling lengths. The larger the distance the smaller the coupling. The individual waveguide mode shapes are somewhat oval. Arrays of 41 and 61 identical waveguides were made, in order that the light will experience an effective infinite system and will not reach the array borders.

The experiment is carried as follows (Fig. 15.3); the power of the input is tuned using a variable filter. A beam sampler picks up a small part of the power to an input power detector. The beam is shaped into an oval shape through cylindrical optics in order to match the individual waveguide mode or, alternatively, into a wider beam if input into a few waveguides is needed. The light is coupled into the sample through a 40× input objective. A glass window, 1 mm thick, is inserted in front of the input objective. By rotating this window, small parallel translations of the beam are transformed by the objective to small angles of incidence at the focal plane, enabling a

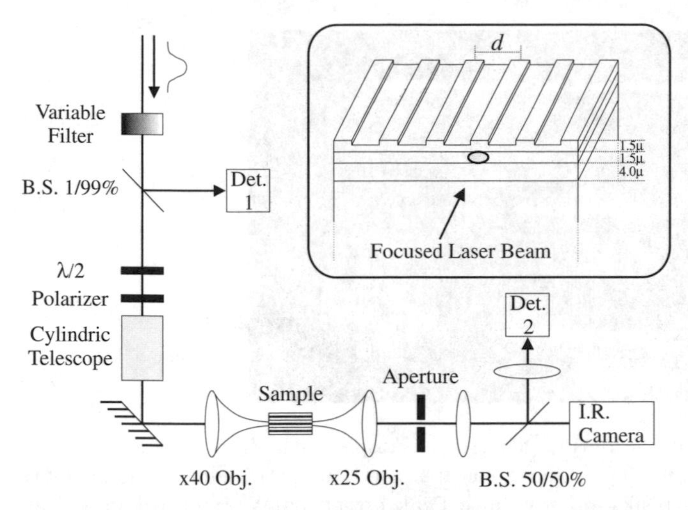

Fig. 15.3. The experimental setup. *Inset*: Schematic drawing of the sample. The sample consists of a $Al_{0.18}Ga_{0.82}As$ core layer and $Al_{0.24}Ga_{0.76}As$ cladding layers grown on top of a GaAs substrate. A few samples where tested with different separations d between the waveguides centers

phase gradient across the beam. After propagating along the sample, the light is collected from the output facet by a second objective. A second beam splitter is sending some of the power to an output power detector. The output objective and another lens image the light from the output facet onto an IR sensitive camera. At the extra focal plane which is formed in between the objective and the lens, a slit can be inserted in order to sample different parts of the image for spectra and auto-correlation measurements. Finally, a computer captures the camera image and the other data for further analysis.

15.3.3 Demonstration of Discrete Solitons

We first describe the basic experiments for the observation of discrete solitons [29]. We used a 6 mm long sample of 41 waveguides. Light was injected into a single central waveguide, as was described before. The first observation is presented in Fig. 15.4 as the output facet images that were captured by the camera. At low power, a wide distribution is obtained, covering about 35 waveguides. This distribution matches what is expected from a 4.2 coupling-lengths long sample (see Fig. 15.1). When the power is increased, we first see the light distribution converging to form a bell-shape. Launching even more power leads to the formation of a confined distribution around the input waveguide, a discrete soliton.

Transverse cross-sections of the output profile for two samples of different coupling lengths are compared in Fig. 15.5. Figure 15.5a,b is from a 1.9 coupling lengths long sample at low and high powers, respectively. Fig-

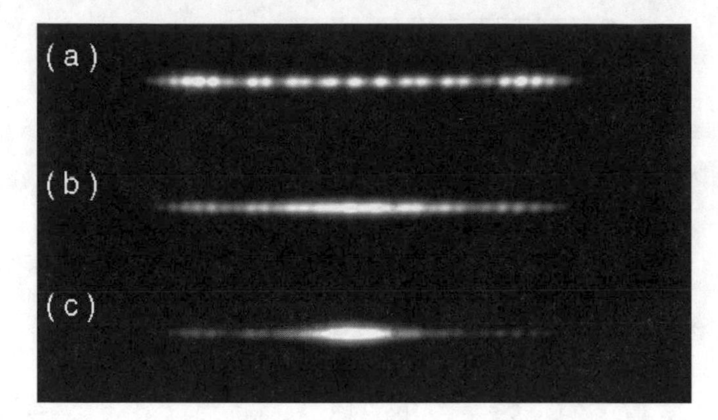

Fig. 15.4. Images of the output facet of a sample with $d = 8\,\mu\text{m}$ for different powers. **a** Peak power 70 W. Linear features are demonstrated: Two main lobes and a few secondary peaks in between. **b** Peak power 320 W. Intermediate power, the distribution is narrowing. **c** Peak power 500 W. A discrete soliton is formed

ure 15.5c,d is from a 3.0 coupling lengths long sample at low and high powers, respectively. The difference in the coupling length is obtained by varying the distance d of adjacent waveguides in the two samples (from 11 µm to 9 µm). The solid curves correspond to the experimentally measured profiles, while the height of the vertical lines stands for the light intensity in each respective waveguide according to a numerical solution of (15.8). The agreement between the experiment and the numerical solution is good, both at low and at high powers. At low power, most of the light energy is pushed out to the profile wings, while characteristic secondary narrow peaks appear in between. The number of these secondary peaks is an indicator to the amount of coupling lengths propagated. When the power is increased (estimated to about 900 W at peak power), light is confined to the vicinity of the central input waveguide.

It is interesting to check the power dependency of the visibility between adjacent peaks in Figs. 15.4 and 15.5. We define the visibility of the nth mode as a function of the intensity values $I(x) = |E(x)|^2$ at the center of two waveguide modes and the intensity value in between them:

$$V_n = 1 - \frac{2I\left[\left(n + \frac{1}{2}\right) d\right]}{I\left[nd\right] + I\left[(n+1)\,d\right]} \,. \tag{15.12}$$

The mode visibility has a minimal value when the light is in-phase between waveguides n and $n + 1$. It is maximal when the light is completely out-of-phase and the intensity is equal in both waveguides. Checking the experimental results of Figs. 15.4 and 15.5 we observe a visibility of about half for all the linear cases. This is in a good agreement with the $\pi/2$ phase relation predicted by the solution of (15.8). The visibility is degraded because

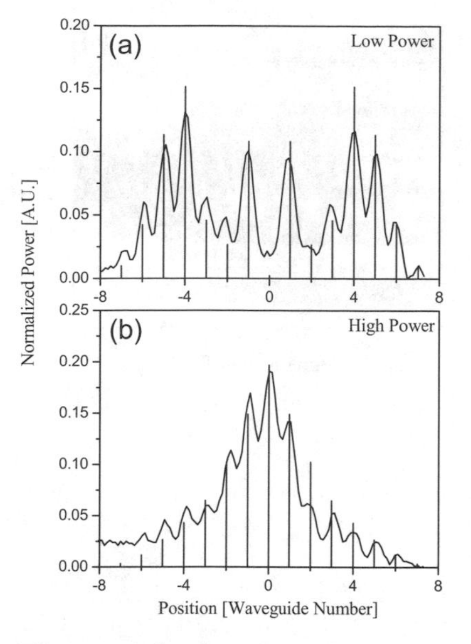

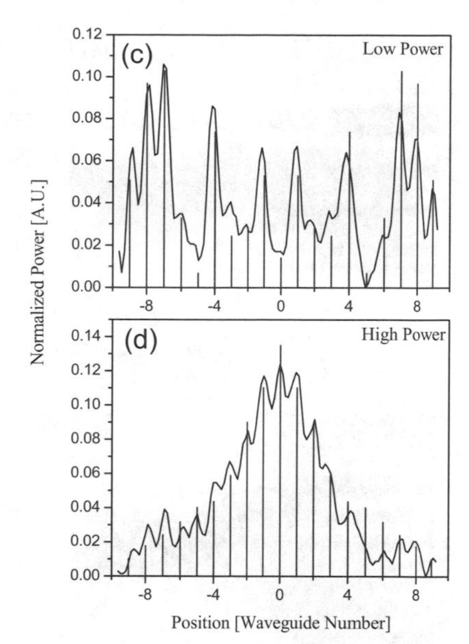

Fig. 15.5. Single waveguide excitation: Experimental and numerical results for samples with $d = 11\,\mu\mathrm{m}$, total propagation distance of 1.9 coupling lengths (**a,b**) and $9\,\mu\mathrm{m}$, total propagation distance of 3.0 coupling lengths (**c,d**). Both experimental results (*solid line*) and numerical results (*vertical lines*) are shown. The integrated power is normalized to unity

of launching light distribution that does not match exactly the central waveguide mode and due to cross-talk between the pixels of the IR camera. When the power is increased, the soliton formation forces a smoother phase relation between light in neighboring waveguides and the mode visibility decreases.

In order to demonstrate a power stability regime of the discrete soliton, we present in Fig. 15.6a–c power evolutions of the output profiles for the two cases of Fig. 15.5 and for the case of Fig. 15.4. Each vertical slice is a power cross-section at a respective power. At low power, the light distribution is broad and most of the light is far from the input waveguide. As the power is increased, light is gradually confined to the center. The important fact to note is that after achieving a certain distribution width, the focusing process almost arrests, and only a slight variation of the width with power is observed. It occurs at about 600 W for all the samples.

15.3.4 Formation of Discrete Solitons from Various Excitations

Until now, we only described discrete solitons that were formed by launching light into a single waveguide. This is equivalent to exciting uniformly all available spatial frequencies, because a single waveguide represents a discrete δ-function excitation. The single waveguide excitation is the case where the

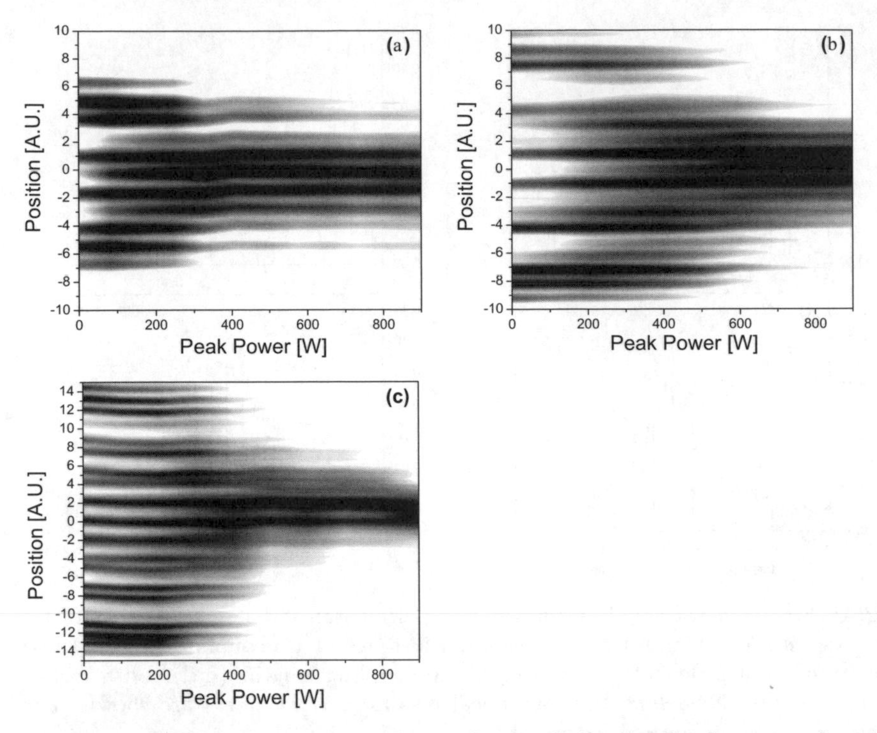

Fig. 15.6. Single waveguide excitation: Output light distributions as a function of the input peak power. Vertical cross-sections are the different power profiles corresponding to each input power. The three frames (**a**–**c**) are taken from samples of 1.9, 3.0 and 4.2 coupling lengths, respectively

discrete nature of the waveguide array is most pronounced. As can be seen in Fig. 15.2, the continuous and discrete diffraction curves have the same curvature for small spatial frequencies. Only when high enough frequencies are excited in the array, discreteness can be pronounced.

Solitons are not affected by such considerations. They are formed in a continuous slab waveguide as well as in an array, independently of their width, as long as enough power is supplied. We present here the results for excitation with various beam widths and shapes. As the launched beam becomes wider and smoother, the narrower is the excited frequency band and the beam experiences diffraction to a smaller extent.

To launch wide input fields, we used arrays with power splitters as input channels. The power splitters were multiple Y-junctions, which split the launched power from one waveguide, almost equally into several waveguides. The output of the splitter was injected into the waveguide array such that each splitter branch is coupled into one waveguide. As a result, the launched light distribution is a discrete rectangular function. It is narrower in its fre-

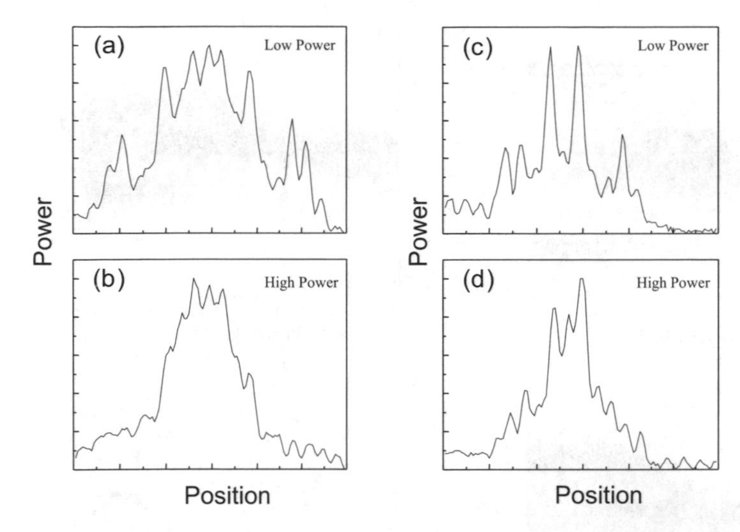

Fig. 15.7. Wide excitation by power splitting: Output profiles of three waveguides wide (**a,b**) and five waveguides wide (**c,d**) excitations. All the samples are of 3.0 coupling lengths

quency content, but due to the sharp distribution edges, it still contains all of the high spatial frequencies.

Figure 15.7 presents the results for low and high power when the power was split into three and five waveguides. The output profiles for a three waveguides input are plotted on Fig. 15.7a,b and for five on Fig. 15.7c,d. The array spacing is 9 μm for both samples, which results in 3.0 coupling-lengths propagation. In the linear, low power cases, we see less expansion due to reduced diffraction of the light. In the single input case the light occupied about 20 waveguides after propagating through the sample, while for the three and five waveguides excitation, it expands to about 17 and 11 waveguides, respectively. For low power, we still see the character of discreteness: some power concentration at the edges and a peaky nature at its center due to out-of-phase relations between the light in neighboring waveguides. Increasing the power causes both distributions to become somewhat narrower. There is a strong effect on the waveguide phases. They are locked together by the nonlinearity as a discrete soliton is formed, and the mode visibility is degraded.

It is also possible to launch a broad light distribution without any sharp features to avoid excitation of high spatial frequencies. These distributions are limited around the center of the diffraction curve, and therefore should behave much like in a continuous system. We inject the light directly into the array after shaping the input beam as a wide ellipse through cylindrical optics. Thc bcam overlaps a few waveguides and the relative power in each of them varies according to the beam gaussian profile. The results of such two

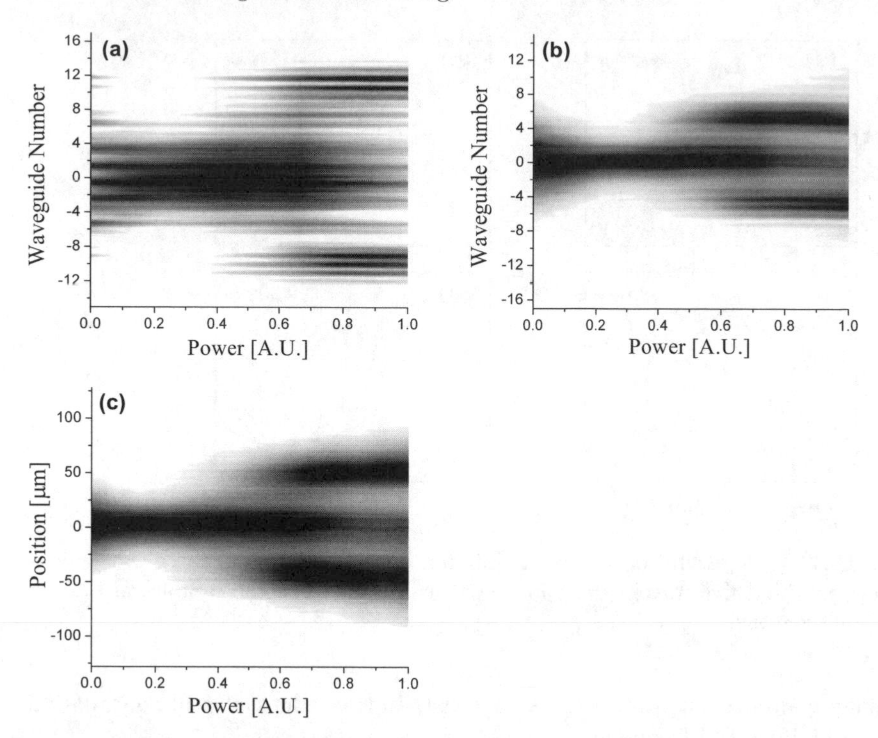

Fig. 15.8. Wide excitation by a wide smooth beam: Output light distributions as a function of the input peak power as in Fig. 15.5. **a** A two waveguides wide excitation. **b** A three waveguides wide excitation. **c** A three waveguides wide excitation of a continuous slab waveguide

experiments are presented in Fig. 15.8. Beams covering about two and three waveguides were tested and their output profiles as a function of input power are shown in Fig. 15.8a,b, respectively. As before, we observe the formation of discrete solitons. However, they now behave more like regular continuous solitons as their profile is bell-shaped and only their width is changing as the power is increased. As with continuous one-dimensional solitons, less power is required to form a wider soliton, therefore the minimal width is reached for weaker intensities than in the single waveguide input case.

Wide discrete solitons exhibit an interesting effect at a power of twice their formation power. Instead of just becoming narrower, when there is enough power for the formation of two solitary beams, the discrete soliton breaks into two. This result has been explained for the continuous case as an instability of the second soliton [51]. The cause for this instability in our case is probably a perturbation to the soliton shape by two- and three-photon absorption. The splitting of a three waveguides wide discrete soliton and of a continuous soliton of a similar width in a planar slab waveguide are compared in Fig. 15.8b,c, respectively. The two cases are clearly very similar, and a three

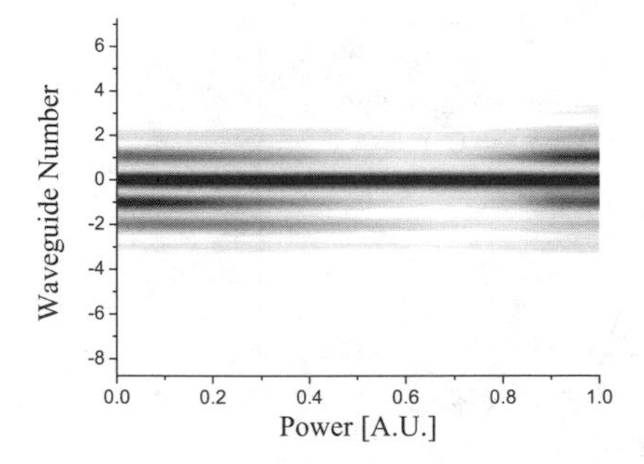

Fig. 15.9. Nonlinear contraction of power into a single waveguide

waveguides excitation is almost indistinguishable from a continuous soliton. Recall that a single waveguide excites all the frequencies of the Brillouin zone (see Fig. 15.2b), between $-\pi$ and π, while a two waveguides wide input would excite only half of this range ($-\pi/2$ to $\pi/2$) which has only one curvature sign. It is not surprising then that by launching a light distribution which is three waveguides wide ($-\pi/3$ to $\pi/3$), even less spectrum is generated, hence the behavior is very similar to continuum.

Using specific conditions of wide excitation and very weak coupling between the waveguides, we achieved contraction of most of the light into a single central waveguide. The output distribution as a function of input power is depicted in Fig. 15.9. By increasing the power, the distribution narrows until all the light is concentrated in a single waveguide. By further increasing the power, the light couples again into more waveguides. It is explained by nonlinear two-photon attenuation and self-phase modulation induced broadening, which are enhanced at the large intensities that are obtained when all the power is concentrated in one waveguide.

In all of the experiments described above, the main reason for differences between the results and the predictions of simple coupled-mode theory (CMT) is the temporal evolution of the pulsed input field [14,16]. In order to examine the pulse effects on the results, we recorded auto-correlation traces of a discrete soliton. A slit was used to select light from the soliton center or from its wing, a few waveguides away from the input center. A comparison between the two results, together with the input pulse auto-correlation is shown in Fig. 15.10a. The pulse length at the soliton center is doubled due to the normal dispersion of AlGaAs and the nonlinear effect of Self-Phase Modulation (SPM) [52]. At the soliton wing, there is an evidence for temporal splitting. The reason for this splitting is the depletion of light from the wings when the stronger spatial focusing occurred near the peak of the pulse. To illustrate this pulse shape, we draw a top view of the pulse in Fig. 15.10b. The spectral broadening of the pulse center is a measurement for the importance of SPM on the temporal evolution. In Fig. 15.10c the spectra of the input pulse and the output pulse center are compared. The difference in widths is

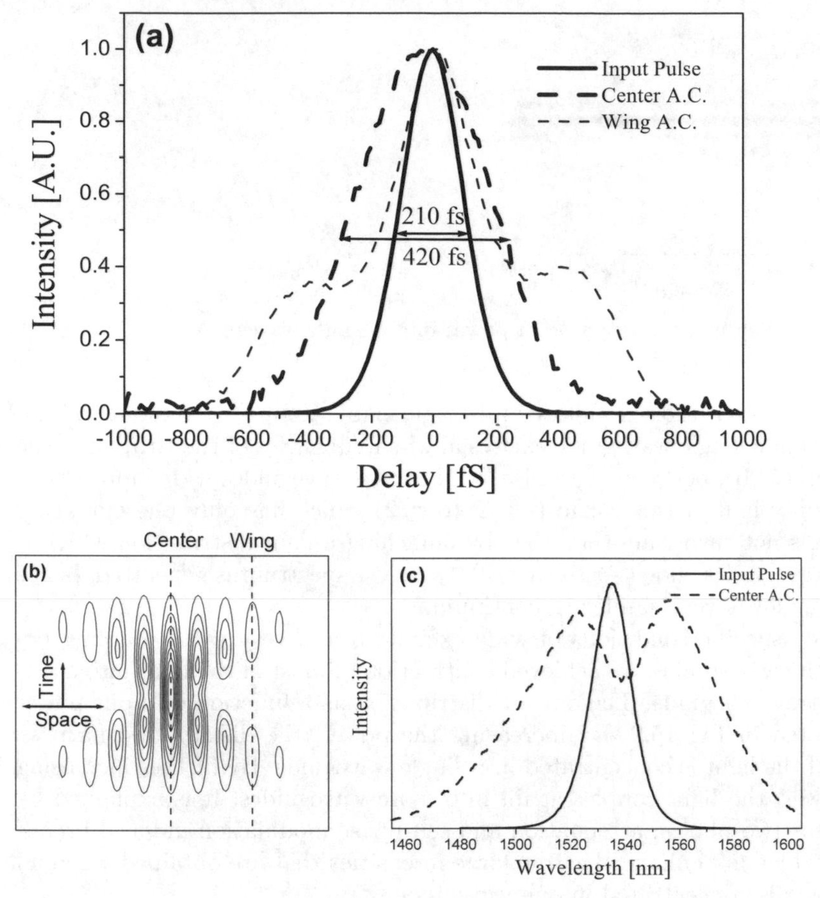

Fig. 15.10. a Temporal auto-correlation of a discrete soliton: the input pulse (*solid line*), the output pulse center (*thick dashed line*) and the output pulse wing (*thin dashed line*). **b** The output pulse sketch. **c** A comparison between the spectra of the input (*solid line*) and the output (*dashed line*) pulses

clear and we can estimate from the oscillatory profile of the output spectrum that SPM introduced a nonlinear phase shift of about $3\pi/2$.

15.3.5 Studies of Discrete Soliton Dynamics

We divide the experiments involving dynamics of discrete solitons in waveguide arrays into two categories [31]. In one type of experiments, a phase gradient is applied along the profile of the input beam, thus giving it transverse kinetic energy (the first term in (15.11)) and forcing the soliton to move across the array. In the other kind of experiments, the beam is launched along the array waveguides while its input position is scanned in between two adjacent waveguides. The experiments of the first category examine the rotational

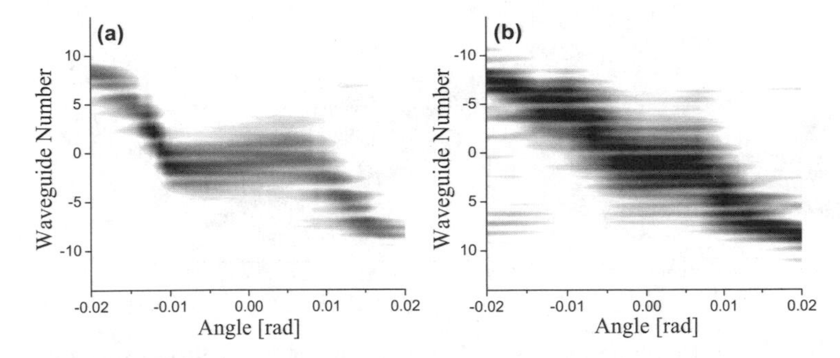

Fig. 15.11. Discrete soliton power steering: Output light distributions as a function of the input angle of **a** two and **b** three waveguides wide initial beams. In the two waveguides wide case, the locking for angles smaller than 0.01 radian is very clear

symmetry breaking in discrete systems, while those of the second category investigate the translational symmetry breaking.

We recorded the output light distribution for various input angles at the power required for launching a soliton. Rotating a glass window positioned in front of the input objective controls the input angle. The results for two- and three-waveguides wide excitations are presented in Fig. 15.11a and b, respectively. Imposing a phase gradient across a single waveguide input has no meaning because phase difference between different waveguides is required for beam steering. Hence a beam coupled to a single guide always move along the waveguides. In the other extreme, the continuous case of a slab waveguide, we expect the output center to move linearly with input angle changes. In the intermediate cases of discrete solitons that extends over a few guides, power dependent locking [8] is expected due to the PN potential [6]. In Fig. 15.11a we note that although small angles are imposed on the input beam, its output center position does not change. Only beyond a critical angle where the kinetic energy overcomes the PN energy barrier, the soliton starts to move across the array. As the input beam becomes broader, the shallower PN potential allows the escape of the discrete soliton at a smaller angle. The smaller locking region in Fig. 15.11b compared to Fig. 15.11a demonstrates this effect.

An interesting result was obtained when the last experiment was repeated with input power which was enough for a second soliton. At a zero angle, the soliton brakes into two, as was shown in Fig. 15.8a–c. When a tilt is introduced, symmetry is broken and more power goes to one of the two parts. The transition of power between the two parts for a two waveguides wide excitation is clearly seen in Fig. 15.12a. In the case of a wider three waveguides beam (Fig. 15.12b), the splitting is even suppressed at large tilts. A single soliton is formed and its center depends linearly on the input angle.

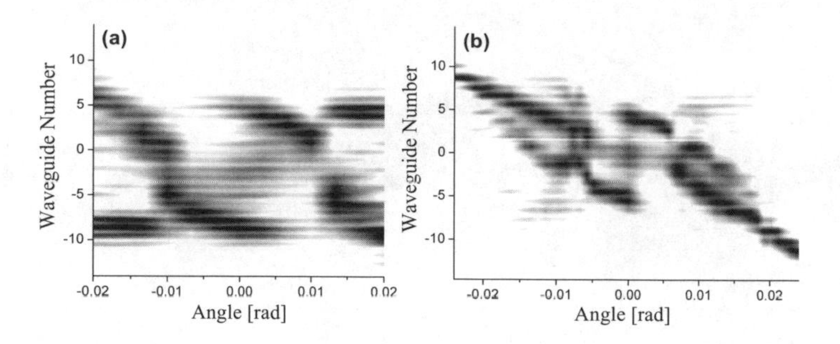

Fig. 15.12. Repeating the experiments of Fig. 15.11 for a two-soliton excitation. The soliton symmetric splitting is broken

Results from experiments of the second kind, with beams with flat phase fronts, are presented in Figs. 15.13 and 15.14. In these experiments the beam is launched along the waveguides direction while its input position is scanned transversally across the array. The sample is positioned on top of a piezo-driven stage and by changing the voltage on the piezo-electric element from zero to 1000 V, the sample is shifted by 12 μm. First, the power was fixed such that a single soliton has been formed. Output profiles for three different input widths of one, two and three waveguides are shown in Fig. 15.13a–c, respectively. We observed an amplification of the lateral translation of the beam, which is induced by discreteness. The nonlinear beam is interacting with the PN potential and as a result it is deflected into large angles. The deflection angle passes zero when the beam is centered on a waveguide (position zero in Fig. 15.13a) and exactly in the middle between two waveguides at ±4.5 μm, as can be expected from symmetry reasons. These two states correspond to the two extrema of the PN potential. In all other input positions, a non-uniform division of the power between two waveguides occurs. After a short propagation distance, the waveguide containing more power acquires more phase shift than the other one. The nonlinearly induced phase difference between the two waveguides causes propagation in an angle towards the waveguide containing more light, as we demonstrated before when we tilted the beam deliberately. The difference between the stable minima and the unstable maxima of the PN potential is manifested in the fast crossing of the output distribution at ±4.5 μm compared to the slow crossing around zero translation. For wider beams, the PN potential is shallower and the effect is reduced. The effect of a two waveguides wide input is small (Fig. 15.13b) while there is none for the three waveguides wide input (Fig. 15.13c).

The effect of input translation on the case where there is enough power for two-solitons is similar to the effect of input rotation. As before, symmetry between the two parts of the beam is broken. Translation has a periodic effect because of the periodicity of the array. Therefore, light is repeatedly concen-

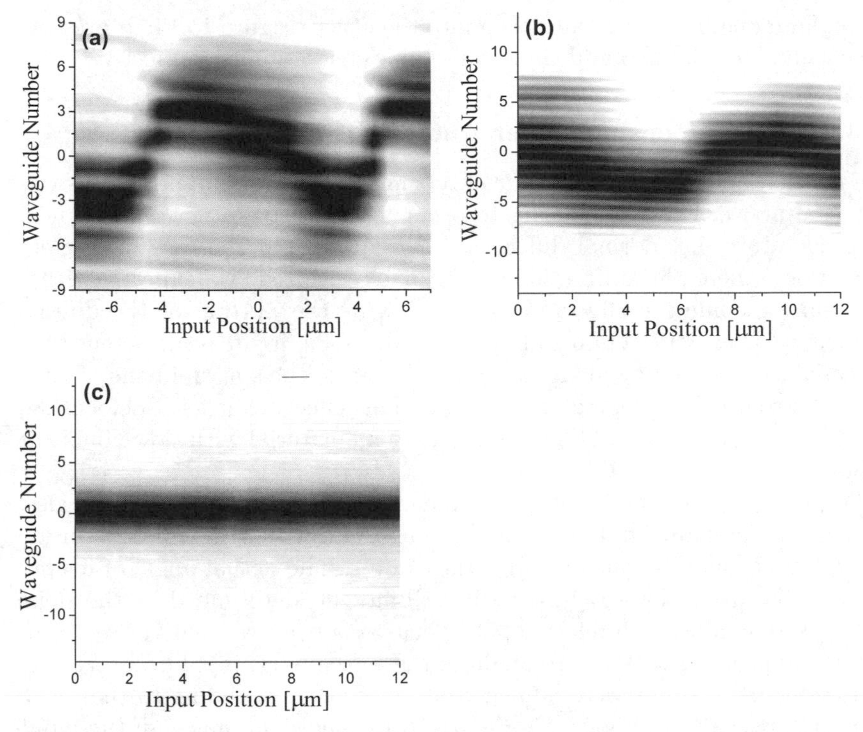

Fig. 15.13. Discrete soliton translation induced steering: **a–c** Output light distributions as a function of the input position for one, two and three waveguide excitation, respectively. Amplified oscillations of the output position are observed. The oscillations are clearly induced by discreteness as they disappear while the excitation becomes wider

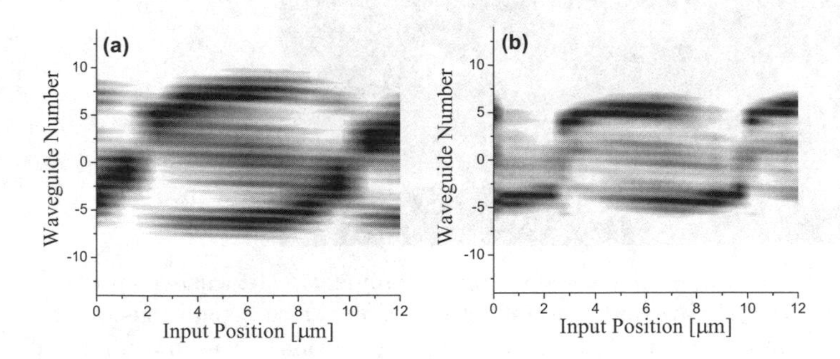

Fig. 15.14. Repeating the experiments of Fig. 15.13 for a two-soliton excitation power. **a,b** Results for inputs of two and three waveguides width, respectively. The soliton symmetric splitting is broken

trating into one beam part and then into the other. Figure 15.14a,b presents these results for the two and three waveguides wide inputs, respectively.

15.3.6 Self-Defocusing Under Anomalous Diffraction

The relation between the nonlinear effects and the diffraction properties was discussed above. For self-focusing to occur in a positive n_2 medium, a negative curvature, i.e. normal diffraction, is required. In the case of discrete diffraction, where the diffraction sign can be inverted, we expect a richer spectrum of nonlinear effects. In particular, at the bottom of the diffraction curve, where diffraction sign is inverted, as the input beam intensity is increased, the excited k_x modes are pushed up into the spectral band, resulting in a stronger broadening of the beam. This effect, called self-defocusing, could be otherwise observed only in negative n_2 material with slow nonlinear response [5].

Applying tilts to the input beam launches different groups of k_x modes. Because of the high refractive index of AlGaAs ($n_0 \approx 3.3$) the angle inside the slab waveguide is much smaller then in air. The actual angle of energy propagation inside the array may still be different, determined by the slope of the discrete diffraction relation [30]. Thus, for a beam around $k_x = \pi/d$ energy is propagating with a zero angle to the waveguides, just like for $k_x = 0$. Results for beams that were launched at angles corresponding to these two cases are presented in Fig. 15.15. Figure 15.15a shows an image of the input beam. Figures 15.15b,c are images of the output facet when $k_x = 0$ for low

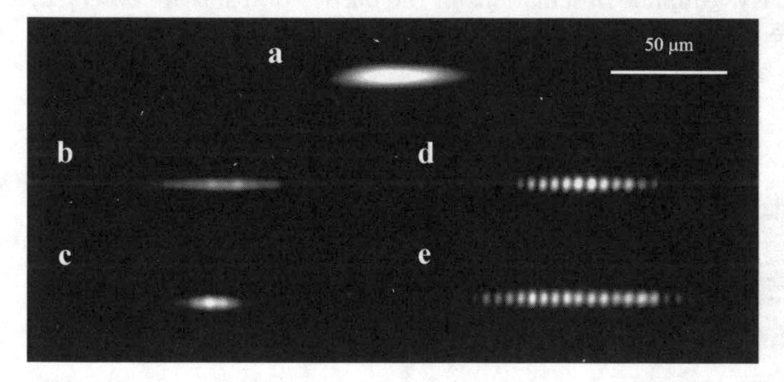

Fig. 15.15. Experimental results showing both nonlinear self-focusing and self-defocusing in an array of waveguides, for slightly different initial conditions. **a** The input beam, ~ 35 mm wide at FWHM. **b** Light distribution at the output facet for normal dispersion in the linear regime. The beam slightly broadens through discrete diffraction. **c** At high power ($I_{\text{peak}} \sim 150$ W), the field shrinks and evolves into a discrete bright soliton. **d** For an anomalous dispersion condition, when the beam is injected at an angle of $2.6 \pm 0.4°$ inside the array, it broadens slightly as in **b**. Note the dark lines between the optical modes resulting from the π-phase flips between adjacent waveguides. **e** When the power is increased ($I_{\text{peak}} \sim 100$ W), the distribution broadens significantly due to self-defocusing

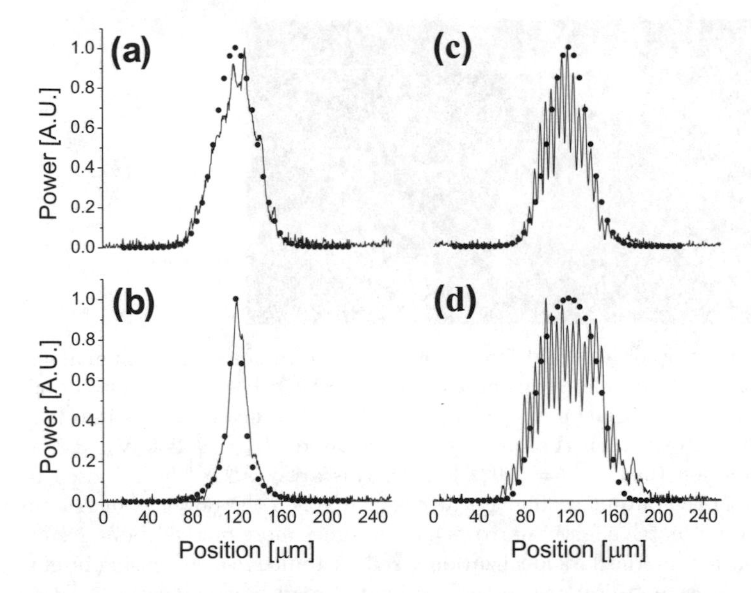

Fig. 15.16. Comparison between cross-sections of the experimental results of Fig. 15.2 (*solid line*) and numerical solutions of coupled mode theory (*solid circles*, each represents the power in a single waveguide). **a,b** Normal discrete diffraction condition, low and high power, respectively. **c,d** Anomalous diffraction condition, low and high power, respectively. Note the difference between nonlinear focusing (**b**) and defocusing (**d**)

and high input powers, respectively. As can be seen, in this condition of normal diffraction, self-focusing which leads to the formation of bright solitons occurs. Figures 15.15d,e are again for low and high power, but when $k_x = \pi/d$ and the diffraction is anomalous. The maximal mode visibility in this case is an indication for the π phase jumps between adjacent guides. When low optical power is injected, the beam is slightly broadened, as in the normal case. The difference is obvious when the power is increased. Instead of focusing, which is expected in such positive n_2 material, the light distribution expands considerably. Cross-sections of the results are compared to a numerical integration of (15.8) (Fig. 15.16). The experimental results prove to be in a very good agreement with this simple theory.

In the anomalous diffraction regime we should expect to observe dark solitons. A dark soliton is a constant illumination in space, with a central dark notch which is a result of a π phase flip [53]. At low powers, the dark zone broadens due to diffraction. If the right conditions are fulfilled, a stable dark notch in a constant background is formed at sufficiently high power – a dark soliton.

Dark solitons can be formed in two cases: the first is when diffraction is normal and n_2 is negative and the second is when diffraction is anomalous

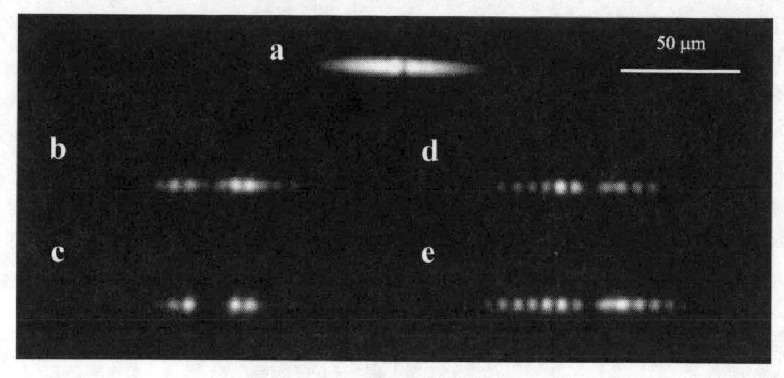

Fig. 15.17. Generation of a dark discrete solitary wave in the case of anomalous diffraction. **a** The input profile, $\sim 40\,\mu\text{m}$ wide at FWHM. **b** For normal diffraction at low power, a notch is visible in the output profile. **c** The beam evolves into two repulsive bright solitons when the intensity is increased ($I_{\text{peak}} \sim 250\,\text{W}$). **d** For anomalous diffraction (beam tilt $= 2.0 \pm 0.4°$ in this array), the "dark" notch is initially present in the output profile (linear case). **e** When the power is increased ($I_{\text{peak}} \sim 250\,\text{W}$), the notch slightly narrows and becomes more marked. As a result of anomalous diffraction, the dark localization is self-sustained in a defocusing bright background, and does not disappear when the beam broadens nonlinearly

and n_2 is positive. As we have the right conditions for the second options, we tried to launch discrete dark solitons in our samples of waveguide arrays [18,19]. A π phase mask was applied on the beam right before entering the sample. The shape of the input beam with the dark notch is presented in Fig. 15.17a. As in Fig. 15.15, we show results of normal incidence ($k_x = 0$) in Fig. 15.15b,c and results of the anomalous diffraction condition ($k_x = \pi/d$) in Fig. 15.15d,e. In the normal regime, the low power notch (Fig. 15.15b) is broader than one waveguide and somewhat blurred. At power that is sufficient to form a bright soliton (Fig. 15.15c), the notch defocuses and becomes broader and more pronounced, while the finite bright background is focusing, as is expected in this regime. On the other hand, in the anomalous case the dark notch is preserved with power (Fig. 15.15e) while the finite bright background undergoes self-defocusing. These results are not a discrete dark soliton in a strictly manner because the background is not a constant illumination. Even more, in some of the results (Fig. 15.15b–d) the bright and dark widths are on a comparable scale. Nevertheless, they show some significant hints for this solitary phenomenon.

15.3.7 Interaction of a Linear Defect State with Discrete Diffraction

There is a well-known way to form a linearly localized state in a discrete system. In an infinite coupled array, when one waveguide has its propagation constant changed to $\beta + \Delta\beta$, a confined mode around this site is created [10,11]. Usually, changing the propagation constant β would also affect the coupling constant to become $C + \Delta C$. The confined defect mode would have the shape:

$$E_n(z) = \frac{E^0}{\rho^{|n|}} \left(1 + \frac{\Delta C}{C}\right) \exp\left(\mathrm{i}\beta_{\mathrm{defect}} z\right) , \tag{15.13}$$

where the defect mode propagation constant is $\beta_{\mathrm{defect}} = \beta + (\rho + 1/\rho)\, C$ and the transversal decay rate ρ is defined as:

$$\rho = \frac{\Delta\beta}{2C} - \sqrt{\left(\frac{\Delta\beta}{2C}\right)^2 + 2\left(1 + \frac{\Delta C}{C}\right)^2 - 1} \, . \tag{15.14}$$

When light is launched into such a linear mode, the high power result depends on the value of β_{defect} compared to the discrete band $\{\beta - 2C < k_z < \beta + 2C\}$. If the defect mode is above the band ($\beta_{\mathrm{defect}} > \beta + 2C$), then the Kerr effect will just raise it higher, without any dramatic consequences. When β_{defect} is in the band, the usual self-focusing is expected, similarly to the discrete soliton case. More interesting case is when the linear defect mode is below the band ($\beta_{\mathrm{defect}} < \beta - 2C$). In the linear regime there is confinement, as outlined before. As the injected power increases, the mode propagation constant enters the optical band, interacting linearly with it and enabling the dispersion of energy into more waveguides. It is the opposite effect of nonlinear self-focusing, but in this case, without the inversion of diffraction.

We tested a sample of waveguide array with a separation of d=9 µm. Each waveguide was 4 µm wide except for the central waveguide ($n = 0$) which was 2.5 µm wide. This difference caused a local change of the parameters such that $\Delta\beta = -1.5C$ and $\Delta C = 0.3C$, resulting in a confined linear defect mode below the optical band ($\beta_{\mathrm{defect}} = \beta - 2.9C$). The calculated decay rate is negative $\rho = -2.5$, forming π phase jumps between adjacent waveguides. The confined mode which was obtained by launching low power light only into the central waveguide is presented in Fig. 15.18a. The π phase jumps give rise to the strong visibility of the three central waveguides modes. The weak wings are formed because the input light did not overlap completely the defect mode, therefore leading to radiation evolving through usual discrete diffraction. As the power is increased (Fig. 15.18b), The defect mode starts to interact with the diffraction band, and light is escaping out. The light distribution is becoming wider from about three waveguides to about seven. The same results are demonstrated numerically in Fig. 15.18c,d.

15.4 Conclusion

In this chapter we have reviewed the progress in the studies of nonlinear waveguide arrays. We have shown how this problem can be treated by a simple discrete model, which predicts many of the key features of linear and

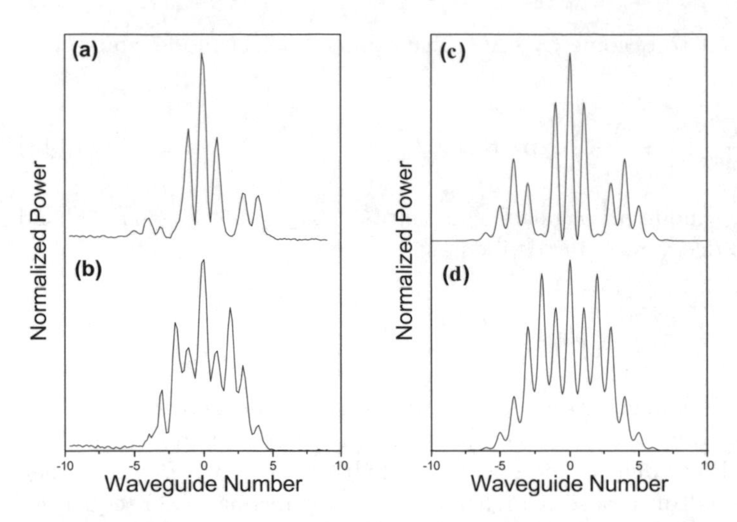

Fig. 15.18. Experimental results of the defect sample for **a** low and **b** high power input. Numerical simulations for the same conditions are presented in **c** and **d**, respectively

nonlinear propagation in such arrays. Furthermore, discrete diffraction offers several new phenomena that stem from the periodicity of the structure. Discrete diffraction can be controlled and managed in a way not possible with continuous diffraction. In particular, the sign and magnitude of diffraction can be varied, much like the parameters of dispersion in optical fibers.

Although we have not discussed any specific applications of the structures or phenomena described here, it is obvious that these waveguide arrays offer several advantages as compared with the more familiar nonlinear slab waveguides. Let us just mention the inherent compatibility with fiber and waveguide devices for input or output, and the possibility to engineer the strength of diffraction for a particular power requirement.

Acknowledgement. The authors wish to thank many of their colleagues who contributed to the work that have been described in this chapter, and primarily to their long-term collaborators Stewart Aitchison and Roberto Morandotti. Among the many others who contributed to different phases of this work are Ulf Peschel, Daniel Mandelik, J.H. Arnold, M. Sorel, G. Pennelli, P. Millar and A.R. Boyd.

References

1. A.L. Jones: J. Opt. Soc. Am. **55**, 261 (1965)
2. S. Somekh, E. Garmire, A. Yariv, H.L. Garvin, and R.G. Hunsperger: Appl. Phys. Lett. **22**, 46 (1973)
3. D.N. Christodoulides and R.I. Joseph: Opt. Lett. **13**, 794 (1988)
4. Y.S. Kivshar and M. Peyrard: Phys. Rev. A **46**, 3198 (1992)
5. Y.S. Kivshar: Opt. Lett. **18**, 1147 (1993)
6. Y.S. Kivshar and D.K. Campbell: Phys. Rev. E **48**, 3077 (1993)
7. W. Królikowsky, U. Trutschel, M. Cronin-Golomb, and C. Schmidt-Hattenberger: Opt. Lett. **19**, 320 (1994)
8. A.B. Aceves, C. De Angelis, S. Trillo, and S. Wabnitz: Opt. Lett. **19**, 332 (1994)
9. A.B. Aceves, C. De Angelis, T. Peschel, R. Muschall, F. Lederer, S. Trillo, and S. Wabnitz: Phys. Rev. E **53**, 1172 (1996)
10. W. Królikowsky and Y.S. Kivshar: J. Opt. Soc. Am. B **13**, 876 (1996)
11. U. Peschel, R. Morandotti, J.S. Aitchison, H.S. Eisenberg, and Y. Silberberg: Appl. Phys. Lett. **75**, 1348 (1999)
12. R. Muschall, C. Schmidt-Hattenberger, and F. Lederer: Opt. Lett. **19**, 323 (1994)
13. I. Relke: Phys. Rev. E **57**, 6105 (1998)
14. A.B. Aceves, C. De Angelis, A.M. Rubenchik, and S.K. Turitsyn: Opt. Lett. **19**, 329 (1994)
15. A.B. Aceves, C. De Angelis, G.G. Luther, and A.M. Rubenchik: Opt. Lett. **19**, 1186 (1994)
16. A.B. Aceves, C. De Angelis, G.G. Luther, A.M. Rubenchik, and S.K. Turitsyn: Physica D **87**, 262 (1995)
17. A.B. Aceves, G.G. Luther, C. De Angelis, A.M. Rubenchik, and S.K. Turitsyn: Phys. Rev. Lett. **75**, 73 (1995)
18. Y.S. Kivshar, W. Królikowsky, and O.A. Chubycalo: Phys. Rev. E **50**, 5020 (1994)
19. M. Johansson and Y. Kivshar: Phys. Rev. Lett. **82**, 85 (1999)
20. S. Darmanyan, A. Kobyakov, E. Schmidt, and F. Lederer: Phys. Rev. E **57**, 3520 (1998)
21. S. Darmanyan, A. Kobyakov, and F. Lederer: JEPT **86**, 682 (1998)
22. Y.S. Kivshar, A.R. Champneys, D. Cai, and A.R. Bishop: Phys. Rev. B **58**, 5423 (1998)
23. W.P. Su, J.R. Schieffer, and A.J. Heeger: Phys. Rev. Lett. **42**, 1698 (1979)
24. A.S. Davydov: Phys. Scr. **20**, 387 (1979)
25. T. Holstein: Ann. Phys. **8**, 325 (1959); Mol. Cryst. Liq. Cryst. **77**, 235 (1981)
26. B. Denardo, B. Galvin, A. Greenfield, A. Larraza, S. Putterman, and W. Wright: Phys. Rev. Lett. **68**, 1730 (1992)
27. P. Marquii, J.M. Bilbaut, and M. Remoissenet: Phys. Rev. E **51**, 6127 (1995)
28. S. Flach and C.R. Willis: Phys. Rep. **295**, 182 (1998)
29. H.S. Eisenberg, Y. Silberberg, R. Morandotti, A.R. Boyd, and J.S. Aitchison: Phys. Rev. Lett. **81**, 3383 (1998)
30. R. Morandotti, U. Peschel, J.S. Aitchison, H.S. Eisenberg and Y. Silberberg, Phys. Rev. Lett. **83**, 2726 (1999)
31. R. Morandotti, U. Peschel, J.S. Aitchison, H.S. Eisenberg, and Y. Silberberg: Phys. Rev. Lett. **83**, 2726 (1999)

32. U. Peschel, T. Pertsch, and F. Lederer: Opt. Lett. **23**, 1701 (1998)
33. G. Lenz, I. Talanina, and C.M. de Sterke: Phys. Rev. Lett **83**, 963 (1999)
34. R. Morandotti, U. Peschel, J.S. Aitchison, H.S. Eisenberg, and Y. Silberberg: Phys. Rev. Lett. **83**, 4756 (1999)
35. T. Pertsch, P. Dannberg, W. Elflein, A. Bräuer, and F. Lederer: Phys. Rev. Lett. **83**, 4752 (1999)
36. H.S. Eisenberg, Y. Silberberg, R. Morandotti, and J.S. Aitchison: Phys. Rev. Lett. **85**, 1863 (2000)
37. A. Yariv: *Quantum Electronics*, 3rd edn. (Wiley & Sons, New York Chichester 1988)
38. S. Kawakami and H.A. Haus: J. Light. Tech. **4**, 160 (1986)
39. P.S.J. Russell: Appl. Phys. B **39**, 231 (1986)
40. P.S.J. Russell: Phys. Rev. A **33**, 3232 (1986)
41. N.W. Ashcroft and N.D. Mermin: *Solid State Physics* (Holt, Rinehart & Winston, New York 1976)
42. V.E. Zakharov and A.B. Shabat: Zh. Eksp. Teor. Fiz. **61**, 118 (1971) [V.E. Zakharov and A.B. Shabat: Sov. Phys. JEPT 34, **62** (1972)]
43. A. Barthelemy, S. Maneuf, and C. Froehly: Opt. Commun. **55**, 201 (1985)
44. J.S. Aitchison, Y. Silberberg, A.M. Weiner, D.E. Leaird, M.K. Oliver, J.L. Jackel, E.M. Vogel, and P.W.E. Smith: J. Opt. Soc. Am. B **8**, 1290 (1990)
45. L.F. Mollenauer, R.H. Stolen, and J.P. Gordon: Phys. Rev. Lett. **45**, 1095 (1980)
46. A.C. Scott and L. Macneil: Phys. Lett. A **98**, 87 (1983)
47. A.B. Aceves, C. De Angelis, A.M. Rubenchik, and S.K. Turitsyn: Opt. Lett. **19**, 329 (1994)
48. C. Schmidt-Hattenberger, U. Trutschel, R. Muschall, and F. Lederer: Opt. Comm. **82**, 461 (1991)
49. P. Millar, J.S. Aitchison, J.U. Kang, G.I. Stegeman, A. Villeneuve, G.T. Kennedy, and W. Sibbett: J. Opt. Soc. Am. B **14**, 3224 (1997)
50. J.S. Aitchison, D.C. Hutchings, J.U. Kang, G.I. Stegeman, and A. Villeneuve: IEEE J. Quant. Elect. **33**, 341 (1997)
51. Y. Silberberg: Opt. Lett. **15**, 1005 (1990)
52. G.P. Agrawal: *Nonlinear Fiber Optics* (Academic Press, San Diego 1995)
53. G.A. Swartzlander Jr., D.R. Andersen, J.J. Regan, Y. Hin, and A.E. Kaplan: Phys. Rev. **66**, 1583 (1991)

16 Nonlinear Localized Modes in 2D Photonic Crystals and Waveguides

S.F. Mingaleev and Y.S. Kivshar

16.1 Introduction

Photonic crystals are usually viewed as an optical analog of semiconductors that modify the properties of light similar to a microscopic atomic lattice that creates a semiconductor band-gap for electrons [1]. It is therefore believed that by replacing relatively slow electrons with photons as the carriers of information, the speed and band-width of advanced communication systems will be dramatically increased, thus revolutionizing the telecommunication industry. Recent fabrication of photonic crystals with a band gap at optical wavelengths from 1.35 µm to 1.95 µm makes this promise very realistic [2].

To employ the high-technology potential of photonic crystals, it is crucially important to achieve a dynamical tunability of their band gap [3]. This idea can be realized by changing the light intensity in the so-called *nonlinear photonic crystals*, having a periodic modulation of the nonlinear refractive index [4]. Exploration of *nonlinear properties* of photonic band-gap (PBG) materials is an important direction of research that opens new applications of photonic crystals for all-optical signal processing and switching, allowing an effective way to create tunable band-gap structures operating entirely with light [5].

One of the important physical concepts associated with nonlinearity is *the energy self-trapping and localization.* In the linear physics, the idea of localization is always associated with disorder that breaks translational invariance. However, during the recent years it was demonstrated that localization can occur in the absence of any disorder and solely due to nonlinearity in the form of *intrinsic localized modes* [6]. A rigorous proof of the existence of time-periodic, spatially localized solutions describing such nonlinear modes has been presented for a broad class of Hamiltonian coupled-oscillator nonlinear lattices [7], but approximate analytical solutions can also be found in many other cases, demonstrating a generality of the concept of *nonlinear localized modes.*

Nonlinear localized modes can be easily identified in numerical molecular-dynamics simulations in many different physical models (see, e.g., [6] for a review), but only very recently the *first experimental observations* of spatially localized nonlinear modes have been reported in mixed-valence transition metal complexes [8], quasi-one-dimensional antiferromagnetic chains [9], and arrays of Josephson junctions [10]. Importantly, very similar types of spatially localized nonlinear modes have been experimentally observed in *macroscopic* mechanical [11] and guided-wave optical [12] systems.

Recent experimental observations of nonlinear localized modes, as well as numerous theoretical results, indicate that nonlinearity-induced localization and spatially localized modes can be expected in physical systems of very different nature. From the viewpoint of possible practical applications, self-localized states in optics seem to be the most promising ones; they can lead to different types of nonlinear all-optical switching devices where light manipulates and controls light itself by varying the input intensity. As a result, the study of *nonlinear localized modes in photonic structures* is expected to bring a variety of realistic applications of intrinsic localized modes.

One of the promising fields where the concept of nonlinear localized modes may find practical applications is the physics of *photonic crystals* [or photonic band gap (PBG) materials] – periodic dielectric structures that produce many of the same phenomena for photons as the crystalline atomic potential does for electrons [1]. Three-dimensional (3D) photonic crystals for visible light have been successfully fabricated only within the past year or two, and presently many research groups are working on creating tunable band-gap switches and transistors operating entirely with light. The most recent idea is to employ nonlinear properties of band-gap materials, thus creating *nonlinear photonic crystals* including those where nonlinear susceptibility is periodic as well [4,13].

Nonlinear photonic crystals (or photonic crystals with embedded nonlinear impurities) create an ideal environment for the generation and observation of nonlinear localized photonic modes. In particular, the existence of such modes for the frequencies in the photonic band gaps has been predicted [14] for 2D and 3D photonic crystals with Kerr nonlinearity. Nonlinear localized modes can be also excited at nonlinear interfaces with quadratic nonlinearity [15], or along dielectric waveguide structures possessing a nonlinear Kerr-type response [16].

In this chapter, we study self-trapping of light and nonlinear localized modes in nonlinear photonic crystals and photonic crystal waveguides. For simplicity, we consider the case of a 2D photonic crystal with embedded nonlinear rods (impurities) and demonstrate that the effective interaction in such a structure is nonlocal, so that the nonlinear effects can be described by a nontrivial generalization of the nonlinear lattice models that include the long-range coupling and nonlocal nonlinearity. We describe several different types of nonlinear guided-wave states in photonic crystal waveguides and analyse their properties [17]. Also, we predict the existence of *stable* nonlinear localized modes (highly localized modes analogous to gap solitons in the continuum limit) in the reduced-symmetry nonlinear photonic crystals [18].

16.2 Basic Equations

Let us consider a 2D photonic crystal created by a periodic lattice of parallel, infinitely long dielectric rods in air (see Fig. 16.1). We assume that the rods are parallel to the x_3 axis, so that the system is characterized by the dielectric constant $\varepsilon(\boldsymbol{x}) = \varepsilon(x_1, x_2)$. As is well known [1], the photonic crystals of this type can possess a complete band gap for the E-polarized (with the electric field $\boldsymbol{E} || \boldsymbol{x}_3$) light propagating in the (x_1, x_2)-plane. The evolution of such a light is governed by the scalar wave equation

$$\nabla^2 E(\boldsymbol{x}, t) - \frac{1}{c^2} \partial_t^2 \left[\varepsilon(\boldsymbol{x}) E \right] = 0 \,, \tag{16.1}$$

where $\nabla^2 \equiv \partial_{x_1}^2 + \partial_{x_2}^2$ and E is the x_3 component of $\boldsymbol{E}$. Taking the electric field in the form $E(\boldsymbol{x}, t) = \mathrm{e}^{-\mathrm{i}\omega t} E(\boldsymbol{x}, t|\omega)$, where $E(\boldsymbol{x}, t|\omega)$ is a slowly varying envelope, i.e. $\partial_t^2 E(\boldsymbol{x}, t|\omega) \ll \omega \partial_t E(\boldsymbol{x}, t|\omega)$, (16.1) reduces to

$$\left[\nabla^2 + \varepsilon(\boldsymbol{x}) \left(\frac{\omega}{c} \right)^2 \right] E(\boldsymbol{x}, t|\omega) \simeq -2\mathrm{i}\varepsilon(\boldsymbol{x}) \frac{\omega}{c^2} \frac{\partial E}{\partial t} \,. \tag{16.2}$$

In the stationary case, i.e. when the r.h.s. of (16.2) vanishes, this equation describes an eigenvalue problem which can be solved, e.g. by the plane waves method [19], in the case of a perfect photonic crystal, for which the dielectric constant $\varepsilon(\boldsymbol{x}) \equiv \varepsilon_p(\boldsymbol{x})$ is a periodic function defined as

$$\varepsilon_p(\boldsymbol{x} + \boldsymbol{s}_{ij}) = \varepsilon_p(\boldsymbol{x}) \,, \tag{16.3}$$

where i and j are arbitrary integers, and

$$\boldsymbol{s}_{ij} = i\boldsymbol{a}_1 + j\boldsymbol{a}_2 \tag{16.4}$$

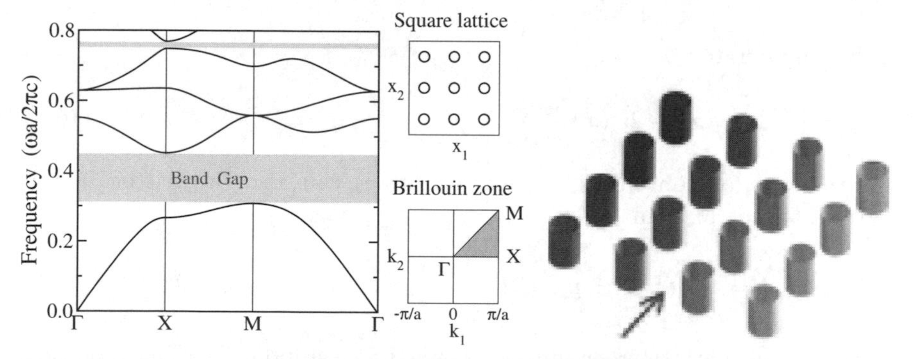

Fig. 16.1. The band-gap structure of the photonic crystal consisting of a square lattice of dielectric rods with $r_0 = 0.18a$ and $\varepsilon_0 = 11.56$ (the band gaps are shaded). The top center inset shows a cross-sectional view of the 2D photonic crystal depicted in the right inset. The bottom center inset shows the corresponding Brillouin zone, with the irreducible zone shaded

is a linear combination of the lattice vectors $\boldsymbol{a}_1$ and $\boldsymbol{a}_2$.

For definiteness, we consider the 2D photonic crystal earlier analyses (in the linear limit) in [20,21]. That is, we assume that cylindrical rods with radius $r_0 = 0.18a$ and dielectric constant $\varepsilon_0 = 11.56$ form a square lattice with the distance a between two neighboring rods, so that $\boldsymbol{a}_1 = a\boldsymbol{x}_1$ and $\boldsymbol{a}_2 = a\boldsymbol{x}_2$. The frequency band structure for this type of 2D photonic crystal is shown in Fig. 16.1 where, using the notations of the solid-state physics, the wave dispersion is mapped onto the Brillouin zone of the so-called *reciprocal lattice* that faces are known as Γ, M, and X. As follows from Fig. 16.1, there exists a large (38%) band gap that extends from the lower cut-off frequency, $\omega = 0.302 \times 2\pi c/a$, to the upper band-gap frequency, $\omega = 0.443 \times 2\pi c/a$. If the frequency of a low-intensity light falls into the band gap, the light cannot propagate through the photonic crystal and is reflected.

16.3 Defect Modes: The Green Function Approach

One of the most intriguing properties of photonic band gap crystals is the emergence of exponentially localized modes that may appear within the photonic band gaps when a defect is embedded into an otherwise perfect photonic crystal. The simplest way to create a defect in a 2D photonic crystal is to introduce an additional defect rod with the radius r_d and the dielectric constant $\varepsilon_d(\boldsymbol{x})$. In this case, the dielectric constant $\varepsilon(\boldsymbol{x})$ can be presented as a sum of periodic and defect-induced terms, i.e.

$$\varepsilon(\boldsymbol{x}) = \varepsilon_p(\boldsymbol{x}) + \varepsilon_d(\boldsymbol{x}) \,,$$

and, therefore, (16.2) takes the form

$$\left[\nabla^2 + \left(\frac{\omega}{c}\right)^2 \varepsilon_p(\boldsymbol{x})\right] E(\boldsymbol{x}, t|\omega) = -\hat{\mathcal{L}} E(\boldsymbol{x}, t|\omega) \,, \qquad (16.5)$$

where the operator

$$\hat{\mathcal{L}} = \left(\frac{\omega}{c}\right)^2 \varepsilon_d(\boldsymbol{x}) + 2\mathrm{i}\varepsilon(\boldsymbol{x}) \frac{\omega}{c^2} \frac{\partial}{\partial t} \qquad (16.6)$$

is introduced for convenience. Equation (16.5) can also be written in the equivalent integral form

$$E(\boldsymbol{x}, t|\omega) = \int \mathrm{d}^2\boldsymbol{y}\, G(\boldsymbol{x}, \boldsymbol{y}|\omega) \hat{\mathcal{L}} E(\boldsymbol{y}, t|\omega) \,, \qquad (16.7)$$

where $G(\boldsymbol{x}, \boldsymbol{y}|\omega)$ is the Green function defined, in a standard way, as a solution of the equation

$$\left[\nabla^2 + \left(\frac{\omega}{c}\right)^2 \varepsilon_p(\boldsymbol{x})\right] G(\boldsymbol{x}, \boldsymbol{y}|\omega) = -\delta(\boldsymbol{x} - \boldsymbol{y}) \,. \qquad (16.8)$$

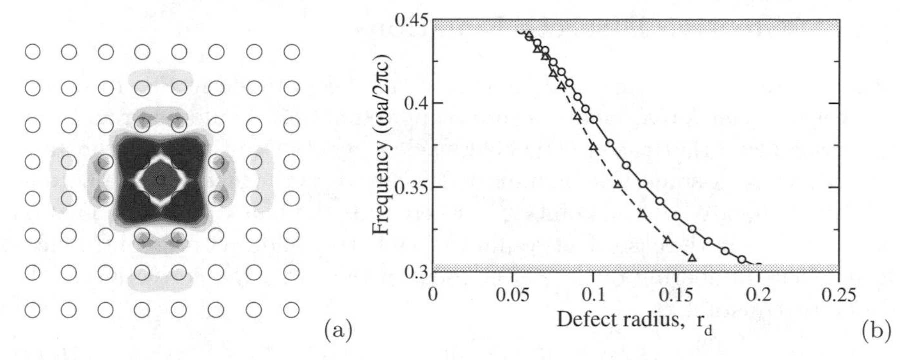

Fig. 16.2. a Electric field structure of a linear localized mode supported by a single defect rod with radius $r_d = 0.1a$ and $\varepsilon_d = 11.56$ in a square-lattice photonic crystal with $r_0 = 0.18a$ and $\varepsilon_0 = 11.56$. The rod positions are indicated by circles and the amplitude of the electric field is indicated by color. **b** Frequency of the defect mode as a function of the radius r_d: calculated precisely from (16.7) (*full line with circles*) and approximately from (16.15) (*dashed line with triangles*)

General properties of the Green function of a perfect 2D photonic crystal are described in more details in [19]. Here, we notice that the Green function is *symmetric*, i.e.

$$G(\boldsymbol{x}, \boldsymbol{y}|\omega) = G(\boldsymbol{y}, \boldsymbol{x}|\omega)$$

and *periodic*, i.e.

$$G(\boldsymbol{x} + \boldsymbol{s}_{ij}, \boldsymbol{y} + \boldsymbol{s}_{ij}|\omega) = G(\boldsymbol{x}, \boldsymbol{y}|\omega)\,,$$

where $\boldsymbol{s}_{ij}$ is defined by (16.4).

The Green function can be calculated by means of the Fourier transform

$$G(\boldsymbol{x}, \boldsymbol{y}|\omega) = \int_{-\infty}^{\infty} \mathrm{d}t\, e^{\mathrm{i}\omega t} G(\boldsymbol{x}, \boldsymbol{y}, t) \tag{16.9}$$

applied to the time-dependent Green function governed by the equation

$$\left[\nabla^2 - \varepsilon_p(\boldsymbol{x})\partial_t^2\right] G(\boldsymbol{x}, \boldsymbol{y}, t) = -\delta(t)\delta(\boldsymbol{x} - \boldsymbol{y})\,, \tag{16.10}$$

which has been solved by the finite-difference time-domain method [22].

Now that we have calculated the Green function, we can figure out the defect states solving (16.7) directly. For example, Fig. 16.2a shows a defect mode created by introducing a single defect rod with the radius $r_d = 0.1a$ and dielectric constant $\varepsilon_d = 11.56$ into the 2D photonic crystal shown in Fig. 16.1. Although direct numerical solution of the integral equation (16.7) remains possible even in the case of a few defect rods, it becomes severely limited by the current computer facilities as soon as we increase the number of the defect rods and start investigation of the line defects (waveguides) and their branches. Thus, looking for new approximate numerical techniques which could combine reasonable accuracy, flexibility, and power to forecast new effects is an issue of the key importance.

16.4 Effective Discrete Equations

Studying the electric field distribution of the defect mode in Fig. 16.2a, one can suggest that a reasonably accurate approximation should be provided by the assumption that the electric field inside the defect rod remains constant. Indeed, let us assume that nonlinear defect rods embedded into a photonic crystal are located at the points $\boldsymbol{x}_m$, where m is the index (or a combination of two indices in the case of a two-dimensional array of defect rods) introduced for explicit numbering of the defect rods. In this case, the correction to the dielectric constant is

$$\varepsilon_d(\boldsymbol{x}) = \left\{\varepsilon_d^{(0)} + |E(\boldsymbol{x},t|\omega)|^2\right\} \sum_m \theta(\boldsymbol{x} - \boldsymbol{x}_m) , \tag{16.11}$$

where

$$\theta(\boldsymbol{x}) = \begin{cases} 1 , & \text{for} \quad |\boldsymbol{x}| \le r_d , \\ 0 , & \text{for} \quad |\boldsymbol{x}| > r_d . \end{cases} \tag{16.12}$$

The second term in (16.11) takes into account a contribution due to the Kerr nonlinearity (we assume that the electric field is scaled with the nonlinear susceptibility, $\chi^{(3)}$). Assuming, as we discussed above, that the electric field $E(\boldsymbol{x},t|\omega)$ inside the defect rods is almost constant, one can derive, by substituting (16.11) into (16.7) and averaging over the cross-section of the rods [16], an approximate *discrete nonlinear equation*

$$i\sigma\frac{\partial}{\partial t}E_n - E_n + \sum_m J_{n-m}(\omega)(\varepsilon_d^{(0)} + |E_m|^2)E_m = 0 , \tag{16.13}$$

for the amplitudes of the electric field $E_n(t|\omega) \equiv E(\boldsymbol{x}_n,t|\omega)$ inside the defect rods. The parameter σ and the coupling constants

$$J_n(\omega) = \left(\frac{\omega}{c}\right)^2 \int_{r_d} d^2\boldsymbol{y}\, G(\boldsymbol{x}_0, \boldsymbol{x}_n + \boldsymbol{y}|\omega) \tag{16.14}$$

are determined in this case by the Green function $G(\boldsymbol{x},\boldsymbol{y}|\omega)$ of the perfect photonic crystal.

To check the accuracy of the approximation provided by (16.13), we solved it in the linear limit for the case of a single defect rod. In this case (16.13) is reduced to the equation

$$J_0(\omega_d) = 1/\varepsilon_d^{(0)} , \tag{16.15}$$

from which one can obtain an estimation for the frequency ω_d of the localized defect mode. As is seen from Fig. 16.2b, the mode frequency calculated in the framework of this approximation is in a good agreement with that calculated directly from (16.7), provided the defect radius r_d is small enough. Even for $r_d = 0.15a$ an error introduced by the approximation does not exceed 5%. It lends a support to the validity of (16.13) allowing us to use it hereafter for studying nonlinear localized modes.

16.5 Nonlinear Waveguides in 2D Photonic Crystals

One of the most promising applications of the PBG structures is a possibility to create a novel type of optical waveguides. In conventional waveguides such as optical fibers, light is confined by *total internal reflection* due a difference in the refractive indices of the waveguide core and cladding. One of the weaknesses of such waveguides is that creating of bends is difficult. Unless the radius of the bend is large compared to the wavelength, much of the light will be lost. This is a serious obstacle for creating "integrated optical circuits", since the space required for large-radius bends is unavailable.

The waveguides based on the PBG materials employ a *different physical mechanism*: the light is guided by a line of coupled defects which possess a localized defect mode with frequency inside the band gap. The simplest photonic-crystal waveguide can be created by a straight line of defect rods, as shown in Fig. 16.3. Instead of a single localized state of an isolated defect, a waveguide supports propagating states (guided modes) with the frequencies in a narrow band located inside the band gap of a perfect crystal. Such guided modes have a periodical profile along the waveguide, and they decay exponentially in the transverse direction, see Fig. 16.3. That is, photonic crystal waveguides operate in a manner similar to resonant cavities, and the light with guiding frequencies is forbidden from propagating in the bulk. Because of this, when a bend is created in a photonic crystal waveguide, the light remains trapped and the only possible problem is that of reflection. However, as was predicted numerically [20,21] and then demonstrated in microwave [23] and optical [24] experiments, it is still possible to get very high transmission efficiency for nearly all frequencies inside the gap.

As we show in Refs. [25,26], the equation (16.3) can accurately describe the dispersion and transmission properties of the photonic crystal waveguides and their circuits.

To employ the high-technology potential of photonic crystal waveguides, it is crucially important to achieve a tunability of their transmission properties. Nowadays, several approaches have been suggested for this purpose.

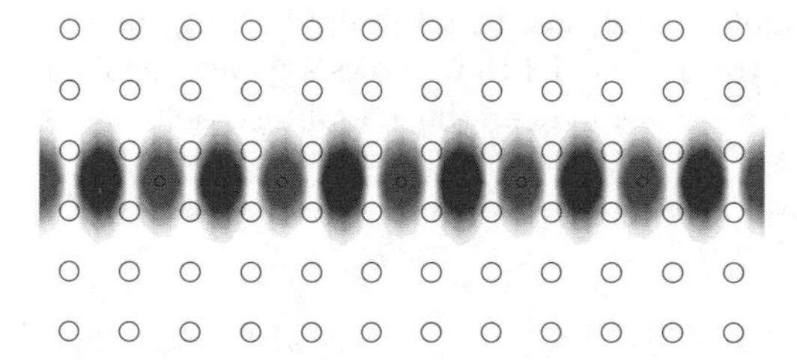

Fig. 16.3. Electric field of the linear guiding mode in a waveguide created by an array of the defect rods. The rod positions are indicated by circles and the amplitude of the electric field is indicated by color

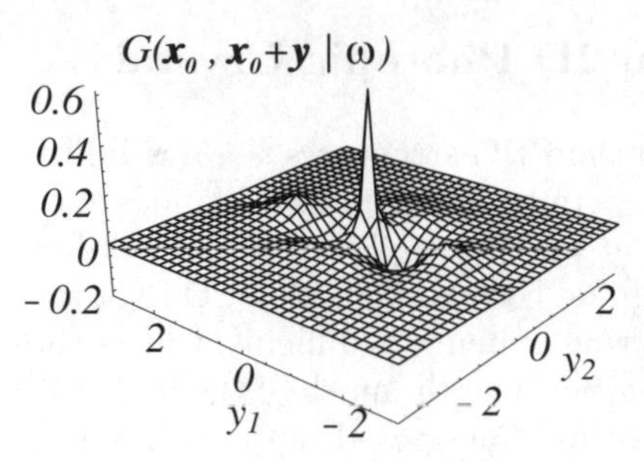

Fig. 16.4. The Green function $G(\boldsymbol{x}_0, \boldsymbol{x}_0 + \boldsymbol{y}|\omega)$ for the photonic crystal shown in Fig. 16.1 ($\boldsymbol{x}_0 = \boldsymbol{a}_1/2$ and $\omega = 0.33 \times 2\pi c/a$)

For instance, it has been recently demonstrated both numerically [27] and in microwave experiments [28], that transmission spectrum of straight and sharply bent waveguides in *quasiperiodic photonic crystals* is rather rich in structure and only some frequencies get near perfect transmission. Another possibility is creation of the *channel drop system* on the bases of two parallel waveguides coupled by the point defects between them. It has been shown [29] that high-Q frequency selective complete transfer can occur between such waveguides by creating resonant defect states of different symmetry and by forcing an accidental degeneracy between them.

However, being frequency selective, the above mentioned approaches do not possess *dynamical tunability* of the transmission properties. The latter idea can be realized by changing the light intensity in the so-called *nonlinear photonic crystal waveguides* [17], created by inserting an additional row of rods made from a Kerr-type nonlinear material characterized by the third-order nonlinear susceptibility $\chi^{(3)}$ and the linear dielectric constant $\varepsilon_d^{(0)}$. For definiteness, we assume that $\varepsilon_d^{(0)} = \varepsilon_0 = 11.56$ and that the nonlinear defect rods are embedded into the photonic crystal along a selected direction $\boldsymbol{s}_{ij}$, so that they are located at the points $\boldsymbol{x}_m = \boldsymbol{x}_0 + m\boldsymbol{s}_{ij}$. As we show below, changing the radius r_d of these defect rods and their location $\boldsymbol{x}_0$ in the crystal, one can create nonlinear waveguides with quite different properties.

As we have already discussed [17,18], the Green function $G(\boldsymbol{x}, \boldsymbol{y}|\omega)$ and, consequently, the coupling coefficients $J_m(\omega)$ are usually highly long-ranged functions. This can be seen directly from Fig. 16.4 that shows a typical spatial profile of the Green function. As a consequence, the coupling coefficients $J_n(\omega)$ calculated from (16.14) decrease slowly with the site number n. For $\boldsymbol{s}_{01}$ and $\boldsymbol{s}_{10}$ directions, the coupling coefficients can be approximated by an exponential function as follows

$$|J_n(\omega)| \approx \begin{cases} J_0(\omega)\,, & \text{for } n = 0\,, \\ J_*(\omega) e^{-\alpha(\omega)|n|}\,, & \text{for } |n| \geq 1\,, \end{cases}$$

where the characteristic decay rate $\alpha(\omega)$ can be as small as 0.85, depending on the values of ω, $\boldsymbol{x}_0$, and r_d, and it can be even smaller for other types of photonic crystals (for instance, for the photonic crystal used in Fig. 16.12 we find $J_m \sim (-1)^m \exp(-0.66m)$ for $m \geq 2$). By this means, (16.13) is a nontrivial long-range generalization of a 2D discrete nonlinear Schrödinger (NLS) equation extensively studied during the last decade for different applications [37]. It allows us to draw an analogy between the problem under consideration and a class of the NLS equations that describe nonlinear excitations in quasi-one-dimensional molecular chains with long-range (e.g. dipole-dipole) interaction between the particles and local on-site nonlinearities [30,31]. For such systems, it was shown that the effect of nonlocal interparticle interaction brings some new features to the properties of nonlinear localized modes (in particular, bistability in their spectrum). Moreover, for our model the coupling coefficients $J_n(\omega)$ can be either unstaggered and monotonically decaying, i.e. $J_n(\omega) = |J_n(\omega)|$, or staggered and oscillating from site to site, i.e. $J_n(\omega) = (-1)^n |J_n(\omega)|$. We therefore expect that effective nonlocality in both linear and nonlinear terms of (16.13) may also bring similar new features into the properties of nonlinear localized modes excited in the photonic crystal waveguides.

16.5.1 Staggered and Unstaggered Localized Modes

As can be seen from the structure of the Green function presented in Fig. 16.4, the case of monotonically varying coefficients $J_n(\omega)$ can occur for the waveguide oriented in the $\boldsymbol{s}_{01}$ direction with $\boldsymbol{x}_0 = \boldsymbol{a}_1/2$. In this case, the frequency of a linear guided mode, that can be excited in such a waveguide, takes a minimum value at $k = 0$ (see Fig. 16.5a), and the corresponding nonlinear mode is expected to be unstaggered.

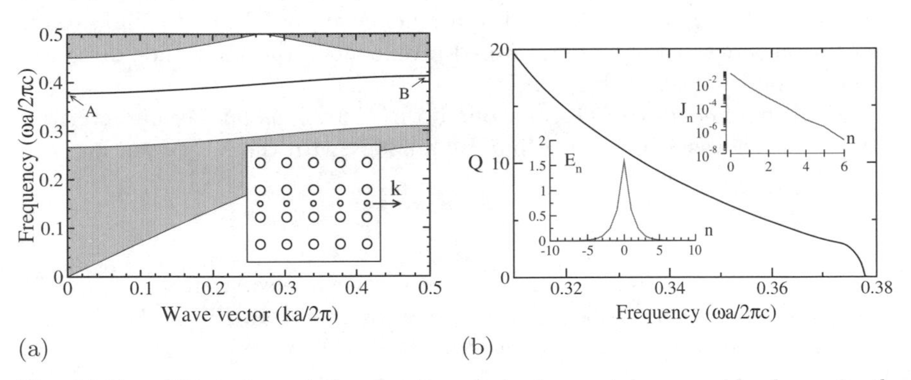

Fig. 16.5. a Dispersion relation for the photonic crystal waveguide shown in the inset ($\varepsilon_0 = \varepsilon_d = 11.56$, $r_0 = 0.18a$, $r_d = 0.10a$). The grey areas are the projected band structure of the perfect 2D photonic crystal. The frequencies at the indicated points are: $\omega_A = 0.378 \times 2\pi c/a$ and $\omega_B = 0.412 \times 2\pi c/a$. **b** Mode power $Q(\omega)$ of the nonlinear mode excited in the corresponding photonic crystal waveguide. The right inset gives the dependence $J_n(\omega)$ calculated at $\omega = 0.37 \times 2\pi c/a$. The left inset presents the profile of the corresponding nonlinear localized mode

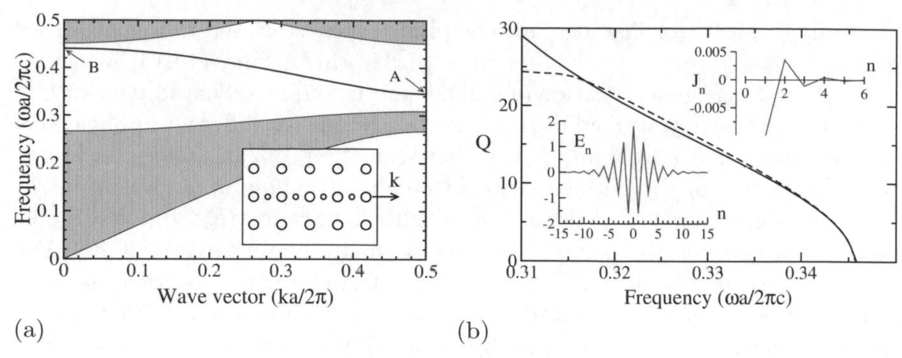

(a) (b)

Fig. 16.6. **a** Dispersion relation for the photonic crystal waveguide shown in the inset ($\varepsilon_0 = \varepsilon_d = 11.56$, $r_0 = 0.18a$, $r_d = 0.10a$). The grey areas are the projected band structure of the perfect 2D photonic crystal. The frequencies at the indicated points are: $\omega_A = 0.346 \times 2\pi c/a$ and $\omega_B = 0.440 \times 2\pi c/a$. **b** Mode power $Q(\omega)$ of the nonlinear mode excited in the corresponding photonic crystal waveguide. Two cases are presented: the case of nonlinear rods in a linear photonic crystal (*solid line*) and the case of a completely nonlinear photonic crystal (*dashed line*). The right inset shows the behavior of the coupling coefficients $J_n(\omega)$ for $n \geq 1$ ($J_0 = 0.045$) at $\omega = 0.33 \times 2\pi c/a$. The left inset shows the profile of the nonlinear mode

We have solved (16.13) numerically and found that nonlinearity can lead to the existence of *guided modes localized in both directions*, i.e. in the direction perpendicular to the waveguide, due to the guiding properties of a channel waveguide created by defect rods, and in the direction of the waveguide, due to the nonlinearity-induced self-trapping effect. Such nonlinear modes exist with the frequencies below the frequency of the linear guided mode of the waveguide, i.e. below the frequency ω_A in Fig. 16.5a, and are indeed unstaggered, with the bell-shaped profile along the waveguide direction shown in the left inset of Fig. 16.5b.

The 2D nonlinear modes localized in both dimensions can be characterized by the mode power which we define, by analogy with the NLS equation, as

$$Q = \sum_n |E_n|^2 . \tag{16.16}$$

This power is closely related to the energy of the electric field in the 2D photonic crystal accumulated in the nonlinear mode. In Fig. 16.5b we plot the dependence of Q on frequency, for the waveguide geometry shown in Fig. 16.5a.

As can be seen from the Green function shown in Fig. 16.4, the case of staggered coupling coefficients $J_n(\omega)$ can be obtained for the waveguide oriented in the $\boldsymbol{s}_{10}$ direction with $\boldsymbol{x}_0 = \boldsymbol{a}_1/2$. In this case, the frequency dependence of the linear guided mode of the waveguide takes the minimum at $k = \pi/a$ (see Fig. 16.6a). Accordingly, the nonlinear guided mode localized along the direction of the waveguide is expected to exist with the frequency below the lowest frequency ω_A of the linear guided mode, with a staggered profile. The longitudinal profile of such a 2D nonlinear localized mode is shown in the left inset in Fig. 16.6b, together with the dependence of the mode power Q on the frequency (solid curve), which in this case is again monotonic.

The results presented above are obtained for linear photonic crystals with nonlinear waveguides created by a row of defect rods. However, we have carried out the same analysis for the general case of *a nonlinear photonic crystal* that is created by rods of different size but made of the same nonlinear material. Importantly, we have found relatively small difference in all the results presented above provided nonlinearity is weak. In particular, for the photonic crystal waveguide shown in Fig. 16.6a, the results for linear and nonlinear photonic crystals are very close. Indeed, for the mode power Q the results corresponding to a nonlinear photonic crystal are shown in Fig. 16.6b by a dashed curve, and for $Q < 20$ this curve almost coincides with the solid curve corresponding to the case of a nonlinear waveguide embedded into a 2D linear photonic crystal.

16.5.2 Stability of Nonlinear Localized Modes

Let us now consider the waveguide created by a row of defect rods which are located at the points $\boldsymbol{x}_0 = (\boldsymbol{a}_1 + \boldsymbol{a}_2)/2$, along a straight line in either the $\boldsymbol{s}_{10}$ or $\boldsymbol{s}_{01}$ directions. The results for this case are presented in Figs. 16.7–16.8. The coupling coefficients J_n are described by a slowly decaying staggered

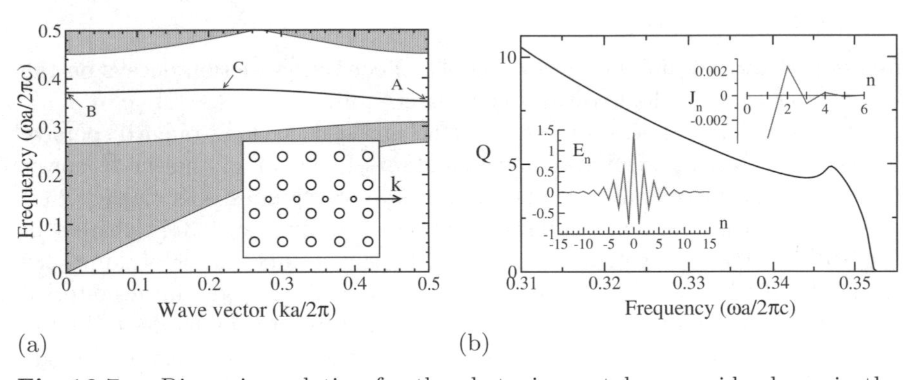

(a) (b)

Fig. 16.7. **a** Dispersion relation for the photonic crystal waveguide shown in the inset ($\varepsilon_0 = \varepsilon_d = 11.56$, $r_0 = 0.18a$, $r_d = 0.10a$). The grey areas are the projected band structure of the perfect 2D photonic crystal. The frequencies at the indicated points are: $\omega_A = 0.352 \times 2\pi c/a$, $\omega_B = 0.371 \times 2\pi c/a$, and $\omega_C = 0.376 \times 2\pi c/a$ (at $k = 0.217 \times 2\pi/a$). **b** Mode power $Q(\omega)$ of the nonlinear mode excited in the corresponding photonic crystal waveguide. The right inset shows the behavior of the coupling coefficients $J_n(\omega)$ for $n \geq 1$ ($J_0 = 0.068$) at $\omega = 0.345 \times 2\pi c/a$. The left inset shows the profile of the corresponding nonlinear mode

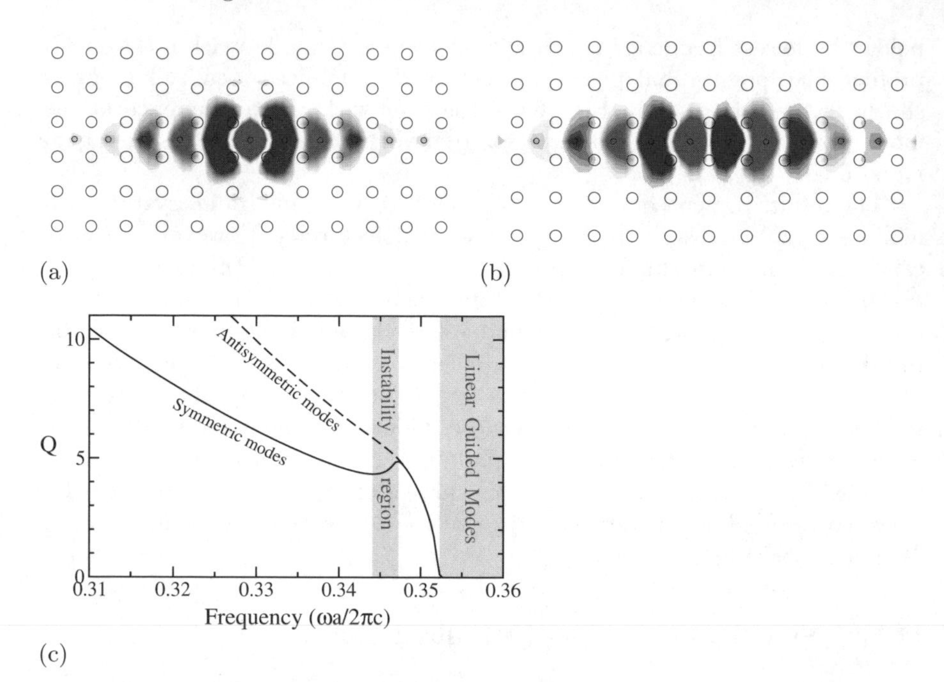

Fig. 16.8. Examples of the **a** symmetric and **b** antisymmetric localized modes. The rod positions are indicated by circles and the amplitude of the electric field is indicated by color (red, for positive values, and blue, for negative values); **c** Power Q vs. frequency dependencies calculated for two modes of different symmetry in the photonic crystal waveguide shown in Fig. 16.7

function of the site number n, so that the effective interaction decays on the scale larger than in the two cases considered above.

It is remarkable that, similar to the NLS models with long-range dispersive interactions [30,31], we find a *non-monotonic* behavior of the mode power $Q(\omega)$ for this type of nonlinear photonic crystal waveguides: specifically, $Q(\omega)$ *increases* in the frequency interval $0.344 < (\omega a/2\pi c) < 0.347$ (shaded in Fig. 16.8c). One can expect that, similar to the results earlier obtained for the nonlocal NLS models [30,31], the nonlinear localized modes in this interval are unstable and will eventually decay or transform into the modes of higher or lower frequency [32]. What counts is that there is an interval of mode power in which *two stable nonlinear localized modes of different widths do coexist.* Since the mode power is closely related to the mode energy, one can expect that the mode energy is also non-monotonic function of ω. Such a phenomenon is known as *bistability*, and in the problem under consideration it occurs as a direct manifestation of the nonlocality of the effective (linear and nonlinear) interaction between the defect rod sites.

Being interested in the mobility of the nonlinear localized modes we have investigated, in addition to the symmetric modes shown in the left inset in

Fig. 16.7b and in Fig. 16.8a, also the *antisymmetric localized modes* shown in Fig. 16.8b. Our calculations show that the power $Q(\omega)$ of the antisymmetric modes always (for all values of ω and all types of waveguides) exceeds that for symmetric ones (see, e.g., Fig. 16.8c). Thus, antisymmetric modes are expected to be unstable and they should transform into a lower-energy symmetric modes.

In fact, the difference between the power of antisymmetric and symmetric modes determines the Peierls-Nabarro barrier which should be overtaken for realizing the mobility of a nonlinear localized mode. One can see in Fig. 16.8c that the Peierls-Nabarro barrier is negligible for $0.347 < (\omega a/2\pi c) < 0.352$ and thus such localized modes should be mobile. However, the Peierls-Nabarro barrier becomes sufficiently large for highly localized modes with $\omega < 0.344 \times 2\pi c/a$ and, as a consequence, such modes should be immobile. Hence, the bistability phenomenon in the photonic crystal waveguides of the type depicted in Figs. 16.7–16.8 opens up fresh opportunities [31] for *switching* between immobile localized modes (used for the energy storage) and mobile localized modes (used for the energy transport).

The foregoing discussions on the mode mobility, based on the qualitative picture of the Peierls-Nabarro barrier, have been established for the discrete *one-dimensional* arrays. It is clear that the *two-dimensional* geometry of photonic crystals under consideration will bring new features into this picture. However, all these issues are still open and would require a further analysis.

16.6 Self-Trapping of Light in a Reduced-Symmetry 2D Nonlinear Photonic Crystal

A low-intensity light cannot propagate through a photonic crystal if the light frequency falls into a band gap. However, it has been recently suggested [14] that in the case of a 2D periodic medium with a Kerr-type nonlinear material, high-intensity light with frequency inside the gap can propagate in the form of *finite energy solitary waves – 2D gap solitons*. These solitary waves were found to be *stable* [14], but the conclusion was based on the coupled-mode equations valid for a *weak modulation* of the dielectric constant $\varepsilon(\boldsymbol{x})$. However, in real photonic crystals the modulation of $\varepsilon(\boldsymbol{x})$ is *comparable to its average value*. Thus, the results of [14] have a limited applicability to the properties of localized modes in *realistic photonic crystals*.

More specifically, the coupled-mode equations are valid if and only if the band gap Δ is vanishingly small, i.e. $\Delta \sim A^2$ where A is an effective amplitude of the mode, that is a small parameter in the multi-scale asymptotic expansions [35]. If we apply this model to describe nonlinear modes in a wider gap (see, e.g., discussions in [35]), we obtain a 2D nonlinear Schrödinger (NLS) equation known to possess *no stable localized solutions*. Moreover, the 2D localized modes described by the coupled-mode equations are expected to possess *an oscillatory instability* recently discovered for a broad class of

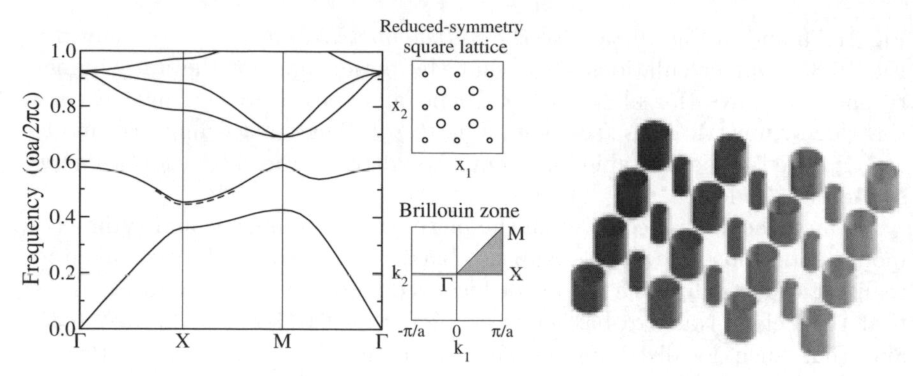

Fig. 16.9. Band-gap structure of the reduced-symmetry photonic crystal with $r_0 = 0.1a$, $r_d = 0.05a$, and $\varepsilon = 11.4$ for both types of rods. Full lines are calculated by the MIT Photonic-Bands program [34] whereas dashed line is found from the effective discrete model. The top center inset shows a cross-sectional view of the 2D photonic crystal depicted in the right inset. The bottom center inset shows the corresponding Brillouin zone

coupled-mode Thirring-like equations [36]. Thus, it is clear that, if nonlinear localized modes do exist in realistic PBG materials, their stability should be associated with *different physical mechanisms* not accounted for by simplified continuum models.

In this section we follow [18] and study the properties of nonlinear localized modes in a 2D photonic crystal composed of *two types of circular rods*: the rods of radius r_0 made from a linear dielectric material and placed at the corners of a square lattice with the lattice spacing a, and the rods of radius r_d made from a nonlinear dielectric material and placed at the center of each unit cell (see right inset in Fig. 16.9). Recently, such *photonic crystals of reduced symmetry* have attracted considerable interest because of their ability to possess *larger absolute band gaps* [33]. The band-gap structure of the reduced-symmetry photonic crystal is shown in Fig. 16.9. As is seen, it possesses two band gaps, first of which extends from $\omega = 0.426 \times 2\pi c/a$ to $\omega = 0.453 \times 2\pi c/a$.

The reduced-symmetry "diatomic" photonic crystal shown in Fig. 16.9 can be considered as a square lattice of the "nonlinear defect rods" of small radius r_d ($r_d < r_0$) embedded into the ordinary single-rod photonic crystal formed by a square lattice of rods of larger radius r_0 in air. The positions of the defect rods can then be described by the vectors $\boldsymbol{x}_{n,m} = n\boldsymbol{a}_1 + m\boldsymbol{a}_2$, where $\boldsymbol{a}_1$ and $\boldsymbol{a}_2$ are the primitive lattice vectors of the 2D photonic crystal. Here, in contrast to the photonic crystal waveguides discussed in the previous section, the nonlinear defect rods are characterized by two integer indices, n and m. However, it is straightforward to extend (16.13) and write an approximate

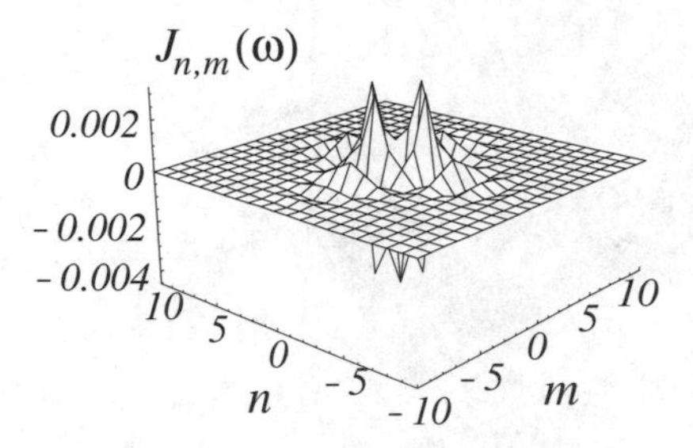

Fig. 16.10. Coupling coefficients $J_{n,m}(\omega)$ for the photonic crystal depicted in Fig. 16.9 (the contribution of the coefficient $J_{0,0} = 0.039$ is not shown). The frequency $\omega = 0.4456$ falls into the first band gap

2D discrete nonlinear equation

$$i\sigma \frac{\partial}{\partial t} E_{n,m} - E_{n,m} + \sum_{k,l} J_{n-k,\, m-l}(\omega)(\varepsilon_d^{(0)} + |E_{k,l}|^2) E_{k,l} = 0\,, \tag{16.17}$$

for the amplitudes of the electric field $E_{n,m}(t|\omega) \equiv E(\boldsymbol{x}_{n,m}, t|\omega)$ inside the defect rods. We have checked the accuracy of the approximation provided by (16.17) solving it in the linear limit, in order to find the band-gap structure associated with linear stationary mode. Since the coupling coefficients $J_{n,m}(\omega)$ in the photonic crystal depicted in Fig. 16.9 are highly long-ranged functions (see Fig. 16.10), one should take into account the interaction between at least 10 neighbors to reach accurate results. As is seen from Fig. 16.9, in this case the frequencies of the linear modes (depicted by a dashed line, with a minimum at $\omega = 0.446 \times 2\pi c/a$) calculated from (16.17) are in a good agreement with those calculated directly from (16.2). It lends a support to the validity of (16.17) and allows us to use it for studying nonlinear properties.

Stationary nonlinear modes described by (16.17) are found numerically by the Newton-Raphson iteration scheme. We reveal the existence of *a continuous family of such modes*, and a typical example [smoothed by continuous optimization for (16.5)] of nonlinear localized mode is shown in Fig. 16.11. In Fig. 16.12, we plot the dependence of the mode power

$$Q(\omega) = \sum_{n,m} |E_{n,m}|^2\,, \tag{16.18}$$

on the frequency ω for the photonic crystal shown in Fig. 16.9. As we have already discussed, this dependence represents a very important characteristic of nonlinear localized modes which allows to determine their stability by means of the Vakhitov-Kolokolov stability criterion: $dQ/d\omega > 0$ for unstable modes (this criterion has been extended [38] to 2D NLS models).

As is well known [38,39], in the 2D discrete cubic NLS equation, only high-amplitude localized modes are stable, whereas no stable modes exist in the continuum limit. For our model, the high-amplitude modes are also stable

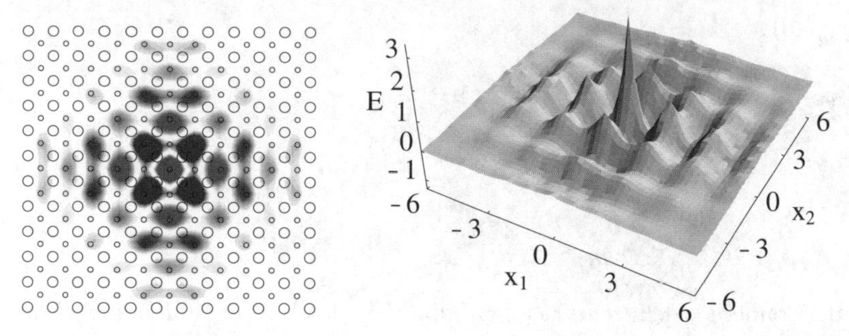

Fig. 16.11. Top (left) and 3D (right) views of a nonlinear localized mode in the first band gap of 2D photonic crystal depicted in Fig. 16.9.

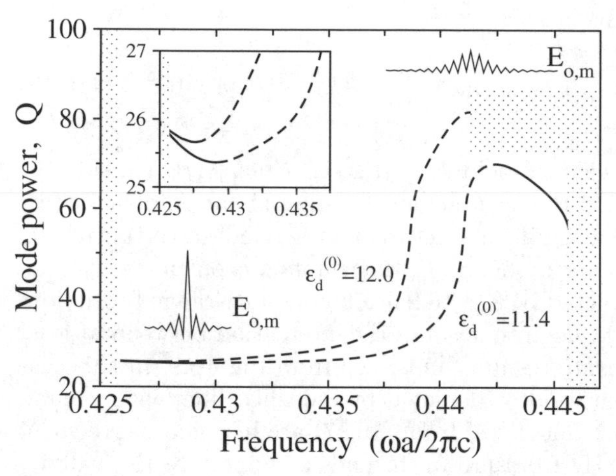

Fig. 16.12. Power Q vs. frequency ω for the 2D nonlinear localized modes in the photonic crystal of Fig. 16.9 with two different $\epsilon_d^{(0)}$. Solid lines – stable modes, dashed lines – unstable modes. Insets show typical profiles of stable modes, and an enlarged part of the power dependence. Grey areas show the lower and upper bands of delocalized modes surrounding the band gap

(see inset in Fig. 16.12), but they are not accessible under realistic conditions: To excite such modes one should increase the refractive index at the mode center in more than 2 times. Thus, for realistic conditions and relatively small values of $\chi^{(3)}$, only low-amplitude localized modes become a subject of much interest since they can be excited in experiment. However, such modes in unbounded 2D NLS models are always unstable and either collapse or spread out [37]. In fact, they can be stabilized by some external forces (e.g., due to interactions with boundaries or disorder [40]), but in this case the excitations are pinned and cannot be used for energy or signal transfer.

Here we reveal that, in a sharp contrast to the 2D discrete NLS models discussed earlier in various applications, the low-amplitude localized modes of (16.17) can be stabilized due to *nonlinear long-range dispersion* inherent to the photonic crystals. It should be emphasized that such stabilization does not occur in the models with only *linear long-range* dispersion [37]. In order to gain a better insight into the stabilization mechanism, we have carried out the studies of (16.13) for the exponentially decaying coupling coefficients $J_{n,m}$. Our results show that the most important factor which determines stability of the low-amplitude localized modes is a ratio of the coefficients at the local nonlinearity ($\sim J_{0,0}$) and the nonlinear dispersion ($\sim J_{0,1}$). If the coupling coefficients $J_{n,m}$ decrease with the distances n and m rapidly, the low-amplitude modes of (16.17) with $\epsilon_d^{(0)} = 11.4$ are essentially stable for $J_{0,0}/J_{0,1} \leq 13$. However, this estimation is usually lowered because the stabilization is favored by the presence of long-range interactions.

It should be mentioned that the stabilization of low-amplitude 2D localized modes is not inherent to all types of nonlinear photonic crystals. On the contrary, the photonic crystals must be *carefully designed* to support *stable low-amplitude nonlinear modes.* For example, in the photonic crystal considered above such modes are stable at least for $11 < \epsilon_d^{(0)} < 12$, however they become unstable for $\epsilon_d^{(0)} \geq 12$ (see Fig. 16.12). The stability of these modes can also be controlled by varying r_d, r_0, or ϵ_0.

16.7 Concluding Remarks

Exploration of nonlinear properties of PBG materials may open new important application of photonic crystals for all-optical signal processing and switching, allowing an effective way to create tunable band-gap structures operating entirely with light. Nonlinear photonic crystals, and nonlinear waveguides created in the photonic structures with a periodically modulated dielectric constant, create an ideal environment for the generation and observation of nonlinear localized modes.

As follows from our results, nonlinear localized modes can be excited in photonic crystal waveguides of different geometry. For several geometries of 2D waveguides, we have demonstrated that such modes are described by a new type of nonlinear lattice models that include long-range interaction and effectively nonlocal nonlinear response. It is expected that the general features of nonlinear guided modes described here will be preserved in other types of photonic crystal waveguides. Additionally, similar types of nonlinear localized modes are expected in photonic crystal fibers [41] consisting of a periodic air-hole lattice that runs along the length of the fiber, provided the fiber core is made of a highly nonlinear material (see, e.g., [42]).

Experimental observation of nonlinear photonic localized modes would require not only the use of photonic materials with a relatively large nonlinear refractive index (such as AlGaAs waveguide PBG structures [43] or polymer PBG crystals [44], but also a control of the group-velocity dispersion and band-gap parameters. The latter can be achieved by employing the surface coupling technique [45] that is able to provide coupling to specific points of the dispersion curve, opening up a very straightforward way to access nonlinear effects.

Acknowledgments. The authors are indebted to O. Bang, K. Busch, P.L. Christiansen, Y.B. Gaididei, S. John, A. McGurn, C. Soukoulis, and A.A. Sukhorukov for encouraging discussions, and R.A. Sammut for collaboration at the initial stage of this project. The work has been partially supported by the Large Grant Scheme of the Australian Research Council and the Performance and Planning Foundation grant of the Institute of Advanced Studies.

References

1. J.D. Joannoupoulos, R.B. Meade, and J.N. Winn, *Photonic Crystals: Molding the Flow of Light* (Princeton University Press, Princeton NJ 1995)
2. J.G. Fleming and S.-Y. Lin: Opt. Lett. **24**, 49 (1999)
3. See, e.g., K. Busch and S. John: Phys. Rev. Lett. **83**, 967 (1999), and discussions therein
4. V. Berger: Phys. Rev. Lett. **81**, 4136 (1998); and also the recent experiment: N.G.R. Broderick, G.W. Ross, H.L. Offerhaus, D.J. Richardson, and D.C. Hanna: Phys. Rev. Lett. **84**, 4345 (2000); S. Saltiel and Y.S. Kivshar: Opt. Lett. **25**, 1204 (2000)
5. S.F. Mingaleev and Yu.S. Kivshar, Opt. Photon. News **13**, No. 7, 48 (2002)
6. See, e.g., S. Flach and C.R. Willis: Phys. Rep. **295**, 181 (1998); O.M. Braun and Y.S. Kivshar: Phys. Rep. **306**, 1 (1998), Chap. 6
7. R.S. MacKay and S. Aubry: Nonlinearity **7**, 1623 (1994); see also S. Aubry: Physica D **103**, 201 (1997)
8. B.I. Swanson, J.A. Brozik, S.P. Love, G.F. Strouse, A.P. Shreve, A.R. Bishop, W.Z. Wang, and M.I. Salkola: Phys. Rev. Lett. **82**, 3288 (1999)
9. U.T. Schwarz, L.Q. English, and A.J. Sievers: Phys. Rev. Lett. **83**, 223 (1999)
10. E. Trias, J.J. Mazo, and T.P. Orlando: Phys. Rev. Lett. **84**, 741 (2000); P. Binder, D. Abraimov, A.V. Ustinov, S. Flach, and Y. Zolotaryuk: Phys. Rev. Lett. **84**, 745 (2000)
11. F.M. Russel, Y. Zolotaryuk, and J.C. Eilbeck: Phys. Rev. B **55**, 6304 (1997)
12. H.S. Eisenberg, Y. Silberberg, R. Marandotti, A.R. Boyd, and J.S. Aitchison: Phys. Rev. Lett. **81**, 3383 (1998)
13. A.A. Sukhorukov, Y.S. Kivshar, O. Bang, J. Martorell, J. Trull, and R. Vilaseca: Optics and Photonics News **10** (12) 34 (1999)
14. S. John and N. Aközbek: Phys. Rev. Lett. **71**, 1168 (1993); Phys. Rev. E **57**, 2287 (1998)
15. A.A. Sukhorukov, Y.S. Kivshar, and O. Bang: Phys. Rev. E **60**, R41 (1999)
16. A.R. McGurn: Phys. Lett. A **251**, 322 (1999); Phys. Lett. A **260**, 314 (1999)
17. S.F. Mingaleev, Y.S. Kivshar, and R.A. Sammut: Phys. Rev. E **62**, 5777 (2000)
18. S.F. Mingaleev and Y.S. Kivshar: Phys. Rev. Lett. **86** (June 2001), Phys. Rev. Lett. **86**, 5474 (2001)
19. A.A. Maradudin and A.R. McGurn, in *Photonic Band Gaps and Localization, NATO ASI Series B: Physics*, Vol. 308, Ed. C.M. Soukoulis (Plenum Press, New York 1993) p. 247

20. A. Mekis, J.C. Chen, I. Kurland, S. Fan, P.R. Villeneuve, and J.D. Joannopoulos: Phys. Rev. Lett. **77**, 3787 (1996)
21. A. Mekis, S. Fan, and J.D. Joannopoulos: Phys. Rev. B **58**, 4809 (1998)
22. A.J. Ward and J.B. Pendry: Phys. Rev. B **58**, 7252 (1998)
23. S.-Y. Lin, E. Chow, V. Hietala, P.R. Villeneuve, and J.D. Joannopoulos: Science **282**, 274 (1998)
24. M. Tokushima, H. Kosaka, A. Tomita, and H. Yamada: Appl. Phys. Lett. **76**, 952 (2000)
25. S.F. Mingaleev and Yu. S. Kivshar: Opt. Lett. **27**, 231 (2002)
26. S.F. Mingaleev and Yu. S. Kivshar: J. Opt. Soc. Am. B **19**, 2241 (2002)
27. S.S.M. Cheng, L.M. Li, C.T. Chan, and Z.Q. Zhang: Phys. Rev. B **59**, 4091 (1999)
28. C. Jin, B. Cheng, B. Mau, Z. Li, D. Zhnag, S. Ban, B. Sun: Appl. Phys. Lett. **75**, 1848 (1999)
29. S. Fan, P.R. Villeneuve, J.D. Joannopoulos, and H.A. Haus: Phys. Rev. Lett. **80**, 960 (1998)
30. Y.B. Gaididei, S.F. Mingaleev, P.L. Christiansen, and K.Ø. Rasmussen: Phys. Rev. E **55**, 6141 (1997)
31. M. Johansson, Y.B. Gaididei, P.L. Christiansen, and K.Ø. Rasmussen: Phys. Rev. E **57**, 4739 (1998)
32. See, e.g., the examples for the continuous generalised NLS models, D.E. Pelinovsky, V.V. Afanasjev, and Y.S. Kivshar: Phys. Rev. E **53**, 1940 (1996)
33. C.M. Anderson and K.P. Giapis: Phys. Rev. Lett. **77**, 2949 (1996); Phys. Rev. B **56**, 7313 (1997)
34. S.G. Johnson, http://ab-initio.mit.edu/mpb/
35. Y.S. Kivshar, O.A. Chubykalo, O.V. Usatenko, and D.V. Grinyoff: Int. J. Mod. Phys. B **9**, 2963 (1995)
36. I.V. Barashenkov, D.E. Pelinovsky, and E.V. Zemlyanaya: Phys. Rev. Lett. **80**, 5117 (1998); A. De Rossi, C. Conti, and S. Trillo: Phys. Rev. Lett. **81**, 85 (1998)
37. See, e.g., V.K. Mezentsev, S.L. Musher, I.V. Ryzhenkova, and S.K. Turitsyn: JETP Lett. **60**, 829 (1994); S. Flach, K. Kladko, and R.S. MacKay, Phys. Rev. Lett. **78**, 1207 (1997); P.L. Christiansen et al.: Phys. Rev. B **57**, 11303 (1998)
38. E.W. Laedke et al.: JETP Lett. **62**, 677 (1995)
39. E.W. Laedke, K.H. Spatschek, S.K. Turitsyn, and V.K. Mezentsev: Phys. Rev. E **52**, 5549 (1995); Y.B. Gaididei, P.L. Christiansen, K.Ø. Rasmussen, and M. Johansson: Phys. Rev. B **55**, R13365 (1997)
40. Y.B. Gaididei, D. Hendriksen, P.L. Christiansen, and K.Ø. Rasmussen: Phys. Rev. B **58**, 3075 (1998)
41. T.A. Birks, J.C. Knight, and P.S.J. Russell: Opt. Lett. **22**, 961 (1997)
42. B.J. Eggleton, P.S. Westbrook, R.S. Windeler, S. Spälter, and T.A. Strasser: Opt. Lett. **24**, 1460 (1999)
43. P. Millar et al.: Opt. Lett. **24**, 685 (1999); A.A. Helmy et al.: Opt. Lett. **25**, 1370 (2000)
44. S. Shoji and S. Kawata: Appl. Phys. Lett. **76**, 2668 (2000)
45. V.N. Astratov et al.: Phys. Rev. B **60**, R16255 (1999)

Index

Printing (Computer to Plate): Saladruck Berlin
Binding: Stürtz AG, Würzburg